The SPINAL CORD *from* GESTATIONAL WEEK 4 *to the* 4TH POSTNATAL MONTH

ATLAS OF HUMAN CENTRAL NERVOUS SYSTEM DEVELOPMENT SERIES

Shirley A. Bayer *and* **Joseph Altman**

VOLUME 1

The Spinal Cord from Gestational Week 4 to the 4th Postnatal Month

VOLUME 2

The Brain During the Third Trimester

VOLUME 3

The Brain During the Second Trimester

VOLUME 4

The Brain During the Late First Trimester

VOLUME 5

The Brain During the Early First Trimester

The SPINAL CORD *from* GESTATIONAL WEEK 4 *to the* 4TH POSTNATAL MONTH

Shirley A. Bayer *and* **Joseph Altman**

CRC PRESS

Boca Raton London New York Washington, D.C.

Library of Congress Cataloging-in-Publication Data

Bayer, Shirley A. (Shirley Ann), 1940-
Atlas of human central nervous system development / Shirley A. Bayer, Joseph Altman.
p. ; cm.
Includes bibliographical references.
Contents: v. 1. The spinal cord from gestational week 4 to the fourth postnatal month -- v. 2. The brain during the third trimester -- v. 3. The brain during the second trimester-- v. 4.The brain during the late first trimester -- v. 5. The brain during the early first trimester.
ISBN 0-8493-1420-8 (alk. paper)
1. Central nervous system--Growth--Atlases. 2. Developmental neurophysiology--Atlases. I. Altman, Joseph, 1925-. II. Title.
[DNLM: 1. Brain--growth & development--Atlases. 2. Spinal Cord--growth & development--Atlases. 3. Brain--embryology--Atlases. 4. Spinal Cord--embryology--Atlases. WL 17 B357a 2002]
QM451 .B394 2002

2002025938

Direct all inquiries to CRC Press LLC, 2000 N.W. Corporate Blvd., Boca Raton, Florida 33431.

Visit the CRC Press Web site at www.crcpress.com

International Standard Book Number 0-8493-1420-8
Library of Congress Card Number 2002025938
Printed in the United States of America 1 2 3 4 5 6 7 8 9 0
Printed on acid-free paper

PREFACE

This book begins an Atlas Series on the Development of the Human Central Nervous System. We start with a volume devoted exclusively to the spinal cord for two reasons: (1) The spinal cord has similar structures throughout its length so a complete picture of its development at any specific age requires only a few different levels–and in early stages, only a single level is needed. (2) A transverse cutting plane is all that is necessary to show spinal cord structures. Thus, a single volume can feature *all* the stages of spinal cord development. In contrast, many different levels of the *brain* must be shown in a variety of cutting planes to get an overview of its structure at *each stage of development.* Four additional volumes (in preparation) in the Atlas Series will present a comprehensive picture of the developing human brain cut in frontal, sagittal, and horizontal planes. The second volume will feature the third trimester; the third volume will feature the second trimester; and finally, volumes 4 and 5 will feature the late and early parts of the first trimester, respectively.

Our recent work on the developing spinal cord (Altman, J., and S. A. Bayer, *Development of the Human Spinal Cord. An Interpretation Based on Experimental Studies in Animals*, Oxford University Press, 2001) and this Atlas are our first publications linking developmental studies of the central nervous system in animals to humans. From the early 1960s to the present, our work has featured the development of the rat nervous system at the cell and tissue levels. We use ^{3}H–thymidine autoradiography to gather a large database that maps the stem cell mosaic in the neuroepithelium, the times of origin of neuronal populations within the nervous system, and the migratory pathways of neuronal populations as they settle in the maturing nervous system. Our initial foray into human central nervous system development began in the late 1980s with a request from Serge Duckett to contribute a chapter on human central nervous system development (Bayer, S.A., J. Altman, R. J. Russo, and X. Zhang, Embryology, In: *Pediatric Neuropathology*, S. Duckett, Ed., Lea & Febiger, Philadelphia, PA, 1995, pp. 54-107). It was while we were working on that chapter that we realized how little is known about the development of the human central nervous system. An up-to-date interpretation simply does not exist. The best work was published in German (Hochstetter, F.: *Beiträge zur Entwicklungsgeschichte des menschlichen Gehirns. I. Teil.* Vienna and Leipzig, Franz Deuticke, 1919) and has never been translated into English. There is a striking similarity between the developing human and rat central nervous systems. But Hochstetter did his pioneering work on development when even the mature anatomy of the nervous system was not well understood. A re-examination of human central nervous system development is needed to link what has been learned from animal studies to normal development in humans. Medical textbooks of human embryology are in need of an update on central nervous system development. In addition, the many disorders resulting from abnormal neural development can be better understood if normal development is better known.

We decided to apply our knowledge of rat nervous system development to humans by directly examining histological sections of normal human specimens. The National Museum of Health and Medicine in the Armed Forces Institute of Pathology (AFIP), Walter Reed Hospital, Washington D.C. has the Carnegie Collection of human embryos and the Yakovlev Collection of human fetuses; while we were there we accidentally found the excellent Minot Collection of human embryos. Unfortunately, attempts to obtain funding for research at the AFIP were unsuccessful, so we decided to finance the research ourselves. During several trips to Washington D.C. in 1996 and mostly during an extended stay in the summer of 1997, we examined and took over 10,000 photographs of the best preserved normal specimens in the Yakovlev, Carnegie, and Minot collections. All of the film developing was done by us. Technicians were employed to scan each negative and create high resolution digitized files. These files are the sources for a comprehensive database on the development of the human central nervous system at the cell and tissue levels. This material is being analyzed in our Laboratory of Developmental Neurobiology that is equipped with photographic instruments, high resolution scanners, printers, and Apple computers capable of handling large files and running 3-dimensional reconstruction software. This Atlas Series is the result of our efforts.

It is our hope that this first volume will update knowledge of the developing human spinal cord and provide new insights into the steps involved in the morphogenesis of the mature spinal cord. All Companion Plates in this Atlas are presented in a "user friendly" format so that the reader can view unlabeled and labeled photographs on facing pages. The first Companion Plate (2A-2B) shows the human spinal cord at the 4th gestational week at its most primitive stage consisting mainly of neuroepithelial stem cells (NEP cells). The last set of Companion Plates (104A-104B to 117A-117B) shows a spinal cord nearing maturity in the 4th postnatal month. What lies in between these Companion Plates are "freeze frames" of the developmental record. The large sections on 3-dimensional reconstructions provide holistic views of the developing spinal cord at critical stages when maturing structures first appear during the first and early second trimesters (Figures 2 through 32). A Glossary at the end of the book gives a brief definition of each structure that is labeled in the Plates and Figures.

Shirley A. Bayer
Joseph Altman

DEDICATION

We dedicate this book to the memory of Dr. Raymond J. Russo, an enthusiastic Professor of Biology, Indiana University–Purdue University in Indianapolis. He introduced us to the marvelous graphic capabilities of Macintosh Computers and generously helped us with our first attempts at 3D reconstructions of the developing nervous system.

ACKNOWLEDGMENTS

We thank our friend, Dr. William DeMyer, a pediatric neurologist at Riley Hospital for Children, Indiana University Medical Center, Indianapolis, Indiana for access to his personal library on human central nervous system development, which included a fragile copy of Hochstetter's 1919 publication. We also thank the staff of the National Museum of Health and Medicine at the Armed Forces Institute of Pathology, Walter Reed Hospital, Washington, D.C.: Dr. Adrianne Noe, Director; Archibald J. Fobbs, Curator of the Yakovlev Collection; Elizabeth C. Lockett, and William Discher. We are most grateful to Dr. James M. Petras at the Walter Reed Institute of Research who made his dark room facilities available so that we could develop all the photomicrographs on location rather than in our laboratory in Indiana. Finally, we thank Barbara Norwitz, Billi W. van der Putten, Tanya Li, and Susan Fox at CRC Press for their personal attention to us and for expert help during production of the manuscript.

CONTENTS

FIGURE 1 Dorsal View of the Brain and Spinal Cord in a Newborn Infant

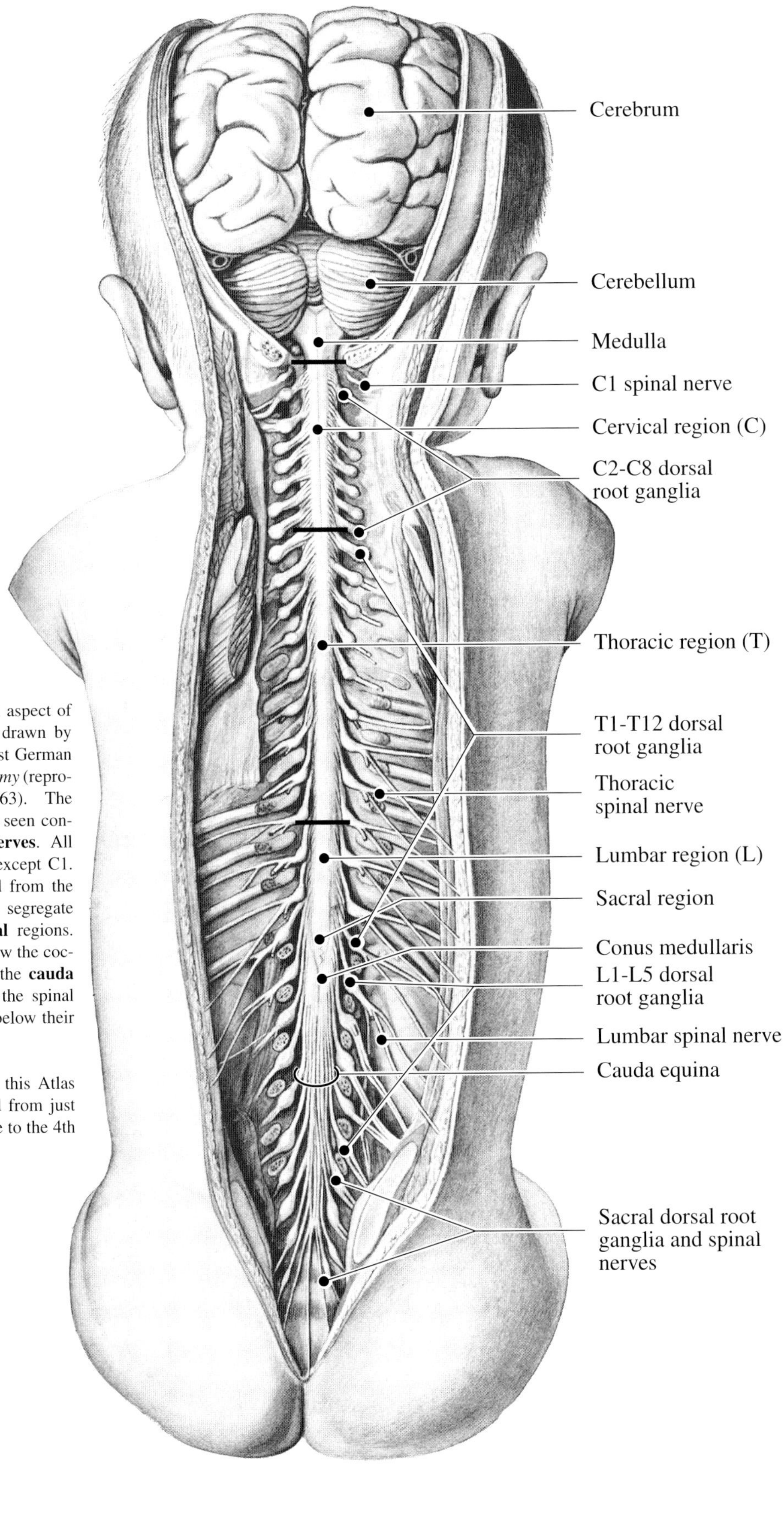

Figure 1. Dissection exposing the dorsal aspect of the brain and spinal cord in a neonate, drawn by Johannes Sobotta and published in the first German edition (1906) of his *Atlas of Human Anatomy* (reproduced from the 8th English edition, 1963). The entire dorsal aspect of the **spinal cord** is seen connected to the dorsal roots of the **spinal nerves**. All spinal nerves have **dorsal root ganglia** except C1. **Top thick line** separates the spinal cord from the **medulla**; **lower thick lines** successively segregate the **cervical, thoracic,** and **lumbosacral** regions. The **conus medullaris** is located just below the coccygeal terminus of the spinal cord, and the **cauda equina** contains the elongated roots of the spinal nerves that exit from the bony spine far below their point of attachment to the spinal cord.

The illustrations that follow throughout this Atlas trace the development of the spinal cord from just after closure of the embryonic neural tube to the 4th postnatal month.

I. Introduction

A. Overview of Atlas Organization

This Atlas focuses on the development of the human spinal cord, and is the first volume in the series, *Atlas of Human Central Nervous System Development*. The largest sections of this Atlas feature photographs of transversely cut spinal cords from normal specimens ranging in age from gestational week (GW) 4.0 up to and including the 4th postnatal month. The first trimester is described in **Part II**, the second in **Part III**, the third in **Part IV**, and the postnatal period in **Part V**. Each specimen or set of specimens is introduced by an *overview plate* that shows thumbnail photographs of all sections in that part of the Atlas at the same scale. The overview plate is followed by *companion plates* designated as **A** and **B** on facing pages. The **A** part of each plate on the left page shows the full contrast high magnification photograph of the specimen without any labels; the **B** part of each plate on the right page shows a low contrast copy of the same photograph with superimposed outlines and unabbreviated labels. **Part VI** presents 3-dimensional reconstructions of the cervical spinal cord during the first trimester (**VIA**) and of motoneuron columns in the ventral horn at cervical, thoracic, and lumbar levels in three specimens at the end of the first trimester and the beginning of the second (**VIB**). In **Part VII**, quantitative summaries of spinal cord development are presented in several graphs along with estimates of timetables for neurogenesis, dates of cell migration and settling, and sequences of myelination in major fiber tracts. A **Glossary** gives brief definitions for each label in the Plates, and defines other terms that are used in notes and figure captions. The concepts presented here are based on Altman and Bayer (2001), a research work that links human spinal cord development to the large body of developmental experiments in animals.

B. Specimen Collections

The 117 Plates in this Atlas are from 20 specimens in three histological collections of human embryos and fetuses housed in the National Museum of Health and Medicine, Armed Forces Institute of Pathology, Washington D.C. Five of the specimens (designated by a **C** prefix) are from the *Carnegie Collection*, which originated in the Department of Embryology of the Carnegie Institution of Washington, under the leadership of Franklin P. Mall (1862-1917), George L. Streeter (1873-1948), and George W. Corner (1889-1981). These specimens were collected over a span of 40 to 50 years and were histologically prepared with a variety of different fixatives, embedding media, cutting planes and histological stains. Initial descriptions and analyses of this material in relation to spinal cord development (most of them in the context of other facets of embryonic development) were published in the early 1900s in *Contributions to Embryology, The Carnegie Institution of Washington*.

Seven of the specimens (designated by an **M** prefix) are from the *Minot Collection*, which is the work of Dr. Charles S. Minot (1852-1914), an embryologist at Harvard University. Throughout his career, Minot collected about 1900 embryos from a variety of species. The 100 human embryos in the collection were probably acquired between 1900 and 1910. From our examination of these specimens and their similar appearance, we presume that they are preserved in the same way although there are no existing records describing fixation procedures. The slides contain information regarding the section number, the section thickness (10 µm), and the stain (aluminum cochineal). Most embryos are cut in the same transverse plane, probably by Minot himself. Because of the constancy in their preparation, the Minot specimens are the source of quantitative data that are used to determine developmental trends during the early first trimester. To our knowledge, the nervous system has never been examined in the Minot specimens, other than by Altman and Bayer (2001).

Eight of the specimens (designated by a **Y** prefix) are from the *Yakovlev Collection*. Over a 40-year span, Paul I. Yakovlev (1894-1983) collected about 1500 normal and abnormal human specimens, ranging in age from the early second trimester of fetal development through old age (Haleem, 1990). All of the specimens are fixed in formalin, embedded in celloidin, and are cut in 35 µm sections. The sections are preserved on two sets of slides, one set is stained for cell bodies using cresylviolet. The other set is stained for myelin using the Loyez modification of Weigert's hematoxylin. Although sections of the brains are consecutively numbered on the slides, only two normal specimens (Y380–62, GW10.5 and Y68–65, GW14), have consecutively numbered sections of the spinal cord; 3D reconstructions of those specimens are presented in **Part VIB**. To our knowledge, this Atlas and the reference work on human spinal cord development (Altman and Bayer, 2001) are the only studies to analyze the spinal cord in the normal specimens of this collection.

C. Methods: Plate Preparation

All sections are photographed using either an Olympus photomicroscope (the Carnegie and Minot specimens) or a Wild photomakroskop (the Yakovlev specimens) using Kodak technical pan black and white negative film #TP442. The film is developed for 6 to 7 minutes in dilution F of Kodak HC–110 developer, followed by stop bath for 30 seconds, Kodak fixer for 5 minutes, Kodak hypo clearing agent for 1 minute, running water rinse for 10 minutes, and a brief rinse in Kodak photo–flo before drying. Each specimen is photographed at the magnification that filled the microscopic field with the largest cross-section of the spinal cord.

The negatives are scanned at 2700 dots per inch (dpi) using a Nikon Coolscan–1000 35 mm film scanner interfaced to a Macintosh PowerMac G3 computer. To bring out all of the subtle shades of gray, the exposures are scanned as color positives rather than as black and white negatives. The software driving the scanner is a plug–in for Adobe Photoshop and generates raw files that are approximately 27 megabytes (mb) each.

The raw files are reversed to positives and are changed to grayscale (reduces file size to 8 to 9 mb). The image resolution is set to 300 dpi. Using the enhancement features built into Adobe Photoshop and the additional features of Extensis Intellihance, adjustments are made to bring out contrast and increase sharpness. A normal-contrast image is in part **A** of each *companion plate* along with a low-contrast copy in part **B**. Labels are placed onto the low-contrast image using Adobe Illustrator. Small copies of each normal-contrast image are shown in the *overview plates*. A small image showing the spinal cord with the surrounding structures of the vertebral column is included in the plates of the first 11 specimens since the spinal cords have not been dissected from the surrounding tissues. For specimens in the late first, second, and third trimesters, the area surrounding the section is removed from the images.

D. Methods: 3D Reconstructions

Cervical Cord in the First Trimester (Part VIA): Eight specimens were used to reconstruct the cervical part of the spinal cord from GW 4.0 to GW10.5, following a 5-step procedure. *First*, photographs of sections are made at regularly spaced intervals through the cervical spinal cord; the negatives are scanned and converted to computer files (as described in Section C above). *Second*, all the files from a single specimen are placed into one large Photoshop file that contains a separate photograph in each layer. By altering the visibility and transparency of these layers the sections are aligned to each other using the middle dorsoventral part of the spinal canal as the reference for alignment. In the living state, there is a curve in the cervical region of the developing cord; that is difficult to reconstruct accurately and often makes interpretation of the rendered images confusing. Therefore to simplify the reconstructions, the centers of the sections are aligned straight. *Third*, Adobe Illustrator is used to outline the following 14 structures in each section: (1) the outside edge of the entire section, (2) the entire neuroepithelium on the left side, (3–5) the dorsal, intermediate, and ventral parts of the neuroepithelium on the right side, (6–7) the roof and floor plates, (8) the undivided gray matter on the left side, (9–11) the dorsal horn, intermediate gray, and ventral horn on the right side, (12) the dorsal funiculus on both sides, (13) the lateral funiculus on both sides, and (14) the ventral funiculus on both sides and the ventral commissure. Since structures are usually symmetrical in most specimens, the outlines from one side are copied to the other side to simplify the images. The outlines from each section are saved in separate Adobe Illustrator eps (encapsulated postscript) files. *Fourth*, the eps files are imported into 3D space (x, y, and z coordinates) using Cinema4DXL (C4D, Maxon Computer, Inc.), a modeling and animation software package. For each section, points on the outlines have unique x–y coordinates and share the same z coordinate. By calculating the distance between sections, the entire array of outlines is stretched out in the z axis. The outlines are segregated into 14 different goups, one for each structure. The C4D loft tool builds a "skin" for each structure by creating a spline mesh of polygons. The polygons start from the x–y points on the first outline with the most anterior z coordinate, to the x–y points on the next outline behind it, and finish with the x–y points on the last outline at the most posterior z coordinate. The polygons are rendered either as completely opaque or as partly transparent surfaces using the C4D ray-tracing engine. Selected parts of the model can be made either invisible or visible during rendering using the various options in C4D. *Fifth*, the rendered images are converted to Adobe Photoshop files, and Adobe Illustrator is used to draw thin lines on some of the surfaces to make the images more easy to understand.

Motor Columns in the Ventral Horn (Part VIB): Using the same steps as described in the preceding paragraph, the entire spinal cord from upper cervical to sacral levels is reconstructed in three specimens at the end of the first trimester and at the beginning of the second trimester (GW8.5, GW10.5, and GW14). The following structural outlines are drawn: (1) the outside edge of the entire section, (2) the gray matter on both sides, (3) the central canal, (4) motoneuron columns in the right ventral horn, (5) motoneuron columns in the left ventral horn. The images are rendered so that the white and gray matter are shown as transparent, glass-like surfaces surrounding the opaque surfaces of the central canal and the motoneuron columns. Because the spinal cord is many times longer than it is wide, the thin motoneuron columns are difficult to see when all axes are rendered at the same scale, especially in the GW14 model. Consequently, the x–y axis is three times larger than the z axis in all models.

E. Terminology

The labels in this Atlas come from a review of the literature (Altman and Bayer, 2001) on experimental studies of anatomical connections in the spinal cord of various mammals. In addition, several atlases of the adult human spinal cord are used (Fix,1987; Roberts et al., 1987; DeArmond et al., 1989; Haines, 2000) and some classical neuroanatomy books are consulted (Ranson and Clark, 1959; Crosby et al., 1962; Truex and Carpenter, 1969; Brodal, 1981). Some of these references use the directional terms *anterior* and *posterior*. Since our orientation is to link the work in humans to experimental work in animals, *dorsal* and *ventral* are the directional terms used throughout this Atlas. For the most part, the labels follow the terminology in these atlases and texts, especially in the gray matter.

However, some parts of the white matter are labeled differently from the standard adult atlases and there are departures from some developmental terminology.

New Subdivisions in the Dorsal Funiculus. The largest components of the dorsal funiculus contain the principal ascending axons of dorsal root ganglion cells in the fasciculus gracilis and the fasciculus cuneatus. However, experimental evidence indicates that the entry zone of the dorsal root fibers and the fibers bordering the medial side of the dorsal horn have a more complex organization.

As the dorsal root axons enter, they first bifurcate into ascending and descending branches (see Figures 12-15 and 12-16 in Truex and Carpenter, 1969; see Chapter 4, Section 4.1.2 in Altman and Bayer, 2001). That part of the dorsal funiculus is labeled the *dorsal root bifurcation zone*. Indeed, that part of the dorsal funiculus is the first to develop, appearing as an indistinct structure around GW4.75 (**Plate 4**) and as a more definite swelling in the white matter by GW5.25 (**Plate 5**). In the developmental literature, that structure is called the oval bundle of His, and labels in the youngest specimens give both names to the structure.

After the dorsal root fibers bifurcate, there is developmental evidence in rats (Kudo and Yamada, 1987) that both the ascending and descending branches give off short, local collaterals (see Figure 143 in Ranson and Clark, 1959; see Figure 12-16 in Truex and Carpenter, 1969; see Figure 2-3 in Brodal, 1981; see Chapter 4, Section 4.2.2 in Altman and Bayer, 2001). These collaterals hug the medial edge of the dorsal horn before they enter the gray matter. For that reason, that part of the dorsal funiculus is called the *dorsal root collateralization zone*. Some of the collaterals invade the gray matter from the top of the dorsal horn in a dorsal bundle that first grows down, then curves upwards to form elaborate terminal arbors in various laminae of the dorsal horn. Other collaterals enter the dorsal horn in a dorsomedial bundle that grows ventrally and forms arbors within the motoneuron columns in the ventral horn (Kudo and Yamada, 1987). Just like the dorsal root bifurcation zone, the dorsal root collateralization zone is an early developing part of the dorsal funiculus. It first appears around GW6.7 (**Plate 9**) and GW6.8 (**Plate 10**).

The latest developing and eventually the largest component of the dorsal funiculus contains the principal ascending axons of the dorsal root ganglia in the traditionally named fasciculus gracilis and fasciculus cuneatus. Most probably, the ascending branch that initially bifurcates and gives off local collaterals in the lateral part of the dorsal funiculus near its point of entry shifts medially as it grows toward the brain to enter these fasciculi (see comments on p. 189 in Ranson and Clark, 1959; see Chapter 4, Section 4.2.3 in Altman and Bayer, 2001). The first evidence that these fasciculi appear is on GW8.5 (**Plates 12, 14-20**).

The Spinocephalic (Spinothalamic) Tract. This tract occupies a crescent-shaped region in the superficial ventral and lateral funiculi. The reason this tract is called the spino*cephalic* tract is because a higher proportion of its fibers terminate in the brainstem rather than the thalamus (Altman and Bayer, 2001). The suffix *cephalic* refers to the brain rather than the thalamus. This tract has been divided into two parts, a ventral and a lateral, but Brodal (1981) maintains that there is no distinction between the two. Haines (2000) calls these tracts the anterolateral system that includes the spino–olivary, spinoreticular, and spinotectal tracts. In addition, some axons terminate in the midbrain central gray and the hypothalamus.

The Intraspinal (Propriospinal) Tract. This tract, which carries local connections between neurons in the spinal cord that do not ascend to the brain, is traditionally called the propriospinal tract or fasciculus. The prefix *proprio-* means "within itself." Unfortunately, that prefix can be confused with a set of proprioceptive fibers that transmit sensory information from muscles and tendons. By that criterion, the spinocerebellar tracts along the margin of the lateral funiculus are also propriospinal tracts because they send proprioceptive information to the cerebellum. To avoid that confusion, we name the propriospinal tract the *intra*spinal tract.

The Overlapping of Tracts. Other than the spinocerebellar tracts, the fiber tracts in the ventral and lateral funiculi of the spinal cord do not have distinct borders (Ranson and Clark, 1959; Crosby et al., 1962; Truex and Carpenter, 1969; Brodal, 1981). The lines drawn in the white matter of the sections in this Atlas designate different densities of the myelin stain or different densities of glia in the cell body stains. They are not the actual borders of fiber tracts, and comments on the **B** parts of several plates indicate that. The ventral fibers of the lateral corticospinal tract overlap with the rubrospinal tract (labeled in only a few Plates). The spinocephalic tracts overlap with the vestibulospinal, spino–olivary (not labeled in any Plate), and spinotectal tracts (not labeled in any Plate). The part of the white matter that is designated as the intraspinal tract in this Atlas is a circumferential zone around the ventral horn, lateral intermediate gray, and lateral dorsal horn that overlaps with the medial longitudinal fasciculus (labeled at cervical levels), the tectospinal tract (labeled at cervical levels), and the lateral reticulospinal tract (only labeled in some Plates). In actuality, the intraspinal tract also extends along the medial surface of the dorsal horn and overlaps with the dorsal root collateralization zone (Fix,1987; Roberts et al., 1987; DeArmond et al., 1989; Haines, 2000). To simplify the plates, the dorsal root collateralization zone is not double-labeled as part of the intraspinal tract.

Developmental Terminology. His (1889) identified three different layers of the embryonic brain and spinal cord: (1) an inner layer surrounding the brain ventricles and spinal canal (*Innerplatte*), (2) a cell-rich mantle layer, and (3) a cell-sparse marginal layer. His further subdivided the inner layer into four plates: a roof plate, a floor plate, and two lateral plates (alar and basal). The sulcus limitans forms the dividing line between the two lateral plates. That terminology is modified as described below.

This Atlas keeps His' designation of the roof and floor plates because these are different cytologically and developmentally from the thick lateral plate that contains the *neuroepithelium*, a term used by Langman et al. (1966). The neuroepithelium contains the stem cells that give rise to neurons and glia. This Atlas modifies His' segregation of the lateral plate by partitioning it into three parts rather than two, the dorsal neuroepithelium (the source of secondary sensory neurons in the dorsal horn), the intermediate neuroepithelium (the source of neurons in the intermediate gray), and ventral neuroepithelium (the source of motoneurons and interneurons in the ventral horn). The sulcus limitans marks that part of the spinal canal that will persist as the central canal and is surrounded by the intermediate neuroepithelium. His' terms mantle and marginal layers are not useful because developing structures should be named according to what they will become at maturity, not be given a different set of names during development. That is a policy followed throughout the *Atlas of Prenatal Rat Brain Development* (Altman and Bayer, 1995). In the spinal cord, the cells that leave the neuroepithelium are specific types of neurons that migrate in a variety of different directions, settle in specific locations, and begin to differentiate in the *primordial gray matter*, not the *mantle layer*. Cell groups in the gray matter are named as soon as they are recognizable; for example, cells in the early ventral horn, intermediate gray, and dorsal horn appear as early as GW4.75 (**Plate 4**). The cell-sparse zone beneath the pial membrane in the developing spinal cord is the *primordial white matter*, not the *marginal layer*. Definite axons appear in the primordial white matter around GW5.25 (**Plate 5**) simultaneously in the ventral and dorsal funiculi; axons appear later in the lateral funiculus.

Two other terms are used throughout this Atlas that refer to developing features not widely known. *Sojourn zone* describes an accumulation of premigratory neurons in the neuroepithelium; the evidence that these are aggregates of postmitotic neurons rather than neuroepithelial cells comes from short survival ^{3}H–thymidine autoradiography (see Figure 3-10 in Altman and Bayer, 2001). The *dorsal root boundary cap* is an aggregate of specialized glial cells adjacent to the external wall of the spinal cord (Altman and Bayer, 1984; see Chapter 4, Section 4.1.3 in Altman and Bayer, 2001). Boundary caps mark spots on the pia where dorsal root fibers penetrate the spinal cord. Indeed, specific boundary caps mark nerve entry or exit points in the developing brain, indicating that they are important developmental structures (see Figure 62 in Altman and Bayer, 1981).

F. References

Altman, J., and S. A. Bayer (1981) *Development of the Cranial Nerve Ganglia and Related Nuclei in the Rat.* (Advances in Anatomy Embryology and Cell Biology, Vol. 74.) Berlin: Springer-Verlag.

Altman, J., and S. A. Bayer (1984) *Development of the Rat Spinal Cord.* (Advances in Anatomy Embryology and Cell Biology, Vol. 85.) Berlin: Springer-Verlag.

Altman, J., and S. A. Bayer (1995) *Atlas of Prenatal Rat Brain Development*. Boca Raton, FL: CRC Press.

Altman, J., and S. A. Bayer (2001) *Development of the Human Spinal Cord. An Interpretation Based on Experimental Studies in Animals.* New York: Oxford University Press.

Brodal, A. (1981) *Neurological Anatomy in Relation to Clinical Medicine*, (3rd edition). New York: Oxford University Press.

Corner, G. W. (1929) A well-preserved human embryo of 10 somites. *Carnegie Institution of Washington, Contributions to Embryology*, 20:81-102.

Crosby, E. C., T. Humphrey, and E. W. Lauer (1962) *Correlative Anatomy of the Nervous System.* New York: The Macmillan Company.

DeArmond, S. J., M. M. Fusco, and M. M. Dewey (1989) *Structure of the Human Brain, A Photographic Atlas,* (3rd edition). New York: Oxford University Press.

Fix, J. D. (1987) *Atlas of the Human Brain and Spinal Cord.* Gaithersburg, MD: Aspen Publishers.

Haines, D. (2000) *Neuroanatomy: An Atlas of Structures, Sections, and Systems,* (5th edition). Philadelphia, PA: Lippincott Williams & Wilkins.

Haleem, M. (1990) *Diagnostic Categories of the Yakovlev Collection of Normal and Pathological Anatomy and Development of the Brain.* Washington, D.C.: Armed Forces Institute of Pathology.

His, W. (1889) Die Neuroblasten und deren Entstehung im embryonalen Mark. *Abhandlungen der königlisches Gesellschaft der Wissenschaften, mathematisch-physikalische Klasse*, 15:313–372.

Kudo, N., and T. Yamada (1987) Morphological and physiological studies of the development of the monosynaptic reflex pathway in the rat lumbar spinal cord. *Journal of Physiology*, 389:441-459.

Langman, J., R. L. Guerrant, and B. G. Freeman (1966) Behavior of neuroepithelial cells during closure of the neural tube. *Journal of Comparative Neurology*, 131:15-26.

Minot, C. S. (1903) *A Laboratory Text-Book of Embryology*. Philadelphia, PA: Blakiston.

Ranson, S. W., and S. L. Clark (1959) *The Anatomy of the Nervous System, its Development and Function,* (10th edition). Philadelphia, PA: W. B. Saunders Company.

Roberts, M., J. Hanaway, and D. K. Morest (1987) *Atlas of the Human Brain in Section*, (2nd edition). Philadelphia, PA: Lea & Febiger.

Sobotta, J. (1963) *Atlas of Human Anatomy*, (8th English Edition). (Volume III, Part II: *Atlas of Neuroanatomy. Central Nervous System, Autonomic Nervous System Eye, Ear, and Skin,* F. H. J. Figge, Ed.) New York: Hafner.

Truex, R. C., and M. B. Carpenter (1969) *Human Neuroanatomy*, (6th edition). Baltimore, MD: Williams and Wilkins Company.

Streeter, G. L., C. H. Heuser, and G. W. Corner (1951) Developmental horizons in human embryos: Description of age groups XIX, XX, XXI, XXII, and XXIII, being the fifth issue of a survey of the Carnegie Collection. *Carnegie Institution of Washington, Contributions to Embryology,* 34:165-196.

II. The First Trimester

A. Cervical levels only: Embryos at GW4.0 to GW8.5

Plate 1 is a survey of sections from the cervical levels of the spinal cord in the youngest 11 specimens. All sections are shown at the same scale. The boxes enclosing each section list: the gestational age in weeks (GW); the crown rump length (CR) in millimeters (mm); the specimen number preceded by **M** (Minot Collection) or **C** (Carnegie Collection); the slide number and section number from the set of slides containing all the sections of that specimen; and the total area of the section in square millimeters (mm^2). Since all Minot specimens have consecutive section numbers, no slide number is given. Full-page normal contrast photographs of each specimen are in **Plates 2A-12A.** Low contrast photographs with superimposed labels and outlines of structural details are in **Plates 2B-12B.** All A-B plates are on facing pages to allow the user to glance back and forth at the unlabeled structure on the left and the labeled one on the right. Plates 2-12 also feature low magnification views of the spinal cord surrounded by other parts of the embryo.

Gross changes in morphology. It is during early stages of embryonic development (GW4.0-GW8.5) that the most dramatic structural changes take place in the spinal cord. On GW4.0 (M714) the most prominent component is the neuroepithelium, the germinal matrix of proliferating stem cells (NEP cells), that surround a slit-shaped spinal canal. The roof and floor plates are cell bridges that close off the spinal canal in the dorsal midline and ventral midline, respectively. Only a few cells are in the differentiating field of future gray matter, and the entire spinal cord is surrounded by a thin layer of primordial white matter. By GW8.5 (C609), the gray matter is the most prominent component of the spinal cord; the ventral horn and the intermediate gray already have complete populations of neurons, and nearly all neuronal populations are complete in the dorsal horn and the central autonomic area. Most major fiber tracts are present in the white matter. The spinal canal remains in a central location (central canal). A remnant of neuroepithelium lines the dorsal part of the central canal, and an ependymal/glial matrix lines the intermediate and ventral part. Roof and floor plates are still visible in the midline.

The sequence of neurogenesis between various components of the gray matter. Between GW4.0 and GW4.75, the ventral part of the neuroepithelium is the largest component. The first neurons outside the ventral neuroepithelium at GW4 are differentiating motoneurons in the ventral horn. Between GW4.75 and GW5.5, ventral horn motoneurons increase as the ventral neuroepithelium recedes. By GW5.5, the center of the ventral neuroepithelium is thinner than its dorsal or ventral borders (see **Plate 6B**). The thinned area may mark the exhaustion of the NEP (stem) cell population that produces motoneurons. Other parts of the ventral neuroepithelium are presumably producing ventral horn interneurons. By GW6.8, the ventral neuroepithelium is becoming an ependymal/glial proliferative matrix that no longer produces neurons. The intermediate neuroepithelium expands rapidly between GW5.25 and GW5.5 prior to growth of the intermediate gray. The intermediate neuroepithelium gradually declines by GW7.25 as the intermediate gray expands. By GW8.5, the intermediate neuroepithelium becomes an ependymal/glial matrix. Between GW5.5 and GW6.5, the dorsal neuroepithelium dramatically increases its size, remains large on GW7.25, and is much smaller on GW8.5. The most rapid growth of the dorsal horn occurs between GW7.25 and GW8.5, coinciding with the shrinkage of the dorsal neuroepithelium (see Chapter 5, Section 5.3, Altman and Bayer, 2001). The dorsal neuroepithelium is characterized by "undulations" on its outer edges. These may be groups of premigratory postmitotic neurons, the postulated "sojourn zones" that have been well documented with ^{3}H–thymidine autoradiography in the spinal cord of rats on embryonic day 15 (see Chapter 3, Figure 3-10 in Altman and Bayer, 2001). These undulations are still present on GW8.5 (**Plate 12**) indicating that a few neurons are still being generated (or have just been generated) by the dorsal neuroepithelium.

The sequence of appearance of various components of the white matter. The oval bundle of His, first definite at GW5.25 (**Plate 5**), is the bifurcation zone of ingrowing axons of the dorsal root ganglion cells, the earliest component of the dorsal funiculus. From that lateral position, the dorsal funiculus grows medially to approach the dorsal midline by GW7.25 (**Plate 11**). By GW8.5, the dorsal funiculus grows downward along the midline, occupying the space vacated by the retreating roof plate and the receding dorsal neuroepithelium (**Plate 12**). The ventral funiculus also appears by GW5.25 in the wedge-like space between the floor plate, the ventral neuroepithelium, and the ventral horn. By GW5.5, axons of the early generated commissural neurons are crossing the midline in the ventral commissure beneath the floor plate. The lateral part of the white matter is still small at GW5.5 in C8998 but is larger at GW5.5 in M2161, indicating the beginning growth of the lateral funiculus. The dorsal, ventral, and lateral funiculi continue to expand throughout the rest of the first trimester (see Chapter 5, Section 5.6, Altman and Bayer, 2001).

PLATE 1

Plates 2A, 2B
GW4.0
CR 4.0 mm

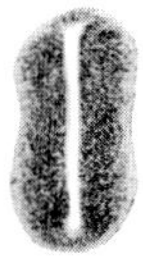

M714
Section 70
Total area:
0.065 mm^2

Plates 3A, 3B
GW4.5
CR 6.3 mm

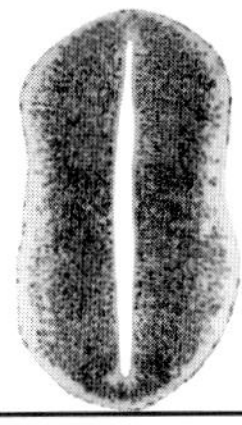

M2300
Section 150
Total area:
0.1795 mm^2

Plates 4A, 4B
GW4.75
CR 6.7 mm

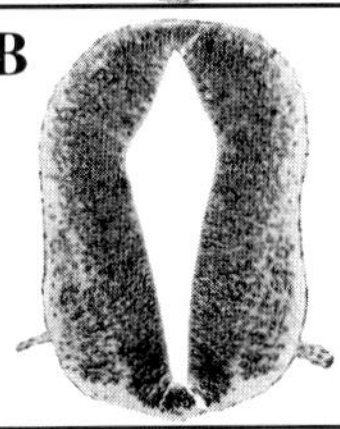

M2285
Section 350
Total area:
0.2324 mm^2

Plates 5A, 5B
GW5.25
CR 8.0 mm

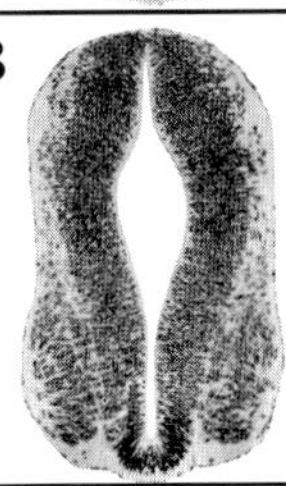

M2065
Section 350
Total area:
0.2478 mm^2

Plates 6A, 6B
GW5.5
CR 11.0 mm

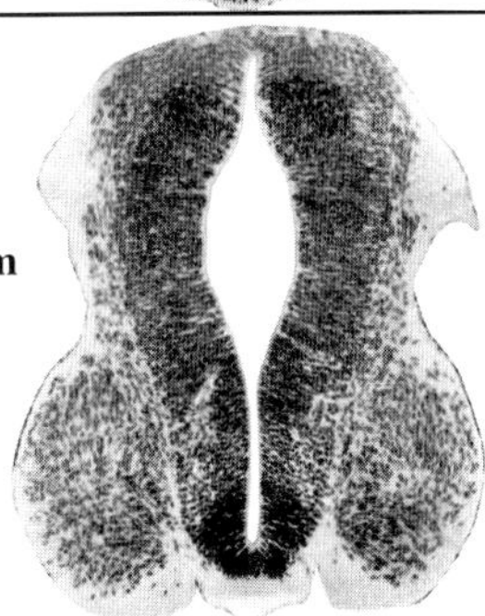

C8998
Slide 19
Section 2
Total area:
0.6447 mm^2

Plates 7A, 7B
GW5.5
CR 11.1 mm

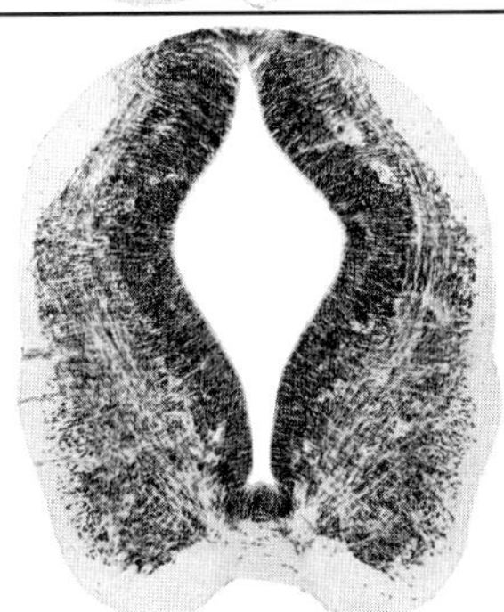

M2161
Section 165
Total area:
0.7530 mm^2

Plates 8A, 8B
GW6.5
CR 14.5 mm

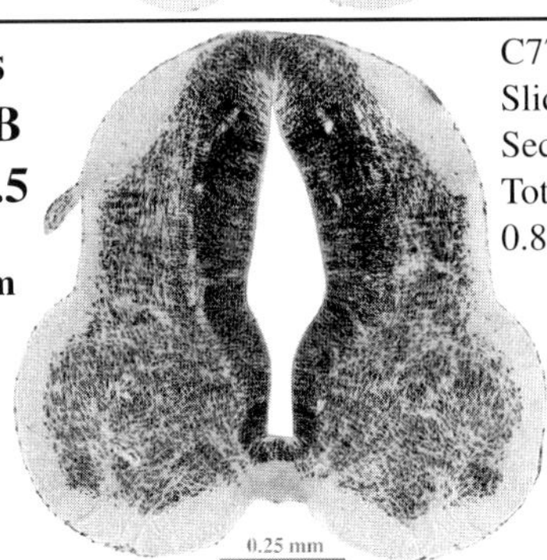

C7707
Slide 34
Section 15
Total area:
0.8541 mm^2

Plates 9A, 9B
GW6.7
CR 17.1 mm

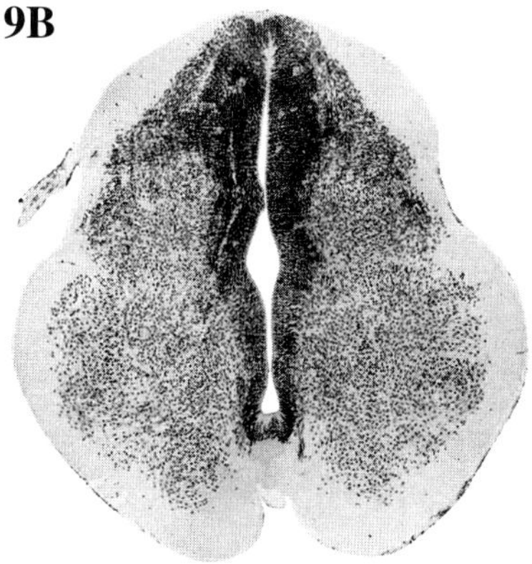

C8965
Slide 29
Section 10
Total area:
1.0092 mm^2

Plates 10A, 10B
GW6.8
CR 17.5 mm

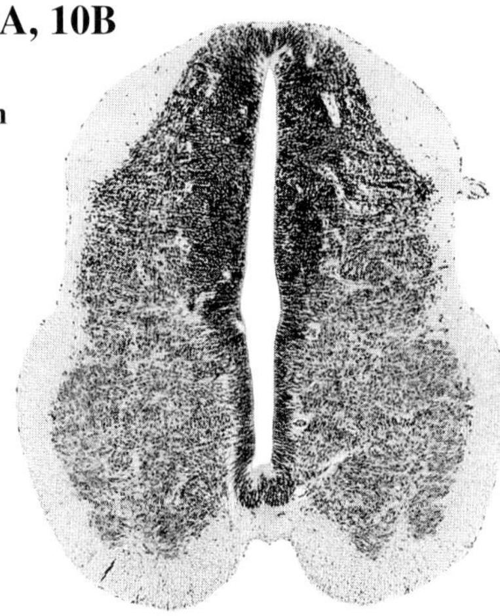

M2155
Section 555
Total area:
1.2977 mm^2

Plates 11A, 11B
GW7.25
CR 22.5 mm

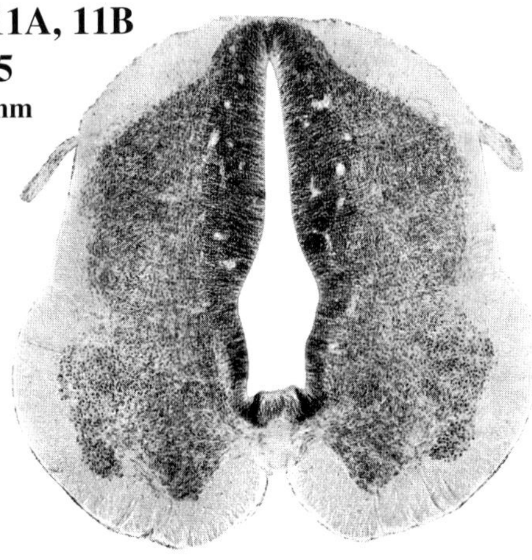

C7254
Slide 34
Section 5
Total area:
1.4555 mm^2

Plates 12A, 12B
GW8.5
CR 32 mm

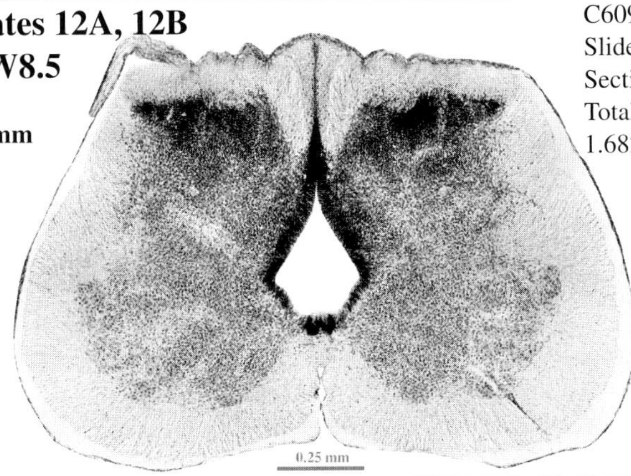

C609
Slide 32
Section 7
Total area:
1.6824 mm^2

PLATE 2A

GW4.0
CR 4.0 mm
M714
Cervical
Cell body stain

Areas (mm^2)

Spinal canal	.0046
Neuroepithelium	.0459
Roof plate	.0021
Floor plate	.0023
Primordial white matter	.0101

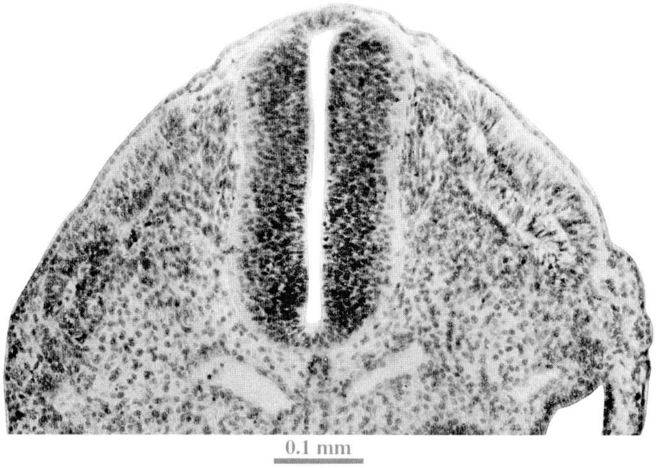

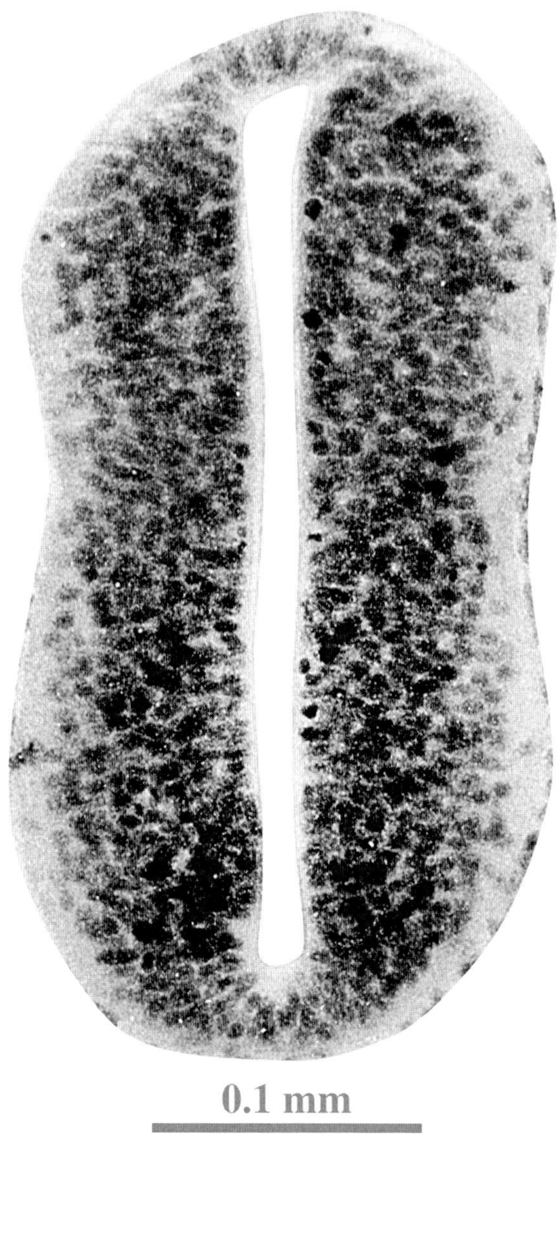

PLATE 2B

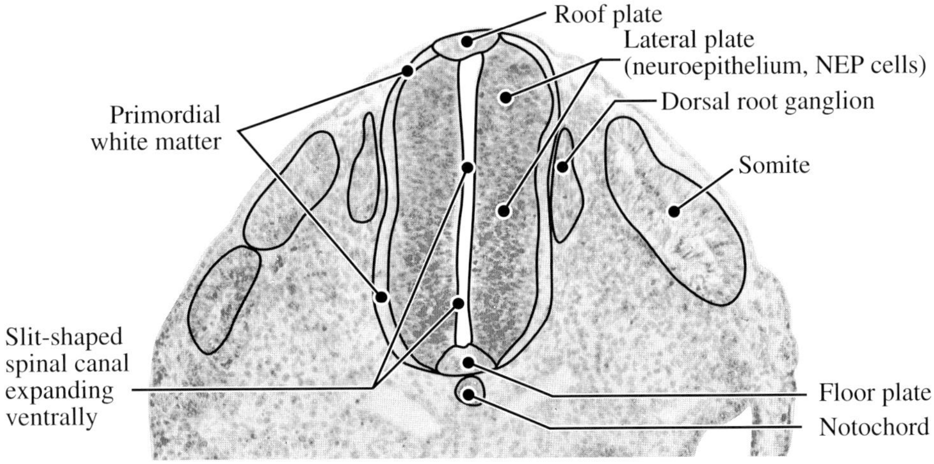

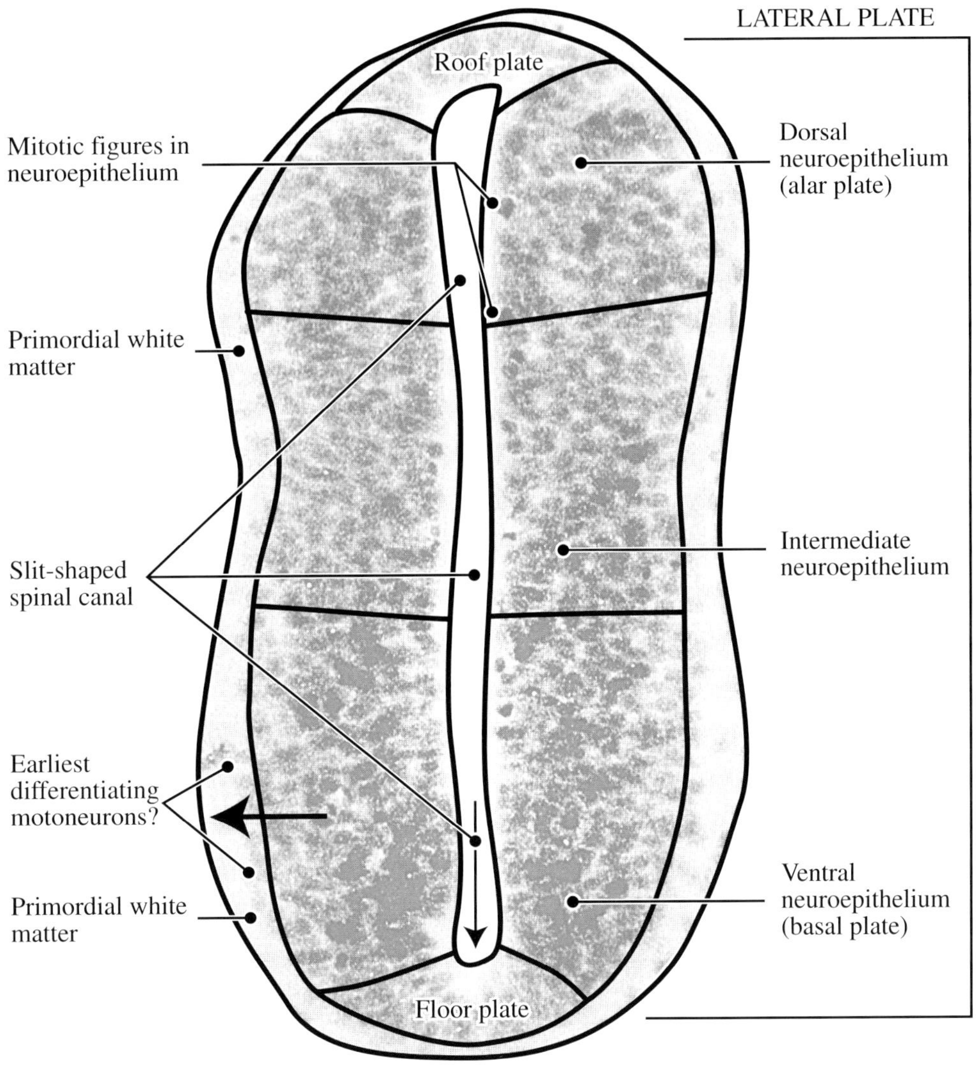

Large arrow extending from the neuroepithelium indicates migrating neurons.
Small arrow in the spinal canal indicates the direction of expansion.

PLATE 3A

GW4.5
CR 6.3 mm
M2300
Cervical
Cell body stain

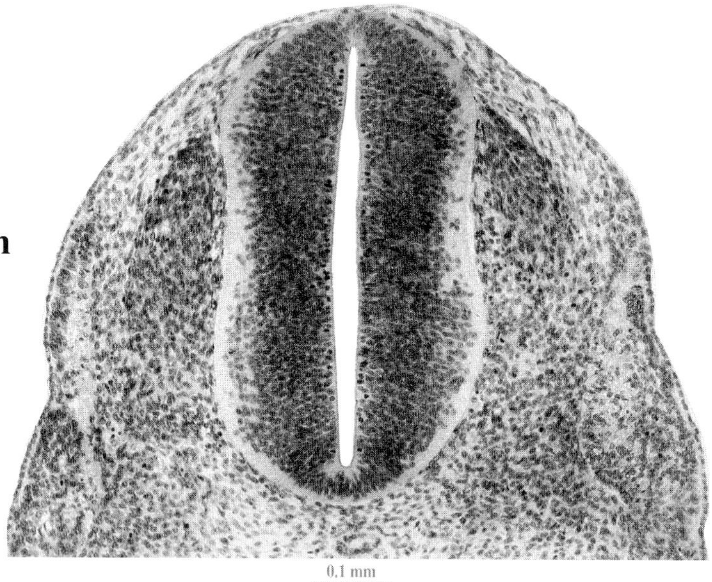

Areas (mm^2)

Spinal canal	.0130
Neuroepithelium	.1163
Roof plate	.0036
Floor plate	.0043
Gray matter	.0098
White matter	.0325

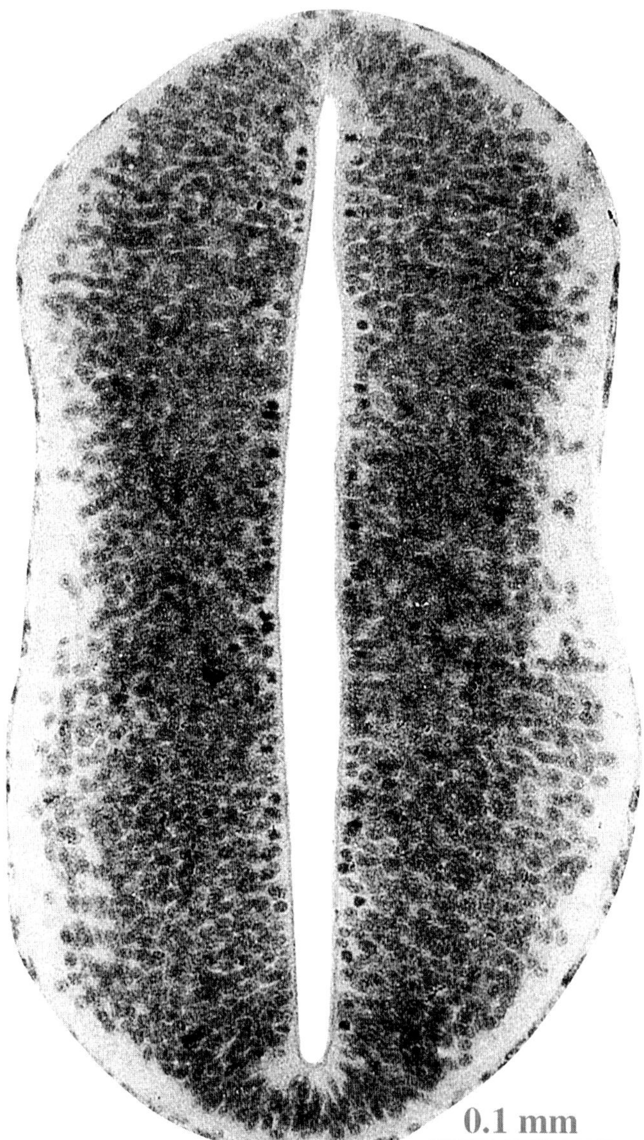

PLATE 3B

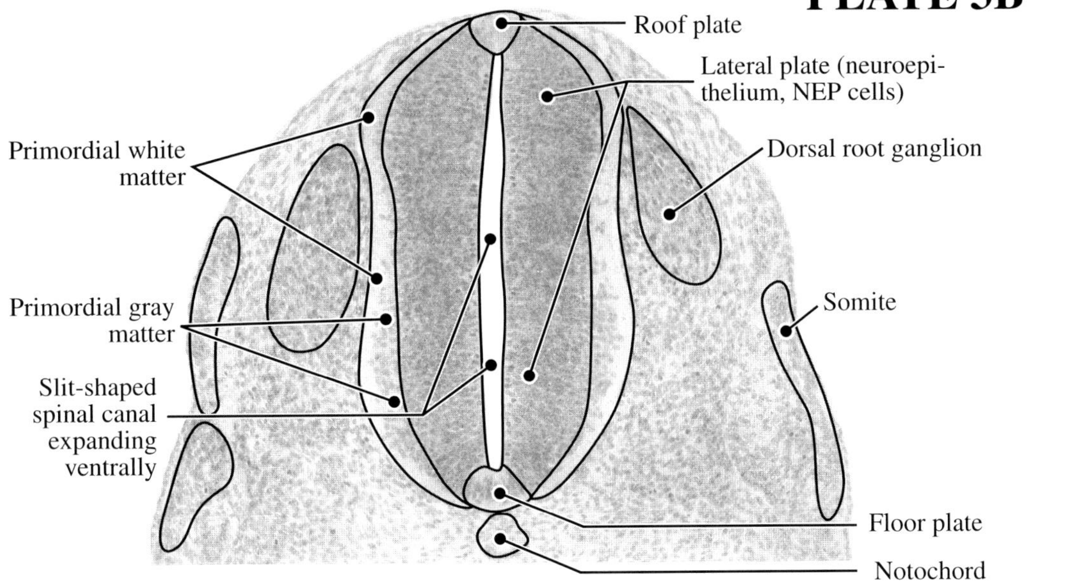

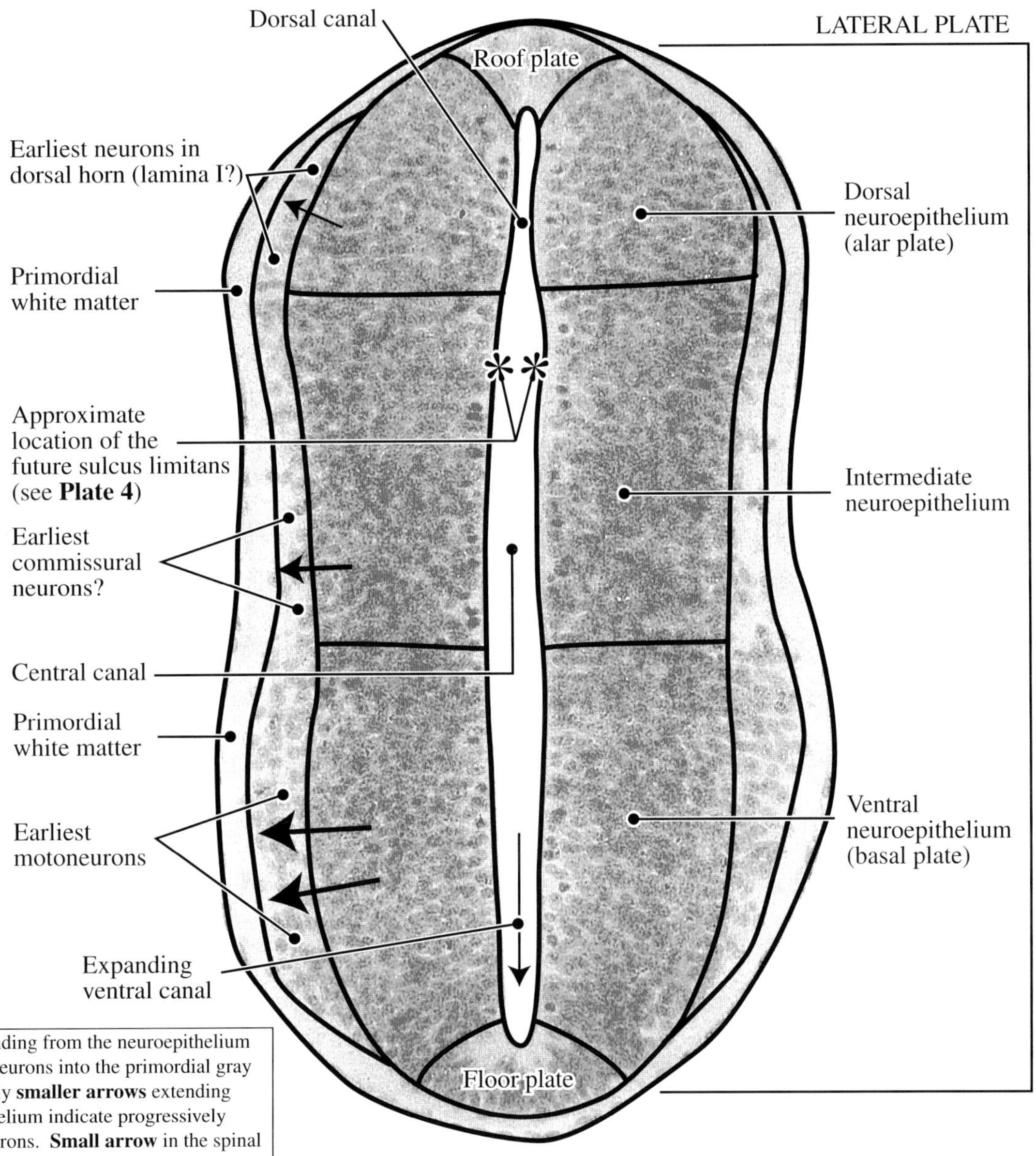

Large arrows extending from the neuroepithelium indicate migrating neurons into the primordial gray matter. Progressively **smaller arrows** extending from the neuroepithelium indicate progressively fewer migrating neurons. **Small arrow** in the spinal canal indicates the direction of expansion.

PLATE 4A

GW4.75
CR 6.7 mm
M2285
Cervical
Cell body stain

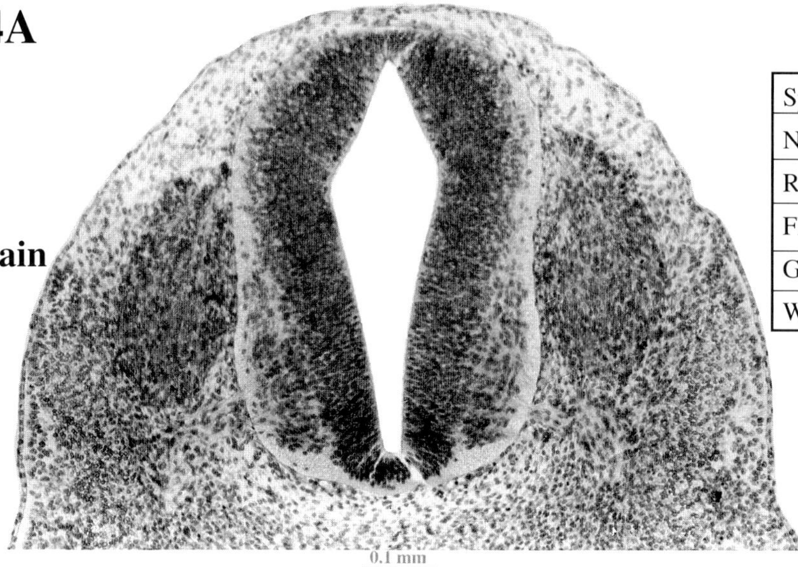

Areas (mm^2)

Spinal canal	.0400
Neuroepithelium	.1080
Roof plate	.0037
Floor plate	.0048
Gray matter	.0476
White matter	.2324

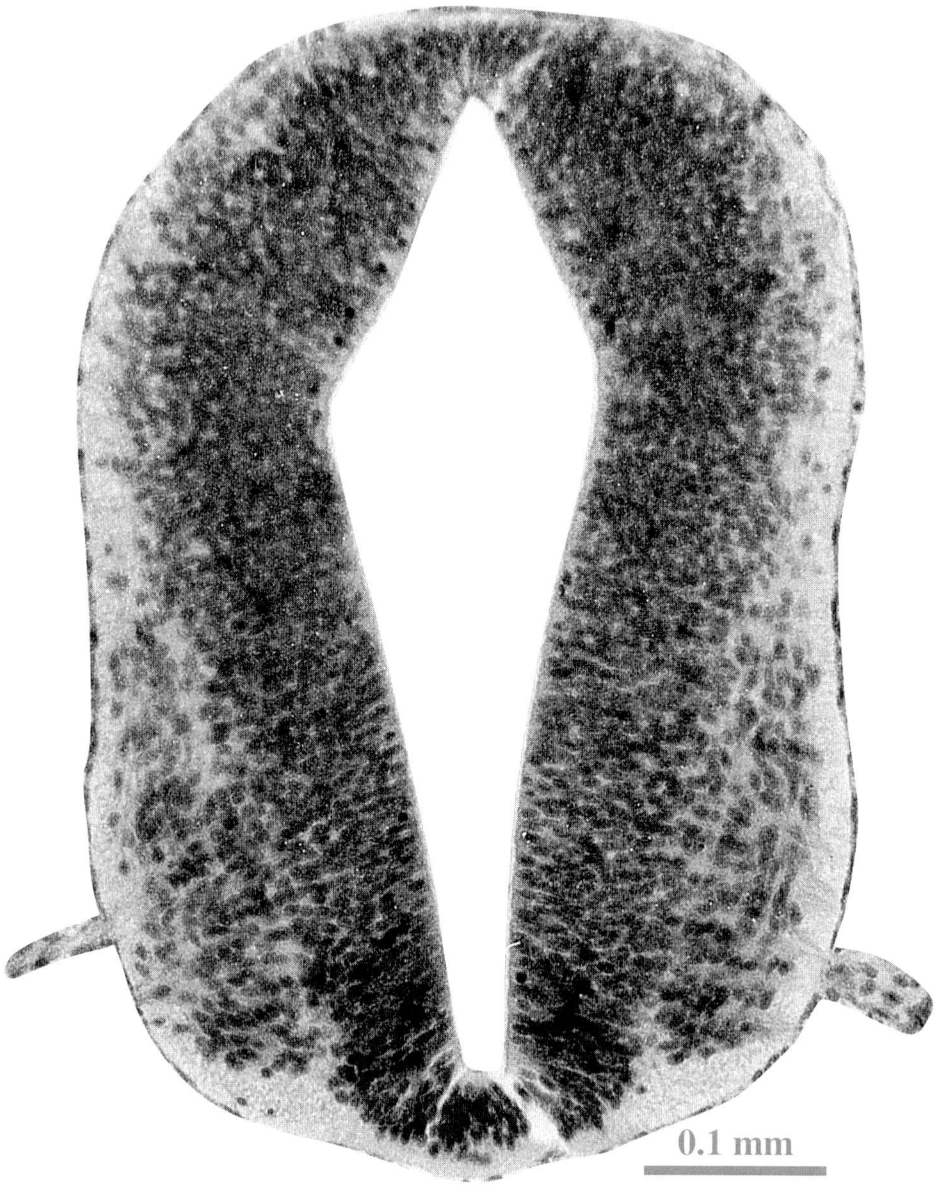

PLATE 4B

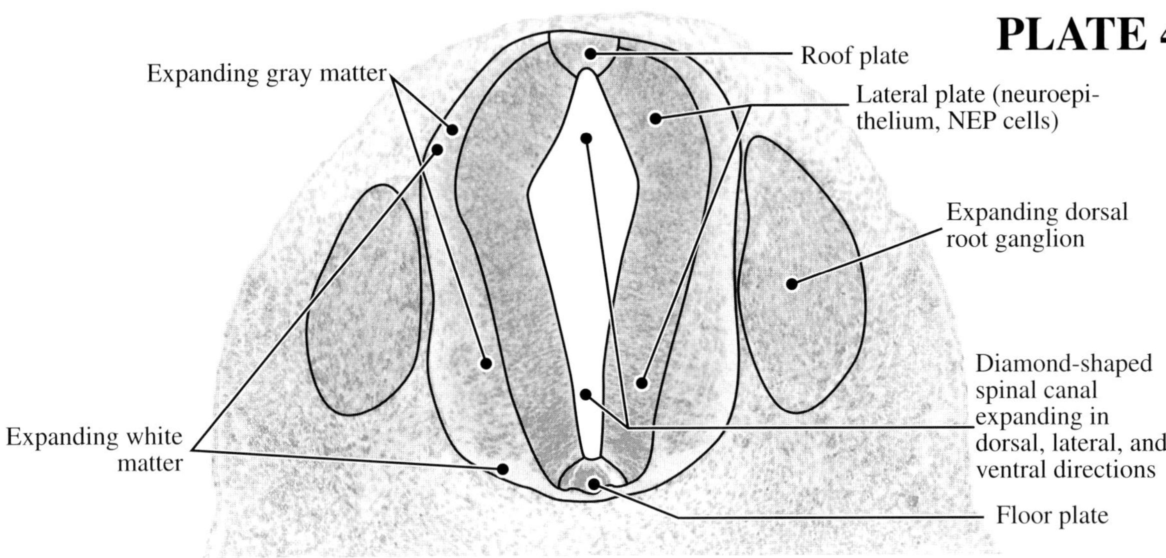

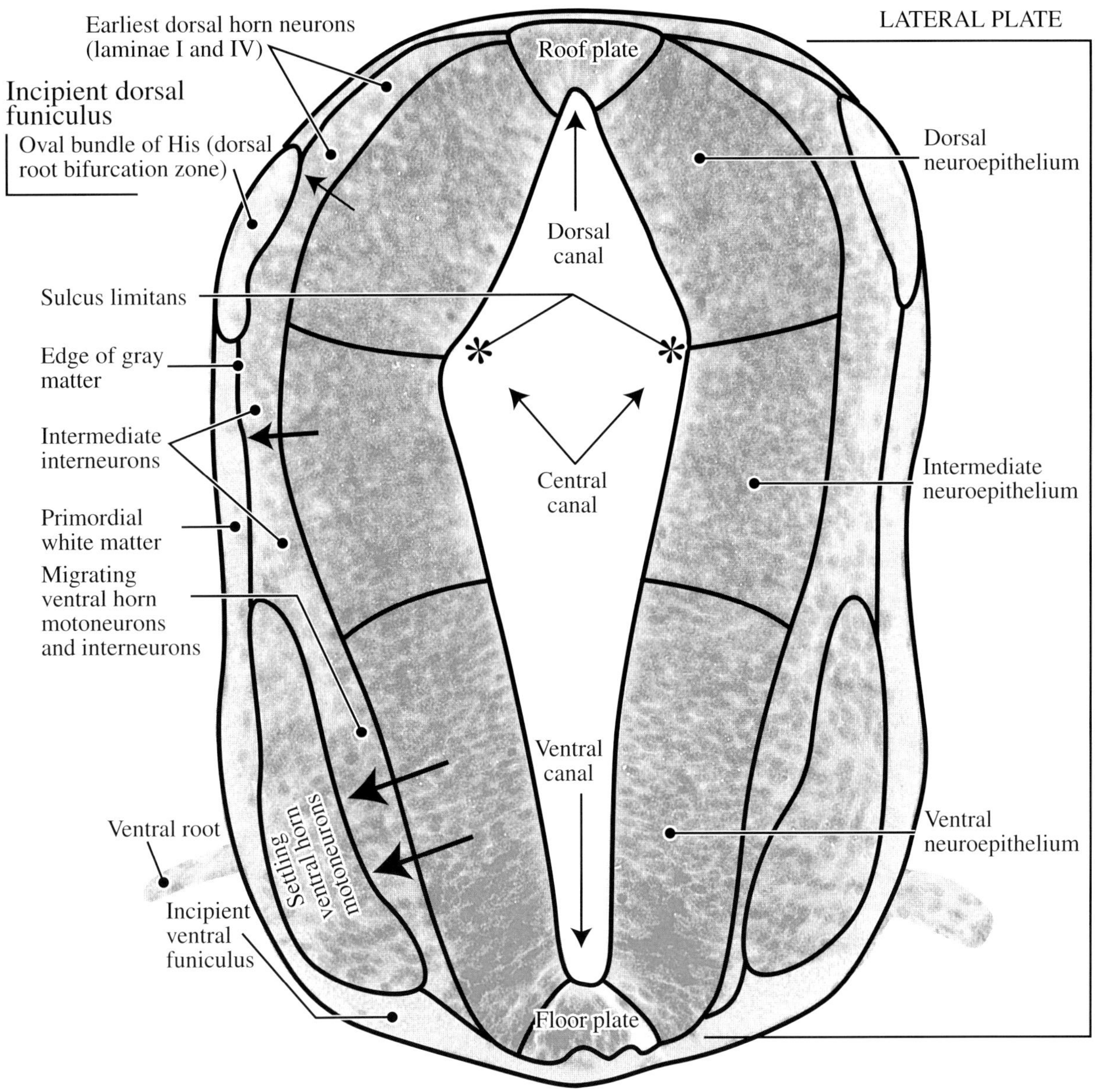

Large arrows extending from the neuroepithelium indicate many migrating neurons. Progressively **smaller arrows** extending from the neuroepithelium indicate progressively fewer migrating neurons. **Arrows** in the spinal canal indicate the direction of expansion.

PLATE 5A

GW5.25
CR 8.0 mm
M2065
Cervical
Cell body stain

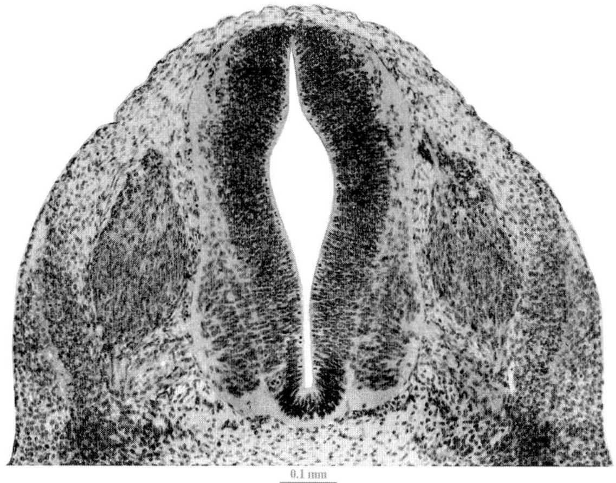

Areas (mm²)

Spinal canal	.0255
Neuroepithelium	.1103
Roof plate	.0012
Floor plate	.0067
Gray matter	.0752
White matter	.0289

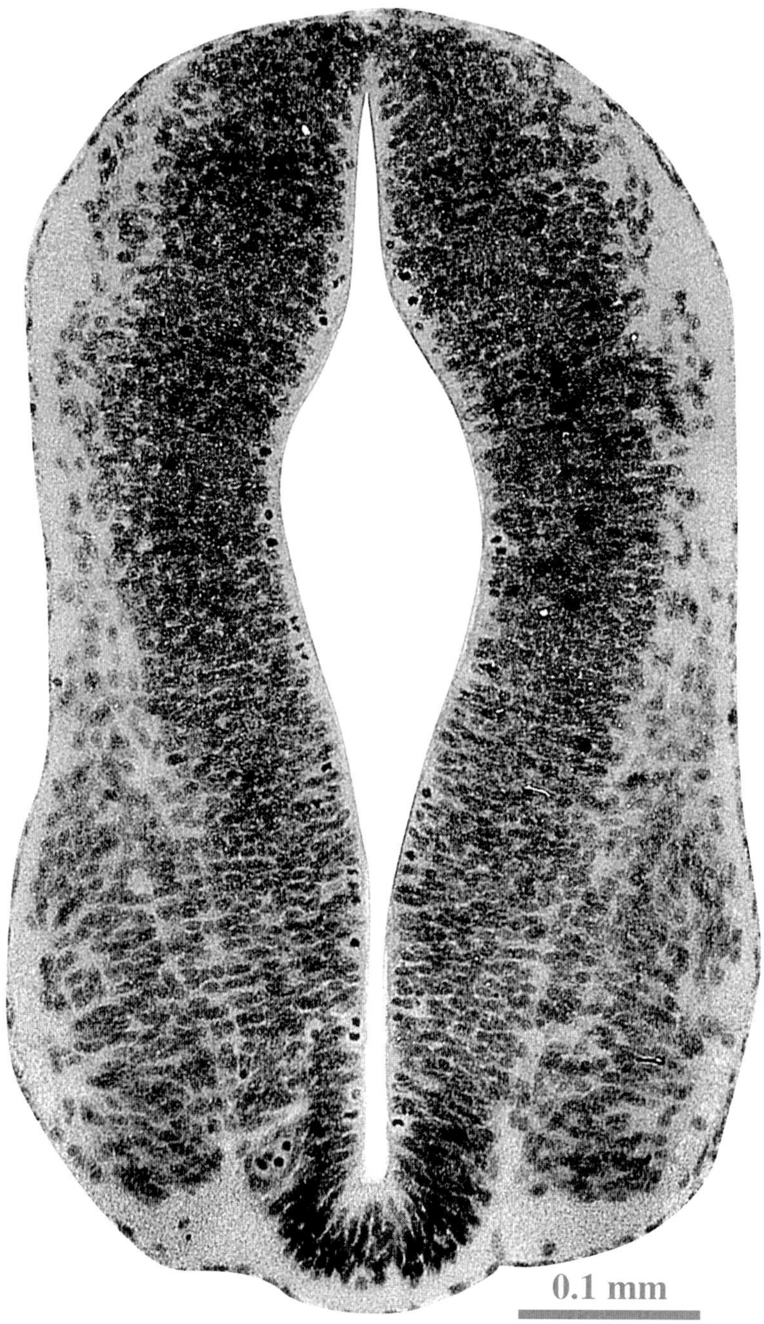

PLATE 5B

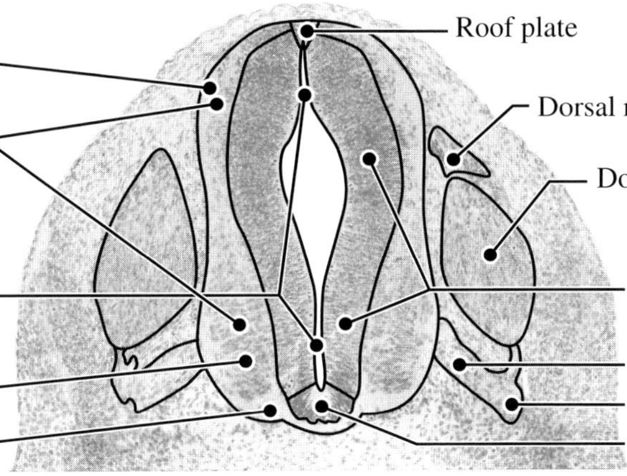

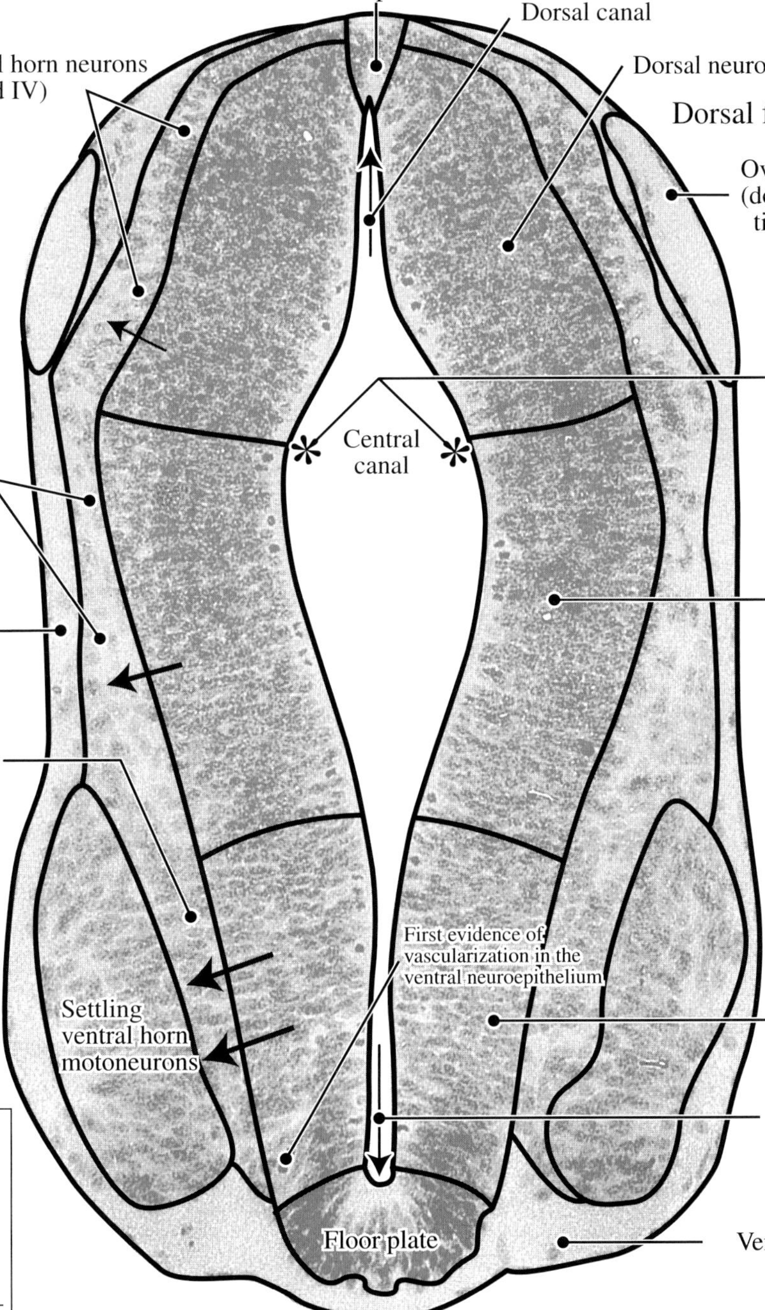

Large arrows extending from the neuroepithelium indicate many migrating neurons. Progressively **smaller arrows** extending from the neuroepithelium indicate progressively fewer migrating neurons. **Arrows** in the spinal canal indicate the direction of expansion.

PLATE 6A

GW5.5
CR 11 mm
C8998
Cervical
Cell body stain

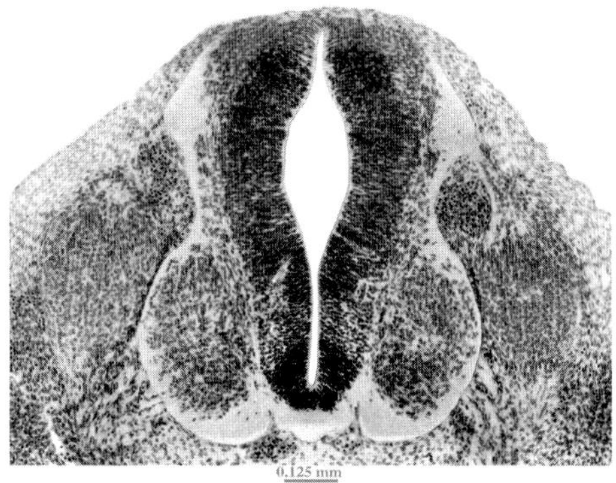

Areas (mm^2)

Spinal canal	.0523
Neuroepithelium	.2376
Roof plate	.0029
Floor plate	.0079
Gray matter	.2522
White matter	.0919

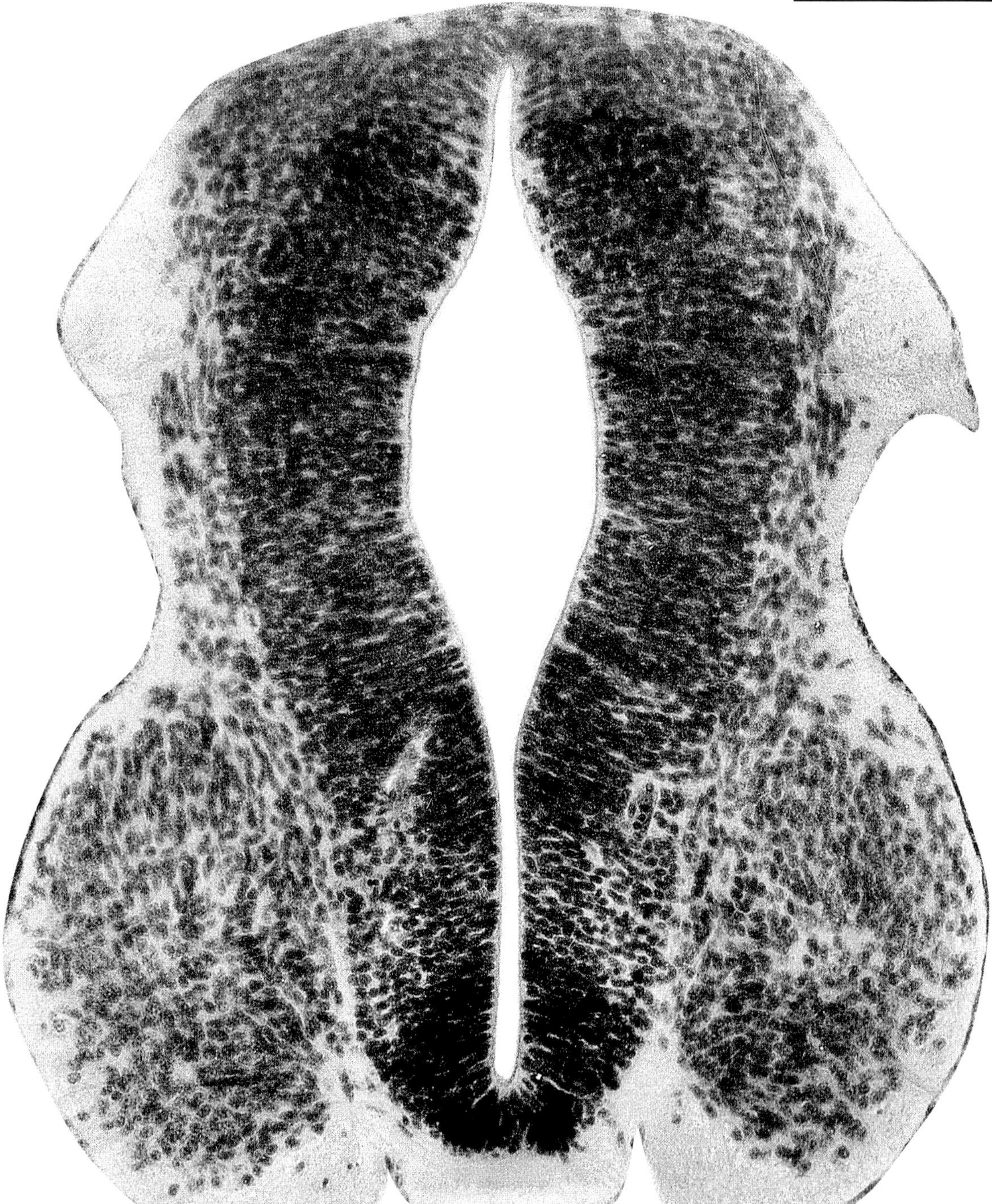

0.125 mm

PLATE 6B

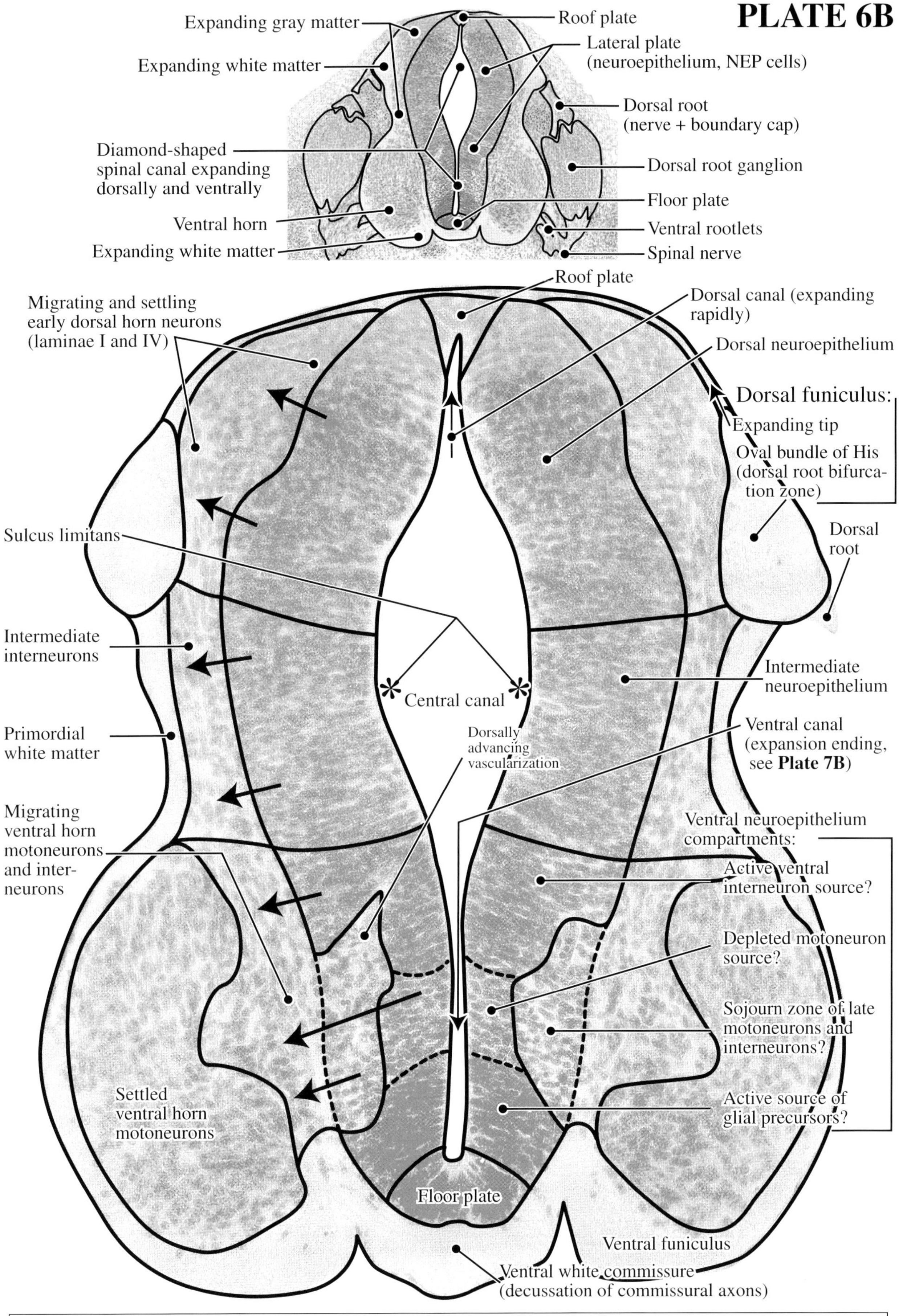

Large arrows extending from the neuroepithelium indicate migrating neurons. **Arrows** in the spinal canal indicate the direction of expansion.

PLATE 7A

GW5.5
CR 11.1 mm
M2161
Cervical
Cell body stain

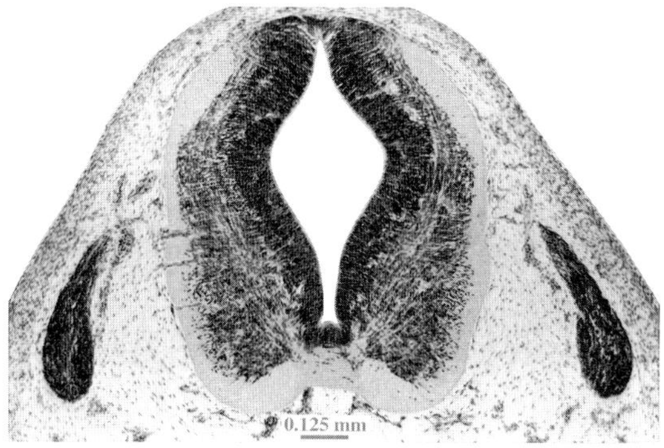

Areas (mm^2)

Spinal canal	.0974
Neuroepithelium	.2160
Roof plate	.0066
Floor plate	.0070
Gray matter	.2842
White matter	.1418

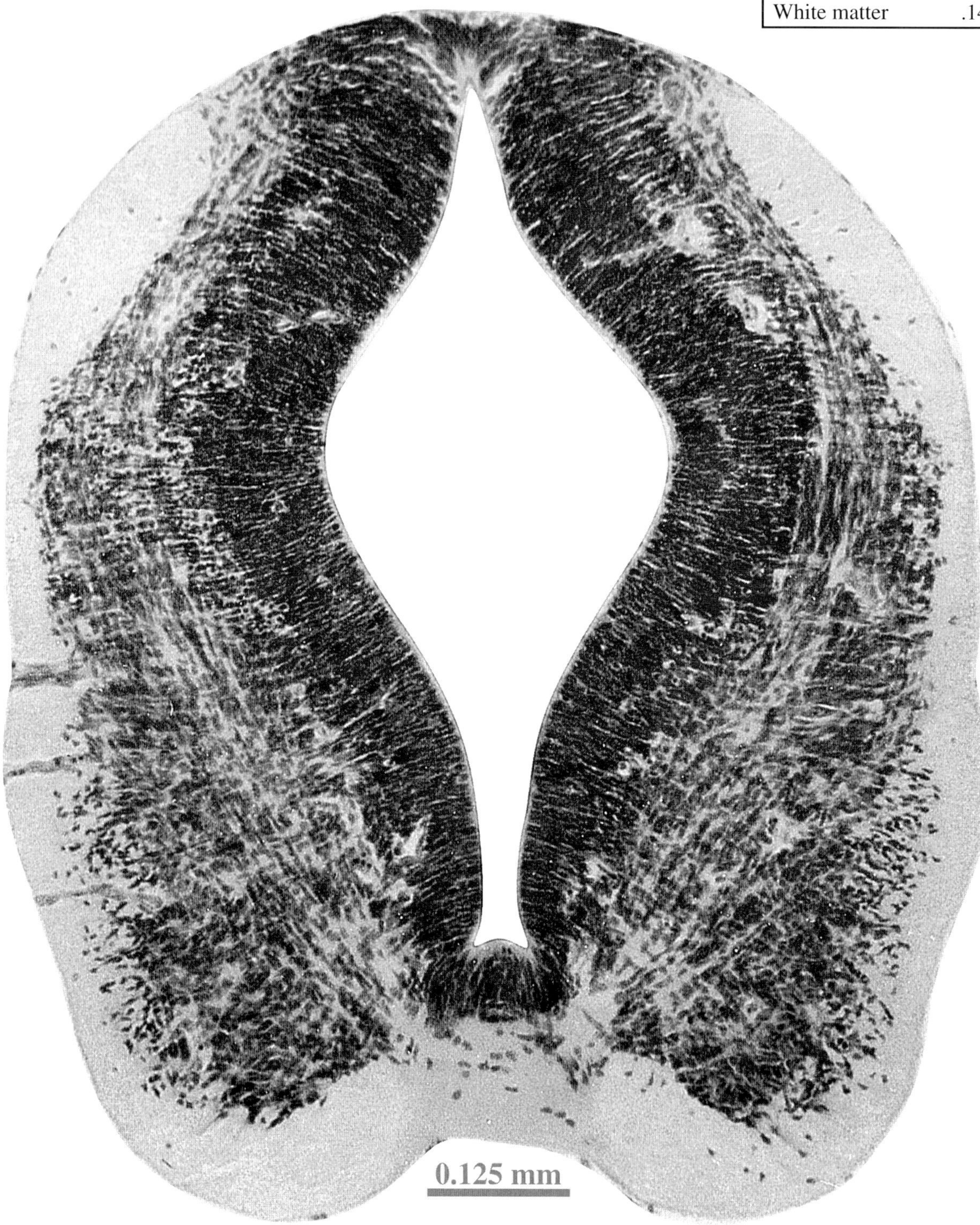

PLATE 7B

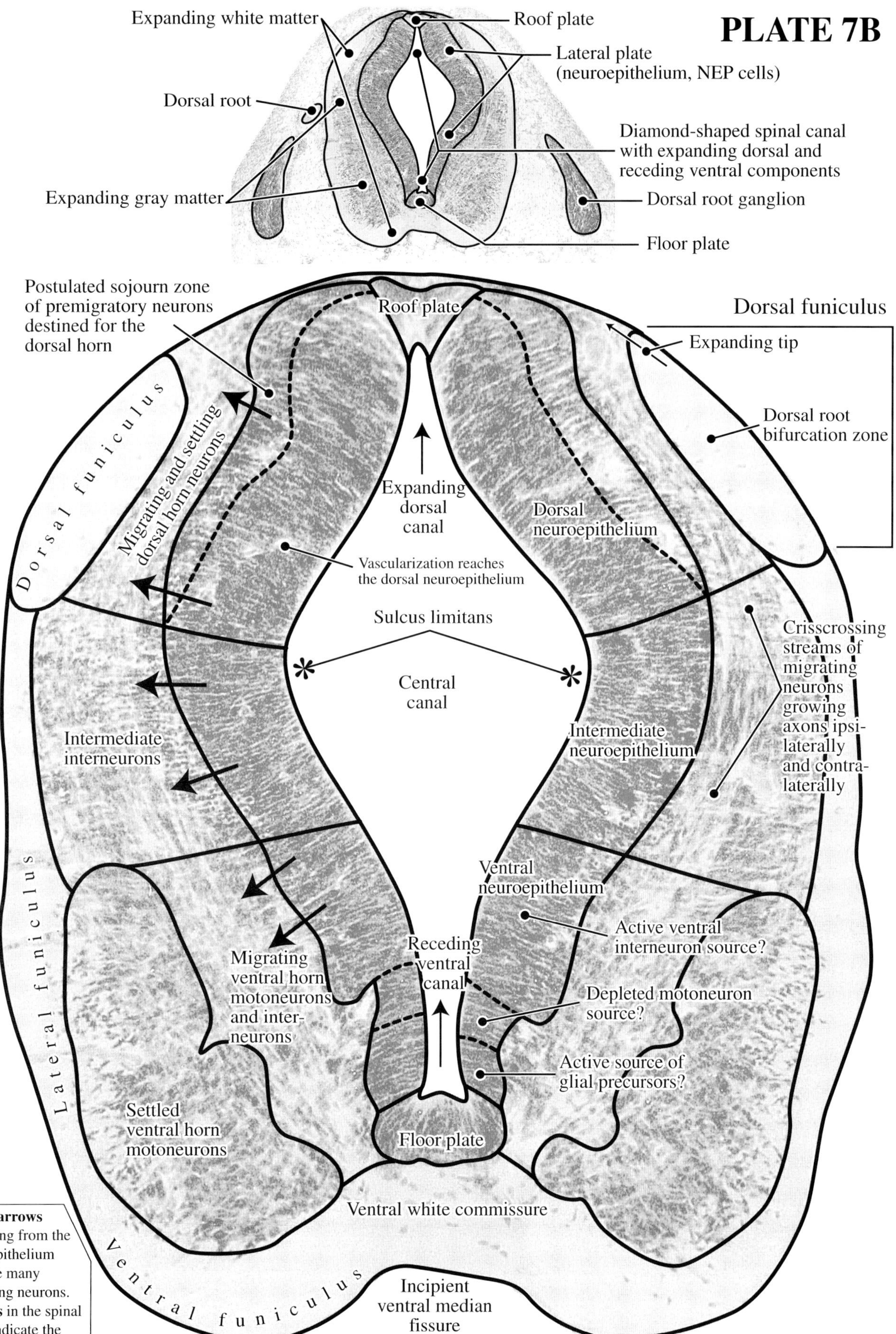

Large arrows extending from the neuroepithelium indicate many migrating neurons. **Arrows** in the spinal canal indicate the direction of expansion or contraction.

PLATE 8A

GW6.5
CR 14.5 mm
C7707
Cervical
Cell body stain

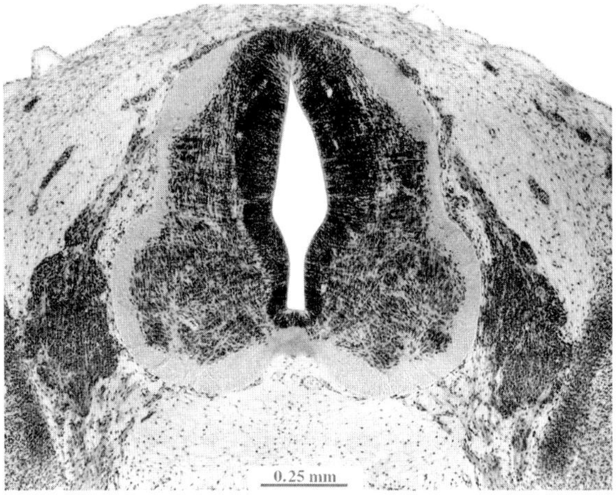

Areas (mm²)

Spinal canal	.0559
Neuroepithelium	.1810
Roof plate	.0136
Floor plate	.0057
Gray matter	.3932
White matter	.2047

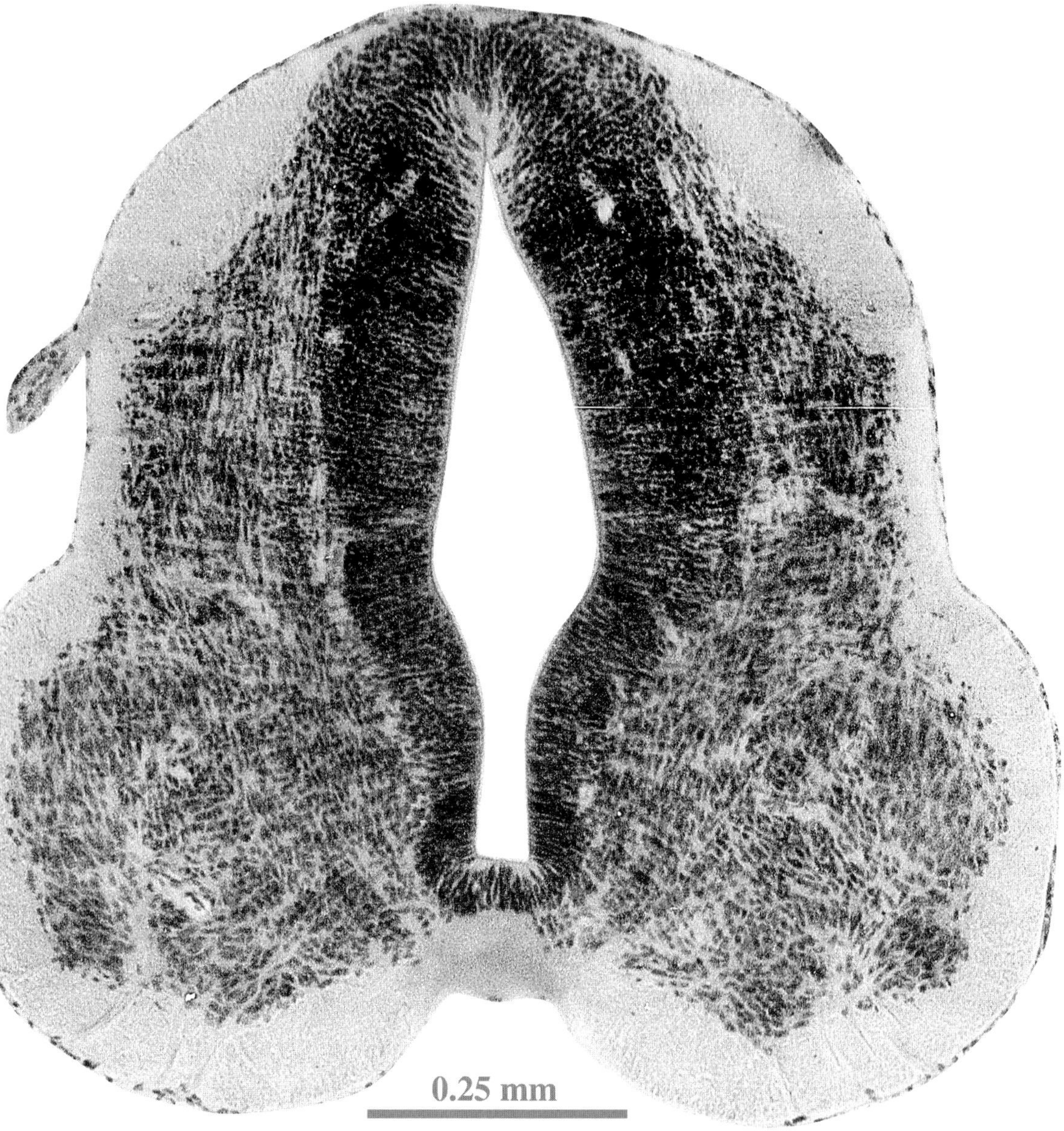

PLATE 8B

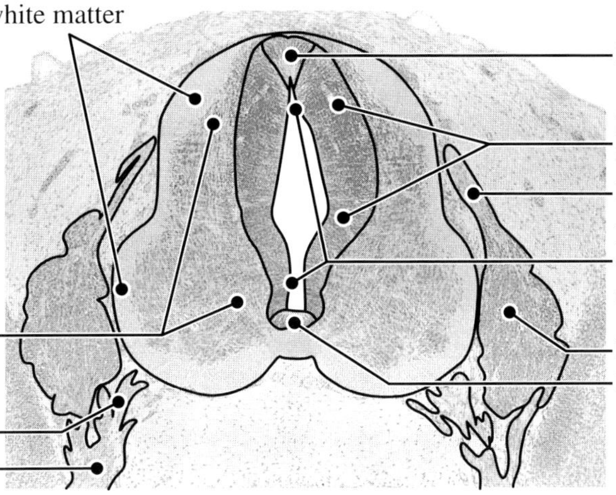

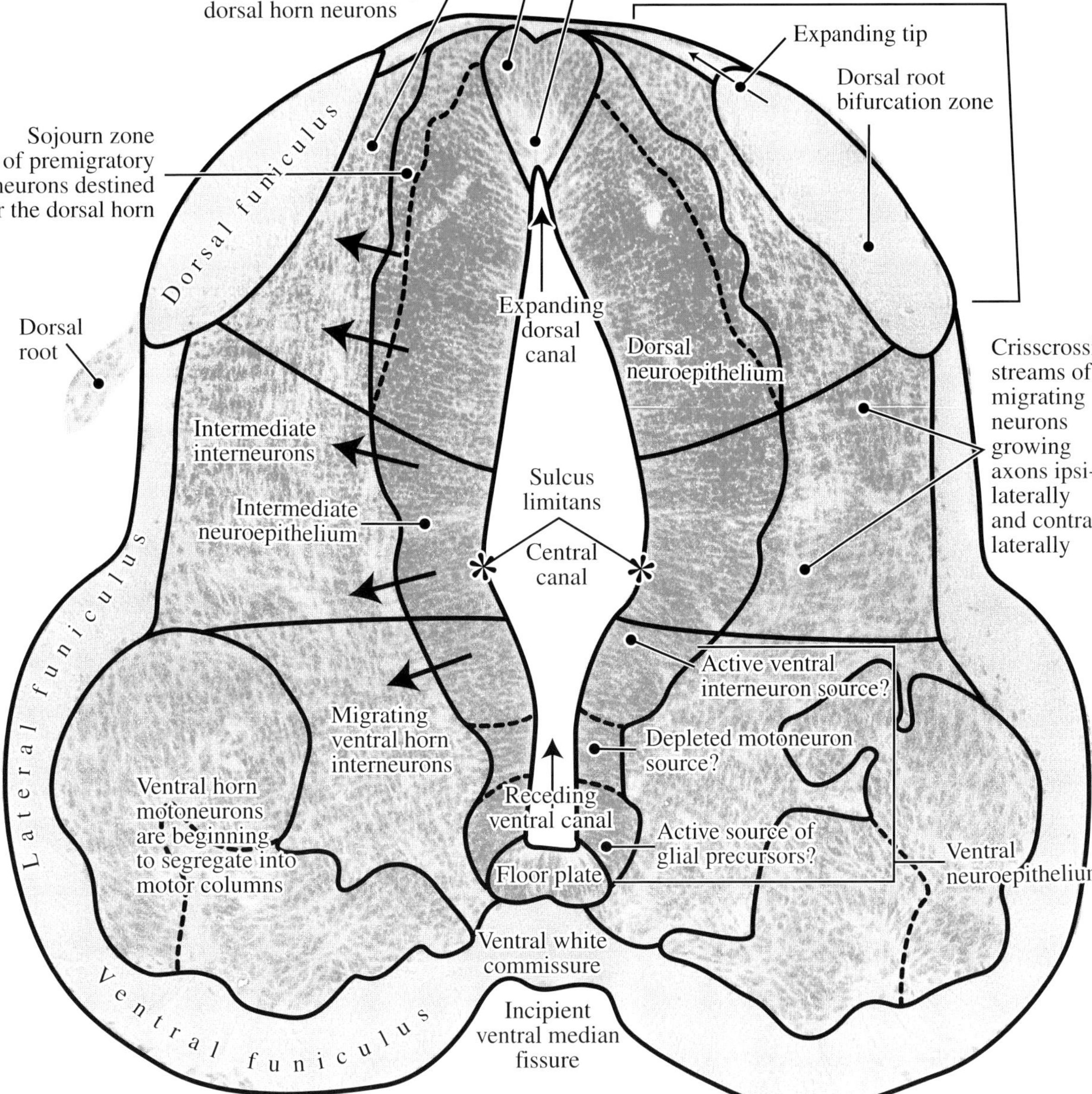

Large arrows extending from the neuroepithelium indicate many migrating neurons.
Arrows in the spinal canal indicate the direction of expansion or contraction.

PLATE 9A

GW6.7
CR 17.1 mm
C8965
Cervical
Cell body stain

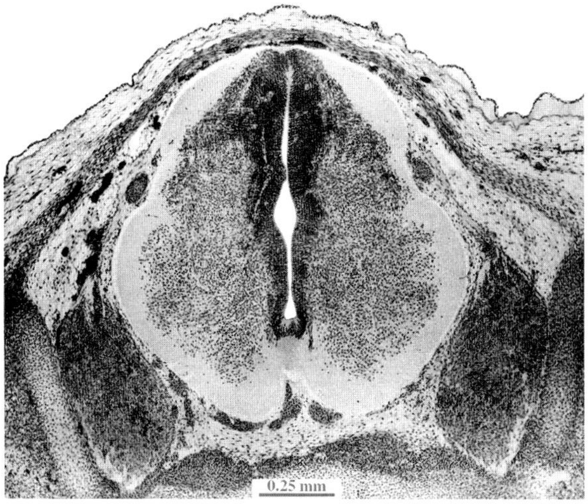

Areas (mm²)

Spinal canal	.0166
Neuroepithelium	.1549
Roof plate	.0069
Floor plate	.0051
Gray matter	.5557
White matter	.2701

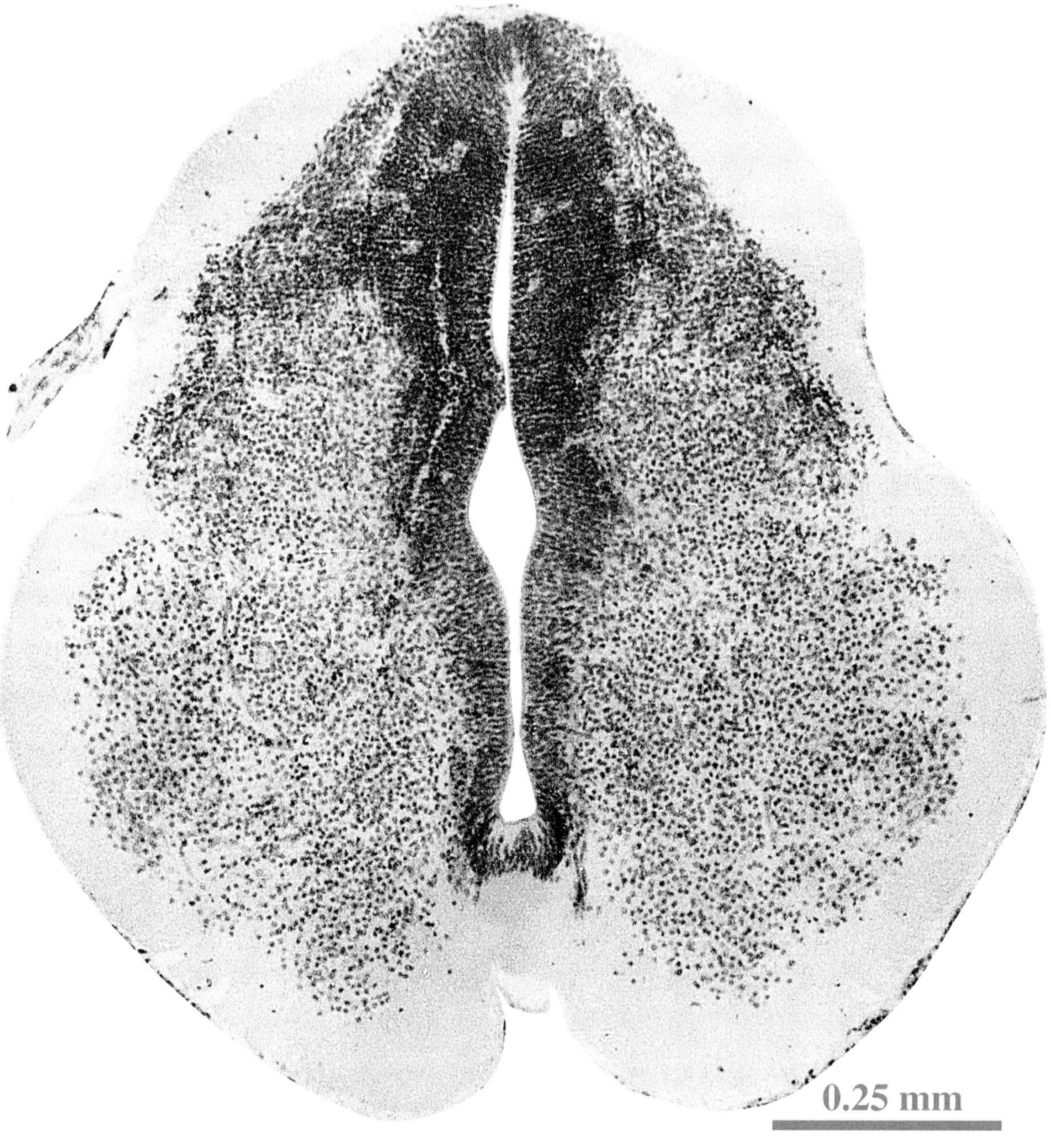

PLATE 9B

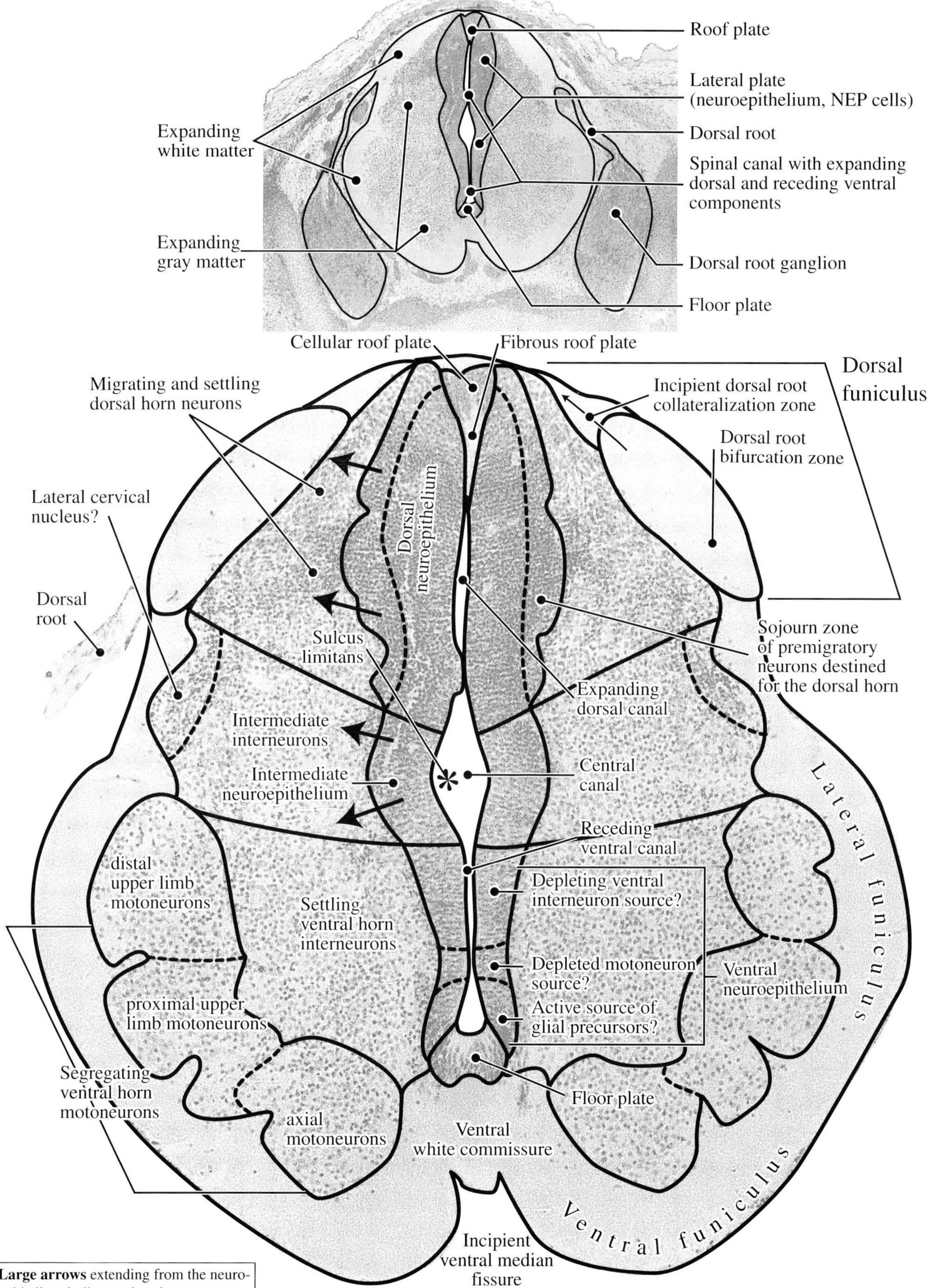

Large arrows extending from the neuroepithelium indicate migrating neurons.

PLATE 10A

GW6.8
CR 17.5 mm
M2155
Cervical
Cell body stain

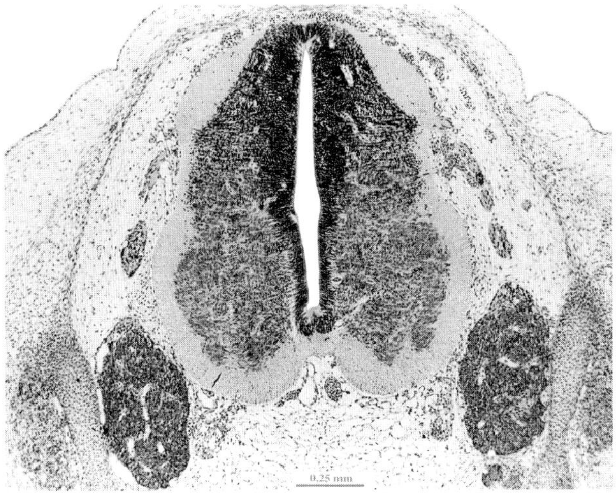

Areas (mm^2)

Spinal canal	.0510
Neuroepithelium	.2350
Roof plate	.0140
Floor plate	.0126
Gray matter	.6475
White matter	.3376

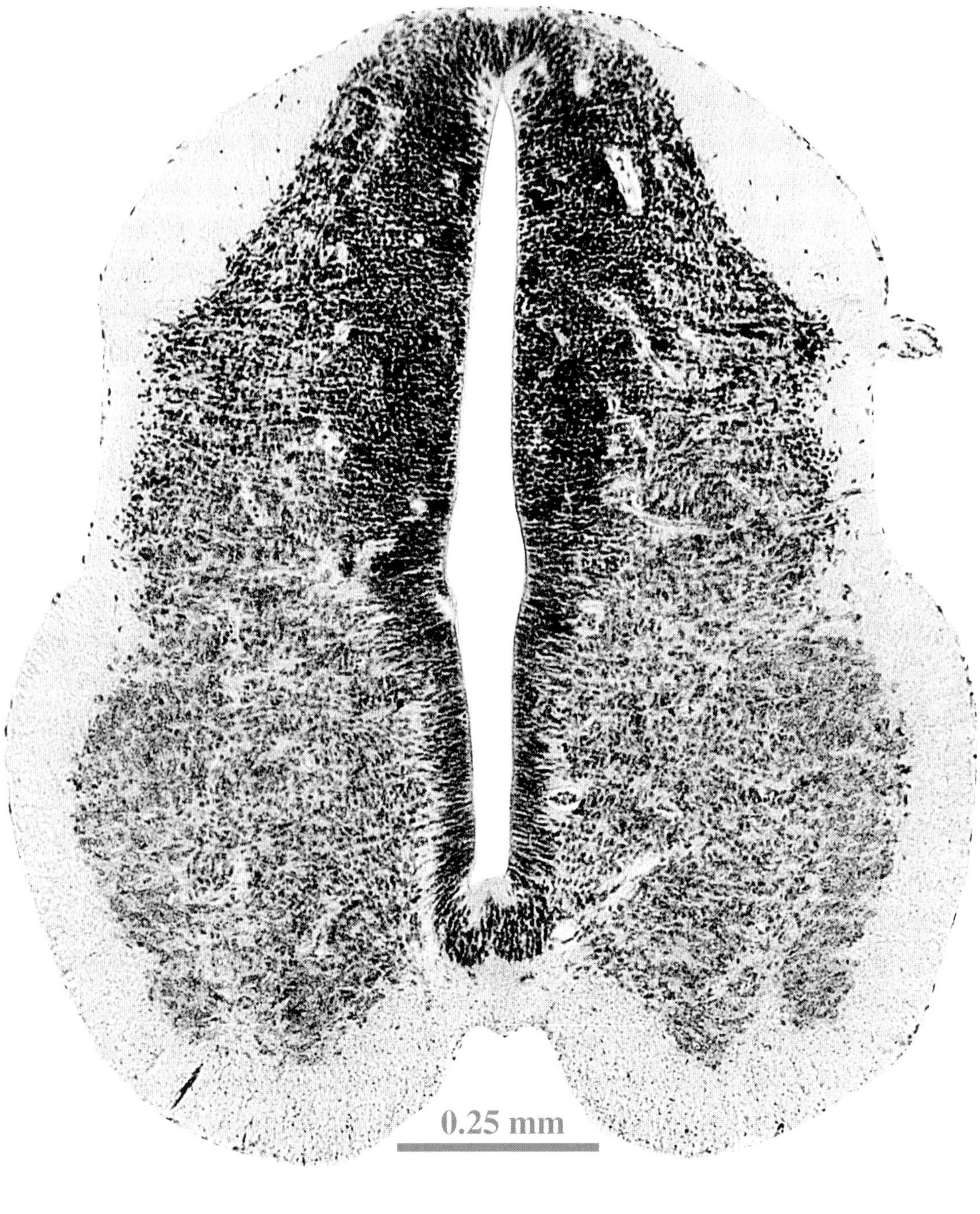

PLATE 10B

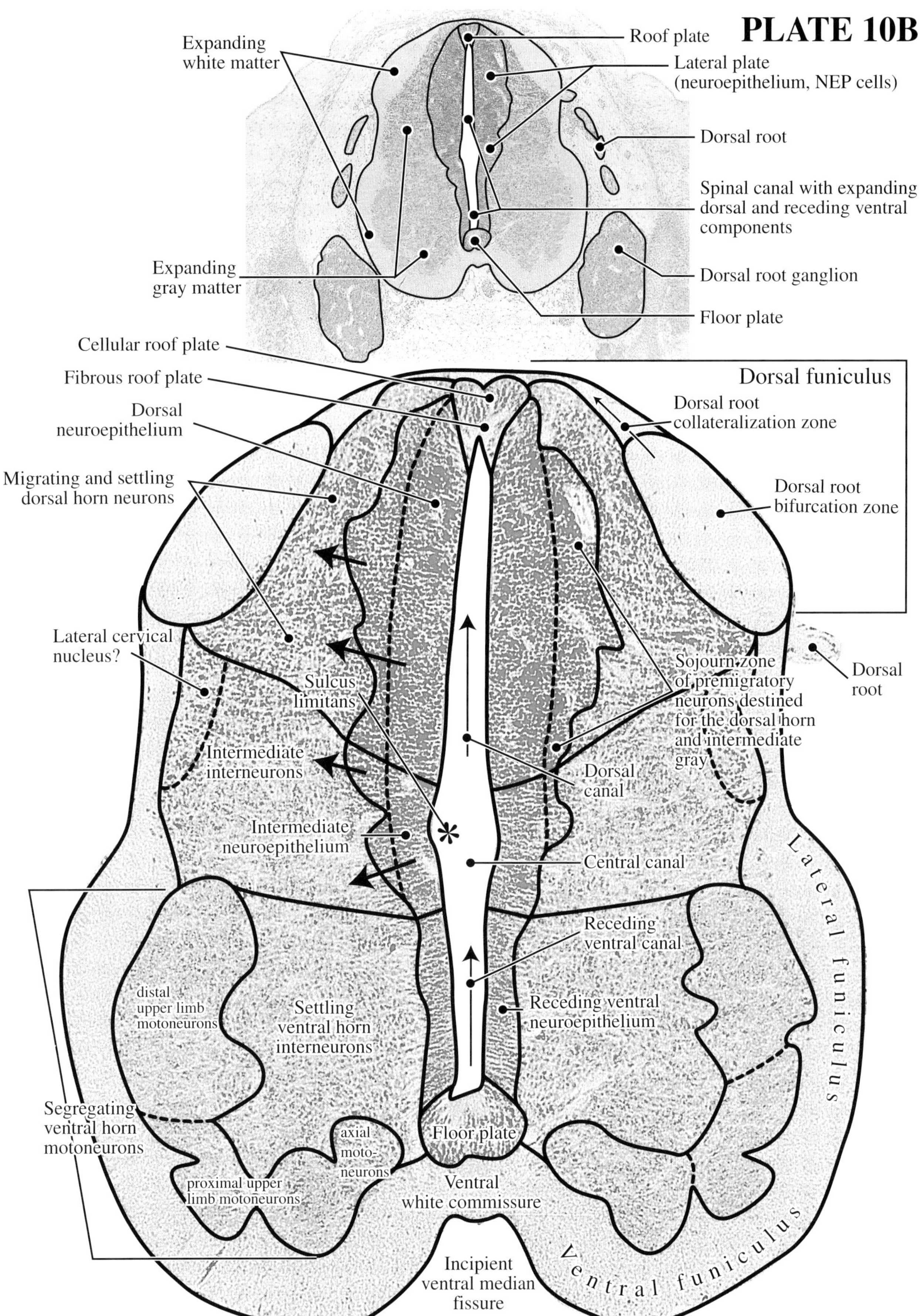

Large arrows extending from the neuroepithelium indicate migrating neurons.
Small arrows in the spinal canal indicate the direction of expansion or contraction.

PLATE 11A

GW7.25
CR 22.5 mm
C7254
Cervical
Cell body stain

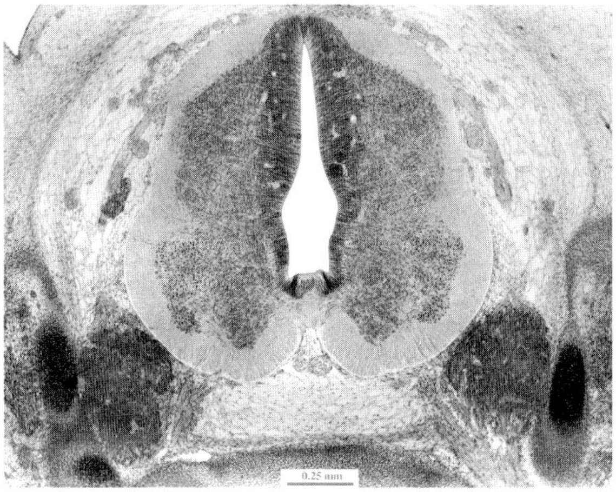

Areas (mm^2)

Spinal canal	.0887
Neuroepithelium	.2207
Roof plate	.0149
Floor plate	.0153
Gray matter	.6798
White matter	.4362

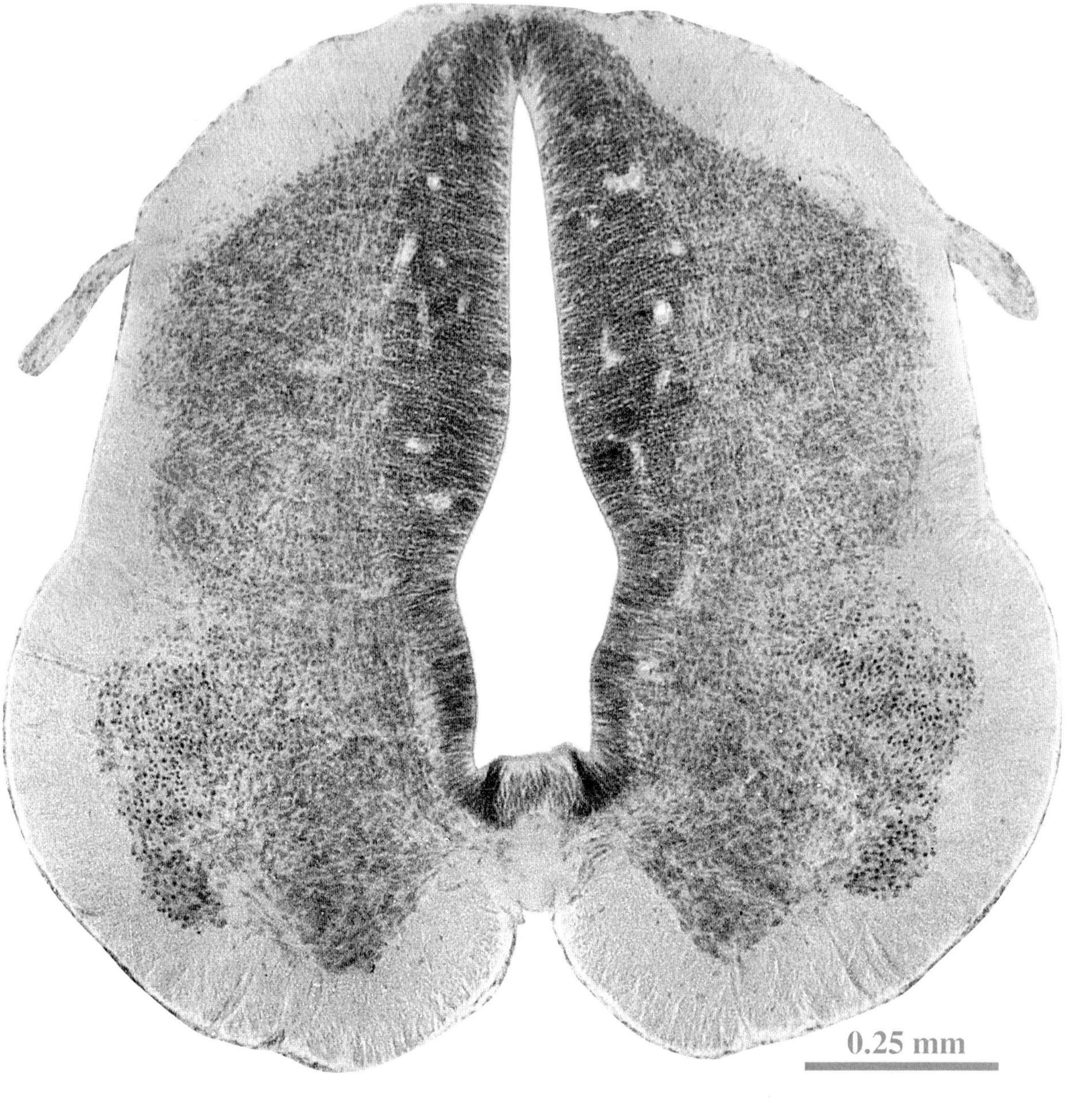

PLATE 11B

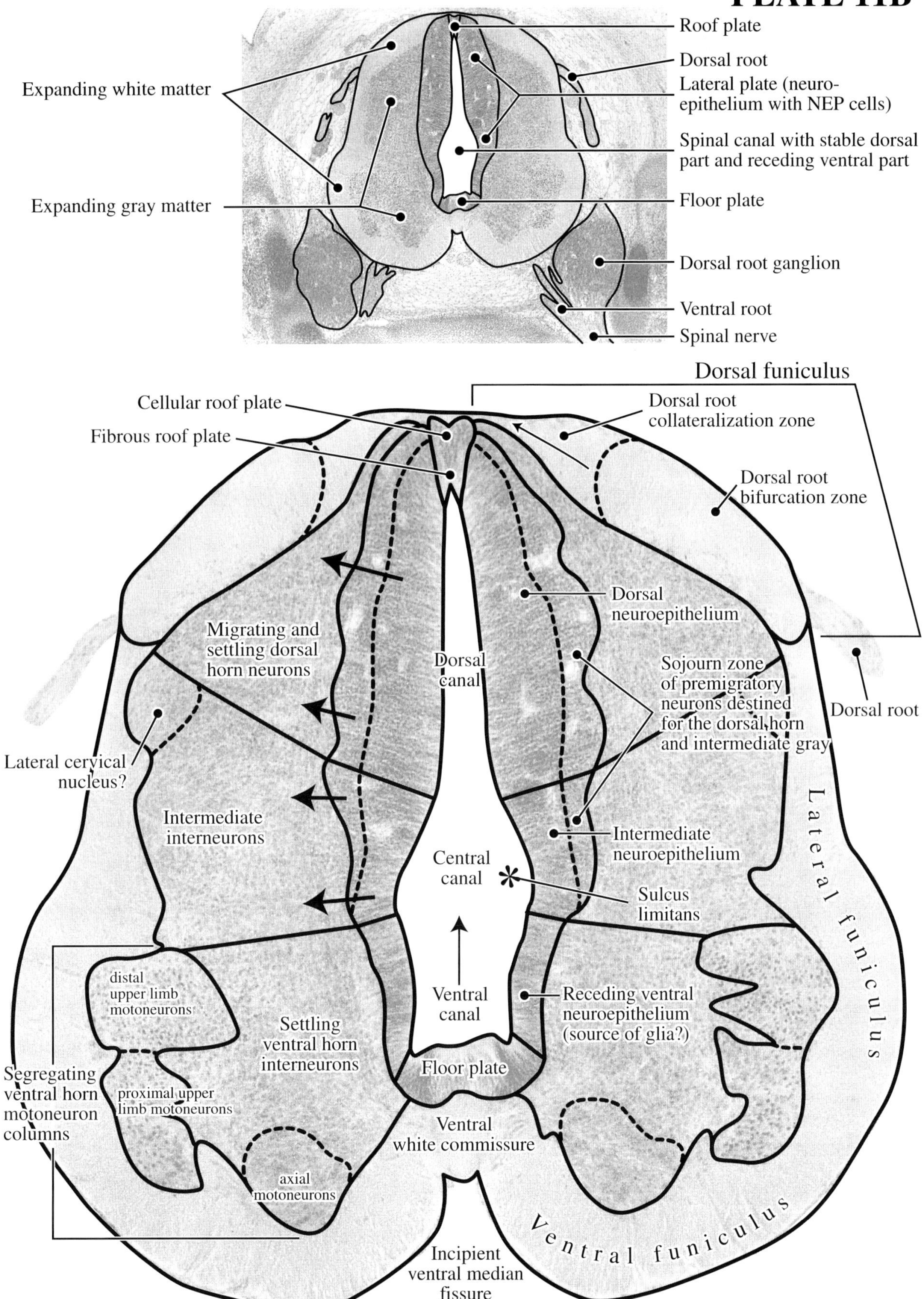

Large arrows extending from the neuroepithelium indicate migrating neurons.
Small arrow in the spinal canal indicates the direction of contraction.

PLATE 12A

GW8.5
CR 32 mm
C609
Cervical
Cell body stain

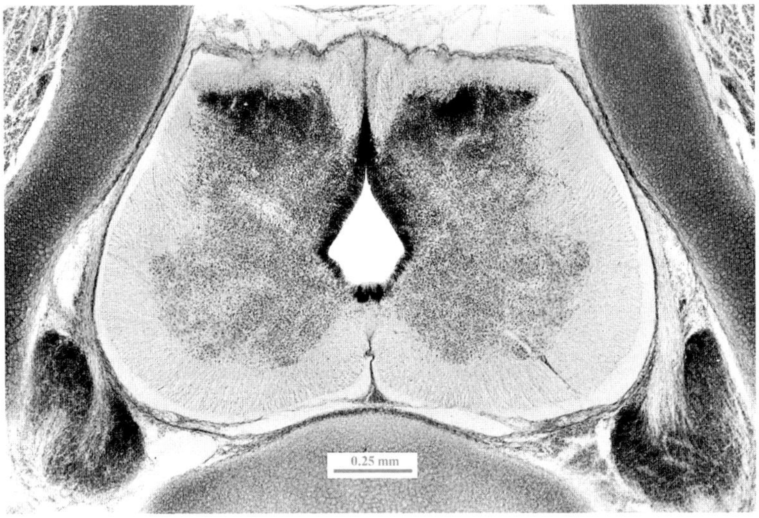

Areas (mm²)

Central canal	.0439
Neuroepithelium	.0470
Roof plate	.0092
Floor plate	.0056
Gray matter	.8925
White matter	.6842

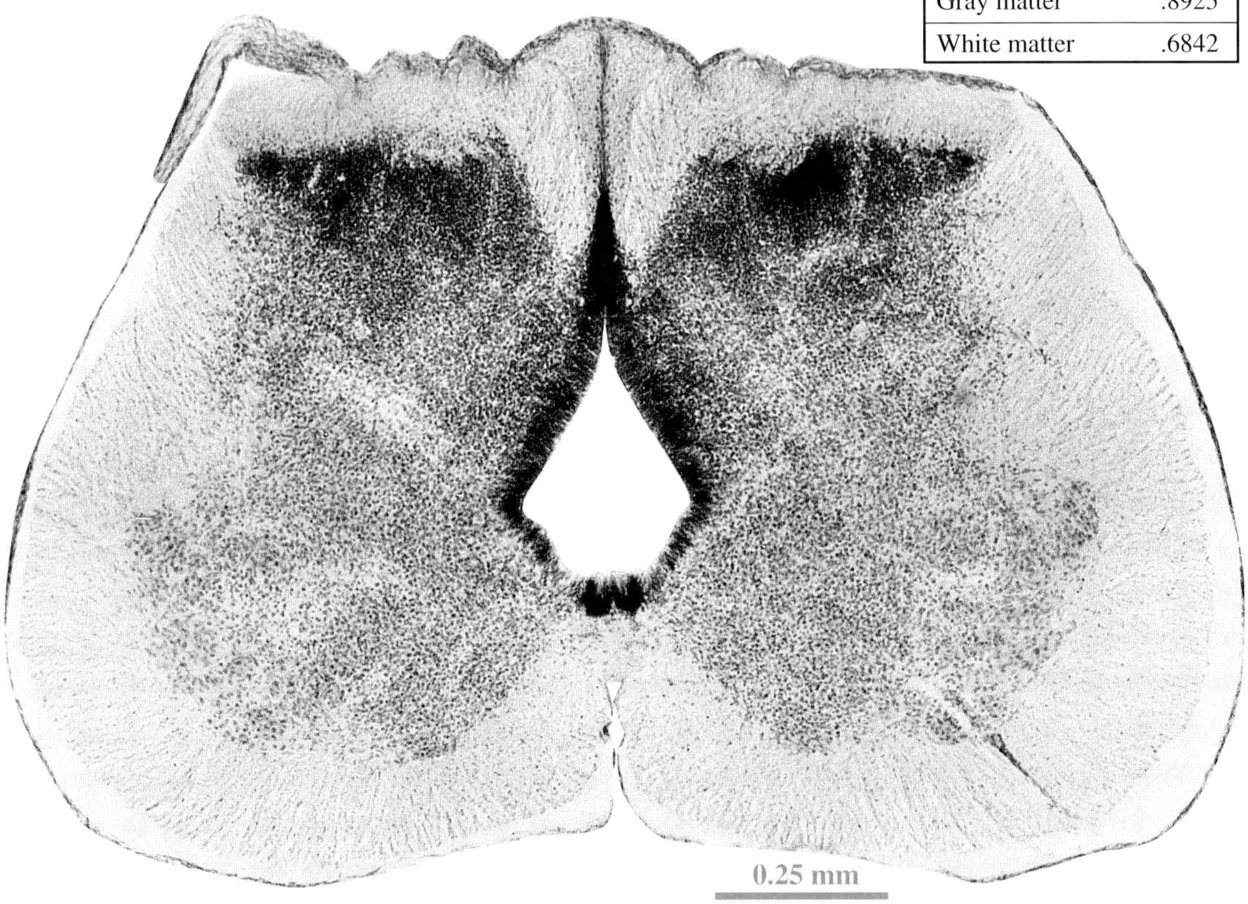

PLATE 12B

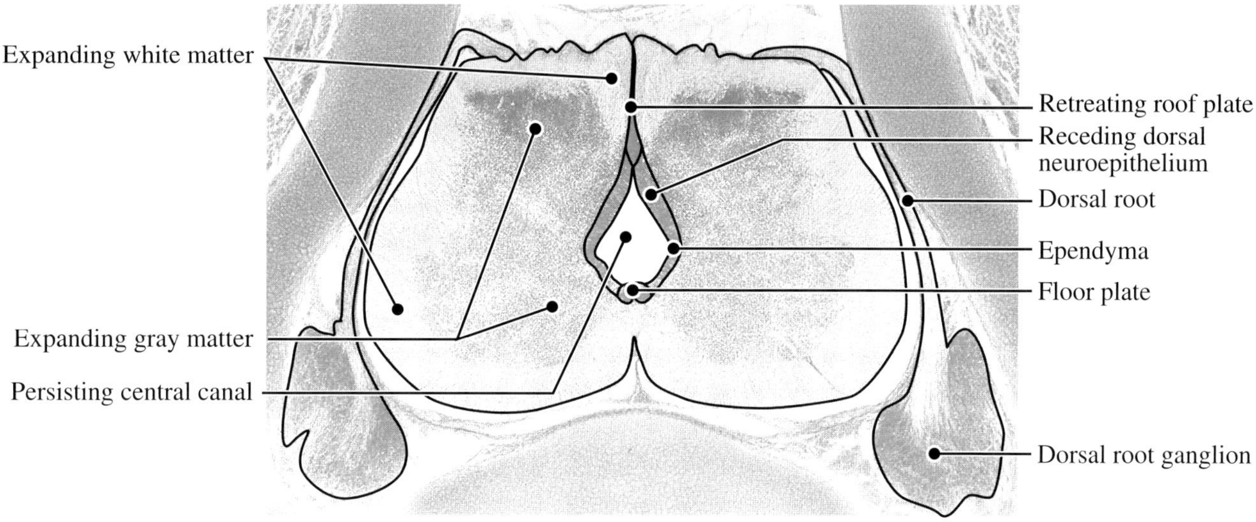

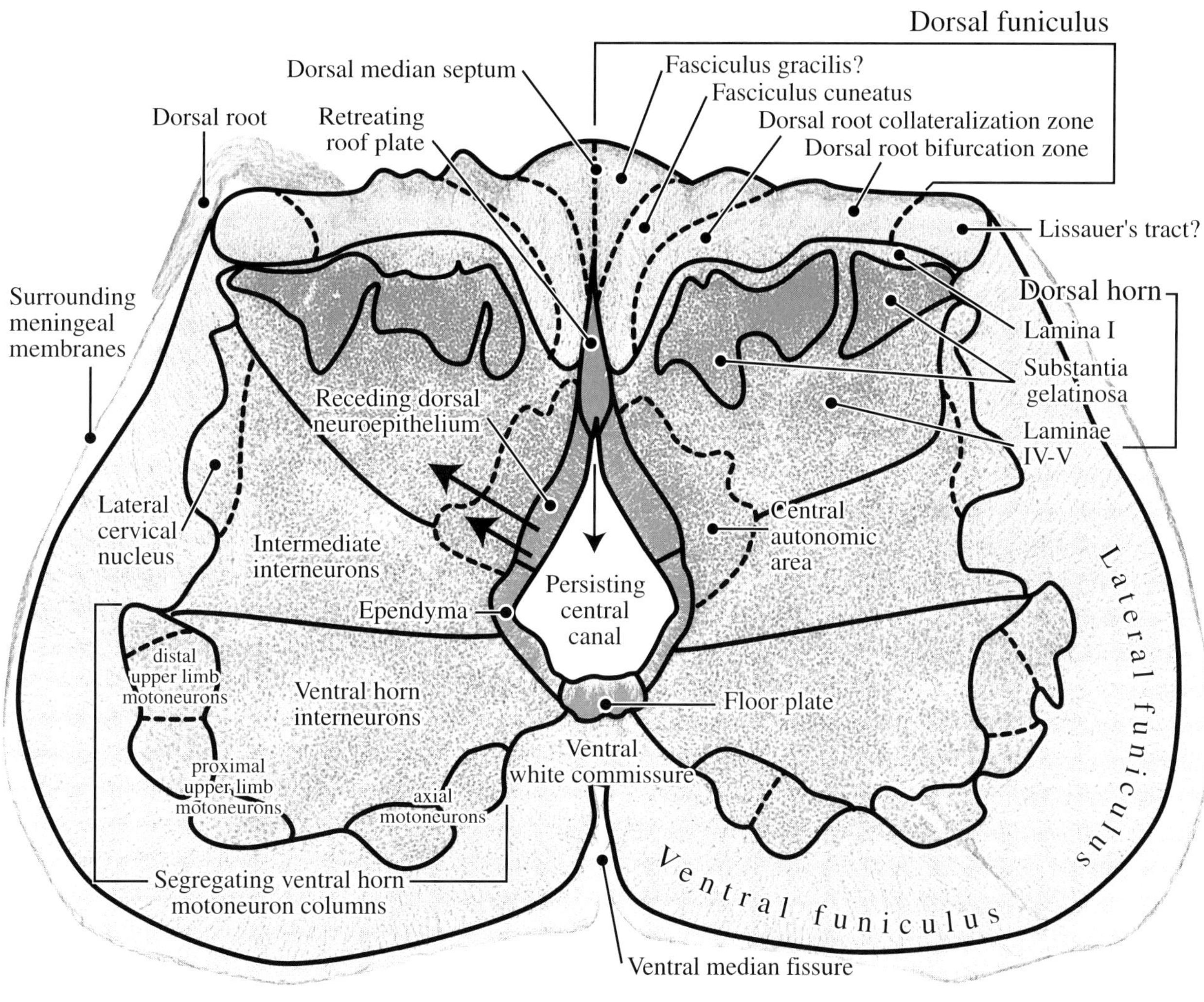

Large arrows extending from the neuroepithelium indicate migrating neurons.
Small arrow in the spinal canal indicates the direction of contraction.

Part II: The First Trimester (continued)

B. Spinal cord of a GW8.5 embryo

Plate 13 is a survey of sections from seven levels of the spinal cord in M2050, a specimen in the Minot Collection with a crown rump length of 36 mm (see Chapter 5 in Altman and Bayer, 2001). All sections are shown at the same scale. The boxes enclosing each section list: the level, ranging from upper cervical to coccygeal; the section number; and the total area of the section in square millimeters (mm^2). Note that the areal measurements are determined *after* fixation, while the crown rump length is measured *before* fixation. The unfixed area of each section could be 40 to 60% larger. Thus, the areal measurements are given only for comparison purposes between levels. Full-page normal contrast photographs of each specimen are in Plates **14A-20A.** Low contrast photographs with superimposed labels and outlines of structural details are in Plates **14B-20B.**

Prior to GW8.5, the spinal cord progressively decreases in size from cervical to coccygeal levels. By GW8.5, this pattern begins to change. First, the cervical enlargement is about the same size as the upper cervical level. Second, regional size differences are also beginning to show. For example, the middle thoracic level is 66% smaller than the cervical enlargement and 30% smaller than the lumbar enlargement. However, the lumbar enlargement is still 50% smaller than the cervical enlargement, and the smallest cross-sectional areas are at sacral and coccygeal levels. These size differences reflect a rostral to caudal gradient in spinal cord maturation.

The proliferating neuroepithelium is declining in this specimen but at all levels an active dorsal neuroepithelium is still present. The basal edges of that neuroepithelium are indistinct as a multitude of cells, presumably young neurons destined for the dorsal horn and the central autonomic area, migrate away to settle. Basal "undulations" are prominent and may be clumps of premigratory young neurons that sojourn in the neuroepithelium just after their generation. The upper part of the intermediate neuroepithelium still appears to be active at low thoracic and lumbar levels, in contrast to cervical levels. At the coccygeal level, even the ventral neuroepithelium appears to be active. Thus, neurons are still being generated in the spinal cord on GW8.5.

The ventral part of the spinal canal is absent except at sacral/coccygeal levels, and the dorsal part of the canal is receding. The central part of the canal will persist into adulthood. Concomitant with the decline of the neuroepithelium, an ependymal layer is appearing as low as lumbar levels.

The gray matter displays maturational changes indicative of the adult spinal cord. In the ventral horn, motoneurons are segregating into columns around the ventral and lateral borders. That is clearly seen in the cervical and lumbar enlargements. The thoracic ventral horn does not have lateral accumulations of motoneurons, but only medial ones. We postulate that the medial columns of motoneurons represent those that supply the axial (skull, neck, and trunk) muscles, while the lateral columns represent those that supply limb muscles. At cervical levels, the intermediate gray has a distinguishable migratory stream of large neurons that may be destined to settle in the lateral cervical nucleus, and an accumulation of large neurons near the central canal that may represent the central cervical nucleus. At thoracic levels, there is only a hint of some large neurons accumulating in Clarke's column (on the right side of the section in **Plates 17A** and **17B**). Throughout all levels of the dorsal horn, there are distinctive clumps of densely packed small cells, presumably settling neurons of laminae II and III (substantia gelatinosa). In addition, migratory streams of cells leaving the dorsal neuroepithelium appear to be heading for the substantia gelatinosa. The central autonomic area can be distinguished as a more dense accumulation of small cells surrounding the central and dorsal parts of the spinal canal.

All components of the white matter are expanding. The most notable changes are occuring in the ventral commissure, ventral funiculus, and dorsal funiculus. The ventral commissure thickens beneath the floor plate, and moves upward as the ventral neuroepithelium and ventral spinal canal disappear. The dorsal funiculus is accumulating axons on either side of the dorsal midline along vertically aligned cells that may connect the retreating roof plate to the pia.

PLATE 13

GW8.5, CR 36 mm, M2050

Plates 14A, 14B
Upper Cervical
Section 730
Total area:
2.8238 mm^2

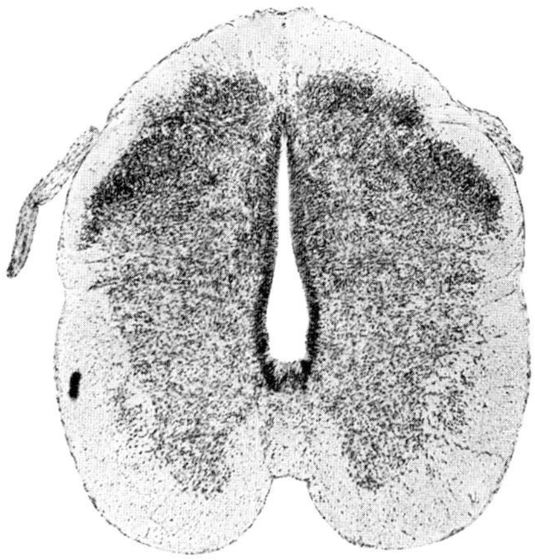

Plates 15A, 15B
Cervical Enlargement
Section 921
Total area:
2.8308 mm^2

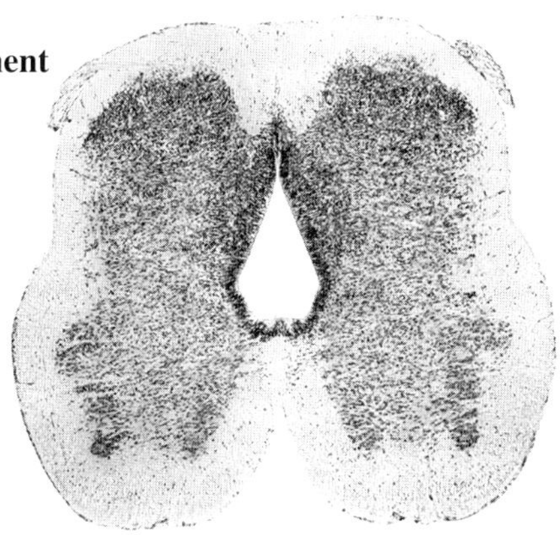

Plates 16A, 16B
Lower Cervical
Section 1113
Total area:
1.7018 mm^2

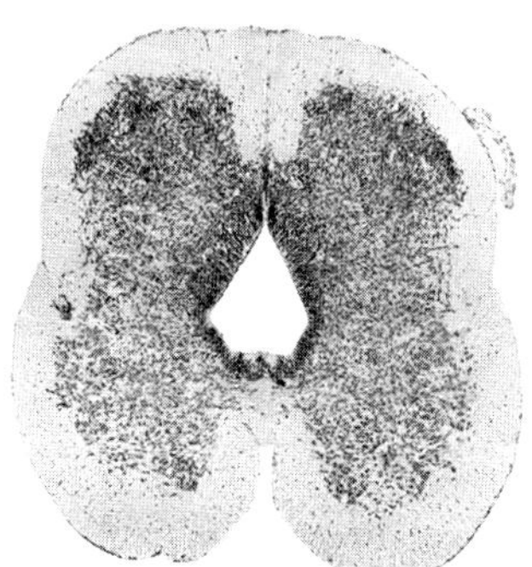

0.5 mm

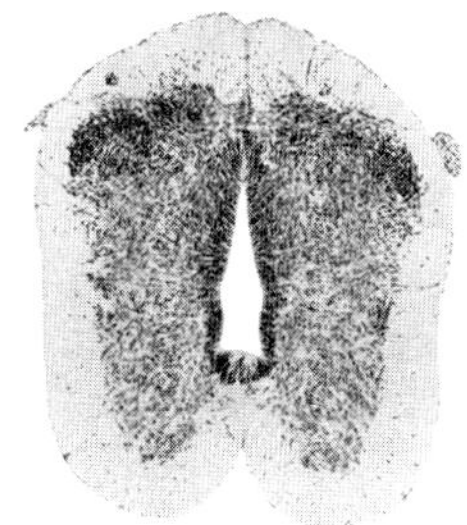

Plates 17A, 17B
Middle Thoracic
Section 1593
Total area:
0.9715 mm^2

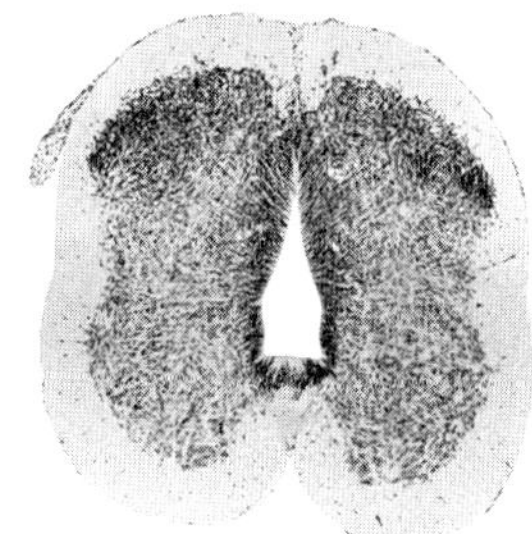

Plates 18A, 18B
Lower Thoracic
Section 1884
Total area:
1.1852 mm^2

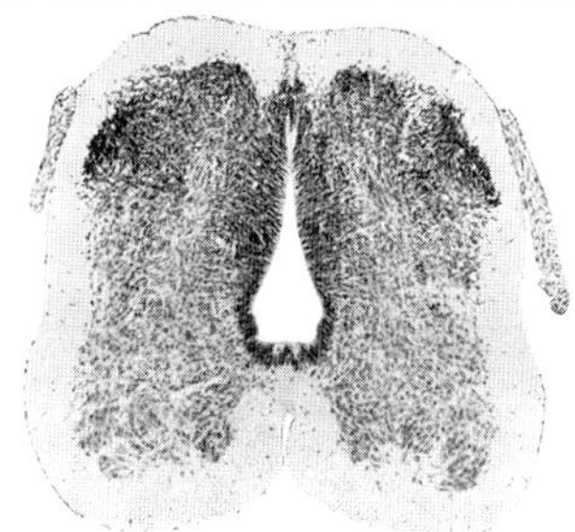

Plates 19A, 19B
Lumbar Enlargement
Section 2134
Total area:
1.3964 mm^2

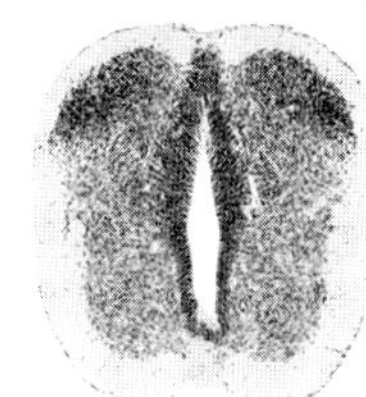

Plates 20A, 20B
(top sections)
Sacral
Section 2607
Total area:
0.6220 mm^2

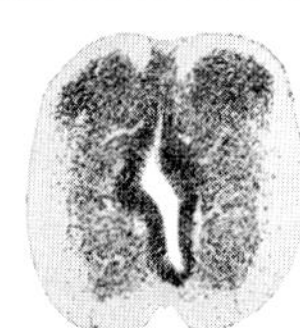

0.5 mm

Plates 20A, 20B
(bottom sections)
Sacral/Coccygeal
Section 2674
Total area:
0.3906 mm^2

PLATE 14A

GW8.5
CR 36 mm
M2050
Upper Cervical
Cell body stain

Areas (mm^2)

Spinal canal	.0622
Neuroepithelium	.1072
Roof plate	.0190
Floor plate	.0209
Gray matter	1.7430
White matter	.8715

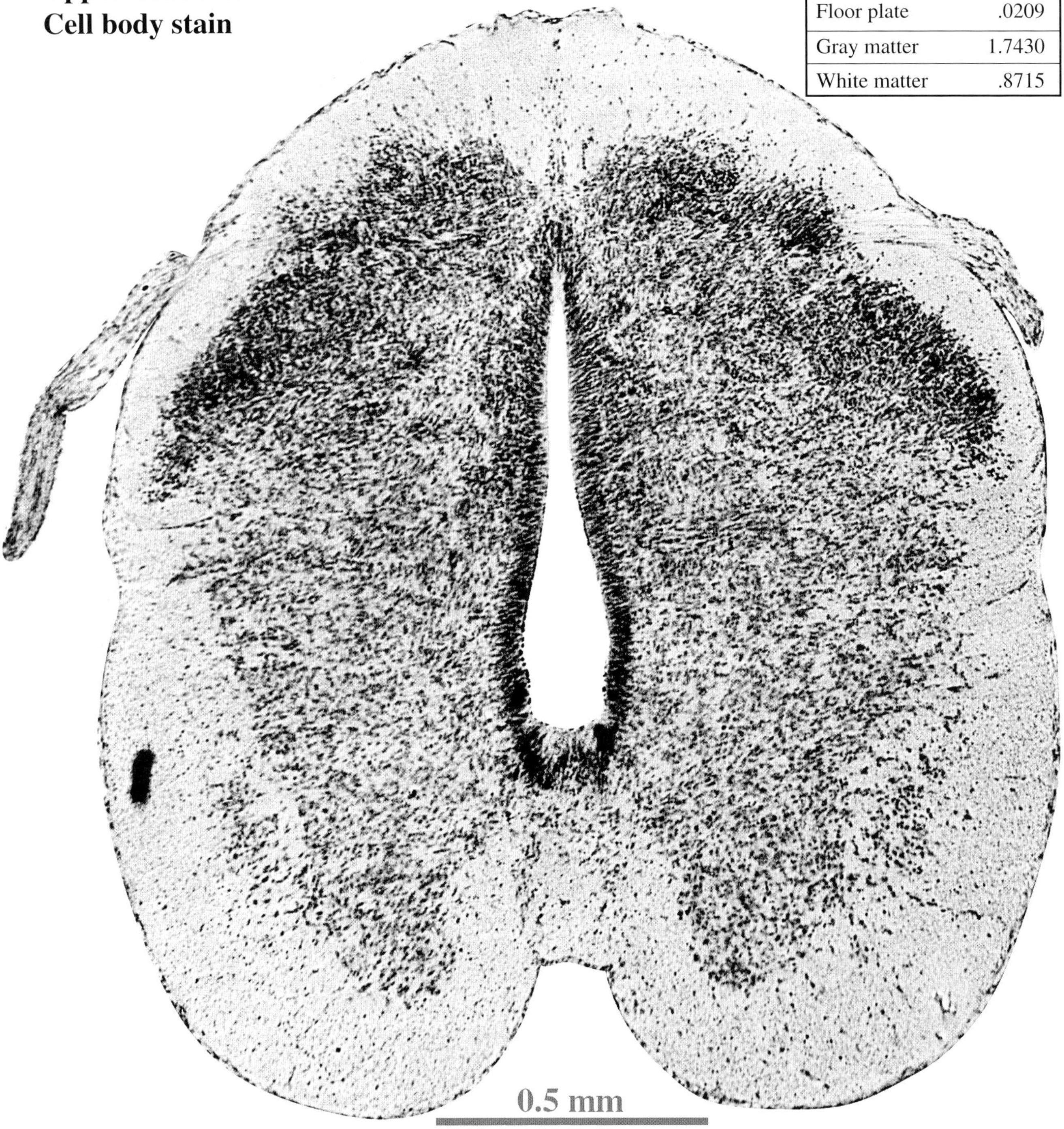

PLATE 14B

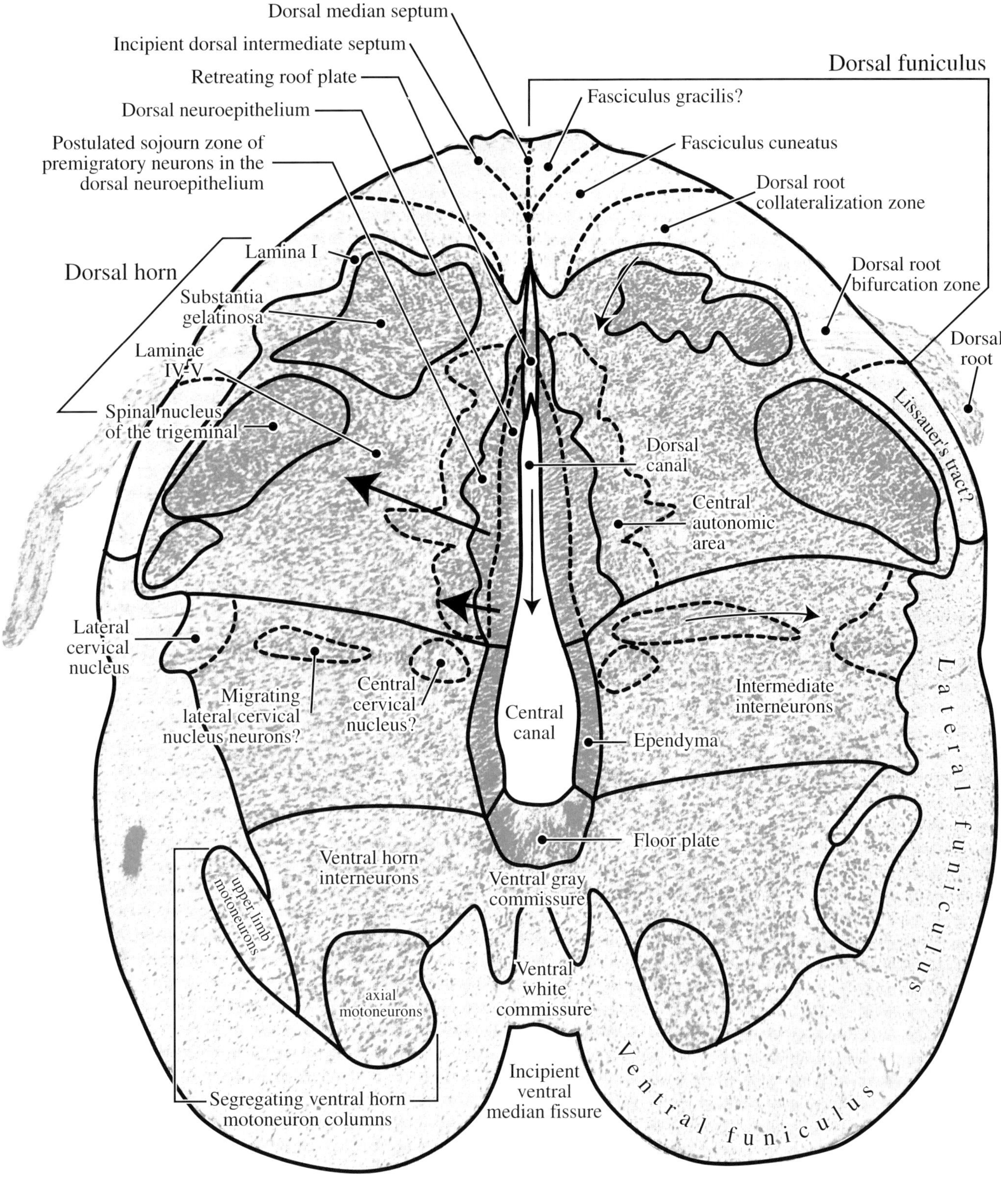
Dorsal median septum
Incipient dorsal intermediate septum
Retreating roof plate
Dorsal neuroepithelium
Postulated sojourn zone of premigratory neurons in the dorsal neuroepithelium
Dorsal funiculus
Fasciculus gracilis?
Fasciculus cuneatus
Dorsal root collateralization zone
Dorsal root bifurcation zone
Dorsal root
Lissauer's tract?
Dorsal horn
Lamina I
Substantia gelatinosa
Laminae IV-V
Spinal nucleus of the trigeminal
Dorsal canal
Central autonomic area
Lateral cervical nucleus
Migrating lateral cervical nucleus neurons?
Central cervical nucleus?
Central canal
Ependyma
Intermediate interneurons
Lateral funiculus
Floor plate
Ventral horn interneurons
Ventral gray commissure
upper limb motoneurons
axial motoneurons
Ventral white commissure
Incipient ventral median fissure
Segregating ventral horn motoneuron columns
Ventral funiculus

PLATE 15A

GW8.5
CR 36 mm
M2050
Cervical Enlargement
Cell body stain

Areas (mm^2)

Spinal canal	.0834
Neuroepithelium	.0966
Roof plate	.0221
Floor plate	.0161
Gray matter	1.5254
White matter	1.0872

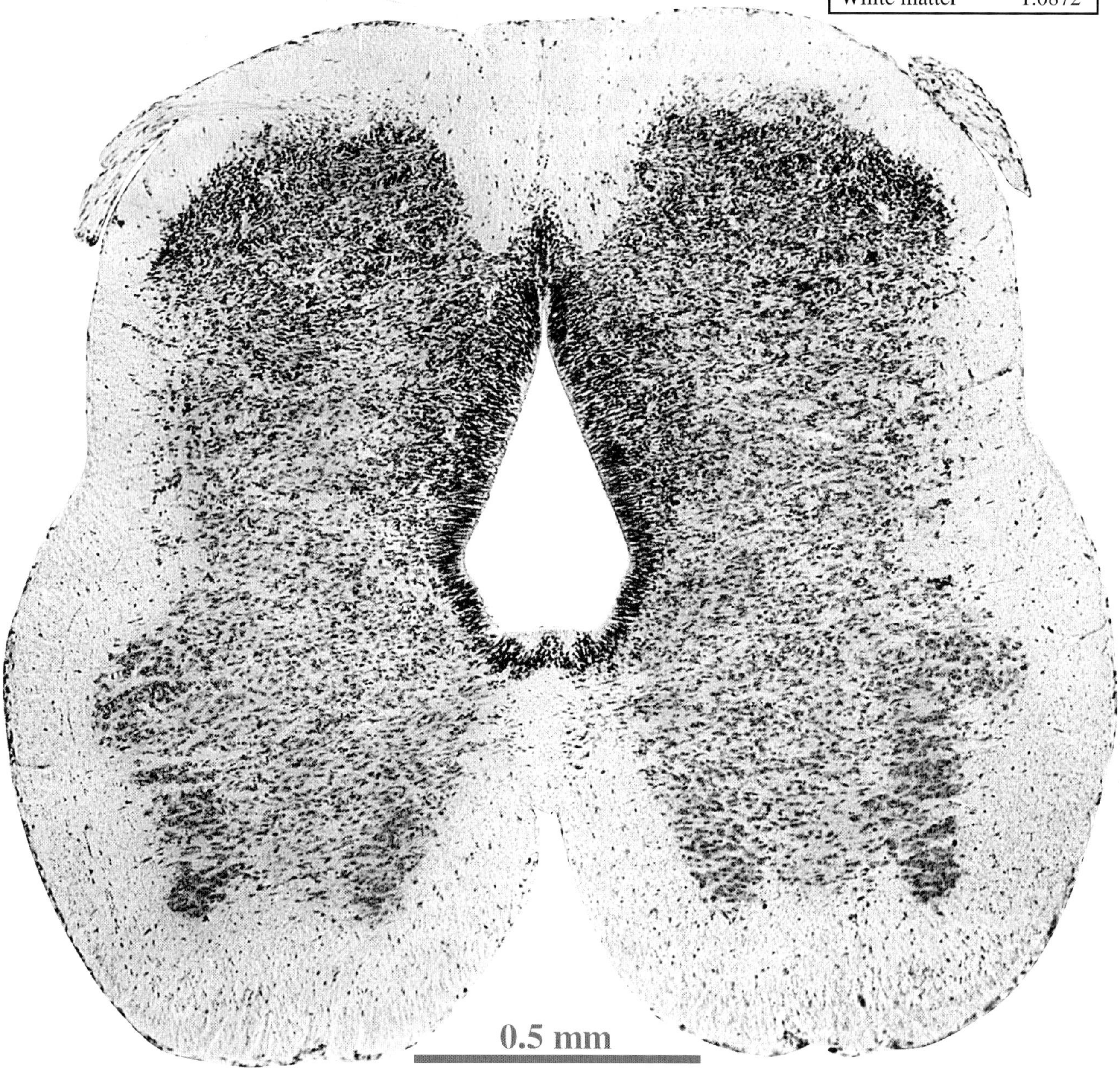

PLATE 15B

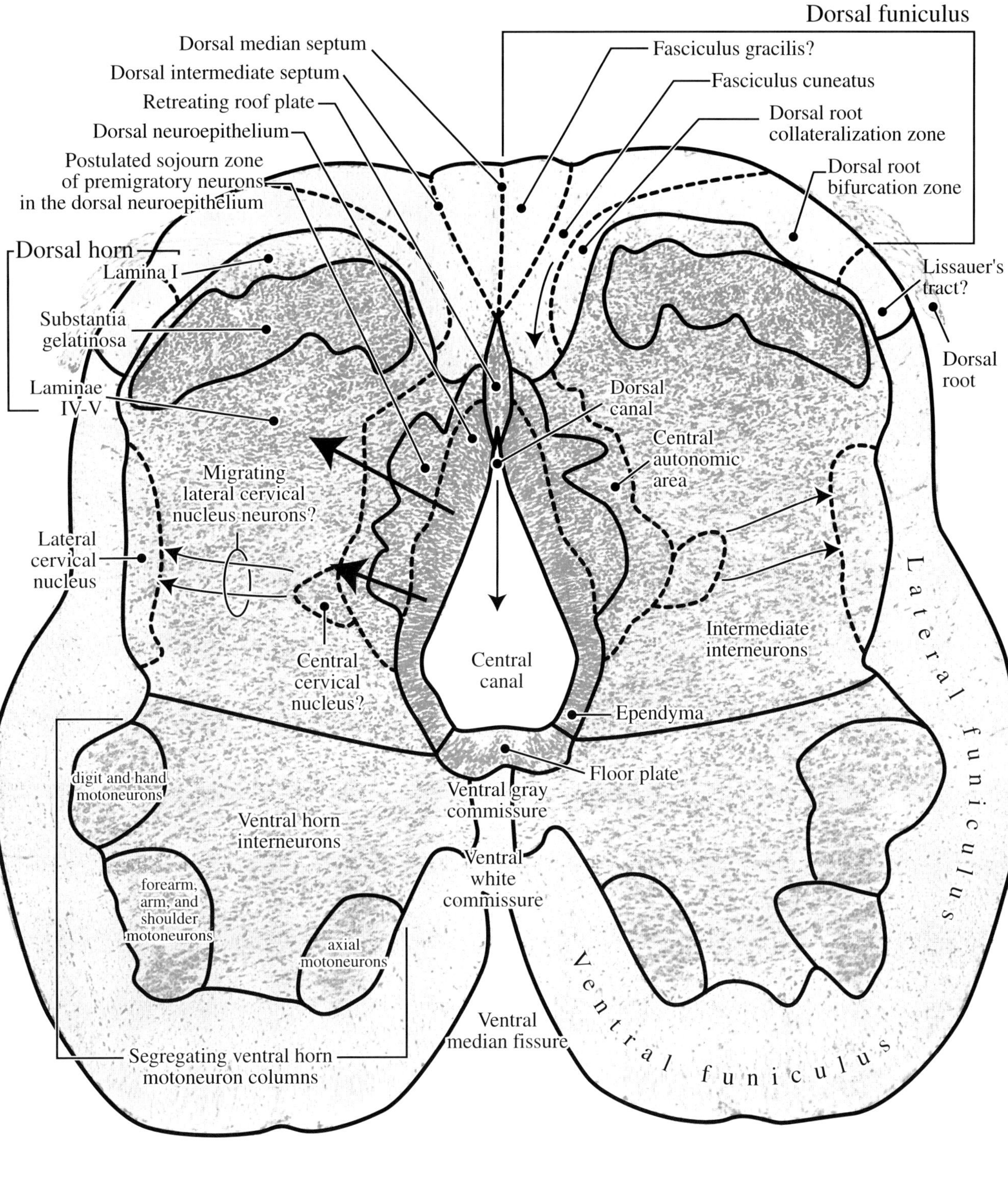
Dorsal funiculus
Dorsal median septum
Dorsal intermediate septum
Retreating roof plate
Dorsal neuroepithelium
Postulated sojourn zone of premigratory neurons in the dorsal neuroepithelium
Fasciculus gracilis?
Fasciculus cuneatus
Dorsal root collateralization zone
Dorsal root bifurcation zone
Dorsal horn
Lamina I
Substantia gelatinosa
Laminae IV-V
Lissauer's tract?
Dorsal root
Dorsal canal
Central autonomic area
Migrating lateral cervical nucleus neurons?
Lateral cervical nucleus
Intermediate interneurons
Central cervical nucleus?
Central canal
Ependyma
Floor plate
Lateral funiculus
digit and hand motoneurons
Ventral gray commissure
Ventral horn interneurons
Ventral white commissure
forearm, arm, and shoulder motoneurons
axial motoneurons
Ventral median fissure
Ventral funiculus
Segregating ventral horn motoneuron columns

PLATE 16A

GW8.5
CR 36 mm
M2050
Lower Cervical
Cell body stain

Areas (mm^2)

Spinal canal	.0557
Neuroepithelium	.0874
Roof plate	.0124
Floor plate	.0156
Gray matter	.8918
White matter	.6389

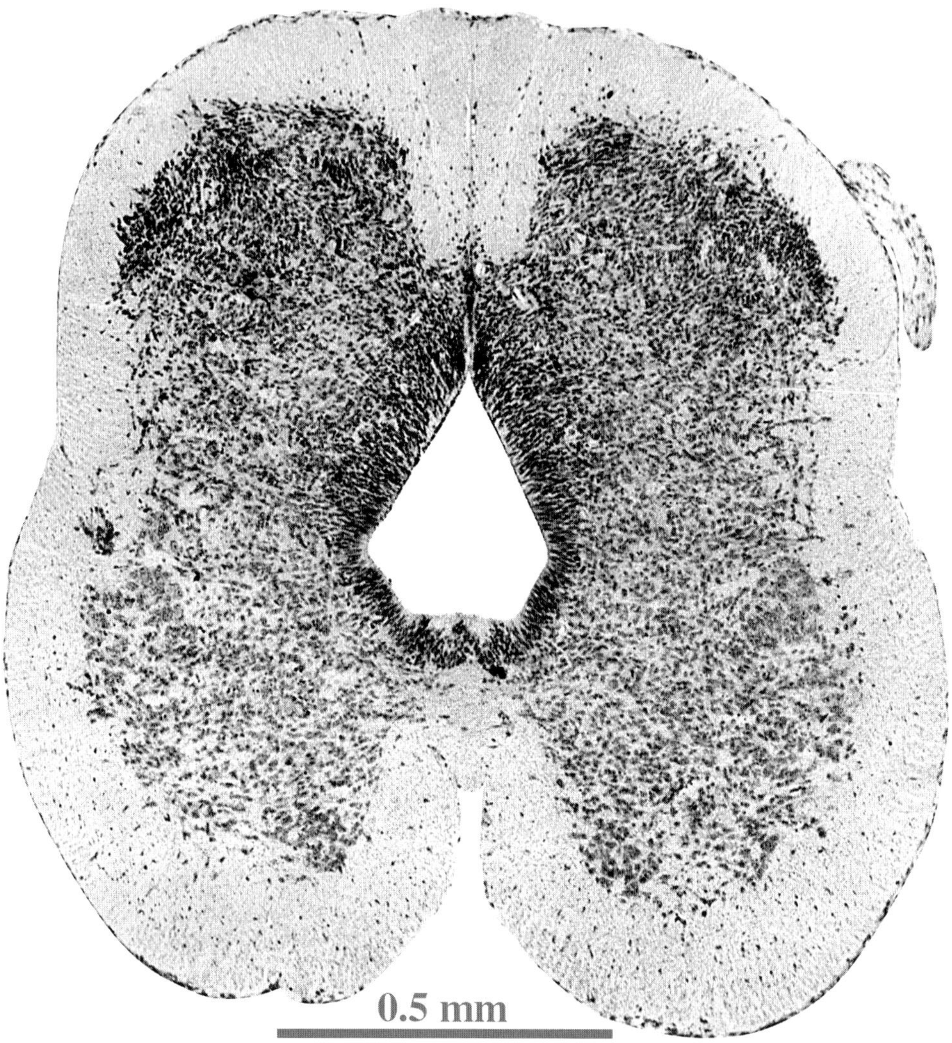

PLATE 16B

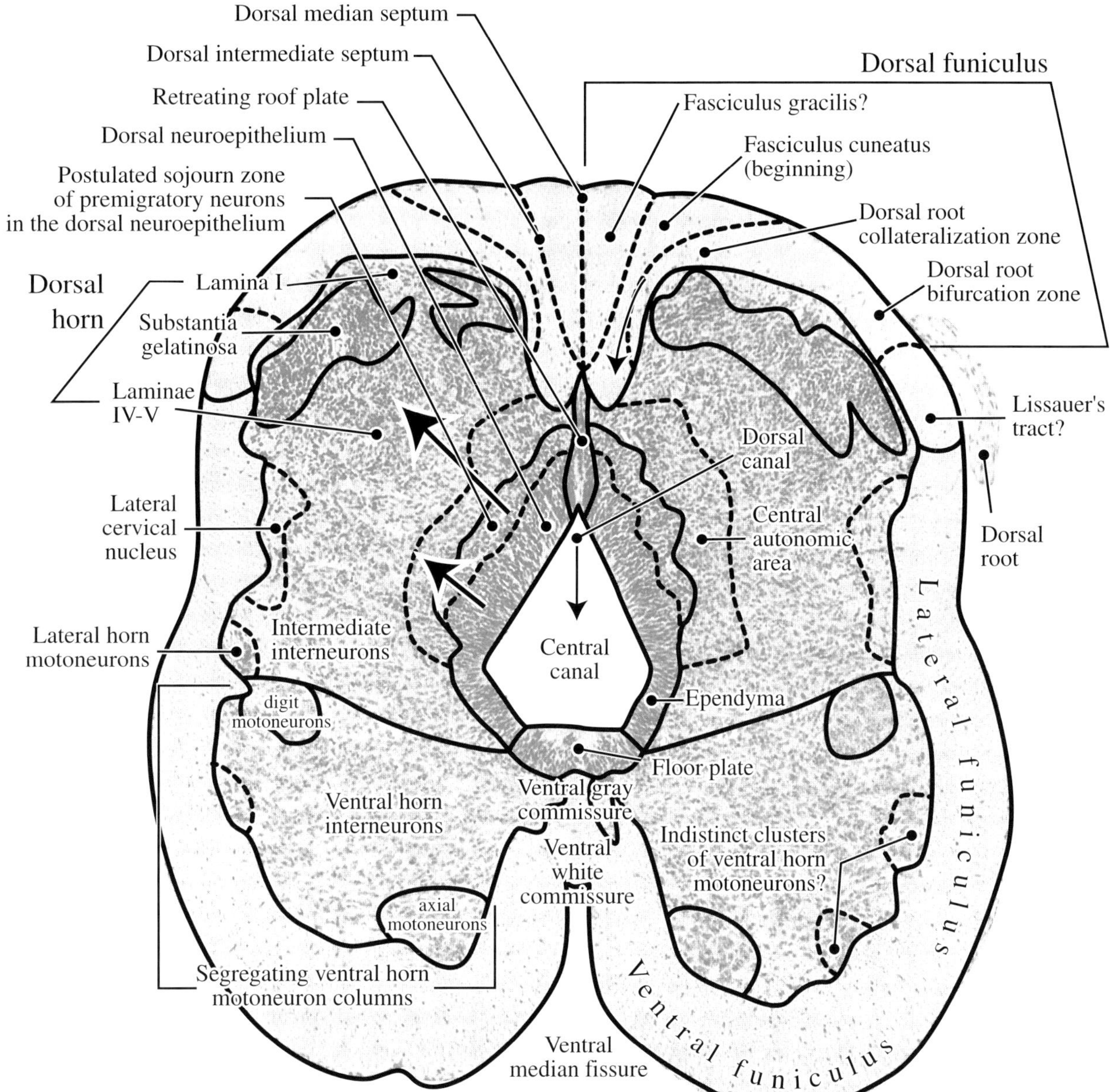
Dorsal median septum
Dorsal intermediate septum
Retreating roof plate
Dorsal neuroepithelium
Postulated sojourn zone of premigratory neurons in the dorsal neuroepithelium
Dorsal funiculus
Fasciculus gracilis?
Fasciculus cuneatus (beginning)
Dorsal root collateralization zone
Dorsal root bifurcation zone
Dorsal horn
Lamina I
Substantia gelatinosa
Laminae IV-V
Lissauer's tract?
Dorsal canal
Lateral cervical nucleus
Central autonomic area
Dorsal root
Lateral horn motoneurons
Intermediate interneurons
Central canal
Lateral funiculus
Ependyma
digit motoneurons
Floor plate
Ventral gray commissure
Ventral horn interneurons
Ventral white commissure
Indistinct clusters of ventral horn motoneurons?
axial motoneurons
Segregating ventral horn motoneuron columns
Ventral funiculus
Ventral median fissure

PLATE 17A

GW8.5
CR 36 mm
M2050
Middle Thoracic
Cell body stain

Areas (mm^2)

Spinal canal	.0238
Neuroepithelium	.0625
Roof plate	.0094
Floor plate	.0110
Gray matter	.4796
White matter	.3852

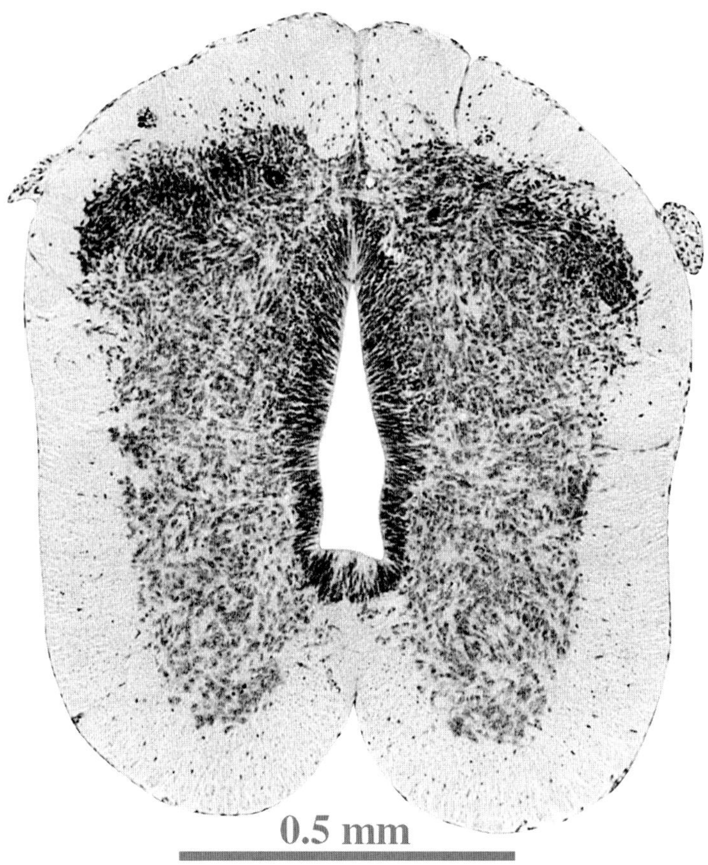

PLATE 17B

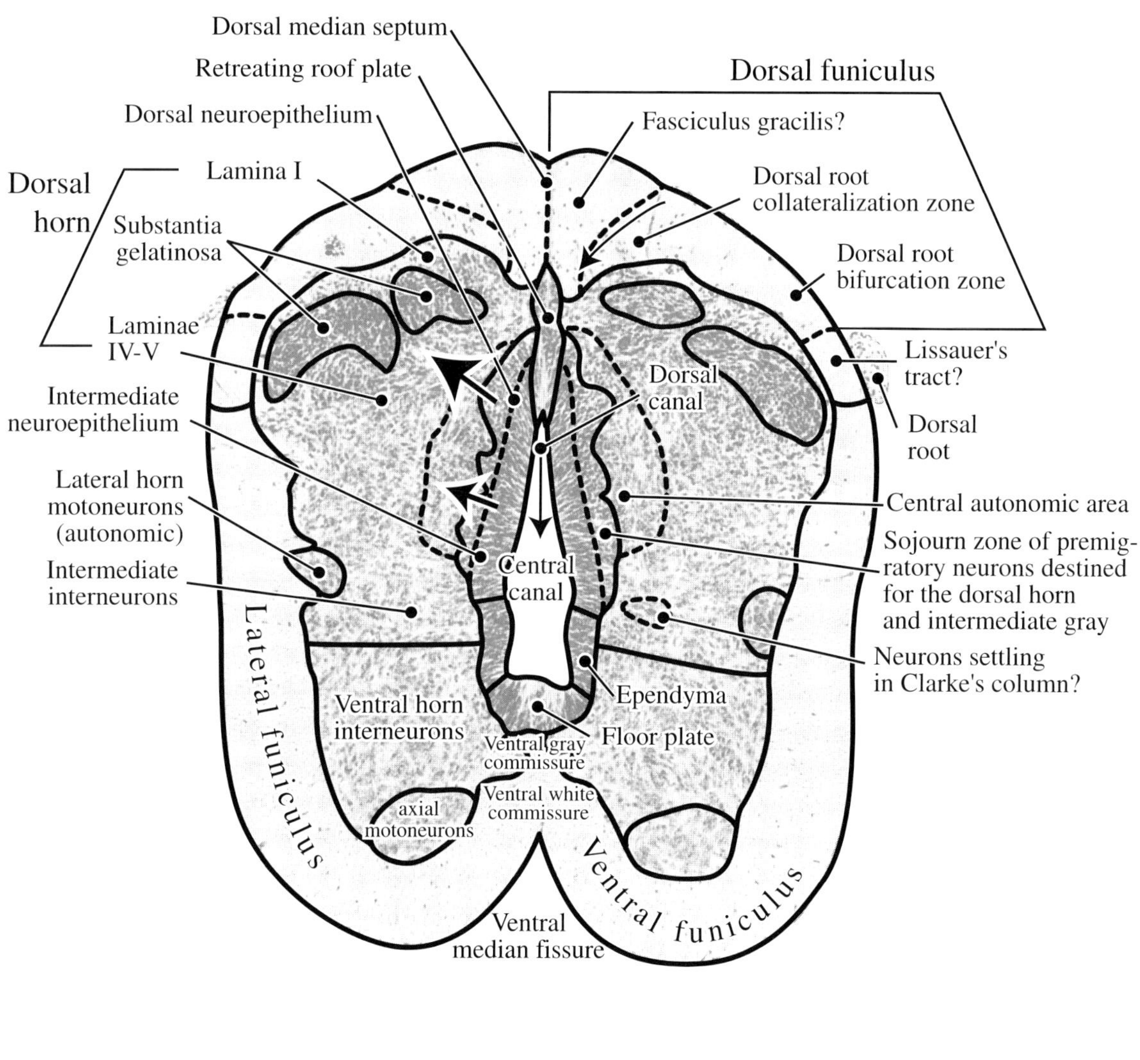
Dorsal median septum
Retreating roof plate
Dorsal funiculus
Dorsal neuroepithelium
Fasciculus gracilis?
Dorsal horn
Lamina I
Dorsal root collateralization zone
Substantia gelatinosa
Dorsal root bifurcation zone
Laminae IV-V
Lissauer's tract?
Dorsal canal
Intermediate neuroepithelium
Dorsal root
Lateral horn motoneurons (autonomic)
Central autonomic area
Sojourn zone of premigratory neurons destined for the dorsal horn and intermediate gray
Intermediate interneurons
Central canal
Neurons settling in Clarke's column?
Lateral funiculus
Ependyma
Ventral horn interneurons
Floor plate
Ventral gray commissure
Ventral white commissure
axial motoneurons
Ventral funiculus
Ventral median fissure

PLATE 18A

GW8.5
CR 36 mm
M2050
Lower Thoracic
Cell body stain

Areas (mm^2)

Spinal canal	.0317
Neuroepithelium	.0857
Roof plate	.0097
Floor plate	.0132
Gray matter	.6133
White matter	.4317

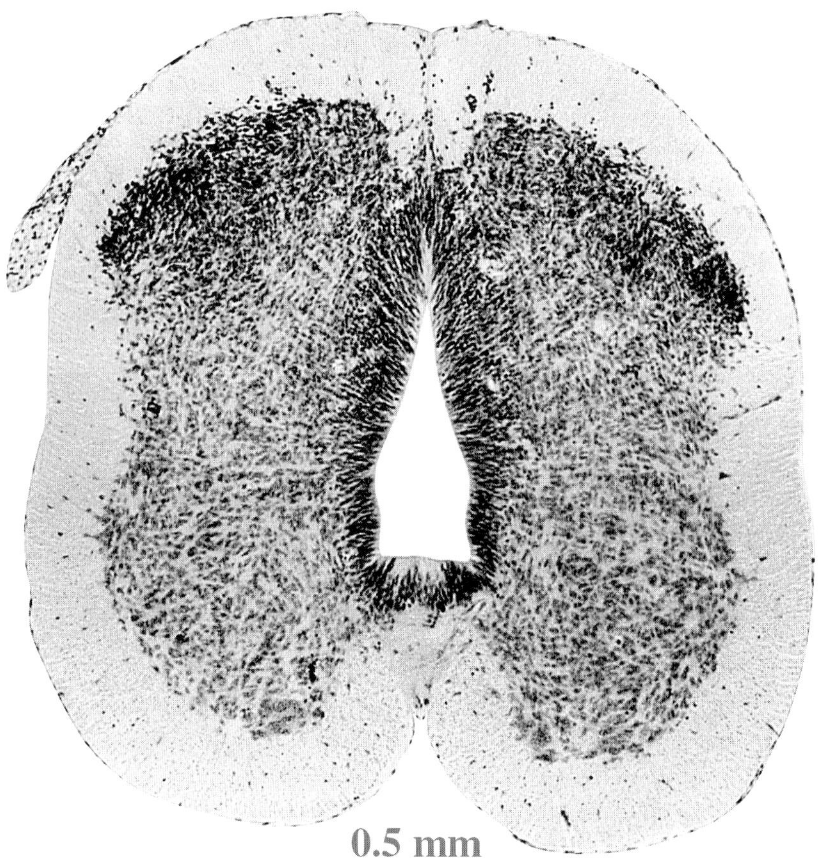

PLATE18B

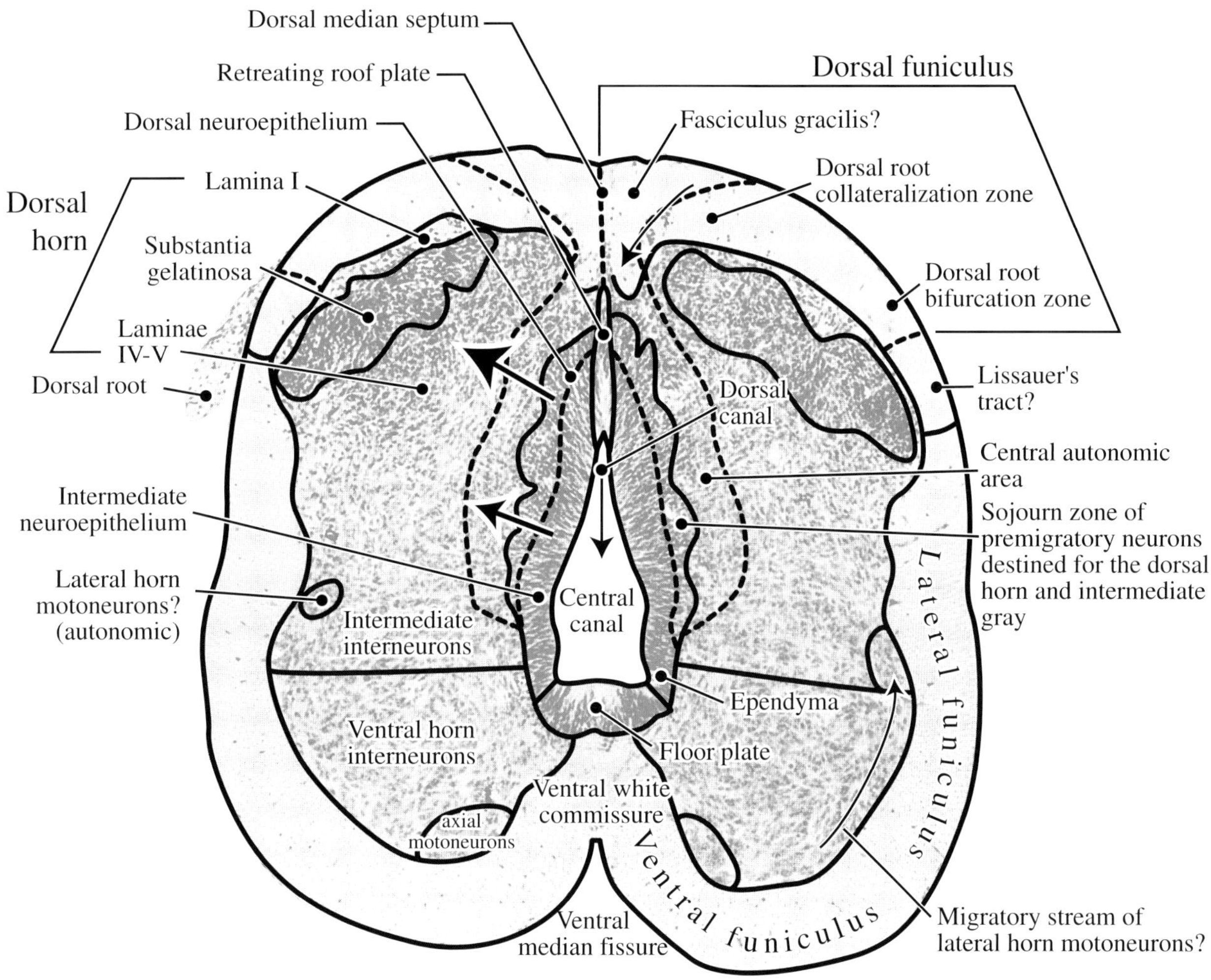

Dorsal median septum
Retreating roof plate
Dorsal neuroepithelium
Dorsal funiculus
Fasciculus gracilis?
Dorsal root collateralization zone
Dorsal horn
Lamina I
Substantia gelatinosa
Laminae IV-V
Dorsal root
Dorsal root bifurcation zone
Lissauer's tract?
Dorsal canal
Central autonomic area
Sojourn zone of premigratory neurons destined for the dorsal horn and intermediate gray
Intermediate neuroepithelium
Lateral horn motoneurons? (autonomic)
Intermediate interneurons
Central canal
Lateral funiculus
Ependyma
Ventral horn interneurons
Floor plate
Ventral white commissure
axial motoneurons
Ventral funiculus
Ventral median fissure
Migratory stream of lateral horn motoneurons?

PLATE 19A

GW8.5
CR 36 mm
M2050
Lumbar Enlargement
Cell body stain

Areas (mm²)

Spinal canal	.0353
Neuroepithelium	.0957
Roof plate	.0238
Floor plate	.0113
Gray matter	.7637
White matter	.4666

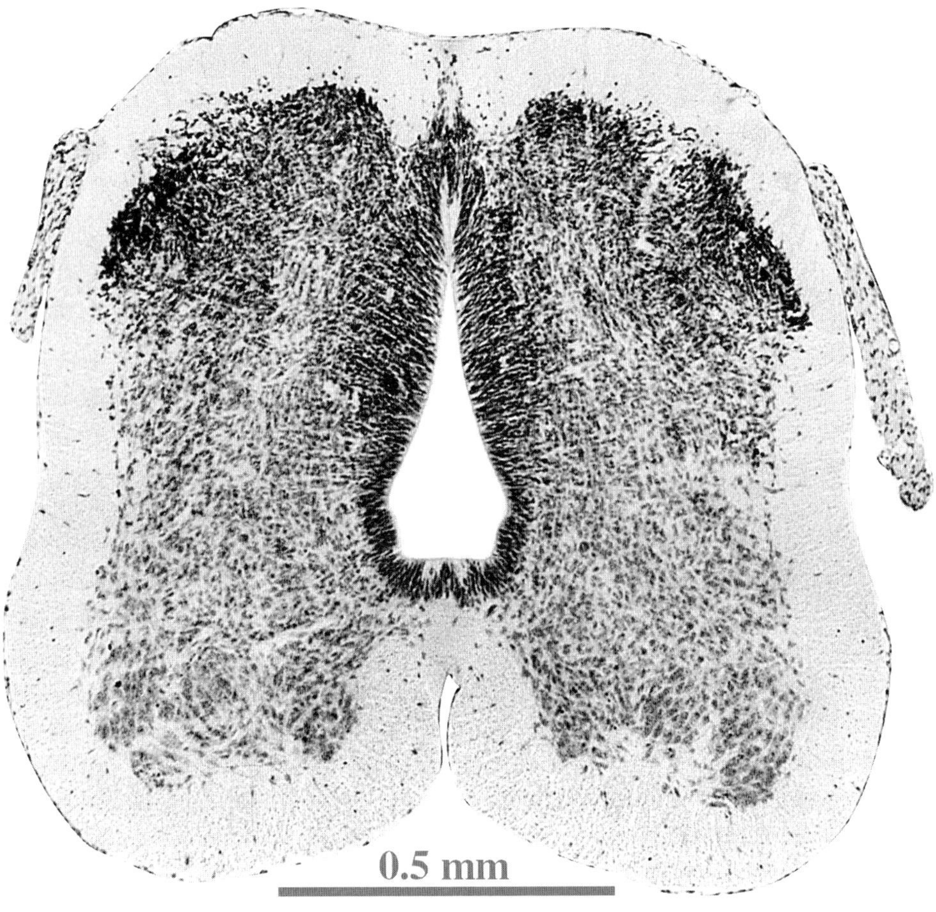

PLATE 19B

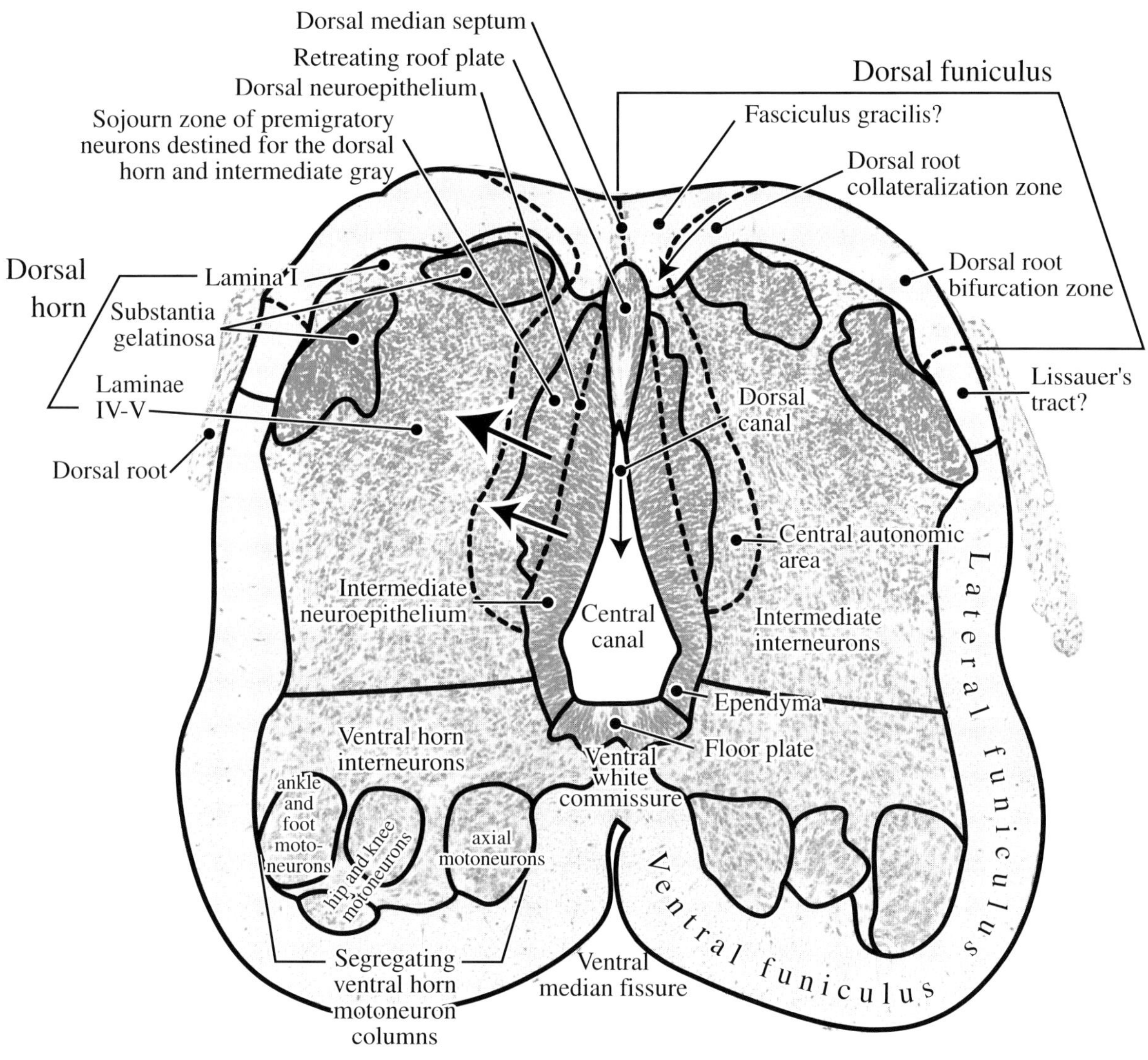
Dorsal median septum
Retreating roof plate
Dorsal neuroepithelium
Sojourn zone of premigratory neurons destined for the dorsal horn and intermediate gray
Dorsal funiculus
Fasciculus gracilis?
Dorsal root collateralization zone
Dorsal horn
Lamina I
Substantia gelatinosa
Laminae IV-V
Dorsal root
Dorsal root bifurcation zone
Lissauer's tract?
Dorsal canal
Central autonomic area
Intermediate neuroepithelium
Central canal
Intermediate interneurons
Ependyma
Floor plate
Ventral horn interneurons
Ventral white commissure
ankle and foot moto-neurons
hip and knee motoneurons
axial motoneurons
Segregating ventral horn motoneuron columns
Ventral median fissure
Ventral funiculus
Lateral funiculus

PLATE 20A

GW8.5
CR 36 mm
M2050
Sacral
Cell body stain

Areas (mm^2)

Spinal canal	.0184
Neuroepithelium	.0647
Roof plate	.0135
Floor plate	.0066
Gray matter	.3275
White matter	.1913

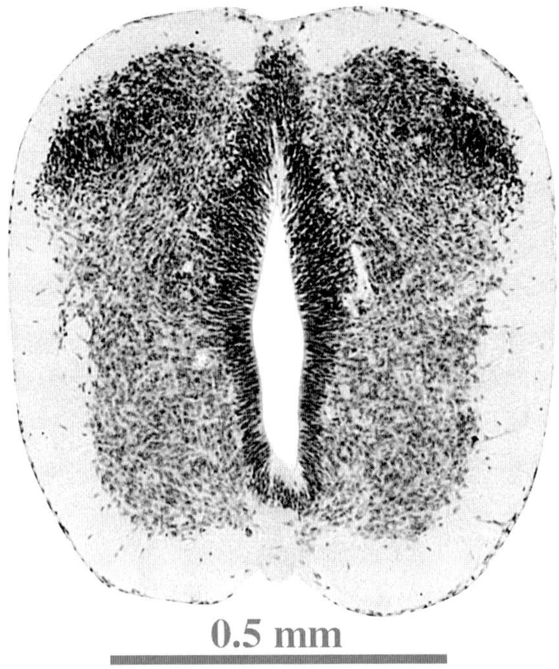

Sacral/Coccygeal
Cell body stain

Areas (mm^2)

Spinal canal	.0089
Neuroepithelium	.0386
Roof plate	.0052
Floor plate	.0029
Gray matter	.1962
White matter	.1389

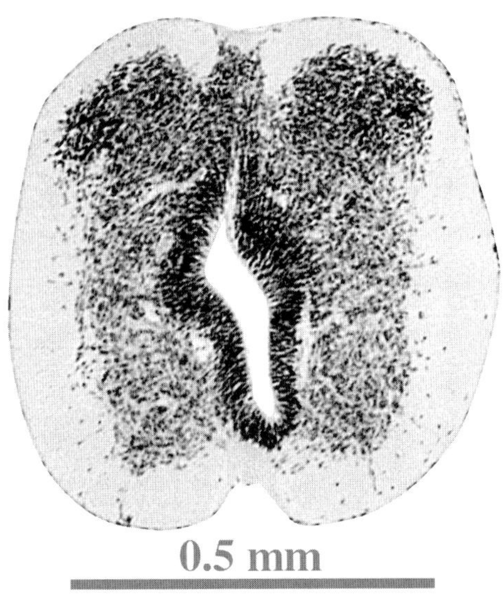

PLATE 20B

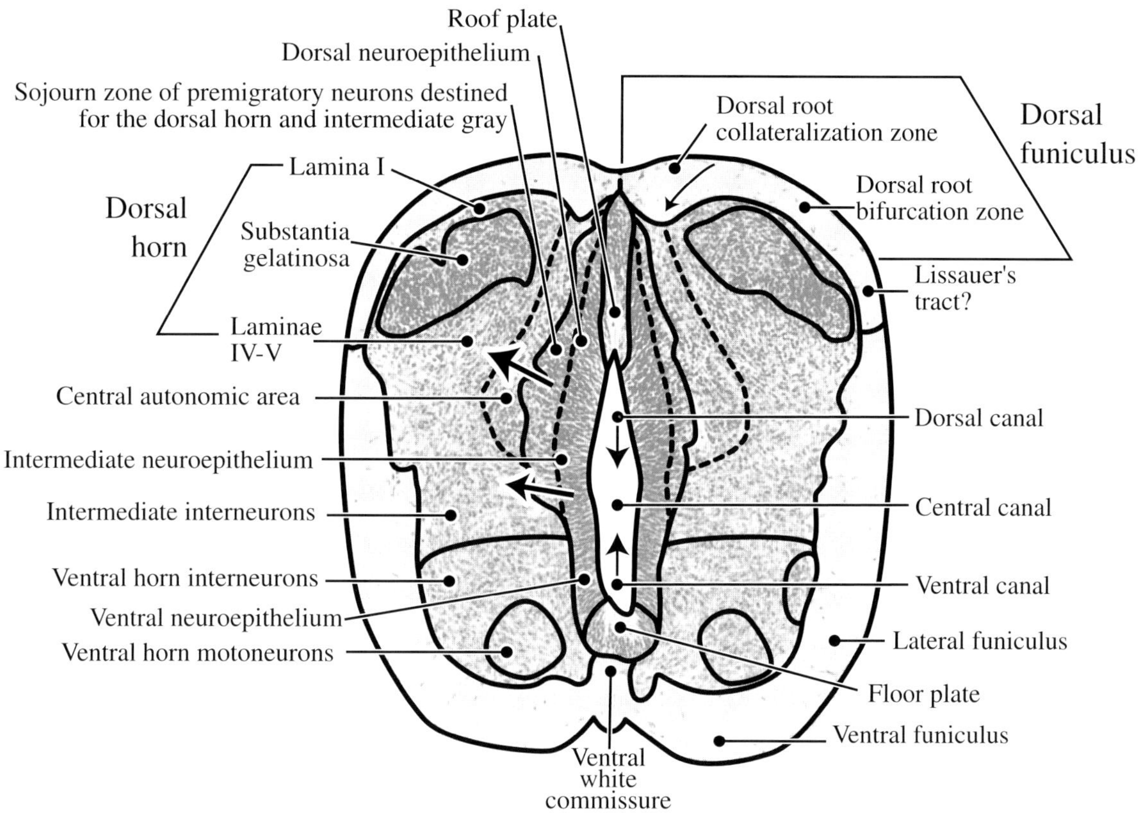
Roof plate
Dorsal neuroepithelium
Sojourn zone of premigratory neurons destined for the dorsal horn and intermediate gray
Dorsal root collateralization zone
Dorsal funiculus
Lamina I
Dorsal horn
Substantia gelatinosa
Dorsal root bifurcation zone
Lissauer's tract?
Laminae IV-V
Central autonomic area
Dorsal canal
Intermediate neuroepithelium
Intermediate interneurons
Central canal
Ventral horn interneurons
Ventral canal
Ventral neuroepithelium
Ventral horn motoneurons
Lateral funiculus
Floor plate
Ventral funiculus
Ventral white commissure

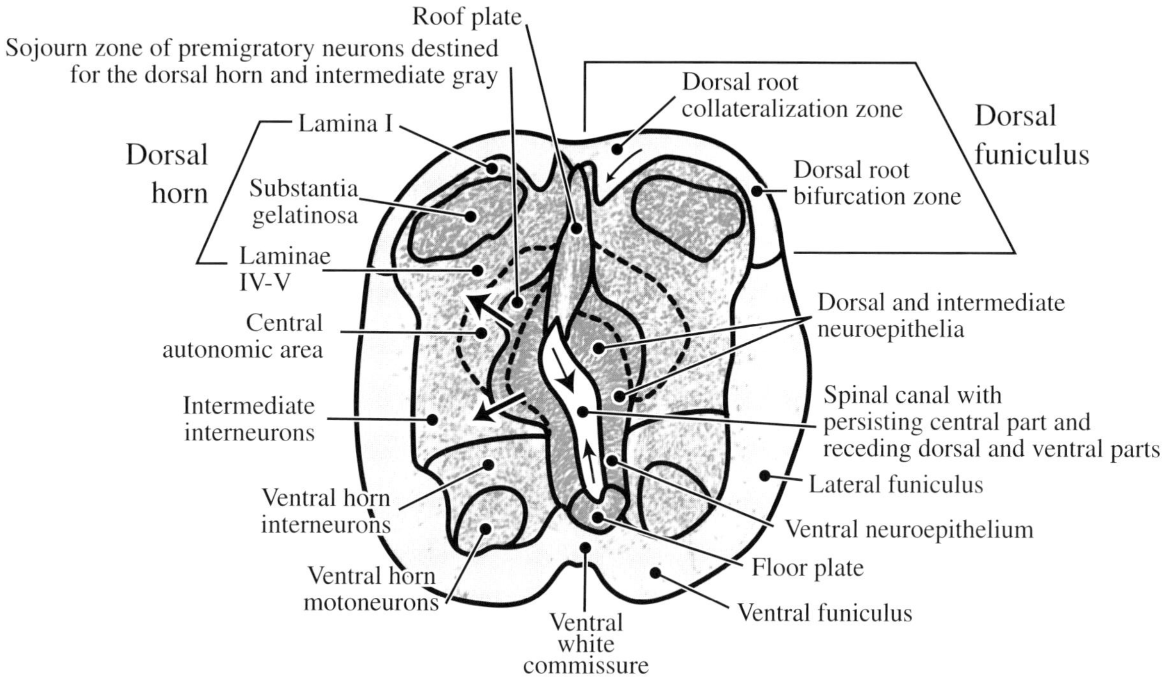
Roof plate
Sojourn zone of premigratory neurons destined for the dorsal horn and intermediate gray
Dorsal root collateralization zone
Dorsal funiculus
Lamina I
Dorsal horn
Dorsal root bifurcation zone
Substantia gelatinosa
Laminae IV-V
Central autonomic area
Dorsal and intermediate neuroepithelia
Intermediate interneurons
Spinal canal with persisting central part and receding dorsal and ventral parts
Lateral funiculus
Ventral horn interneurons
Ventral neuroepithelium
Floor plate
Ventral horn motoneurons
Ventral funiculus
Ventral white commissure

Part II: The First Trimester (concluded)

C. Spinal cord of a GW10.5 embryo

Plate 21 is a survey of sections from seven levels of the spinal cord in Y380-62, a specimen in the Yakovlev Collection with a crown rump length of 56 mm (see Chapter 5 in Altman and Bayer, 2001). All sections are shown at the same scale. The boxes enclosing each section list: the level, ranging from upper cervical to coccygeal; the section number; and the total area of the section in square millimeters (mm^2). Note that the areal measurements are determined *after* fixation, while the crown-rump length is measured *before* fixation. The unfixed area of each section could be 40 to 60% larger. Thus, the areal measurements are given only for comparison purposes between levels. Full-page normal contrast photographs of each specimen are in Plates **22A-28A.** Low contrast photographs with superimposed labels and outlines of structural details are in Plates **22B-28B.**

The size differences between levels seen at GW8.5, reflecting a rostral-to-caudal gradient in spinal cord maturation, persist in this specimen. The lumbar enlargement is 11% smaller than the cervical enlargement, and the smallest cross-sectional area is at the sacral level. However, the lower thoracic level is 37% smaller than the cervical enlargement and 29% smaller than the lumbar enlargement, reflecting regional size differences characteristic of the adult spinal cord.

The dorsal neuroepithelium is no longer present in this specimen. At all levels, the central canal is shaped like a triangle, and is lined by a dense layer of cells, presumably the ependymal layer. However, at the most caudal level (**Plate 28**), a small ventral neuroepithelium may be the source of glia. The lining of the persisting central canal is considered to be composed of ependymal cells.

The gray matter is continuing to show maturational changes indicative of the adult spinal cord. In the ventral horn, the motoneurons are segregating into more discrete motor columns than in the GW8.5 specimen. The thoracic ventral horn has a large accumulation of motoneurons in the medial motor column. At cervical levels, the intermediate gray has a distinguishable lateral cervical nucleus, and an accumulation of large neurons near the central canal that may represent the central cervical nucleus. However, the migratory stream of large neurons that appeared to be moving into the lateral cervical nucleus (seen in the GW8.5 specimen) is no longer evident. The dense groups of cells that appear to be migrating to and settling in the substantia gelatinosa, prominent at GW8.5, are less distinct in this more mature specimen. In the white matter, all components are expanding. The dorsal funiculus continues to fill in with fibers as it descends from the dorsal midline, following the retreating roof plate.

PLATE 21

GW10.5, CR 56mm, Y380-62

Plates 22A, 22B
Lower Medulla/
Upper Cervical
Section 361
Total Area:
3.2032 mm^2

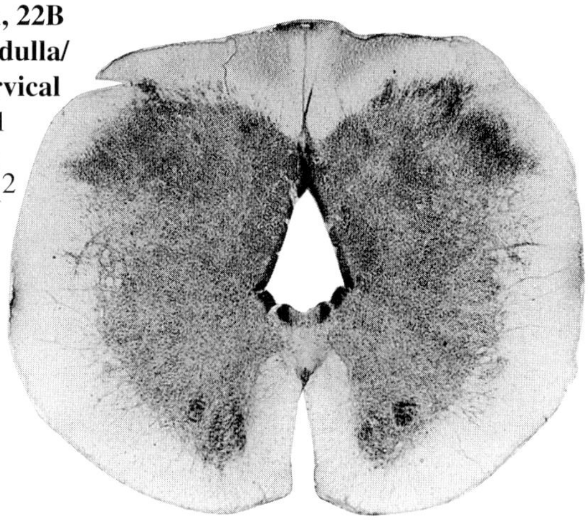

Plates 23A, 23B
Cervical Enlargement
Section 521
Total Area:
1.9640 mm^2

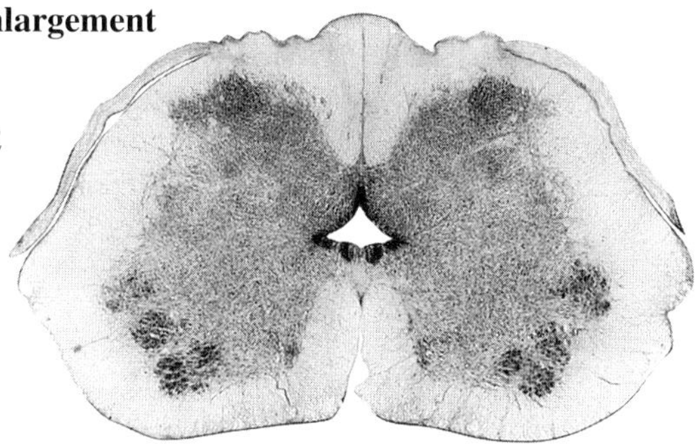

Plates 24A, 24B
Upper Thoracic
Section 671
Total Area:
1.4213 mm^2

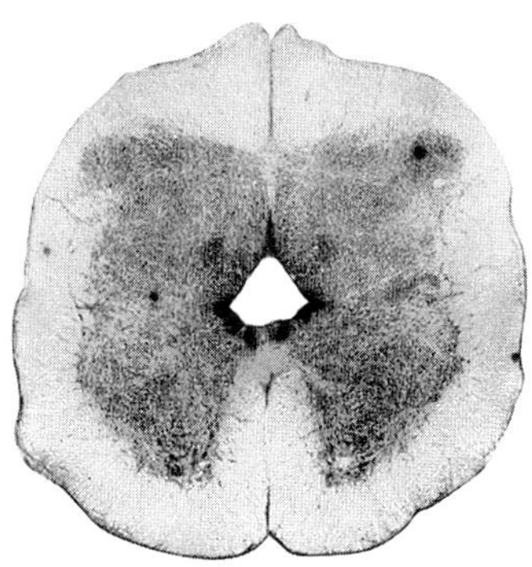

0.5 mm

Plates 25A, 25B
Lower Thoracic
Section 881
Total Area:
1.2365 mm^2

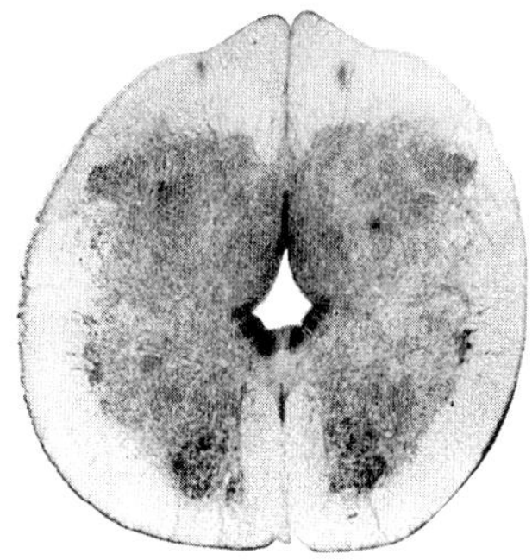

Plates 26A, 26B
Upper Lumbar
Section 911
Total Area:
1.4099 mm^2

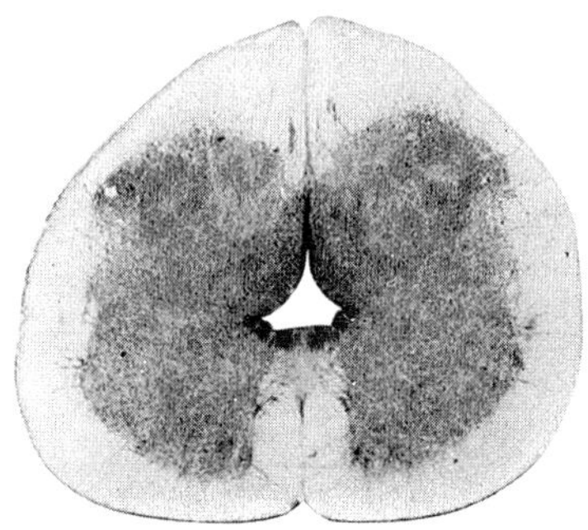

Plates 27A, 27B
Lumbar
Enlargement
Section 1066
Total Area:
1.7510 mm^2

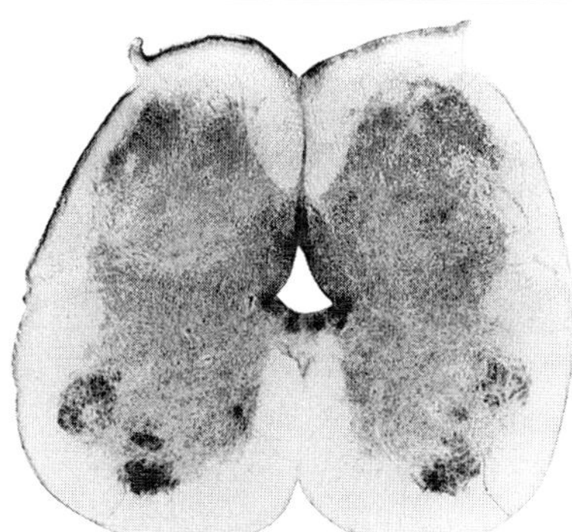

Plates 28A, 28B
Sacral
Section 1211
Total Area:
0.6813 mm^2

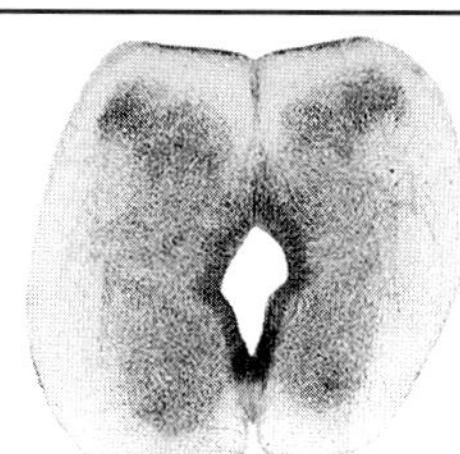

0.5 mm

PLATE 22A

GW10.5
CR 56 mm
Y380-62
Lower Medulla/Upper Cervical
Cell body stain

Areas (mm^2)

Central canal	.0728
Neuroepithelium	.0395
Roof plate	.0199
Floor plate	.0139
Gray matter	1.7140
White matter	1.3431

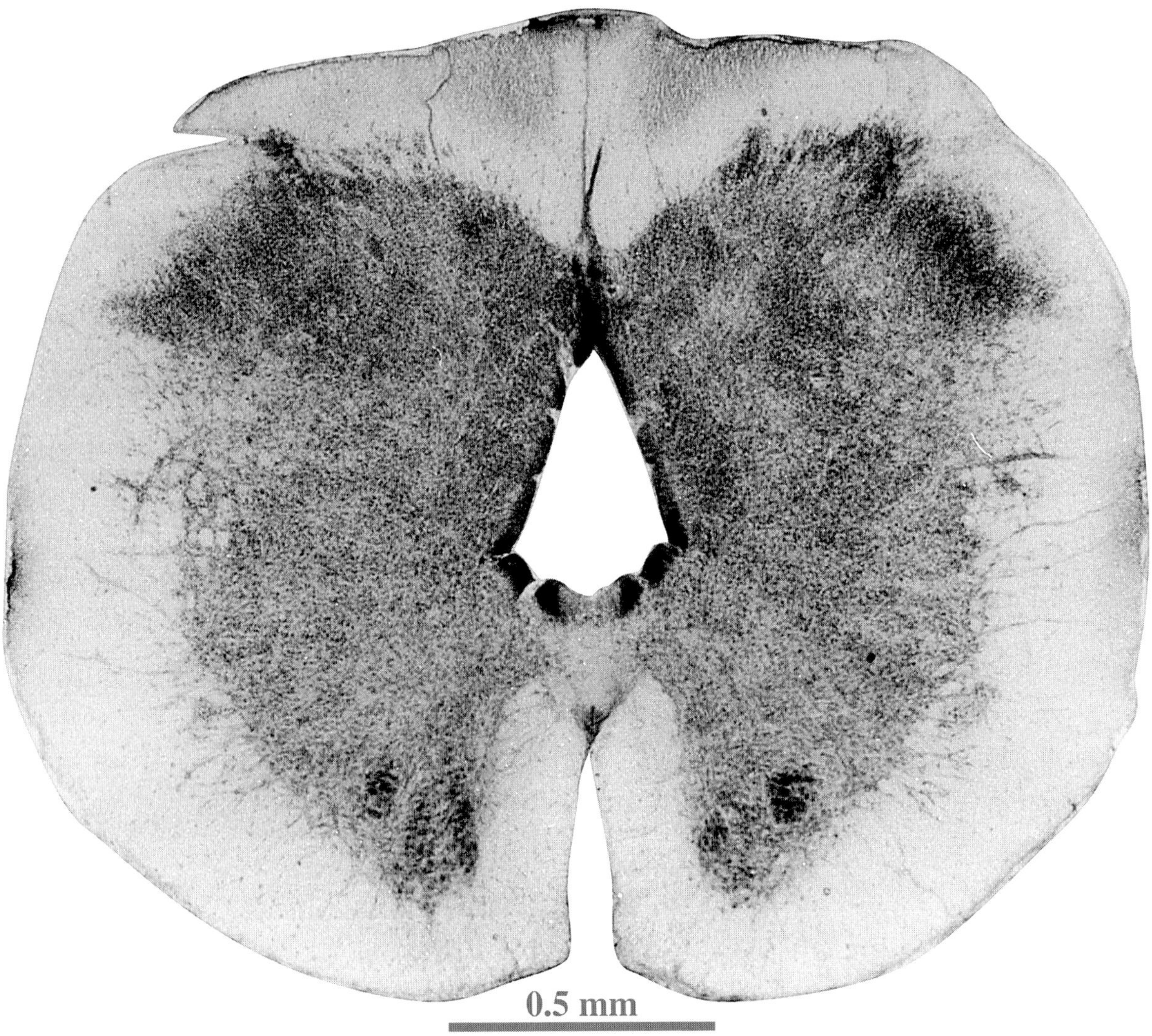

PLATE 22B

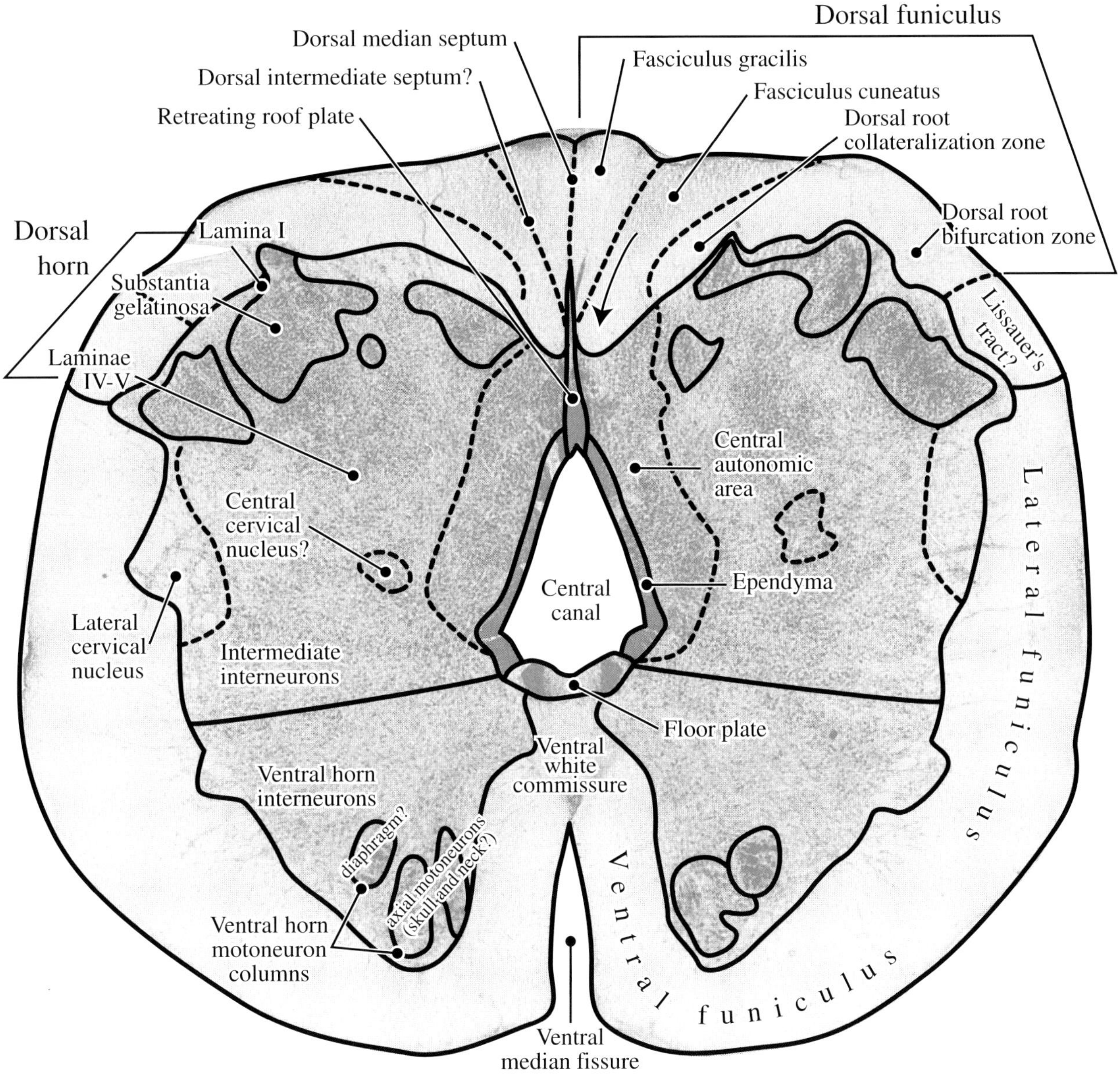
Dorsal funiculus
Dorsal median septum
Fasciculus gracilis
Dorsal intermediate septum?
Fasciculus cuneatus
Retreating roof plate
Dorsal root
collateralization zone
Dorsal root
bifurcation zone
Dorsal
horn
Lamina I
Substantia
gelatinosa
Lissauer's
tract?
Laminae
IV-V
Central
autonomic
area
Central
cervical
nucleus?
Central
canal
Ependyma
Lateral
cervical
nucleus
Intermediate
interneurons
Lateral funiculus
Floor plate
Ventral
white
commissure
Ventral horn
interneurons
diaphragm?
axial motoneurons
(skull and neck?)
Ventral horn
motoneuron
columns
Ventral funiculus
Ventral
median fissure

PLATE 23A

GW10.5
CR 56 mm
Y380-62
Cervical Enlargement
Cell body stain

Areas (mm^2)

Central canal	.0105
Neuroepithelium	.0135
Roof plate	.0133
Floor plate	.0086
Gray matter	1.0320
White matter	.8861

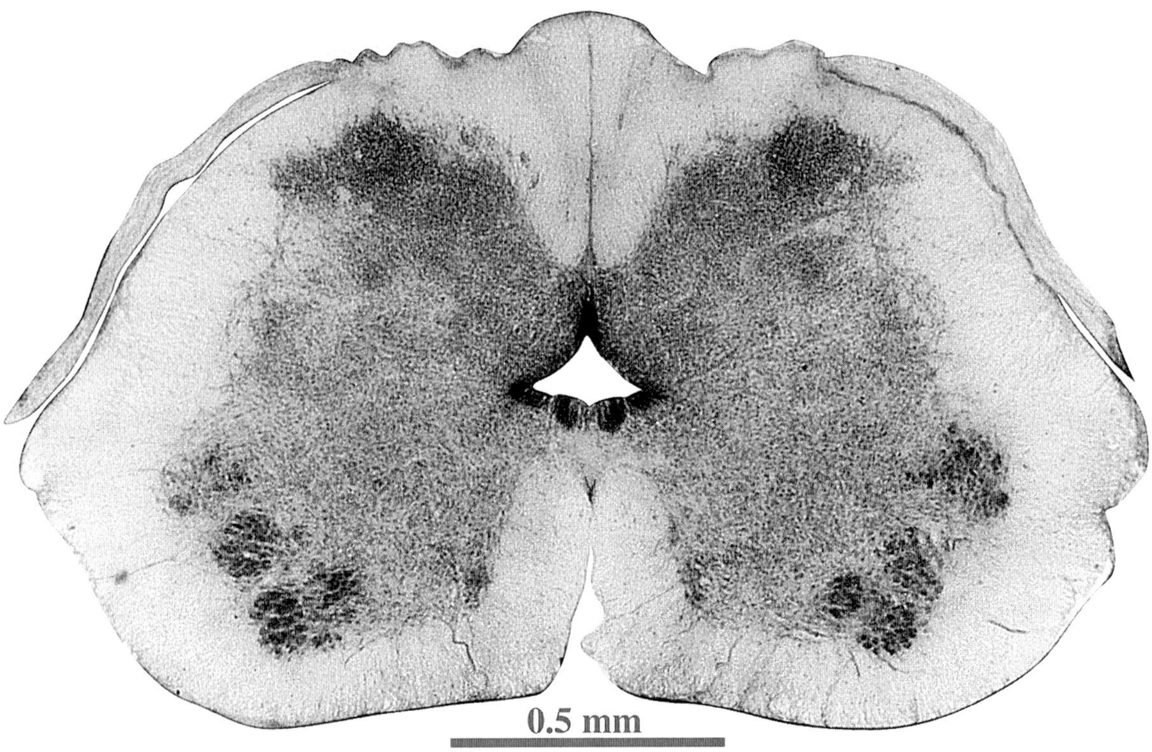

PLATE 23B

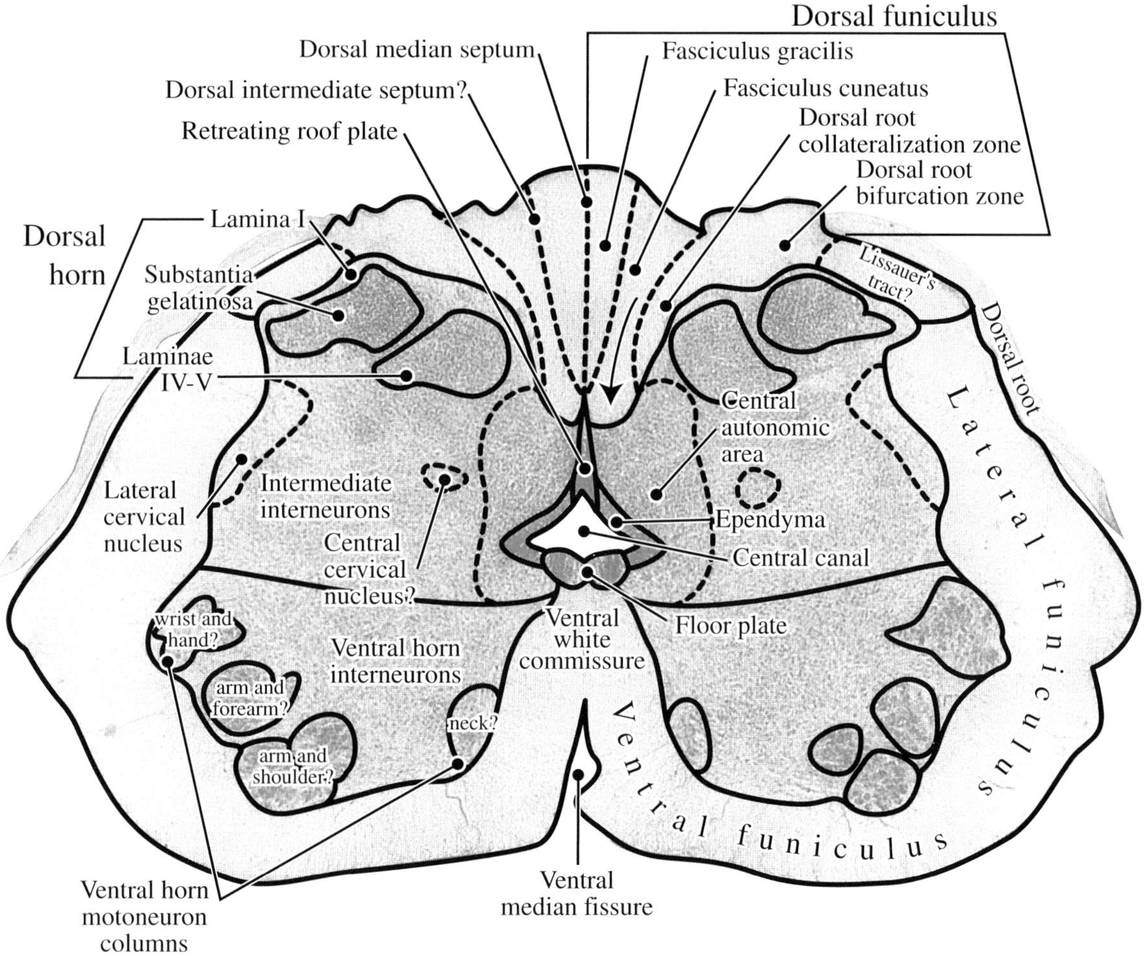
Dorsal funiculus
Dorsal median septum
Fasciculus gracilis
Dorsal intermediate septum?
Fasciculus cuneatus
Retreating roof plate
Dorsal root collateralization zone
Dorsal root bifurcation zone
Dorsal horn
Lamina I
Substantia gelatinosa
Laminae IV-V
Lissauer's tract?
Dorsal root
Central autonomic area
Lateral funiculus
Lateral cervical nucleus
Intermediate interneurons
Central cervical nucleus?
Ependyma
Central canal
Floor plate
Ventral white commissure
wrist and hand?
Ventral horn interneurons
arm and forearm?
neck?
arm and shoulder?
Ventral funiculus
Ventral horn motoneuron columns
Ventral median fissure

PLATE 24A

GW10.5
CR 56 mm
Y380-62
Upper Thoracic
Cell body stain

Areas (mm^2)

Central canal	.0206
Neuroepithelium	.0198
Roof plate	.0143
Floor plate	.0083
Gray matter	.6812
White matter	.6771

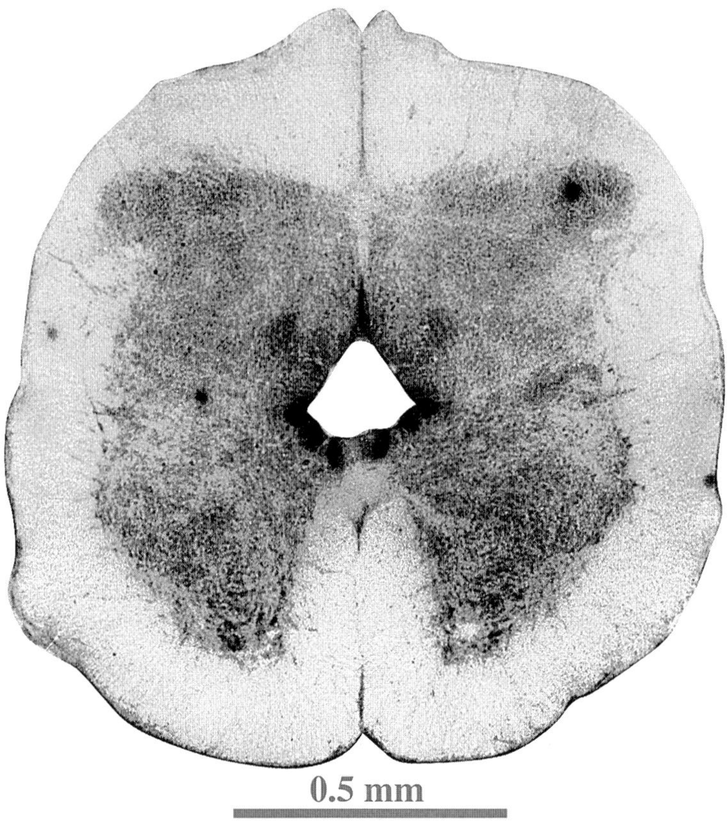

PLATE 24B

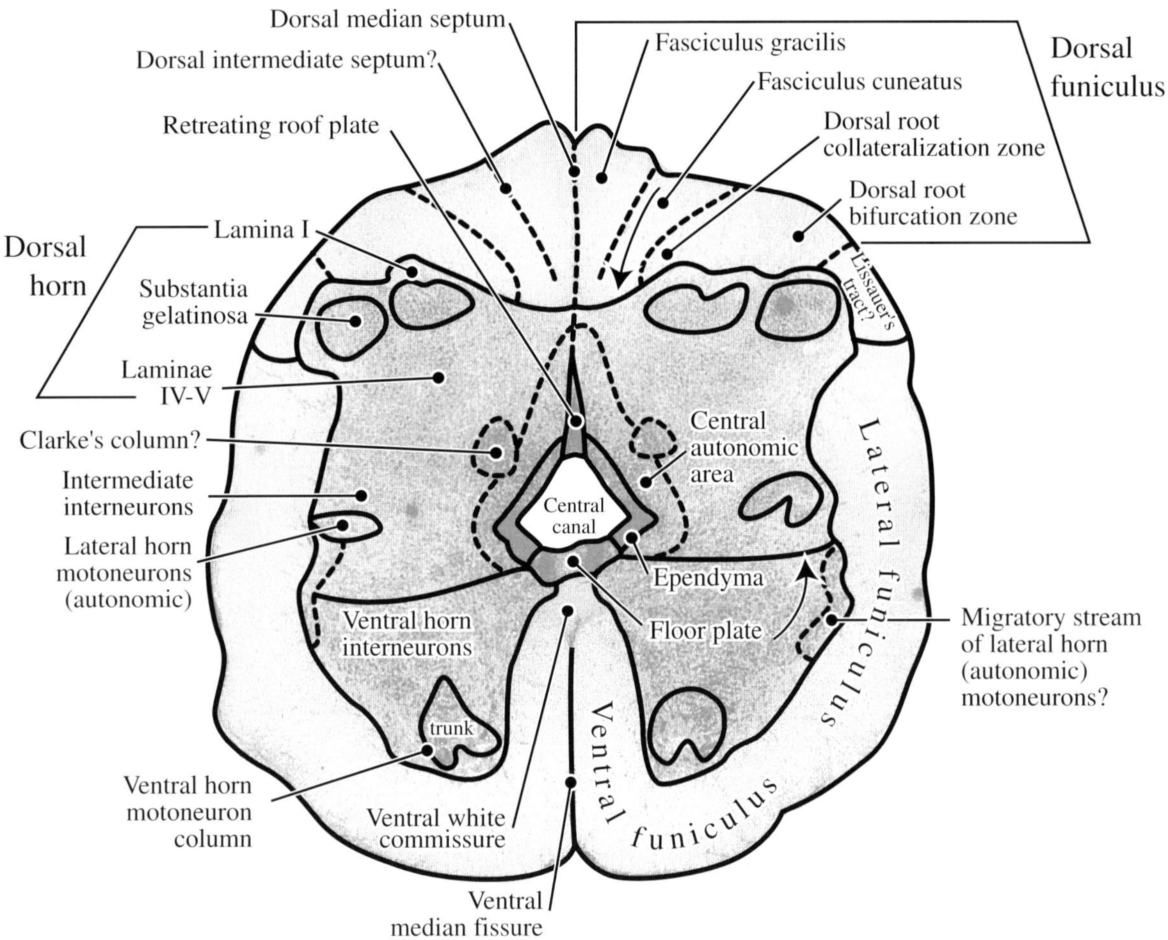
Dorsal median septum
Dorsal intermediate septum?
Retreating roof plate
Fasciculus gracilis
Fasciculus cuneatus
Dorsal funiculus
Dorsal root collateralization zone
Dorsal root bifurcation zone
Lissauer's tract?
Dorsal horn
Lamina I
Substantia gelatinosa
Laminae IV-V
Clarke's column?
Central autonomic area
Intermediate interneurons
Central canal
Lateral horn motoneurons (autonomic)
Ependyma
Lateral funiculus
Floor plate
Migratory stream of lateral horn (autonomic) motoneurons?
Ventral horn interneurons
trunk
Ventral horn motoneuron column
Ventral white commissure
Ventral funiculus
Ventral median fissure

PLATE 25A

GW10.5
CR 56 mm
Y380-62
Lower Thoracic
Cell body stain

Areas (mm^2)

Central canal	.0114
Neuroepithelium	.0207
Roof plate	.0100
Floor plate	.0082
Gray matter	.6527
White matter	.5336

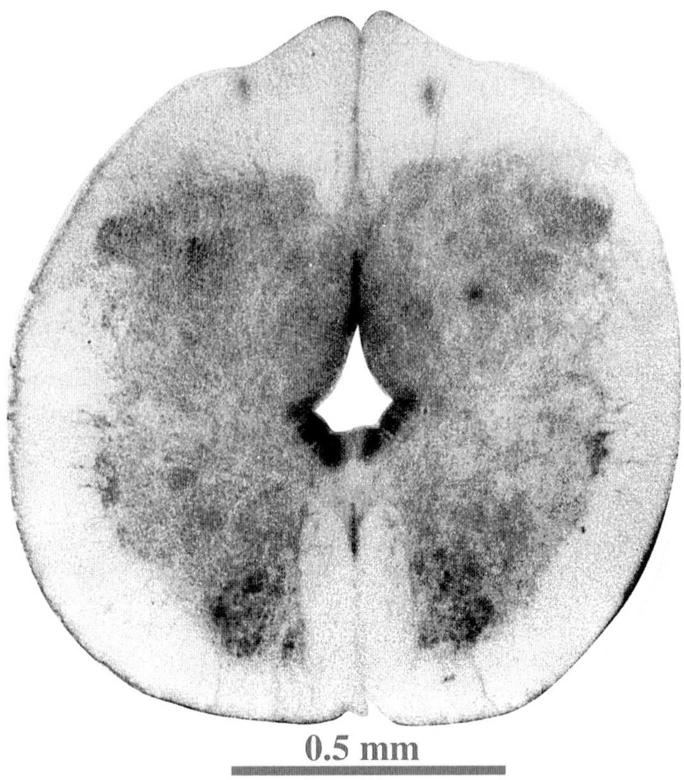

PLATE 25B

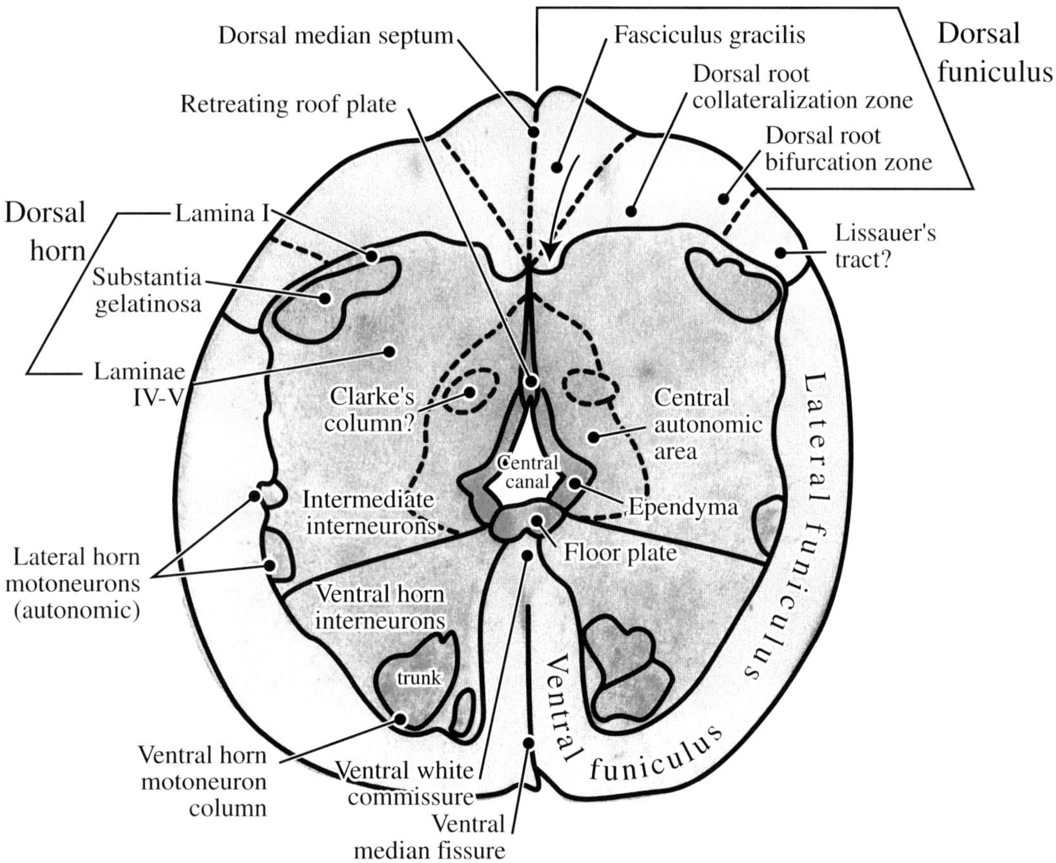
Dorsal median septum
Fasciculus gracilis
Dorsal funiculus
Dorsal root collateralization zone
Retreating roof plate
Dorsal root bifurcation zone
Dorsal horn
Lamina I
Lissauer's tract?
Substantia gelatinosa
Laminae IV-V
Clarke's column?
Central autonomic area
Lateral funiculus
Central canal
Intermediate interneurons
Ependyma
Floor plate
Lateral horn motoneurons (autonomic)
Ventral horn interneurons
trunk
Ventral funiculus
Ventral horn motoneuron column
Ventral white commissure
Ventral median fissure

PLATE 26A

GW10.5
CR 56 mm
Y380-62
Upper Lumbar
Cell body stain

Areas (mm^2)

Central canal	.0129
Neuroepithelium	.0232
Roof plate	.0143
Floor plate	.0091
Gray matter	.7558
White matter	.5992

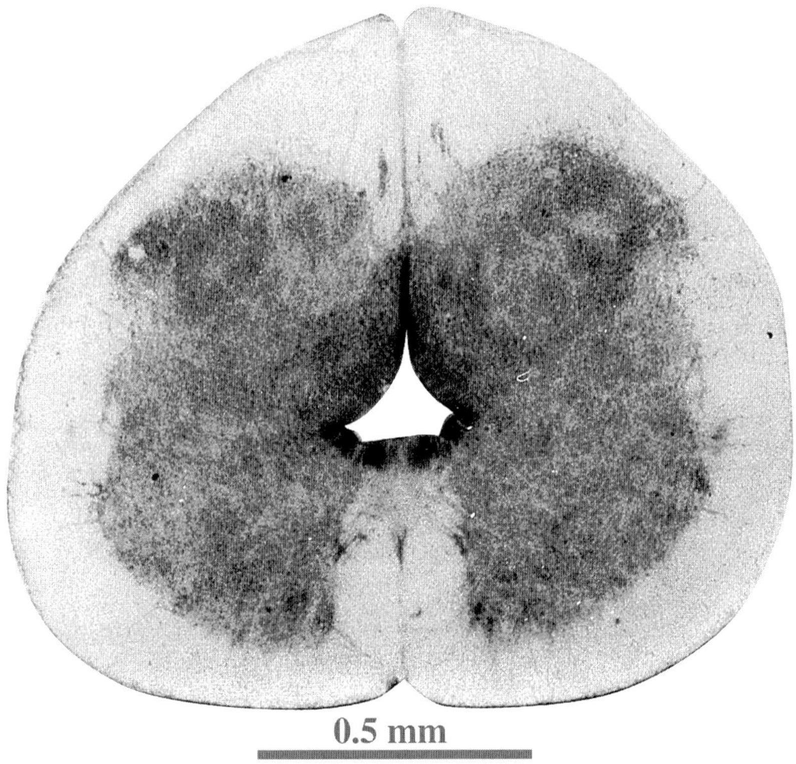

PLATE 26B

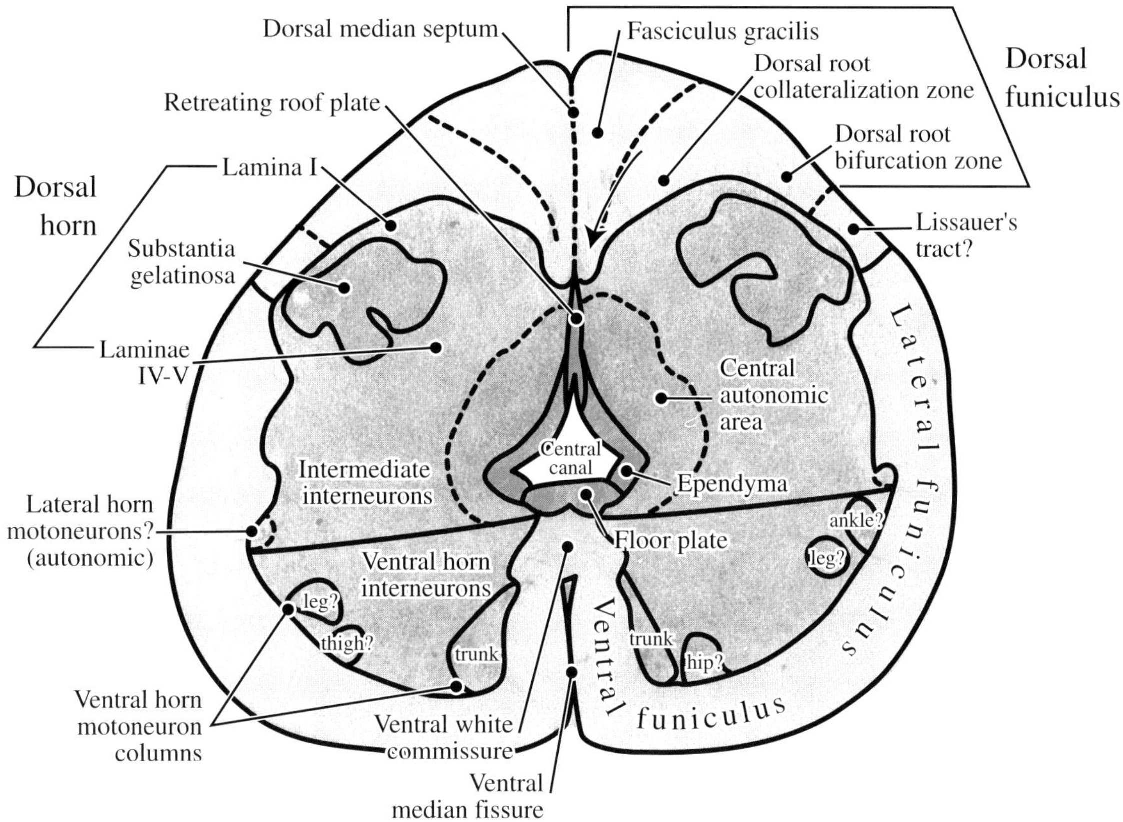
Dorsal median septum
Fasciculus gracilis
Dorsal root collateralization zone
Dorsal funiculus
Retreating roof plate
Dorsal root bifurcation zone
Dorsal horn
Lamina I
Lissauer's tract?
Substantia gelatinosa
Laminae IV-V
Central autonomic area
Lateral funiculus
Central canal
Intermediate interneurons
Ependyma
Lateral horn motoneurons? (autonomic)
ankle?
Floor plate
leg?
Ventral horn interneurons
leg?
thigh?
trunk
trunk
hip?
Ventral funiculus
Ventral horn motoneuron columns
Ventral white commissure
Ventral median fissure

PLATE 27A

GW10.5
CR 56 mm
Y380-62
Lumbar Enlargement
Cell body stain

Areas (mm^2)

Central canal	.0128
Neuroepithelium	.0183
Roof plate	.0128
Floor plate	.0081
Gray matter	.9752
White matter	.7221

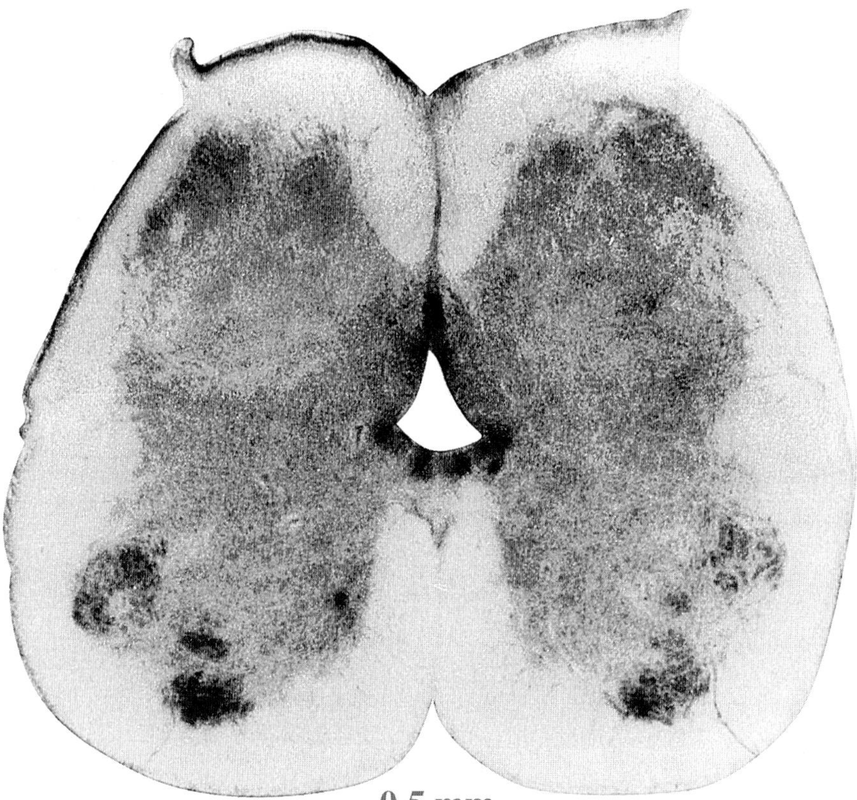

0.5 mm

PLATE 27B

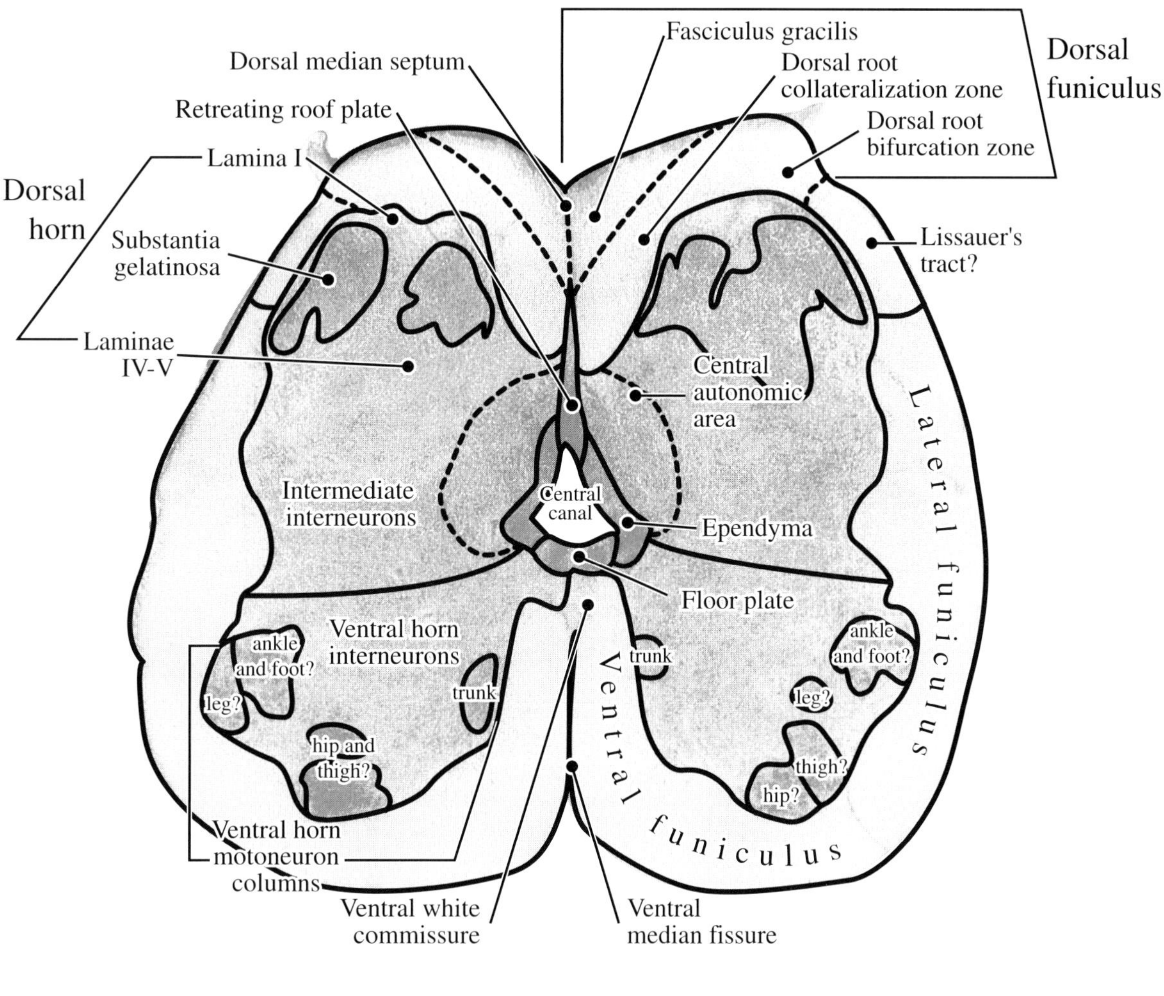
Fasciculus gracilis
Dorsal median septum
Dorsal root collateralization zone
Dorsal funiculus
Retreating roof plate
Dorsal root bifurcation zone
Lamina I
Dorsal horn
Substantia gelatinosa
Lissauer's tract?
Laminae IV-V
Central autonomic area
Lateral funiculus
Intermediate interneurons
Central canal
Ependyma
Floor plate
Ventral horn interneurons
ankle and foot?
leg?
trunk
hip and thigh?
thigh?
hip?
Ventral funiculus
Ventral horn motoneuron columns
Ventral white commissure
Ventral median fissure

PLATE 28A

GW10.5
CR 56 mm
Y380-62
Sacral
Cell body stain

Areas (mm²)

Central canal	.0144
Neuroepithelium	.0311
Roof plate	.0201
Floor plate	.0058
Gray matter	.3385
White matter	.2754

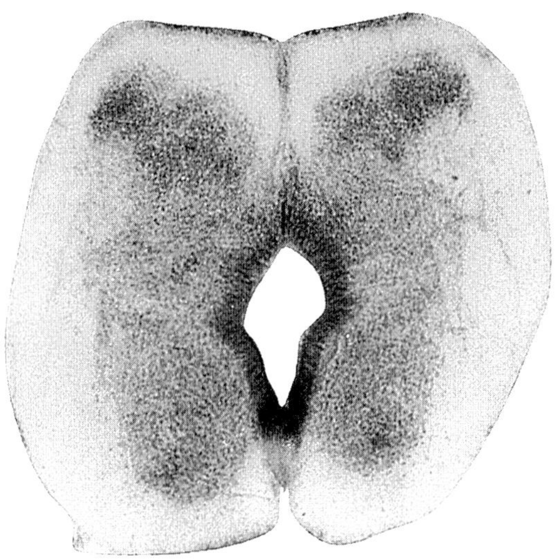

0.5 mm

PLATE 28B

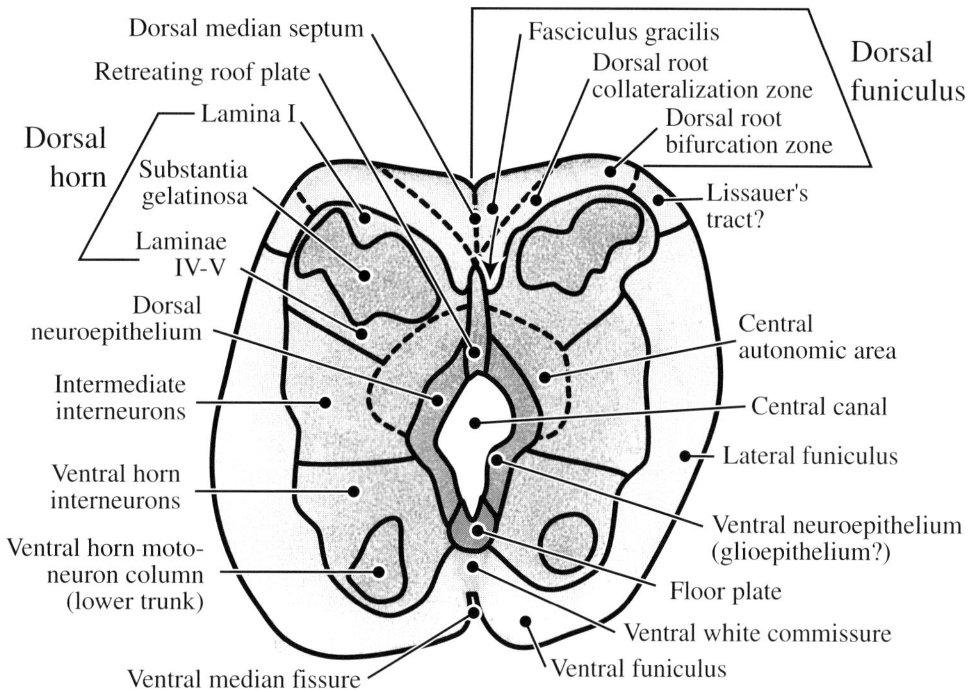

Dorsal median septum
Retreating roof plate
Lamina I
Dorsal horn
Substantia gelatinosa
Laminae IV-V
Dorsal neuroepithelium
Intermediate interneurons
Ventral horn interneurons
Ventral horn moto-neuron column (lower trunk)
Ventral median fissure
Fasciculus gracilis
Dorsal root collateralization zone
Dorsal funiculus
Dorsal root bifurcation zone
Lissauer's tract?
Central autonomic area
Central canal
Lateral funiculus
Ventral neuroepithelium (glioepithelium?)
Floor plate
Ventral white commissure
Ventral funiculus

III. The Second Trimester

A. Spinal cord of a GW14 fetus

Plate 29 is a survey of sections from Y68-65, a specimen in the Yakovlev Collection with a crown-rump length of 108 mm (see Chapters 5 and 6 of Altman and Bayer, 2001). All sections are shown at the same scale. The boxes enclosing each section list: the level from upper cervical to sacral/coccygeal; the section number from the set of slides containing all the sections of that specimen; and the total area (post-fixation) of the section in square millimeters (mm^2). Full-page normal contrast photographs of each specimen are in **Plates 30A-37A**. Low contrast photographs with superimposed labels and outlines of section details are on facing pages in **Plates 30B-37B**.

The outer surface of this specimen has sharp spike-like projections, and the white matter is thin around the gray matter, interpreted to be a shrinkage artifact of histological processing. Cells in the gray matter are not distinct, except for the prominent clumps of large motoneurons in the ventral horn, possibly due to inadequate penetration of fixative. In spite of these histological artifacts, this specimen is illustrated becaue it is the only one in the Yakovlev Collection where sections of the spinal cord are consecutively numbered from cervical to coccygeal levels.

Most of the size differences between levels at GW14 are similar to those in the adult spinal cord. The smallest cross-sectional area is in the middle thoracic level, which is 27% smaller than the sacral level. That the cervical enlargement level is smaller than both the upper cervical and lumbar enlargement levels reflects variability unique to this specimen. Within the gray matter, the most obvious sign of ongoing maturation is the prominent columnar arrangement of motoneurons in the ventral horn. Not only are these columns larger than at GW10.5, but there are more of them. Within the white matter, the dorsal funiculus is deepening further in the dorsal midline, accompanying the retreating roof plate.

PLATE 29

GW14, CR 108 mm, Y68-65

Plates 30A, 30B
Upper Cervical
Section 741
Total area:
4.1972 mm^2

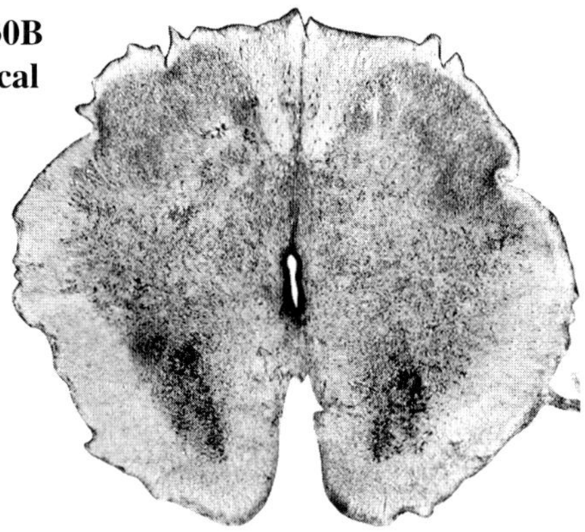

Plates 31A, 31B
Cervical Enlargement
Section 981
Total area:
3.0426 mm^2

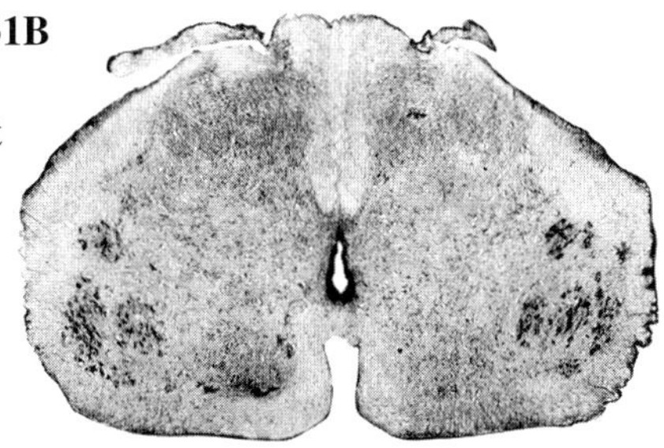

Plates 32A, 32B
Upper Thoracic
Section 1141
Total area:
1.4745 mm^2

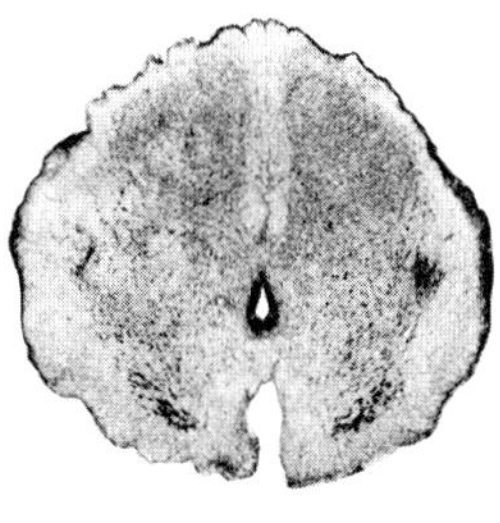

Plates 33A, 33B
Middle Thoracic
Section 1581
Total area:
1.1659 mm^2

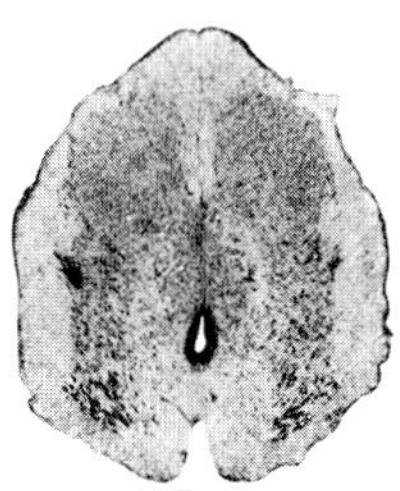

0.5mm

Plates 34A, 34B
Lower Thoracic
Section 1741
Total area:
1.6474 mm^2

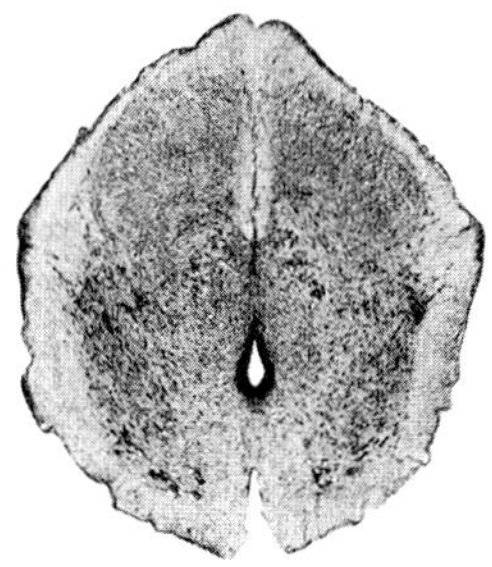

Plates 35A, 35B
Upper Lumbar
Section 1821
Total area:
2.7717 mm^2

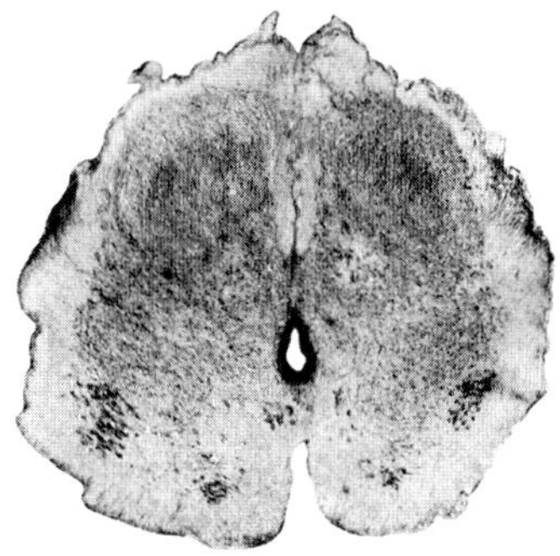

Plates 36A, 36B
Lumbar Enlargement
Section 1901
Total area:
3.8584 mm^2

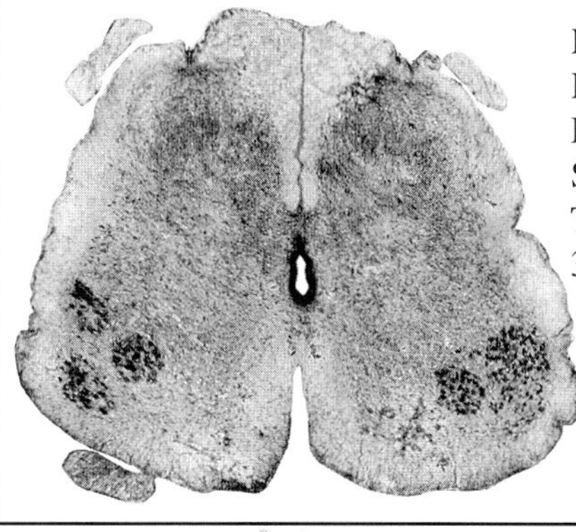

Plates 37A, 37B
Sacral/ Coccygeal
Section 2061
Total area:
1.6071 mm^2

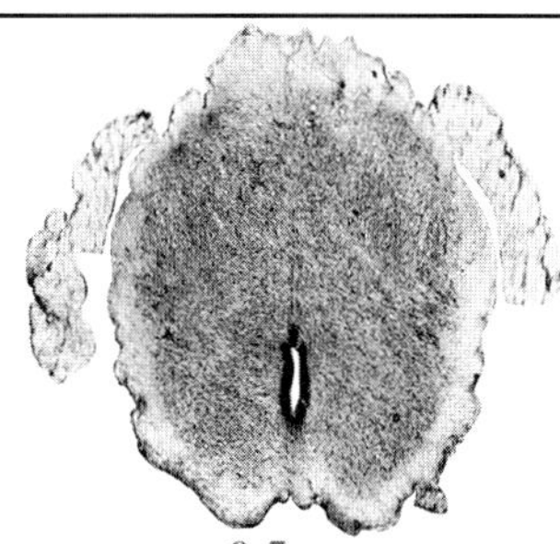

0.5mm

PLATE 30A

GW14
CR 108 mm
Y68-65
Upper Cervical
Cell body stain

Areas (mm^2)

Central canal	.0053
Neuroepithelium	.0182
Roof plate	.0290
Floor plate	.0029
Gray matter	2.5381
White matter	1.6037

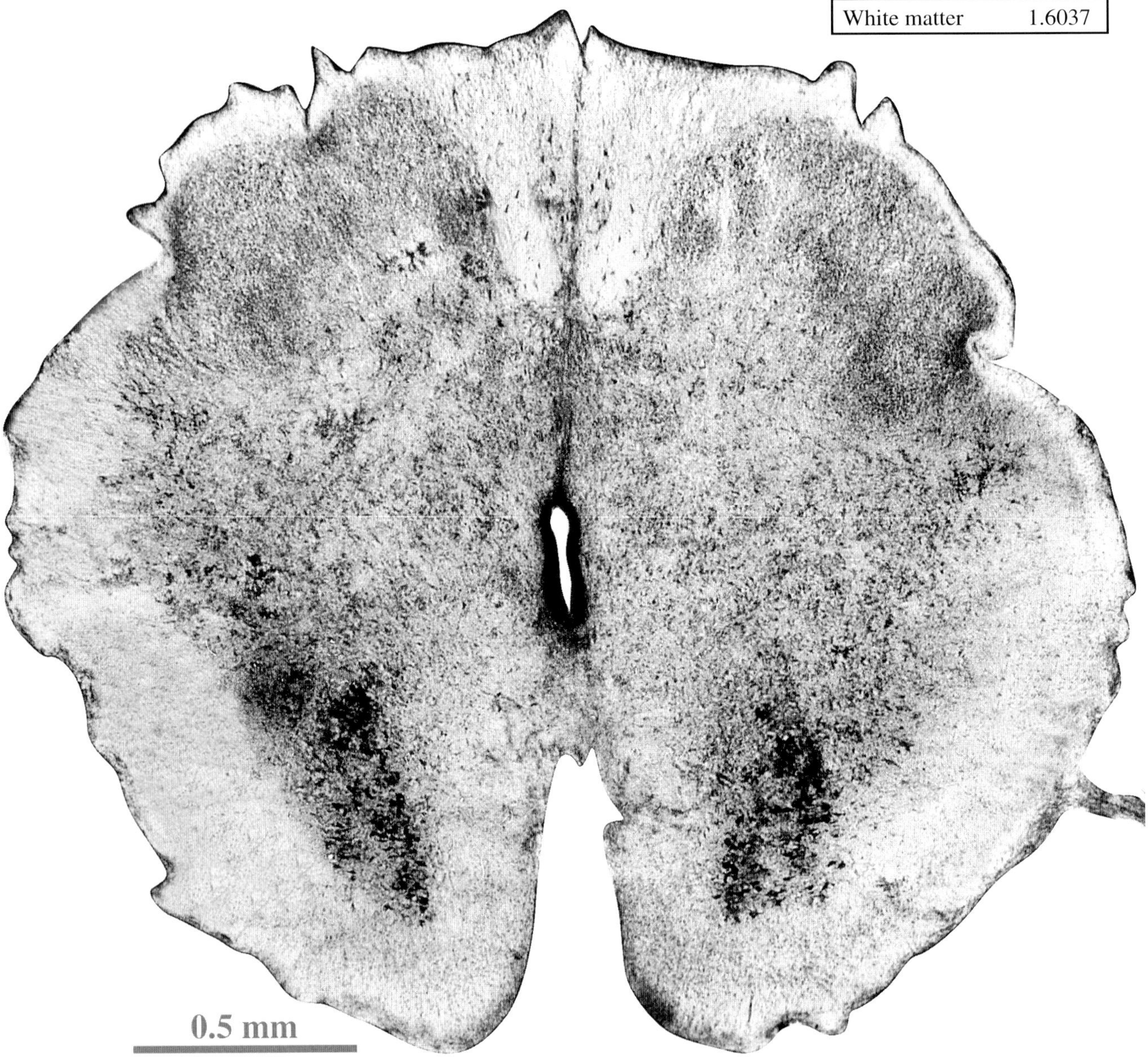

PLATE 30B

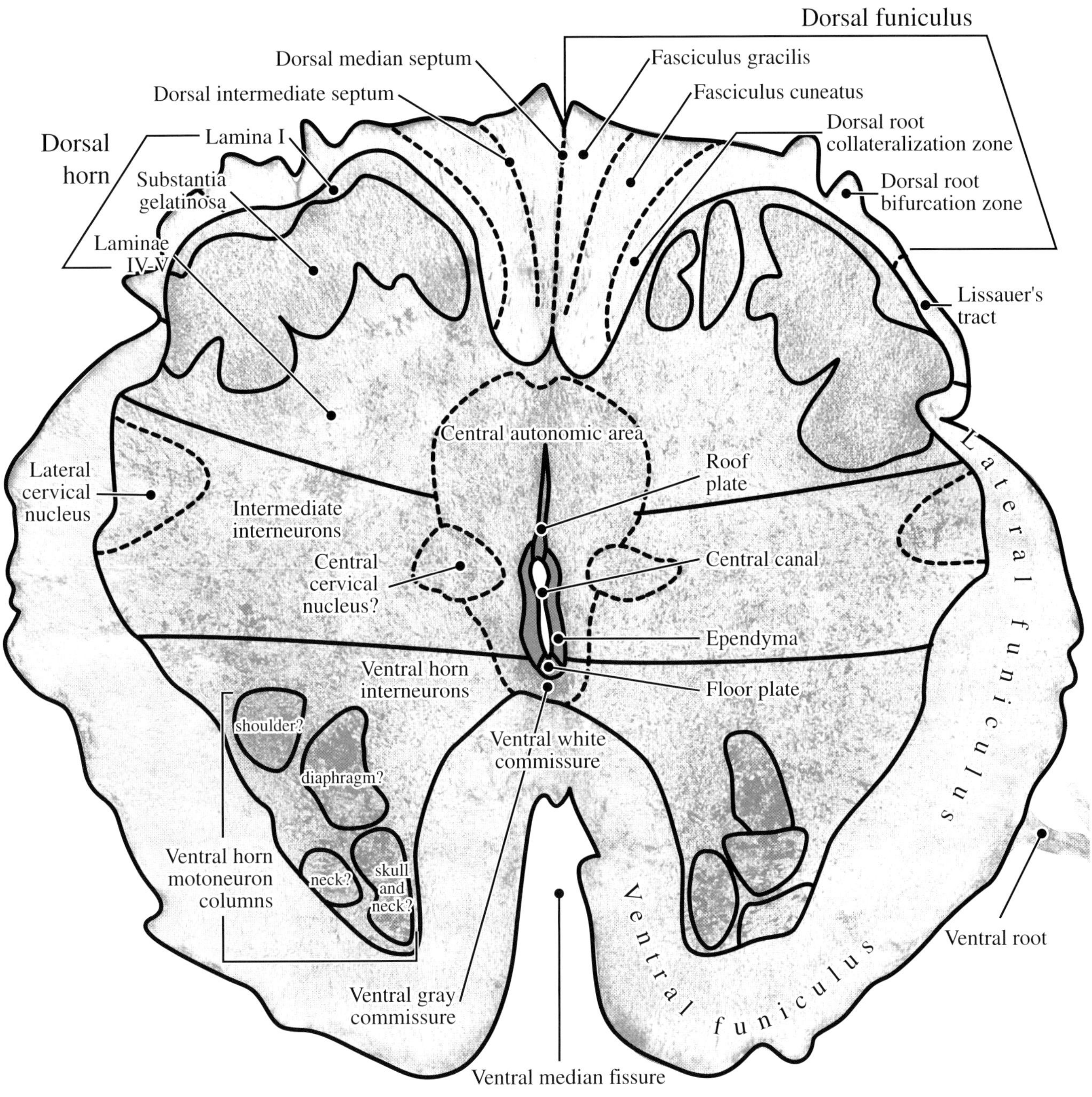
Dorsal funiculus
Dorsal median septum
Fasciculus gracilis
Dorsal intermediate septum
Fasciculus cuneatus
Dorsal horn
Lamina I
Dorsal root collateralization zone
Substantia gelatinosa
Dorsal root bifurcation zone
Laminae IV-V
Lissauer's tract
Central autonomic area
Lateral cervical nucleus
Roof plate
Intermediate interneurons
Central cervical nucleus?
Central canal
Ependyma
Ventral horn interneurons
Floor plate
shoulder?
Ventral white commissure
diaphragm?
Ventral horn motoneuron columns
neck?
skull and neck?
Lateral funiculus
Ventral funiculus
Ventral root
Ventral gray commissure
Ventral median fissure

PLATE 31A

GW14
CR 108 mm
Y68-65
Cervical Enlargement
Cell body stain

Areas (mm^2)

Central canal	.0056
Neuroepithelium	.0127
Roof plate	.0192
Floor plate	.0035
Gray matter	1.9505
White matter	1.0511

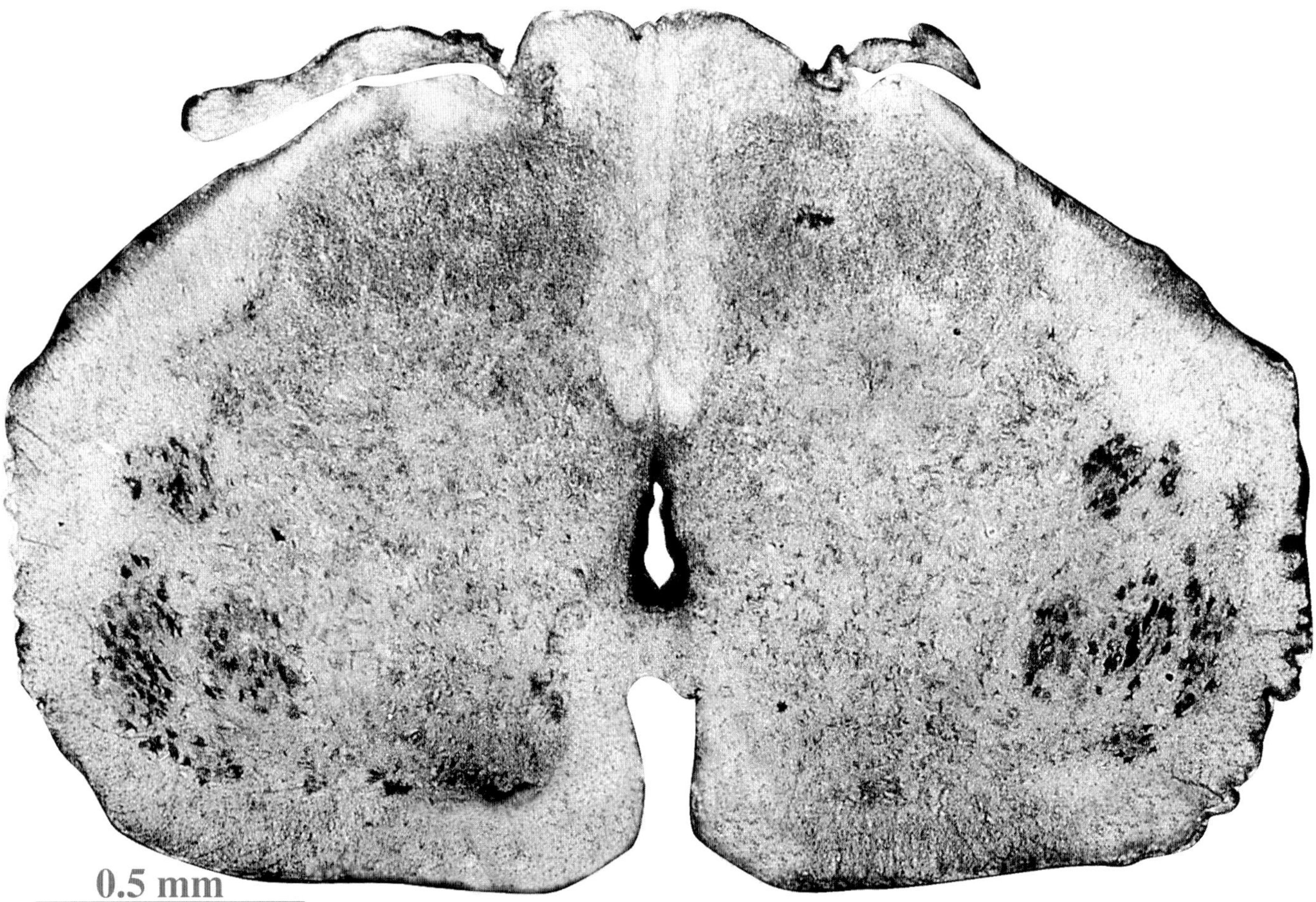

PLATE 31B

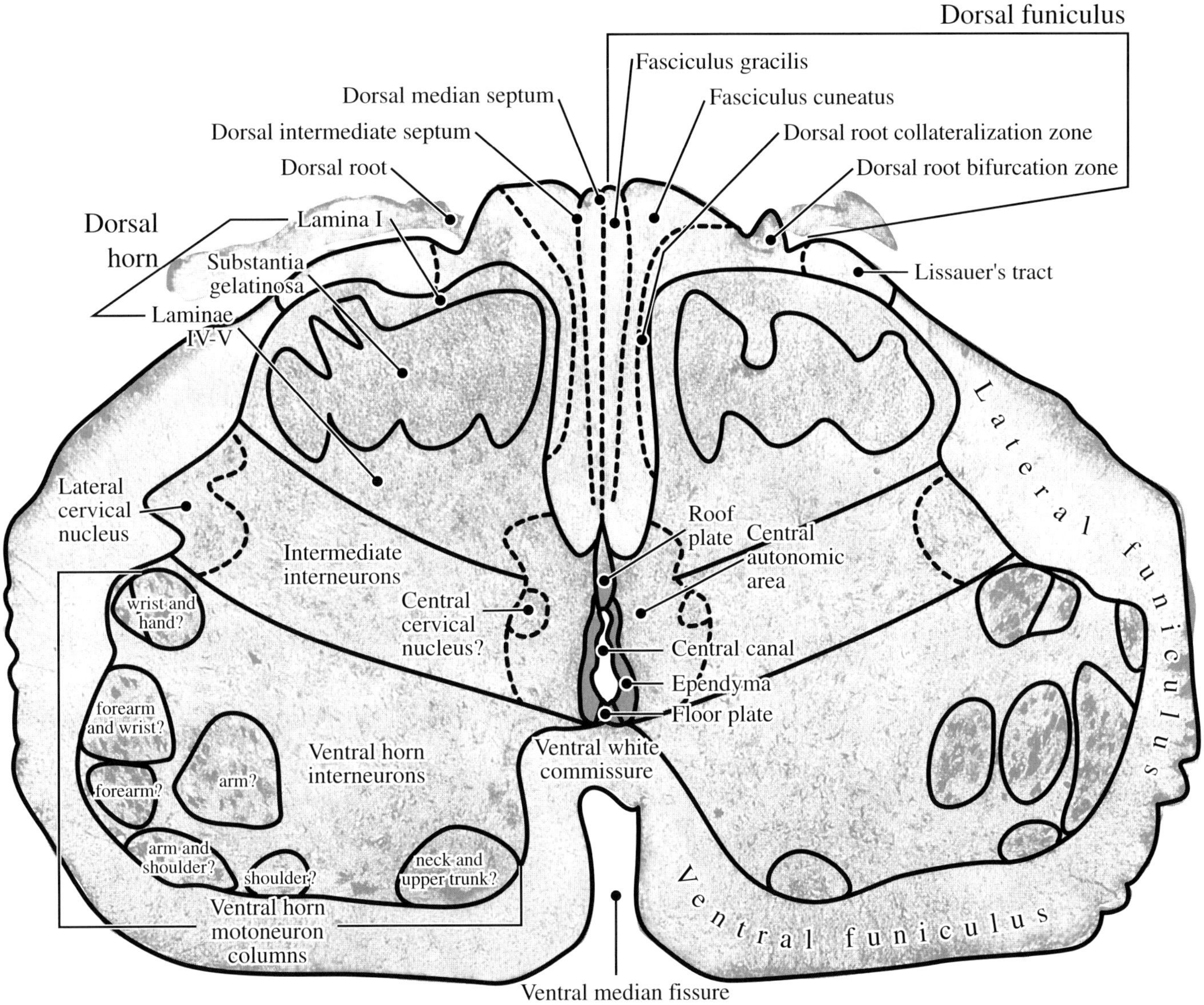
Dorsal funiculus
Fasciculus gracilis
Fasciculus cuneatus
Dorsal median septum
Dorsal intermediate septum
Dorsal root collateralization zone
Dorsal root
Dorsal root bifurcation zone
Dorsal horn
Lamina I
Substantia gelatinosa
Laminae IV-V
Lissauer's tract
Lateral funiculus
Lateral cervical nucleus
Intermediate interneurons
Roof plate
Central autonomic area
Central cervical nucleus?
Central canal
Ependyma
Floor plate
wrist and hand?
forearm and wrist?
forearm?
arm?
arm and shoulder?
shoulder?
neck and upper trunk?
Ventral horn interneurons
Ventral white commissure
Ventral horn motoneuron columns
Ventral funiculus
Ventral median fissure

PLATE 32A

GW14
CR 108 mm
Y68-65
Upper Thoracic
Cell body stain

Areas (mm^2)

Central canal	.0021
Neuroepithelium	.0063
Roof plate	.0135
Floor plate	.0040
Gray matter	.8242
White matter	.6243

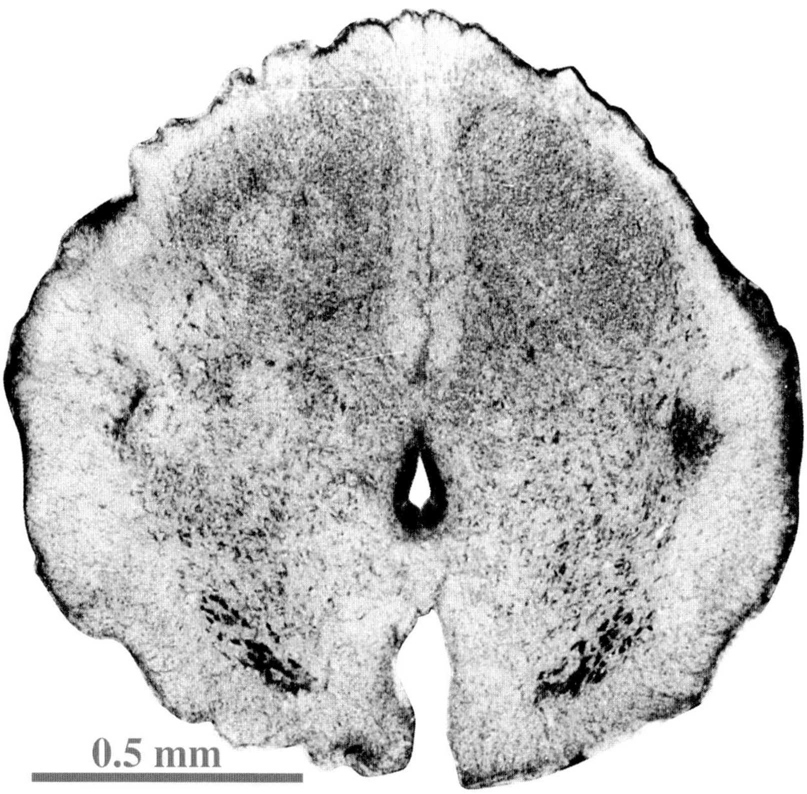

PLATE 32B

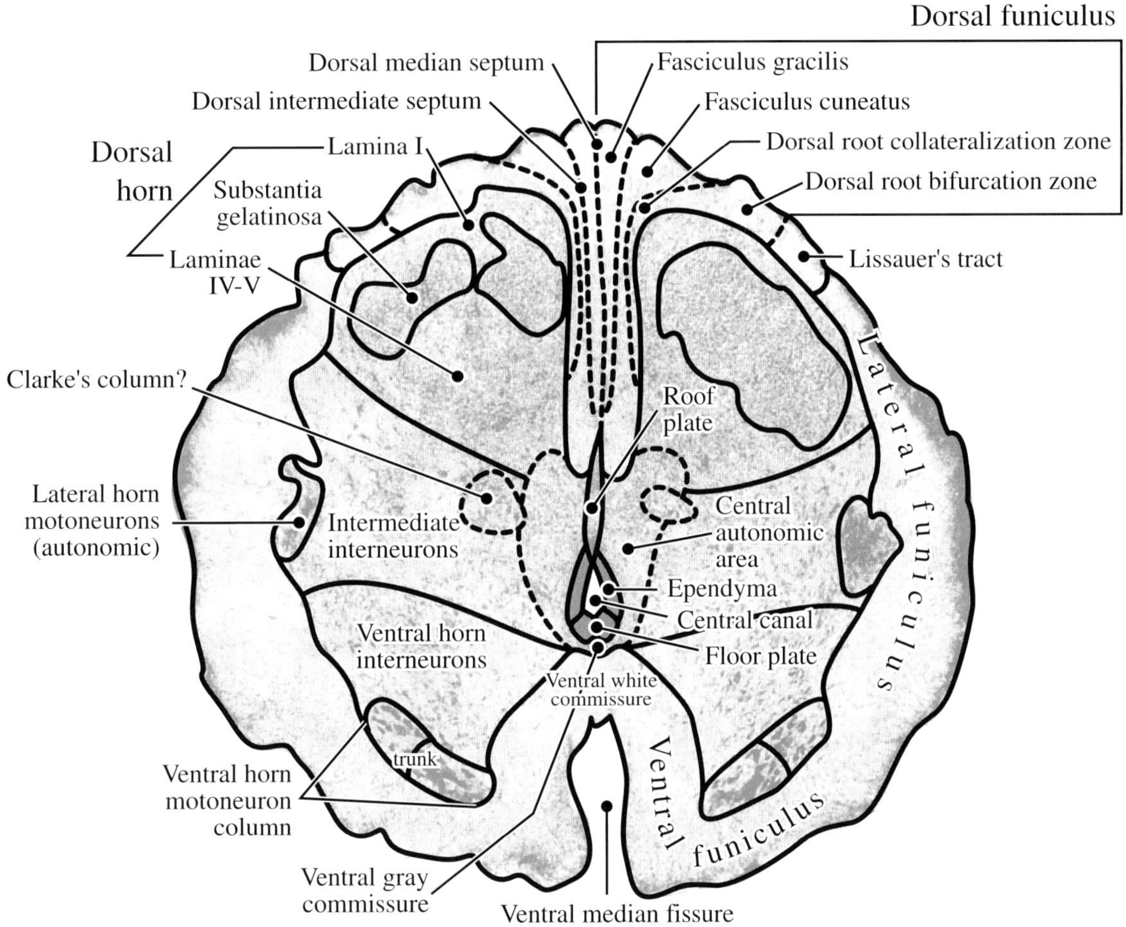
Dorsal funiculus
Dorsal median septum
Fasciculus gracilis
Dorsal intermediate septum
Fasciculus cuneatus
Dorsal horn
Lamina I
Dorsal root collateralization zone
Substantia gelatinosa
Dorsal root bifurcation zone
Laminae IV-V
Lissauer's tract
Clarke's column?
Roof plate
Lateral funiculus
Lateral horn motoneurons (autonomic)
Intermediate interneurons
Central autonomic area
Ependyma
Central canal
Floor plate
Ventral horn interneurons
Ventral white commissure
trunk
Ventral horn motoneuron column
Ventral funiculus
Ventral gray commissure
Ventral median fissure

PLATE 33A

GW14
CR 108 mm
Y68-65
Middle Thoracic
Cell body stain

Areas (mm^2)

Central canal	.0018
Neuroepithelium	.0084
Roof plate	.0152
Floor plate	.0035
Gray matter	.6913
White matter	.4458

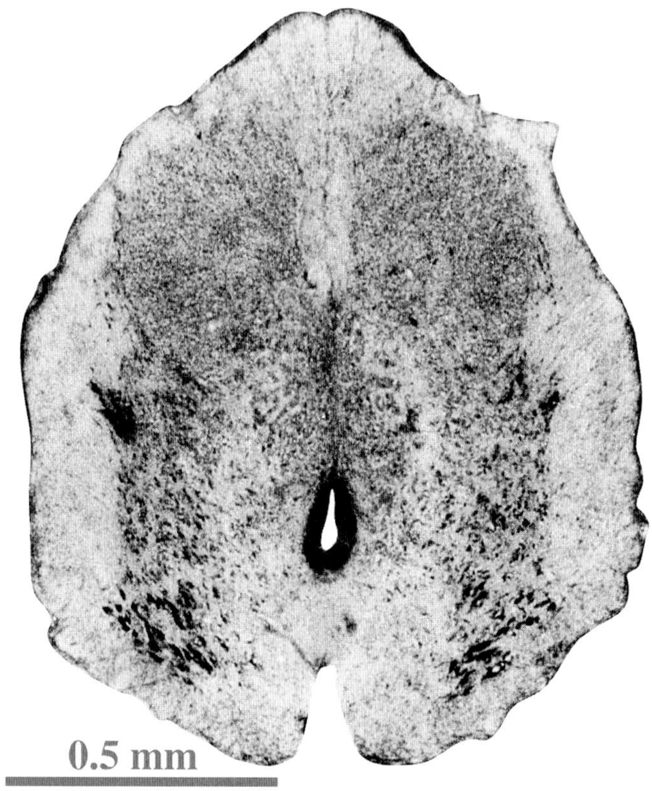

PLATE 33B

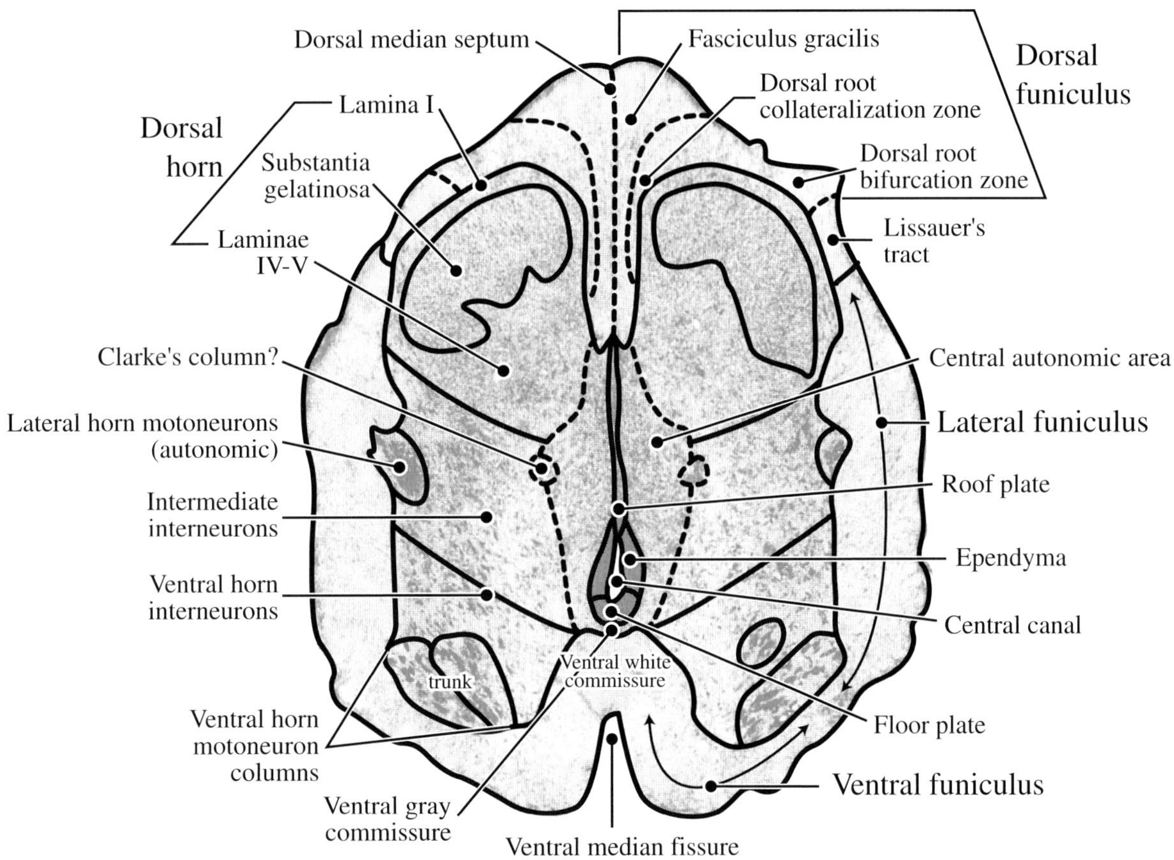
Dorsal median septum
Fasciculus gracilis
Dorsal funiculus
Dorsal root collateralization zone
Dorsal horn
Lamina I
Substantia gelatinosa
Dorsal root bifurcation zone
Lissauer's tract
Laminae IV-V
Clarke's column?
Central autonomic area
Lateral horn motoneurons (autonomic)
Lateral funiculus
Roof plate
Intermediate interneurons
Ependyma
Ventral horn interneurons
Central canal
Ventral white commissure
trunk
Floor plate
Ventral horn motoneuron columns
Ventral funiculus
Ventral gray commissure
Ventral median fissure

PLATE 34A

GW14
CR 108 mm
Y68-65
Lower Thoracic
Cell body stain

Areas (mm^2)

Central canal	.0039
Neuroepithelium	.0128
Roof plate	.0171
Floor plate	.0019
Gray matter	1.0249
White matter	.5868

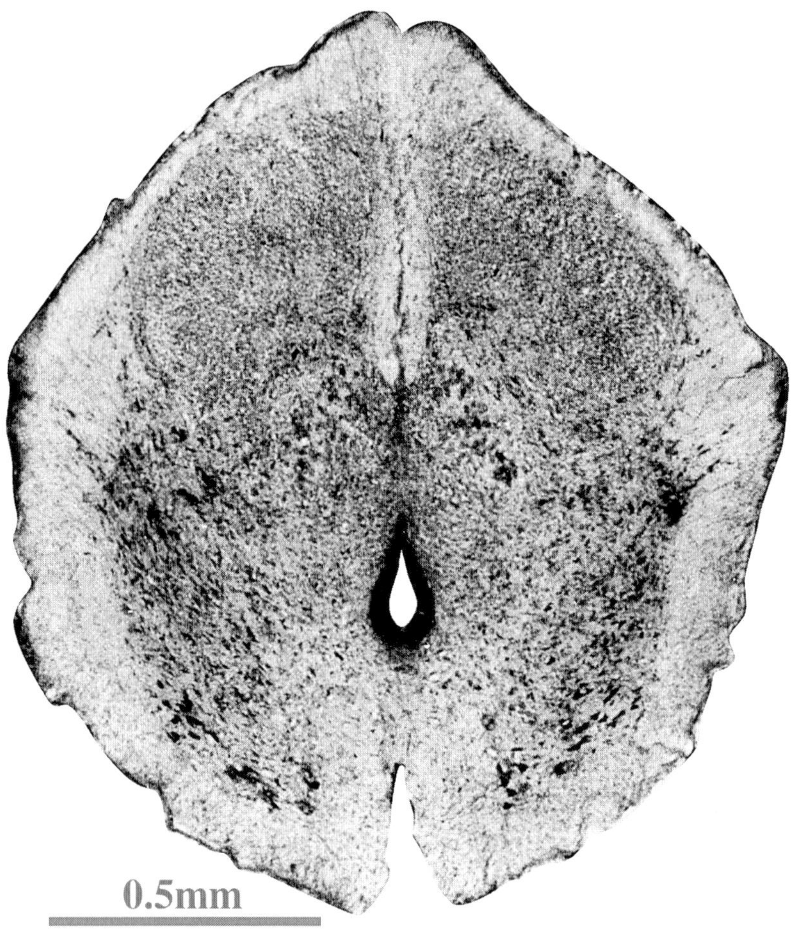

PLATE 34B

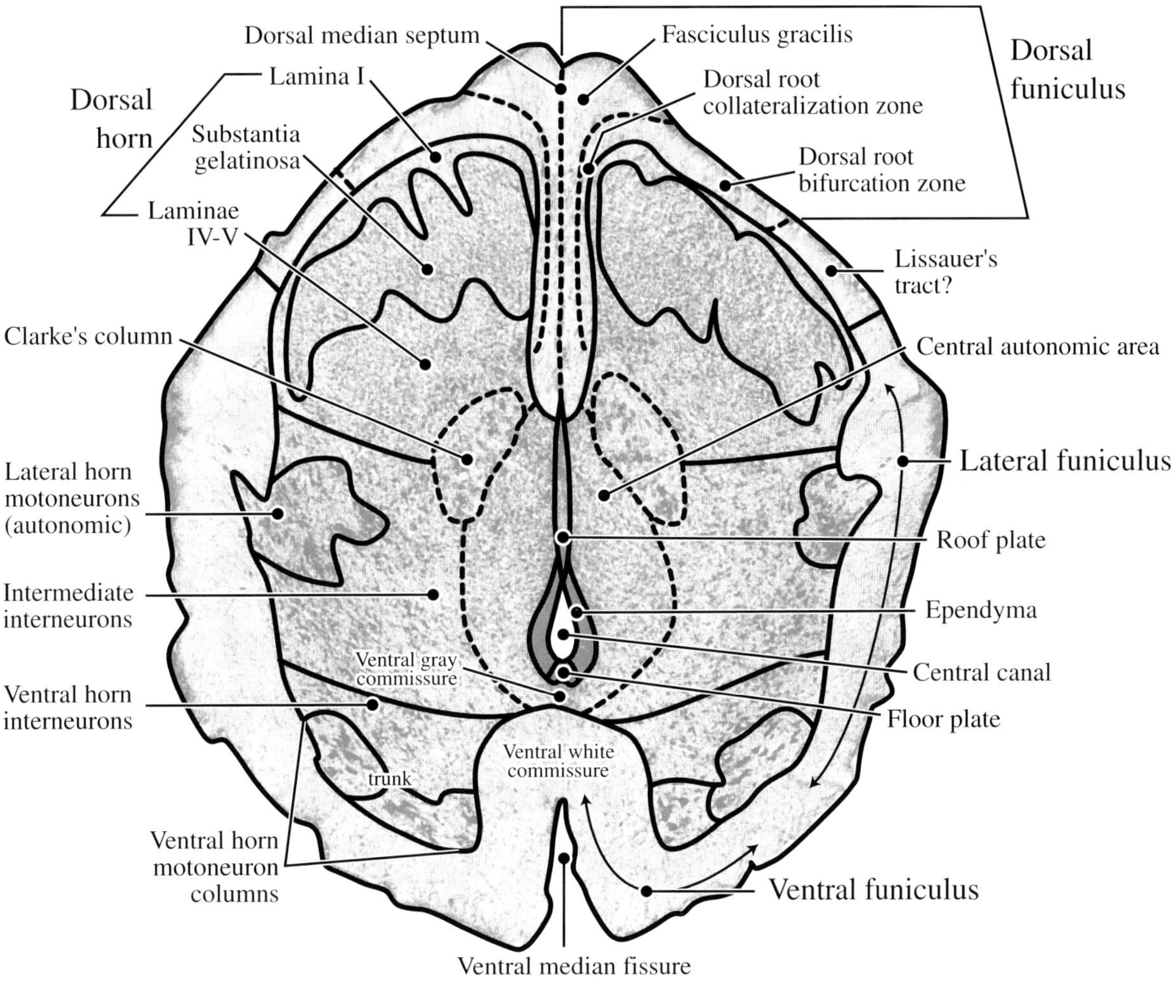
Dorsal median septum
Fasciculus gracilis
Dorsal funiculus
Lamina I
Dorsal horn
Dorsal root collateralization zone
Substantia gelatinosa
Dorsal root bifurcation zone
Laminae IV-V
Lissauer's tract?
Clarke's column
Central autonomic area
Lateral funiculus
Lateral horn motoneurons (autonomic)
Roof plate
Intermediate interneurons
Ependyma
Ventral gray commissure
Central canal
Ventral horn interneurons
Floor plate
Ventral white commissure
trunk
Ventral horn motoneuron columns
Ventral funiculus
Ventral median fissure

PLATE 35A

GW14
CR 108 mm
Y68-65
Upper Lumbar
Cell body stain

Areas (mm^2)

Central canal	.0056
Neuroepithelium	.0140
Roof plate	.0315
Floor plate	.0076
Gray matter	1.6817
White matter	1.0312

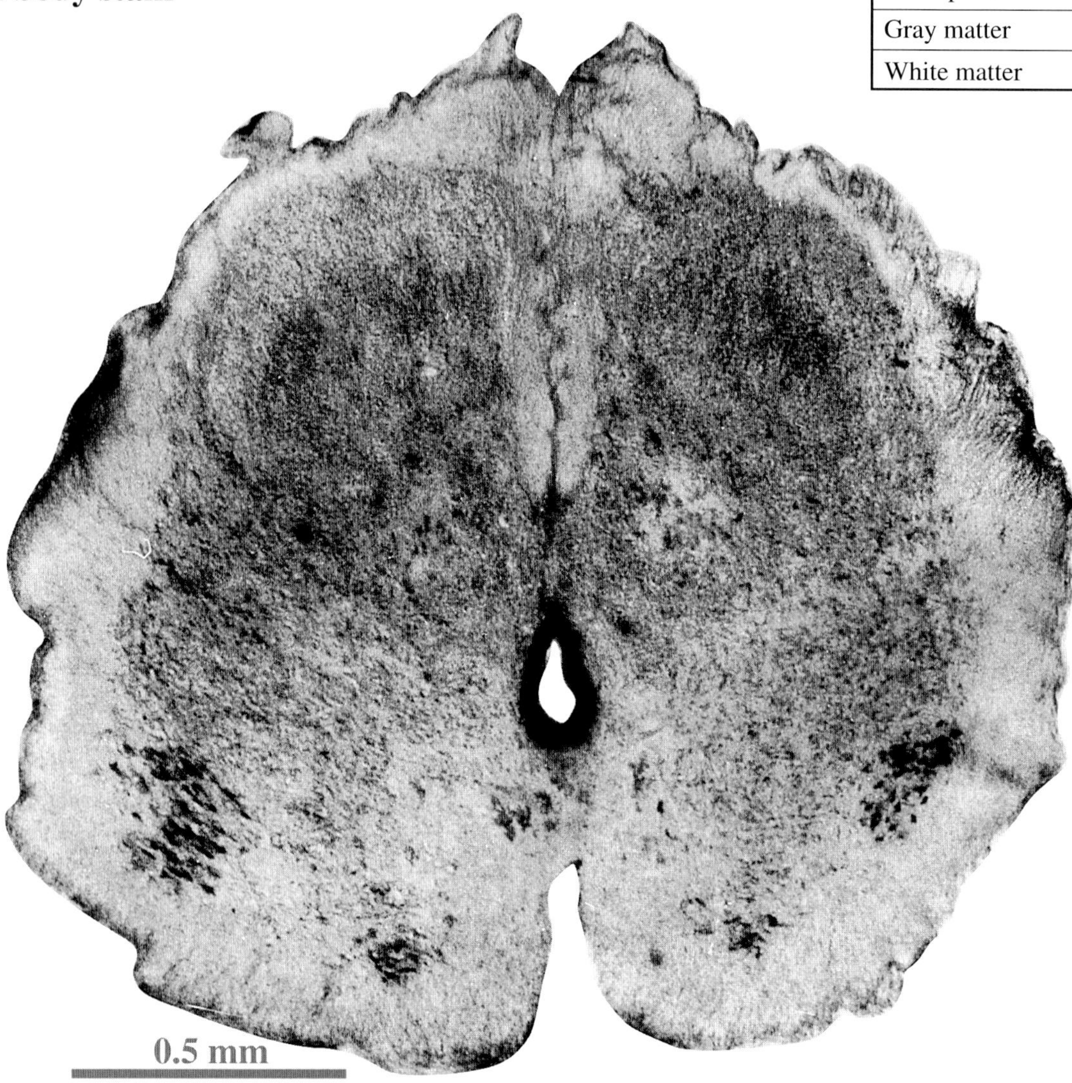

PLATE 35B

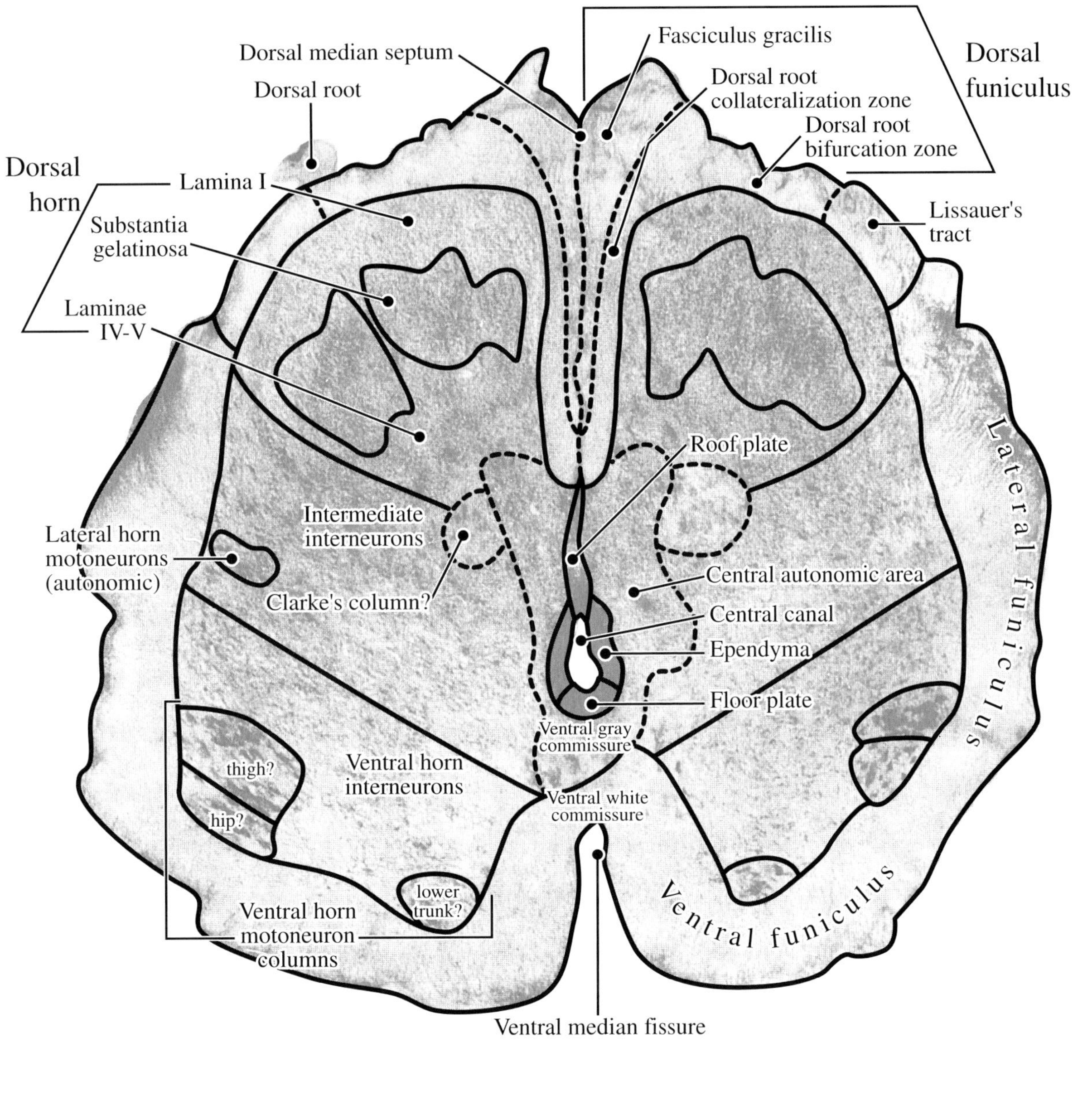
Dorsal median septum
Dorsal root
Fasciculus gracilis
Dorsal root collateralization zone
Dorsal funiculus
Dorsal root bifurcation zone
Dorsal horn
Lamina I
Substantia gelatinosa
Laminae IV-V
Lissauer's tract
Roof plate
Intermediate interneurons
Lateral horn motoneurons (autonomic)
Clarke's column?
Central autonomic area
Central canal
Ependyma
Floor plate
Lateral funiculus
Ventral gray commissure
thigh?
Ventral horn interneurons
Ventral white commissure
hip?
lower trunk?
Ventral horn motoneuron columns
Ventral funiculus
Ventral median fissure

PLATE 36A

GW14
CR 108 mm
Y68-65
Lumbar Enlargement
Cell body stain

Areas (mm^2)

Central canal	.0059
Neuroepithelium	.0118
Roof plate	.0271
Floor plate	.0056
Gray matter	2.5458
White matter	1.2621

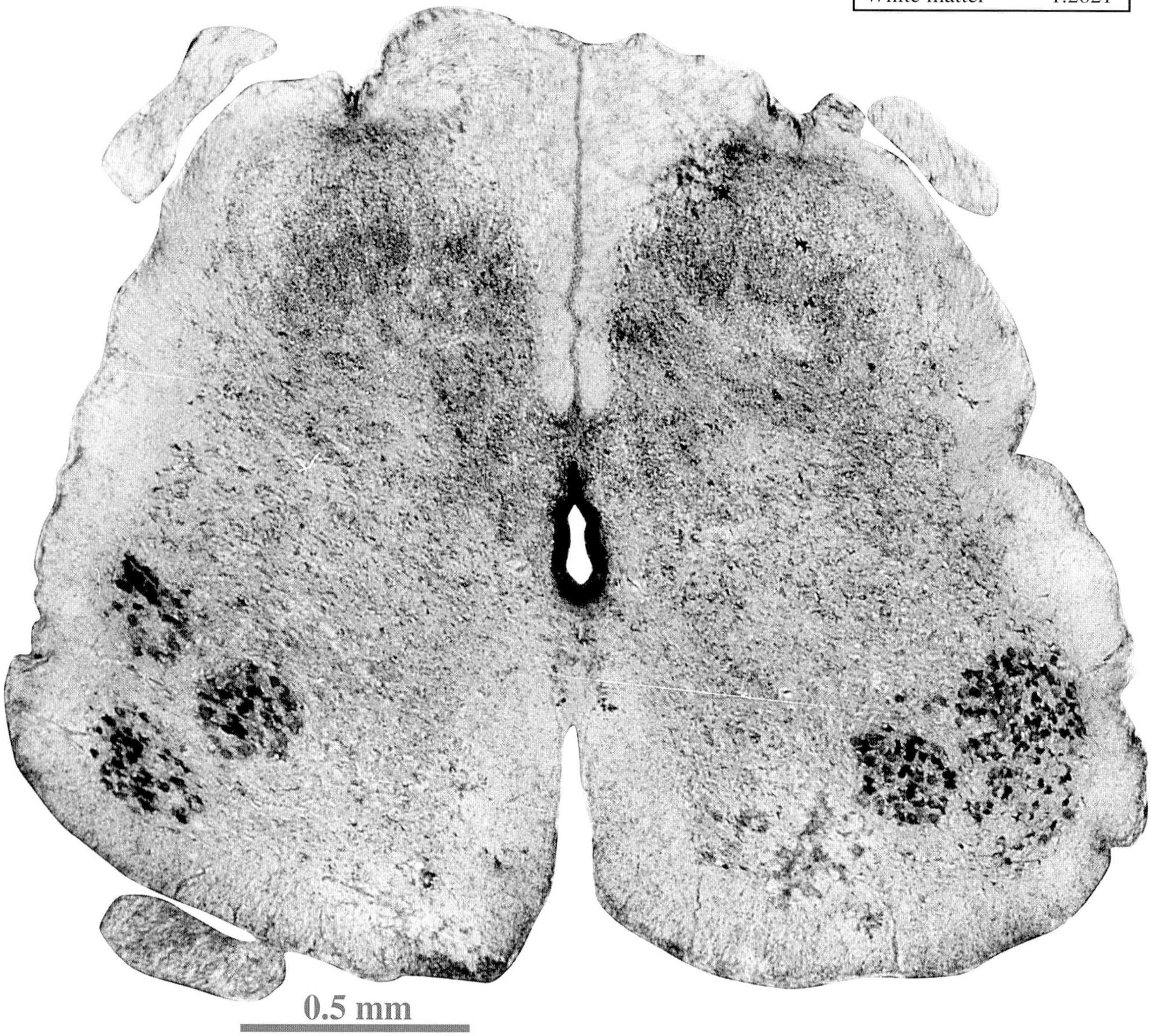

PLATE 36B

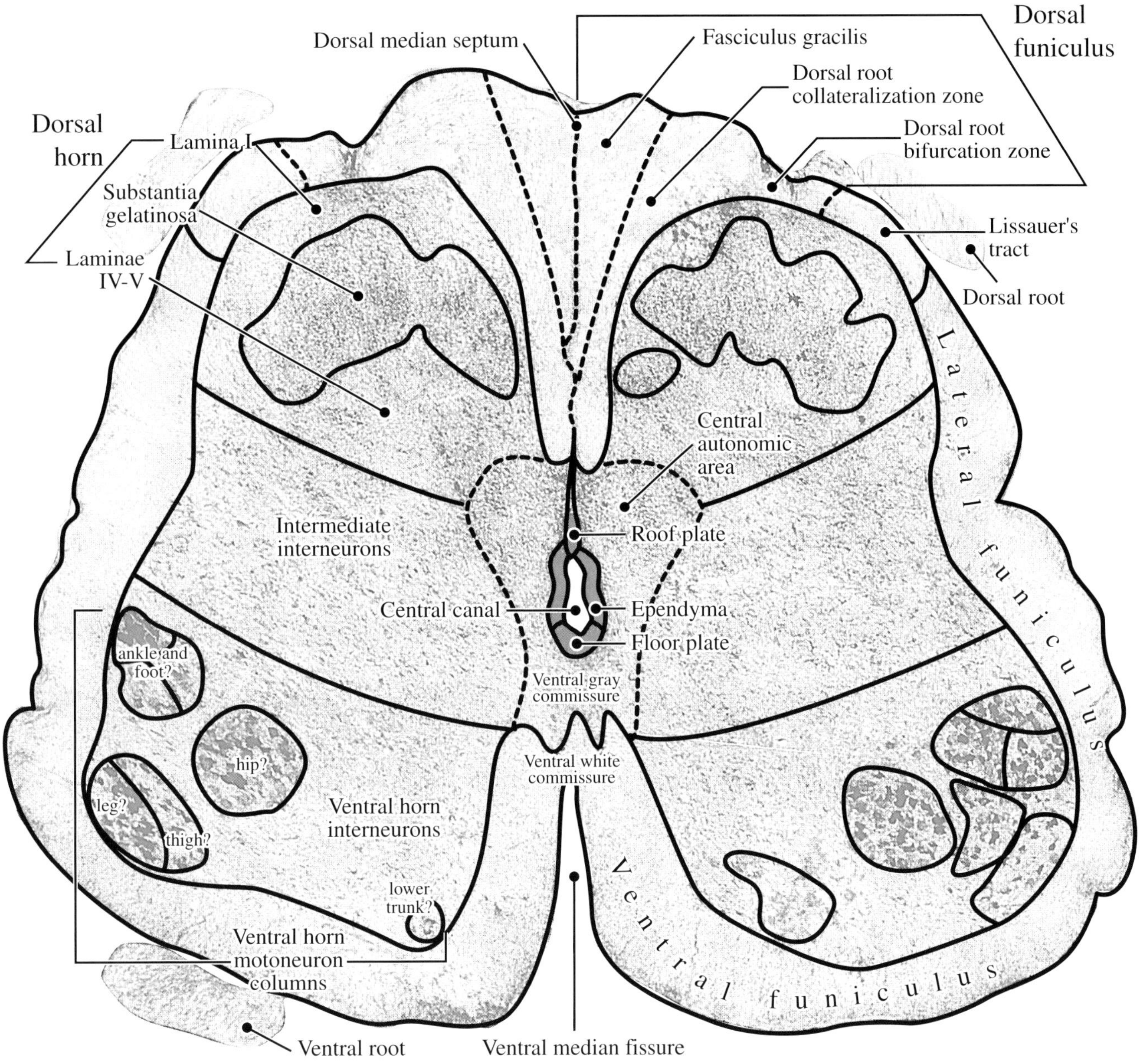
Dorsal funiculus
Dorsal median septum
Fasciculus gracilis
Dorsal root collateralization zone
Dorsal root bifurcation zone
Dorsal horn
Lamina I
Substantia gelatinosa
Laminae IV-V
Lissauer's tract
Dorsal root
Lateral funiculus
Central autonomic area
Intermediate interneurons
Roof plate
Central canal
Ependyma
Floor plate
Ventral gray commissure
ankle and foot?
hip?
leg?
thigh?
Ventral white commissure
Ventral horn interneurons
lower trunk?
Ventral horn motoneuron columns
Ventral funiculus
Ventral root
Ventral median fissure

PLATE 37A

GW14
CR 108 mm
Y68-65
Sacral/Coccygeal
Cell body stain

Areas (mm^2)

Central canal	.0045
Neuroepithelium	.0134
Roof plate	.0254
Floor plate	.0057
Gray matter	1.1322
White matter	.4258

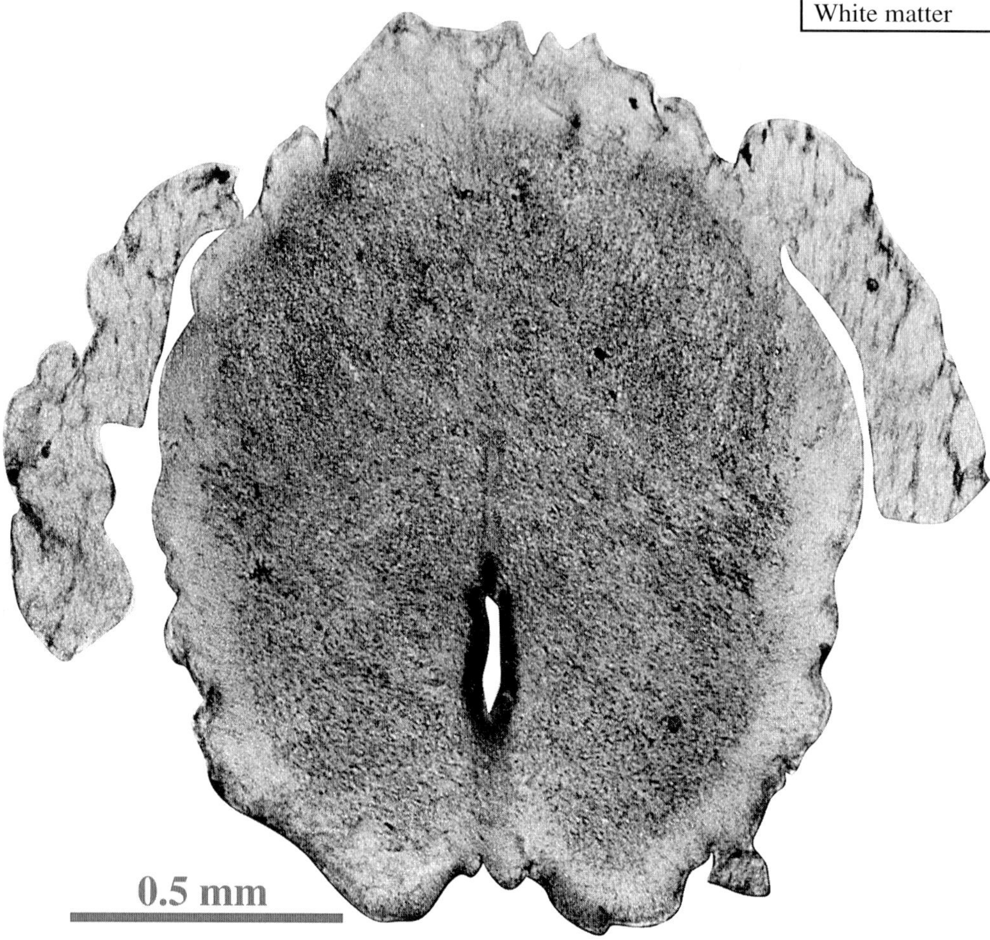

PLATE 37B

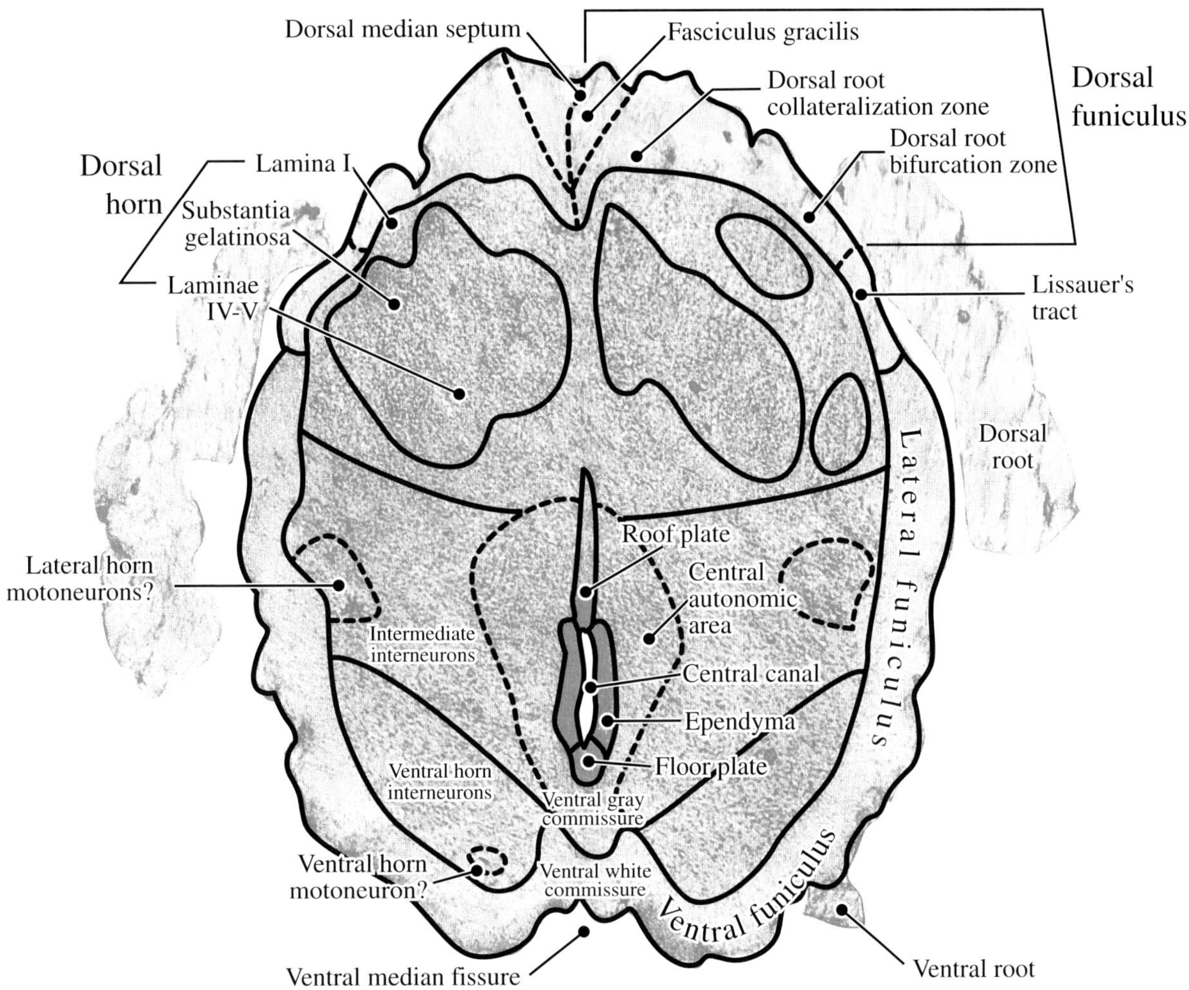

Dorsal median septum
Fasciculus gracilis
Dorsal root collateralization zone
Dorsal funiculus
Dorsal root bifurcation zone
Dorsal horn
Lamina I
Substantia gelatinosa
Laminae IV-V
Lissauer's tract
Dorsal root
Lateral funiculus
Roof plate
Central autonomic area
Lateral horn motoneurons?
Intermediate interneurons
Central canal
Ependyma
Floor plate
Ventral horn interneurons
Ventral gray commissure
Ventral horn motoneuron?
Ventral white commissure
Ventral funiculus
Ventral root
Ventral median fissure

Part III: The Second Trimester (continued)

B. Spinal cord of a GW19 fetus

Plate 38 is a survey of sections from Y52-61, a specimen in the Yakovlev Collection with a crown-rump length of 130 mm. All sections are shown at the same scale. The boxes enclosing each section list the approximate level and the total area (post-fixation) in square millimeters (mm^2). Full-page normal contrast photographs of each specimen are in **Plates 39A-45A.** Low contrast photographs with superimposed labels and outlines of structural details are in **Plates 39B-45B.** In this specimen, the section numbers are not given because they are placed on large glass plates without any numbers. Twenty sections were photographed ranging from upper cervical to sacral levels. Low sacral and coccygeal levels were not preserved. To determine the approximate level, the 20 photographs were intuitively arranged using features such as the appearance and progressively smaller rostral-to-caudal size of the ventral and lateral corticospinal tracts.

The most notable characteristic of this specimen is the dense accumulation of glial cells in specific regions of the white matter. These are assumed to be glial cells proliferating prior to myelination (myelination gliosis). The myelination of fiber tracts in the spinal cord follows a progressive sequence (see Chapter 6, Section 6.3 in Altman and Bayer, 2001). Differences in the concentration of glia allow several major fiber tracts to be tentatively identified by this age (**Table III B-1**). The lateral and ventral corticospinal tracts (the last to myelinate) stand out as very sparse regions. Other regions in the ventral and lateral funiculi have different densities of proliferative glia in bands and clumps. A densely populated outer band at the cervical level (only in the lateral funiculus) is postulated to be the spinocerebellar tracts. A sparsely populated band just inside that (the outermost band in the ventral funiculus) is postulated to be the vestibulospinal tract and the spinocephalic tracts. A very dense to dense inner band adjacent to the ventral gray matter is postulated to contain the medial longitudinal fasciculus, the tectospinal tract (only at the cervical levels), and the intraspinal (propriospinal) tracts. Both of these fiber tracts contain density gradients of proliferating glia (see notes beneath **Table III B-1**). In the dorsal funiculus, the dorsal root collateralization and bifurcation zones contain very dense to dense proliferative glia all the way down to the lumbar enlargement. At cervical and middle thoracic levels, the cuneate fasciculus can be distinguished from the gracile fasciculus by the greater concentration of proliferating glia. However, the gracile fasciculus, especially its deep part, contains proliferating glia that is most dense in the lumbar enlargement and gradually declines through thoracic levels. It is least dense at cervical levels, except for the deep wedge. Generally, the lower concentration of proliferating glia in the white matter at lumbosacral levels (*column 4*, **Table III B-1**) reflects the gradient of maturation from upper cervical to coccygeal levels.

Within the gray matter, columns of motoneurons continue to show progressive segregation in the ventral horn. The accumulation of lateral horn (autonomic) motoneurons is obvious at thoracic levels, and possibly at the sacral level. Large neurons in Clarke's column are evident at the low thoracic level, and may also be present at the middle thoracic level.

Table III B-1: Density of proliferating glia in the white matter at GW19

Name	Cervical	Thoracic	Lumbosacral
DORSAL ROOT	Sparse	---	Sparse
VENTRAL ROOT	Very dense	---	Very dense
DORSAL FUNICULUS:			
dorsal root bif. zone	Dense	Dense	Sparse
dorsal root col. zone	Very dense	Very dense	Dense*
deep fas. gracilis	Dense	Very dense	---
superficial fas. gracilis	Very sparse	Sparse	Sparse
deep fas. cuneatus	Very dense	Dense	---
superficial fas. cuneatus	Sparse	Very sparse	---
Lissauer's tract	Very sparse	Very sparse	Very sparse
LATERAL and VENTRAL FUNICULI:			
lat. cortricospinal tract	Very sparse	Very sparse	Very sparse
ven. cortricospinal tract	very sparse	very sparse	---
rubrospinal tract	Dense	---	---
spinocerebellar tracts	Dense	Sparse	---
ven. commissure	Very dense	Dense	Sparse
intraspinal tracts**	Gradient†	Gradient†	--
spinocephalic tracts***	Gradient††	Gradient††	--
med. long. fasiculus	Very dense	---	---
tectospinal tract	Very dense	---	---
vestibulospinal tract	Sparse	Sparse	---

* Sparse at the most caudal level (**Plate 45**).

** Overlaps with the medial longitudinal fasciculus, the tectospinal tract, and the lateral reticulospinal tract.

† Very dense medial to the ventral horn, dense around the remaining ventral horn, sparse around the intermediate gray and lateral dorsal horn.

*** Contains anterior and lateral parts; overlaps with the vestibulospinal tract, the spinotectal tract, and the spino-olivary tract.

†† Deep parts (adjacent to intraspinal tracts) are more dense than superficial parts (adjacent to the ventrolateral pial membrane).

PLATE 38

GW19, CR 130 mm, Y52-61

Plates 39A, 39B
Upper Cervical
Total area:
4.1202 mm^2

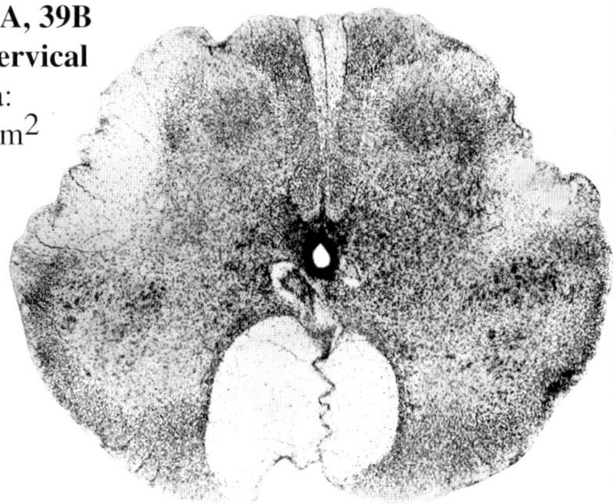

Plates 40A, 40B
Cervical Enlargement
Total area:
3.7511 mm^2

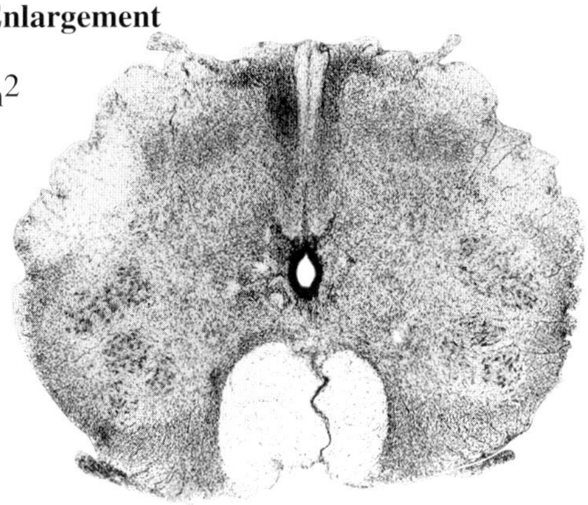

Plates 41A, 41B
Middle Thoracic
Total area:
1.3996 mm^2

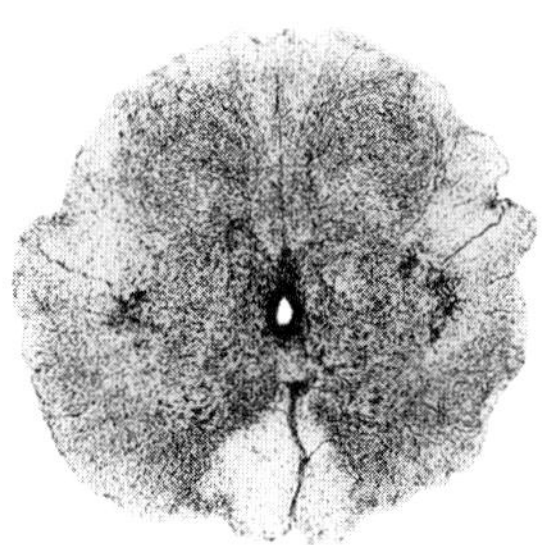

0.5 mm

Plates 42A, 42B
Lower Thoracic
Total area:
1.4929 mm^2

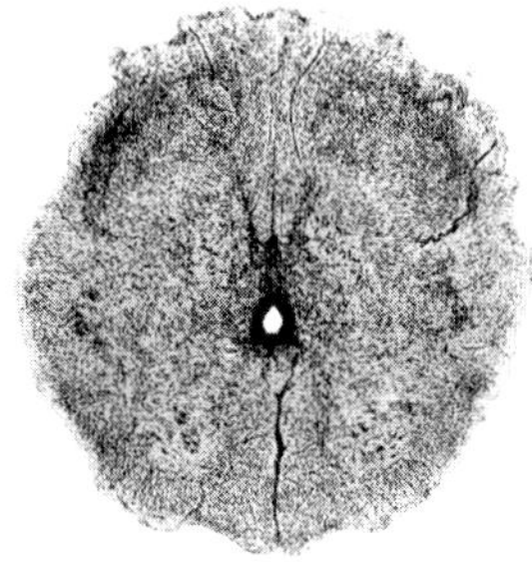

Plates 43A, 43B
Upper Lumbar
Total area:
2.2128 mm^2

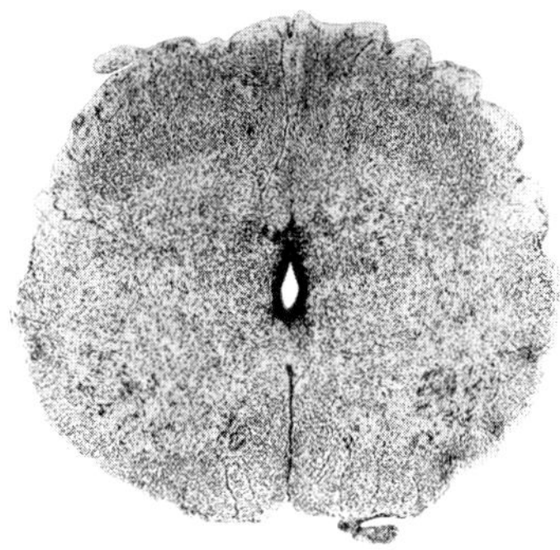

Plates 44A, 44B
Lumbar Enlargement
Total area:
2.6480 mm^2

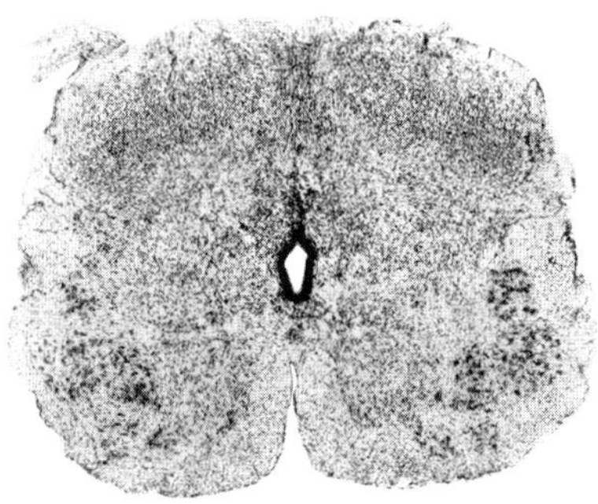

Plates 45A, 45B
Lumbosacral
Total area:
2.0164 mm^2

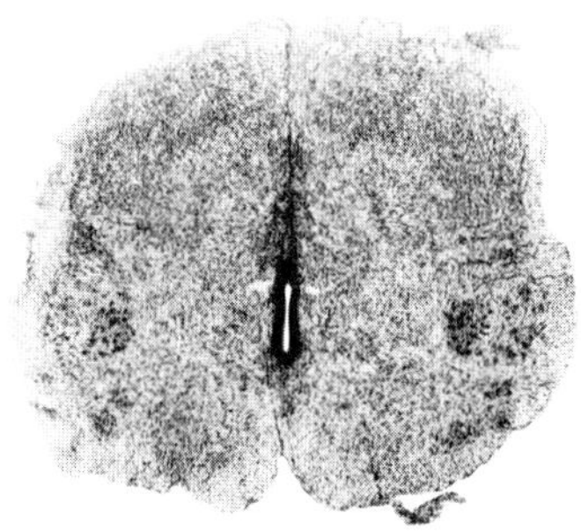

0.5 mm

PLATE 39A

GW19
CR 130 mm
Y52-61
Upper Cervical
Cell body stain

Areas (mm^2)

Central canal	.0050
Neuroepithelium	.0170
Roof plate	.0152
Floor plate	.0054
Gray matter	1.8505
White matter	2.2271

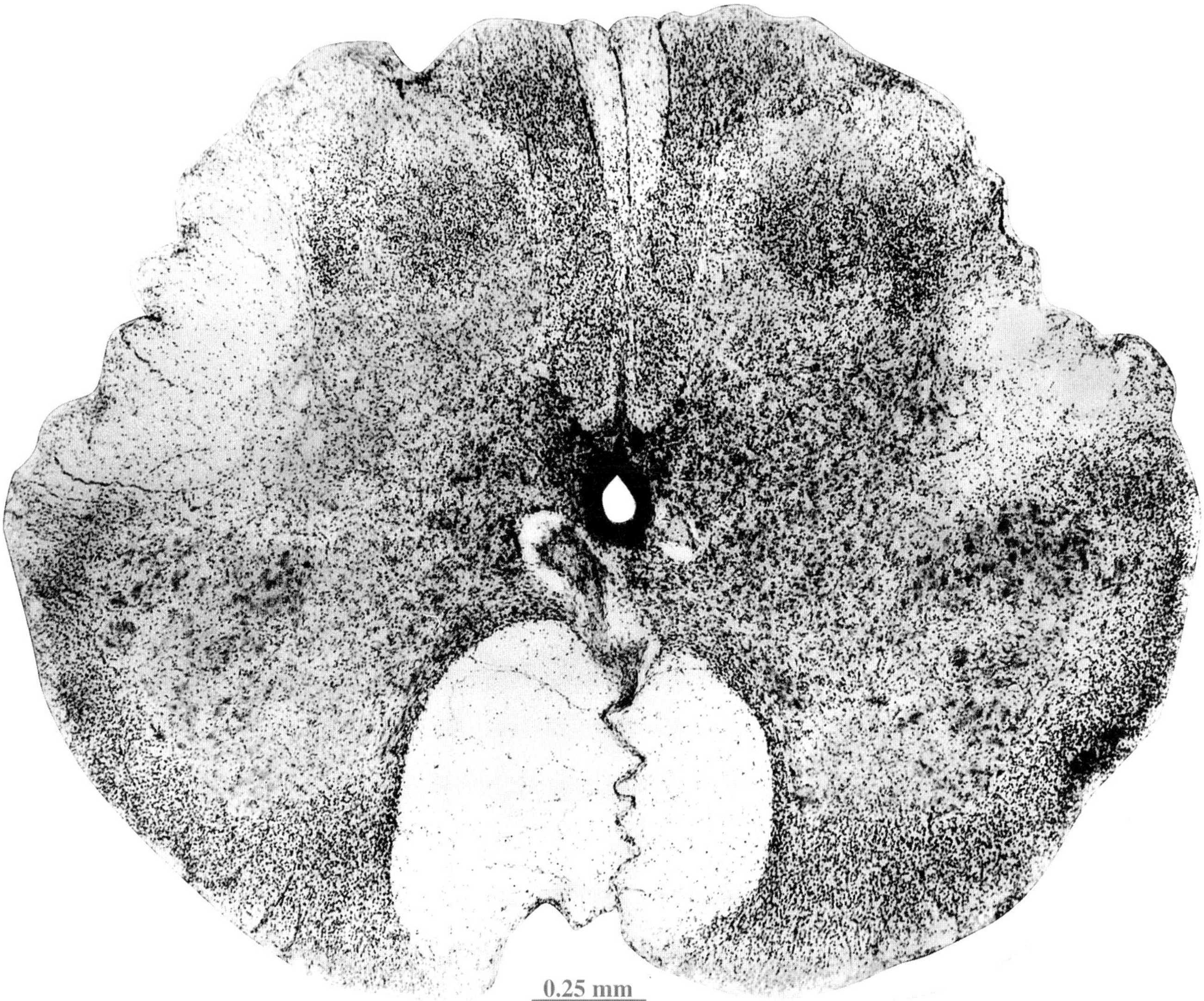

See Table III B-1 for the list of glial densities in each fiber tract. Examples of glial densities in each category are labeled in the lower left corner of this plate.

PLATE 39B

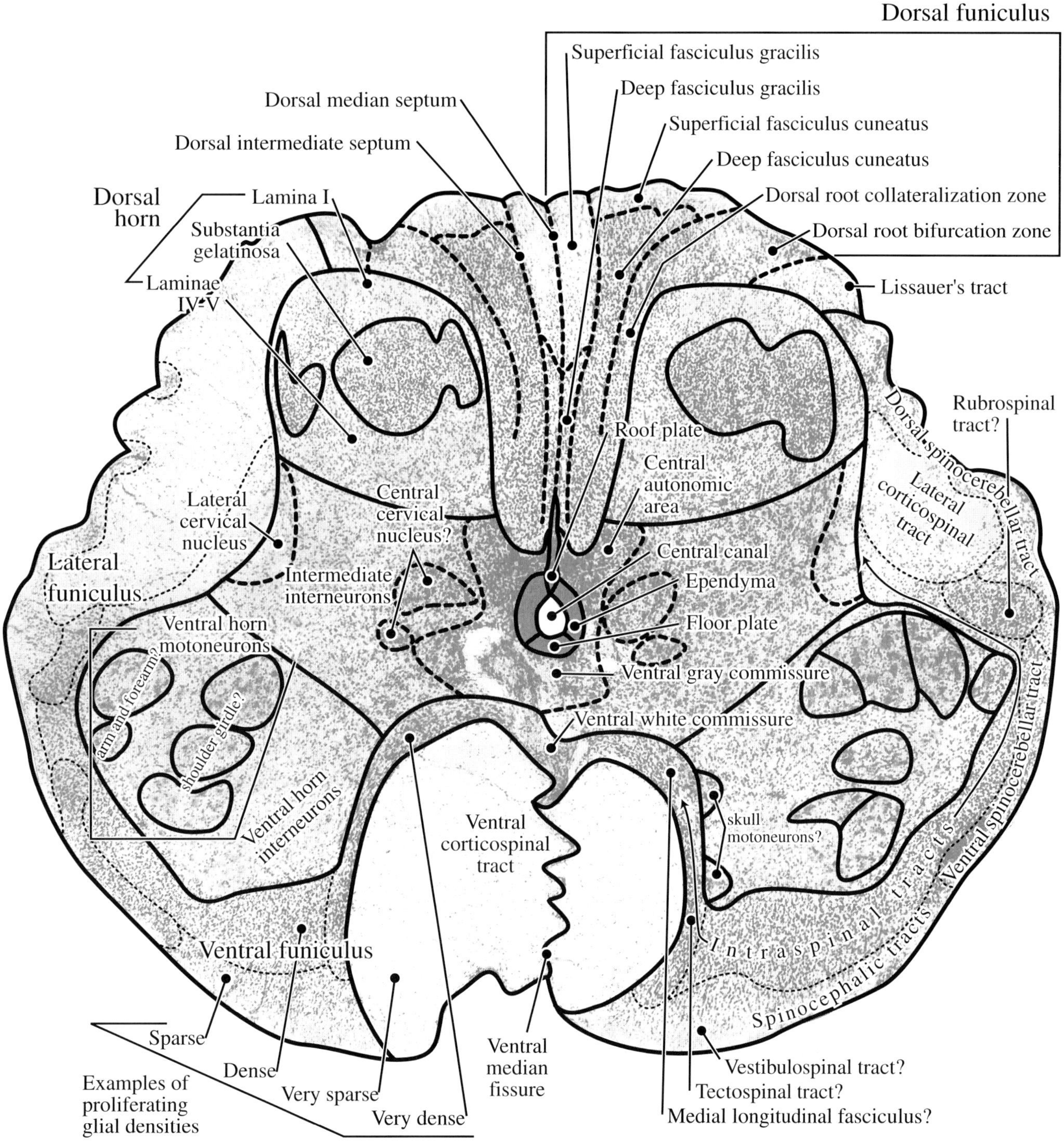

The fine lines in the lateral and ventral funiculi segregate regions of differing densities of proliferating glia, not the borders of fiber tracts. These lines may correspond to the borders of some fiber tracts, such as the spinocerebellar and the lateral corticospinal tracts. Other lines run across fiber tracts, such as the very dense region adjacent to the ventral corticospinal tract that includes the medial longitudinal fasciculus, the tectospinal tract, and the medial part of the intraspinal tracts. Many of the fiber tracts in the ventral and lateral funiculi overlap and do not have distinct borders.

PLATE 40A

GW19
CR 130 mm
Y52-61
Cervical Enlargement
Cell body stain

Areas (mm^2)

Central canal	.0069
Neuroepithelium	.0172
Roof plate	.0154
Floor plate	.0037
Gray matter	1.8661
White matter	1.8419

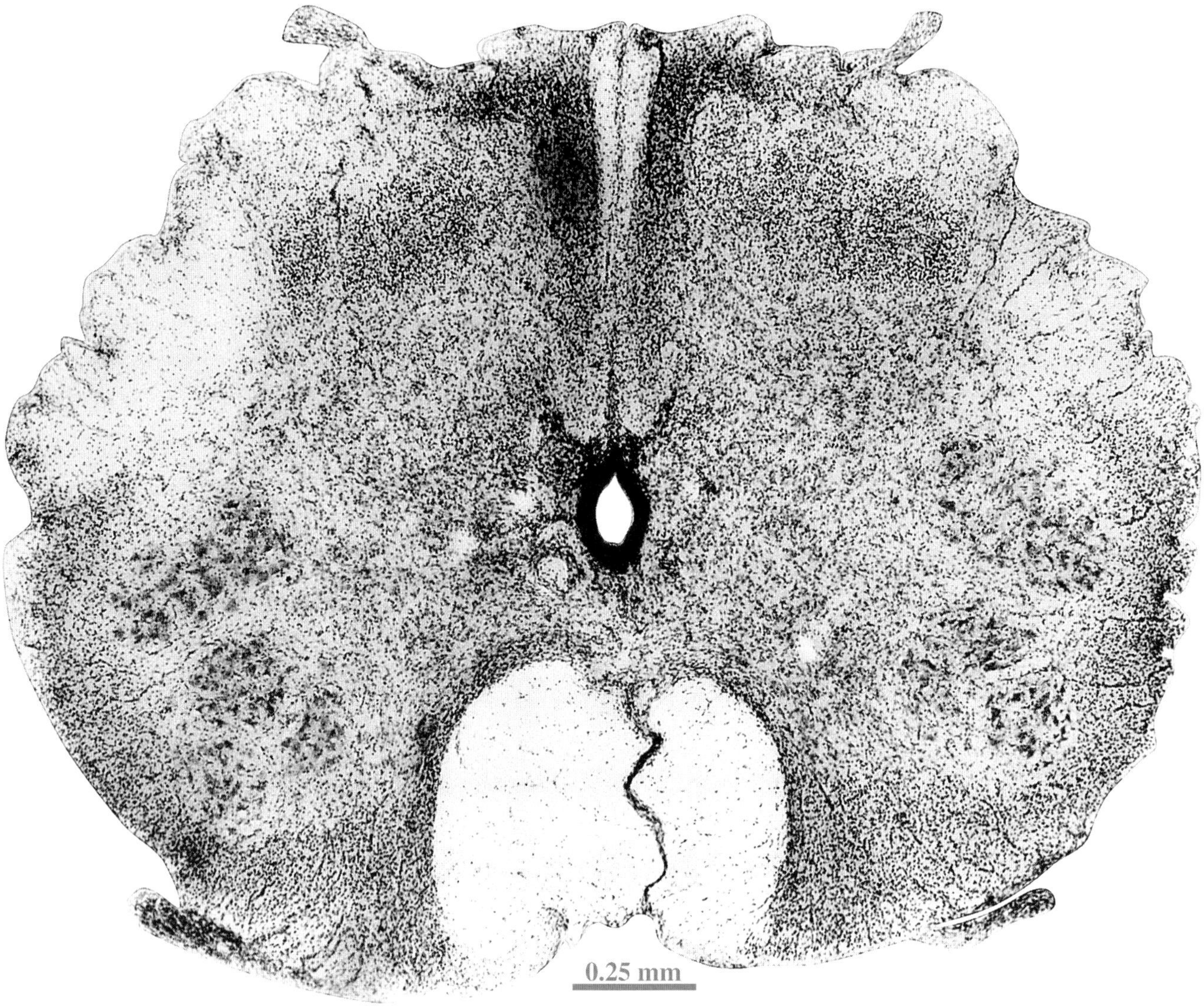

See Table III B-1 for the list of glial densities in each fiber tract. Examples of glial densities in each category are labeled in the lower left corner of Plate 39B.

PLATE 40B

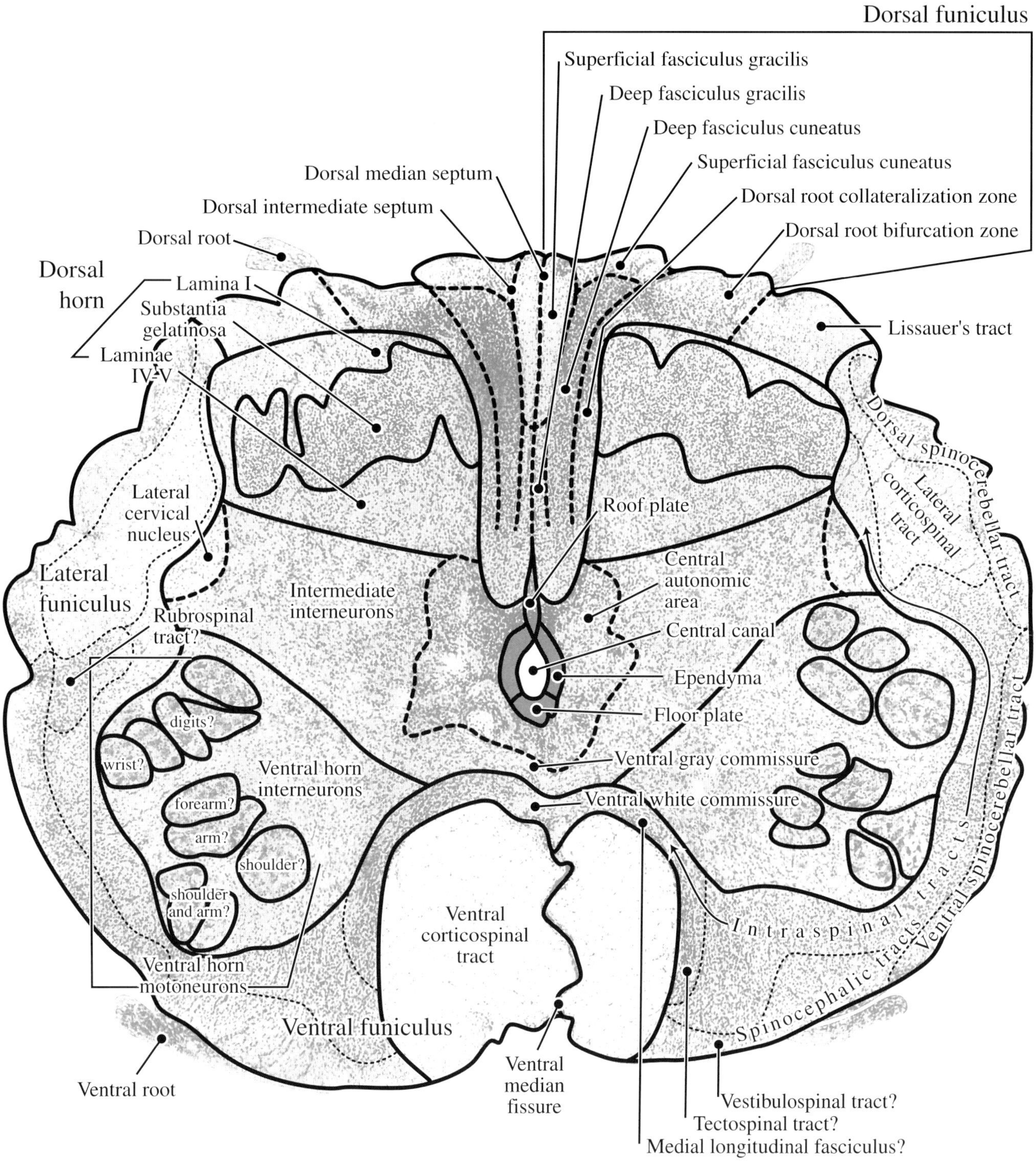

Note the larger lateral and ventral corticospinal tracts on one side of the spinal cord and the smaller ones on the other (see also **Plates 39** and **41**). Axons in the lateral corticospinal tract cross the midline, while axons in the ventral corticospinal tract remain ipsilateral. The *larger lateral* corticospinal tract (crossed component) is linked to the *smaller ventral* corticospinal tract (uncrossed component) and vice versa. It is unknown which is the true right or left side. This specimen is the only one in the Atlas to show such a pronounced asymmetry.

PLATE 41A

GW19
CR 130 mm
Y52-61
Middle Thoracic
Cell body stain

Areas (mm^2)

Central canal	.0019
Neuroepithelium	.0092
Roof plate	.0156
Floor plate	.0029
Gray matter	.6000
White matter	.7701

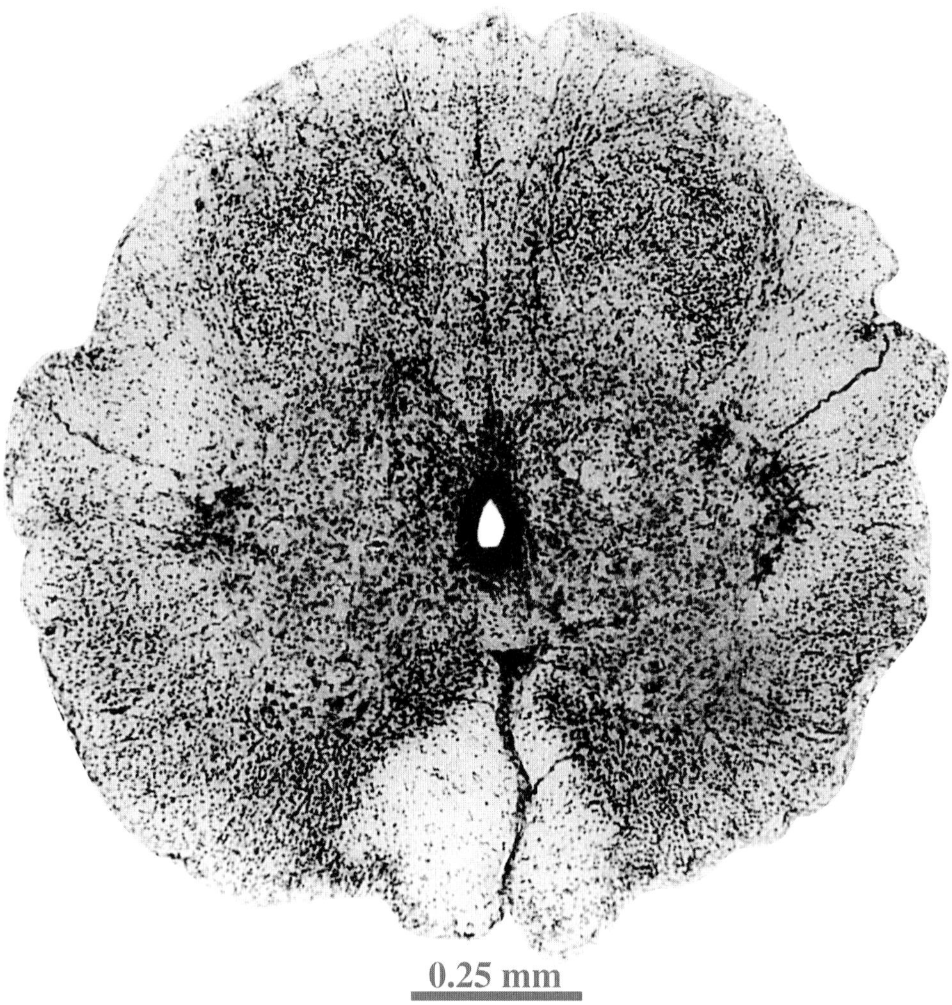

See Table III B-1 for the list of glial densities in each fiber tract. Examples of glial densities in each category are labeled in the lower left corner of Plate 39B.

PLATE 41B

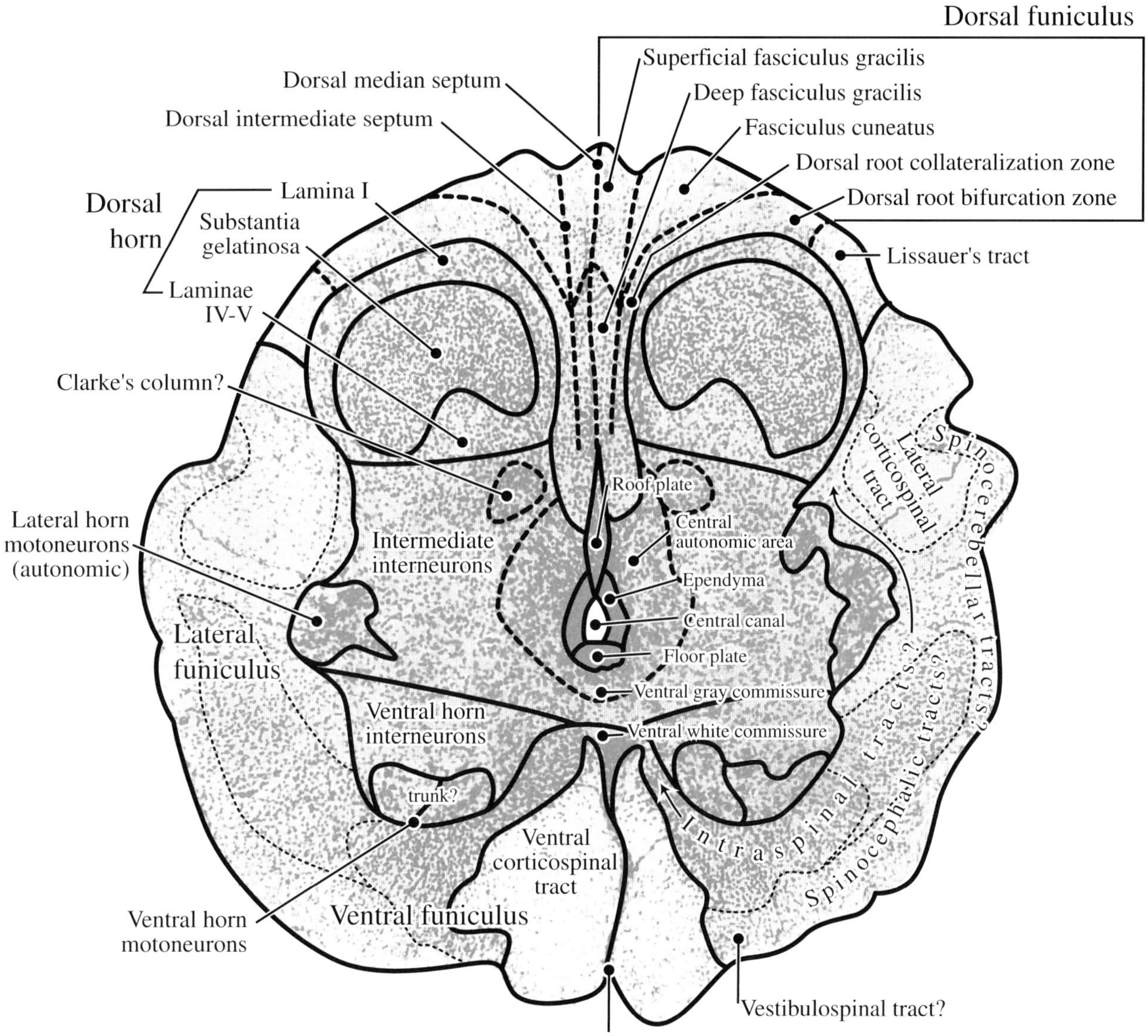

The fine lines in the lateral and ventral funiculi segregate regions of differing densities of proliferating glia, not the borders of fiber tracts.

PLATE 42A

GW19
CR 130 mm
Y52-61
Lower Thoracic
Cell body stain

Areas (mm^2)

Central canal	.0025
Neuroepithelium	.0103
Roof plate	.0170
Floor plate	.0031
Gray matter	.7999
White matter	.6600

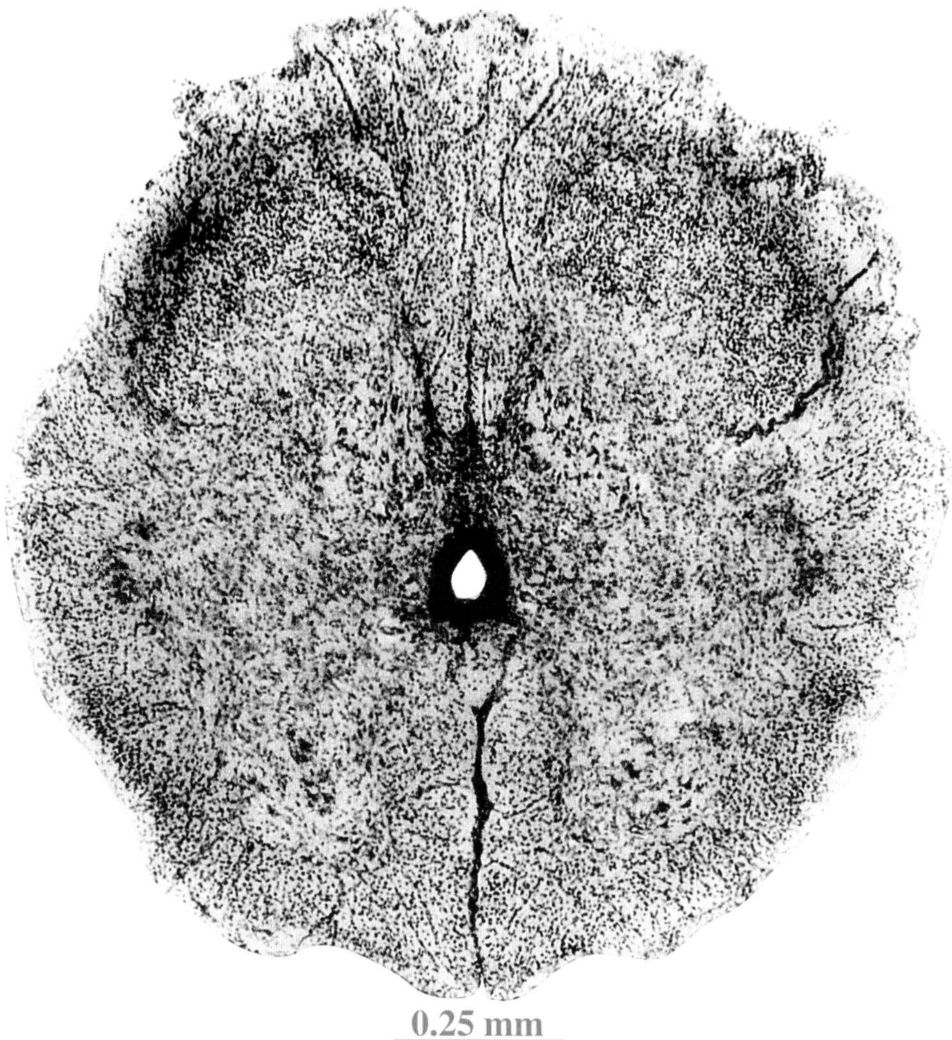

See Table III B-1 for the list of glial densities in each fiber tract. Examples of glial densities in each category are labeled in the lower left corner of Plate 39B.

PLATE 42B

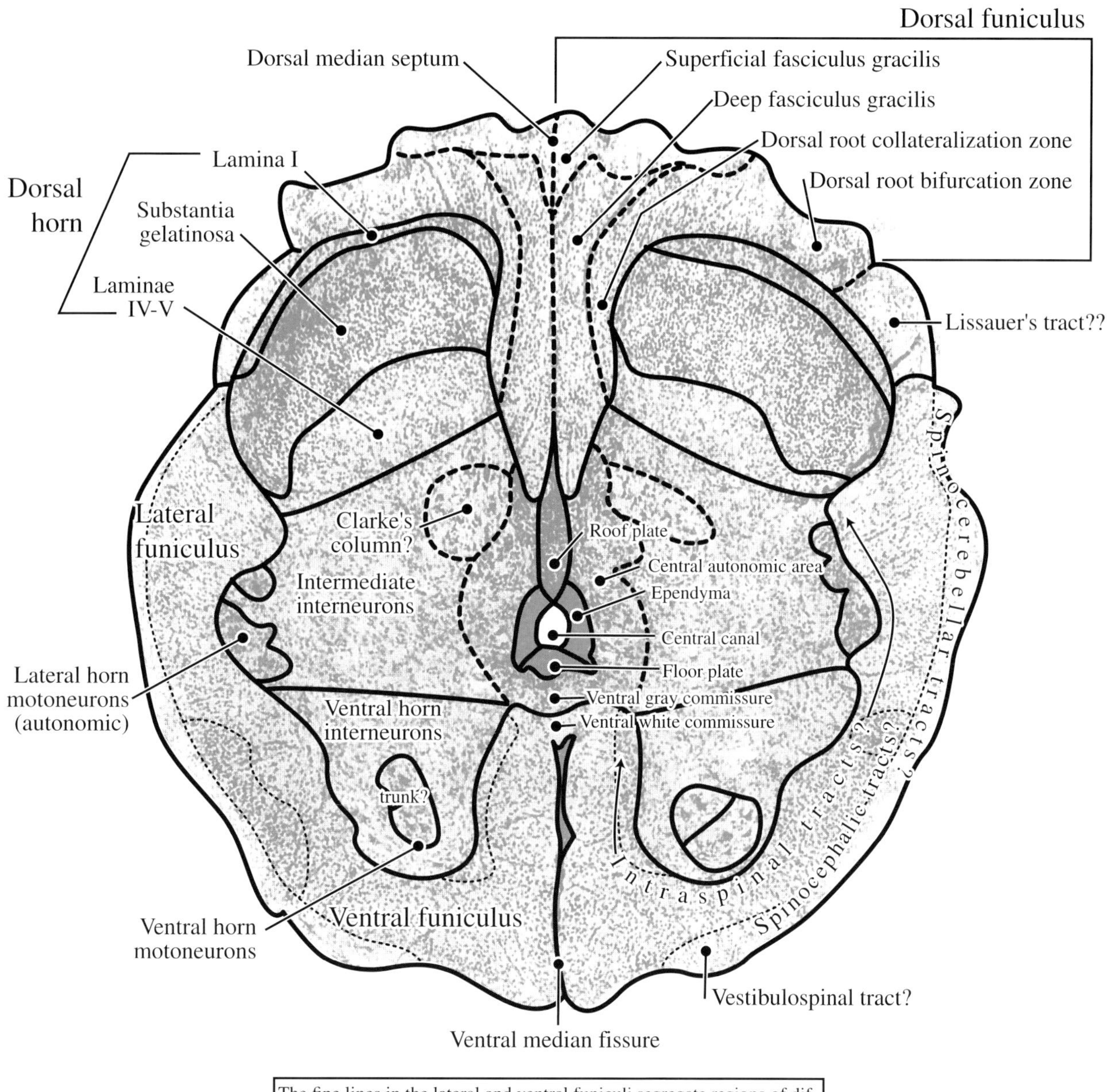

The fine lines in the lateral and ventral funiculi segregate regions of differing densities of proliferating glia, not the borders of fiber tracts.

PLATE 43A

GW19
CR 130 mm
Y52-61
Upper Lumbar
Cell body stain

Areas (mm^2)

Central canal	.0049
Neuroepithelium	.0152
Roof plate	.0175
Floor plate	.0031
Gray matter	1.3115
White matter	.8607

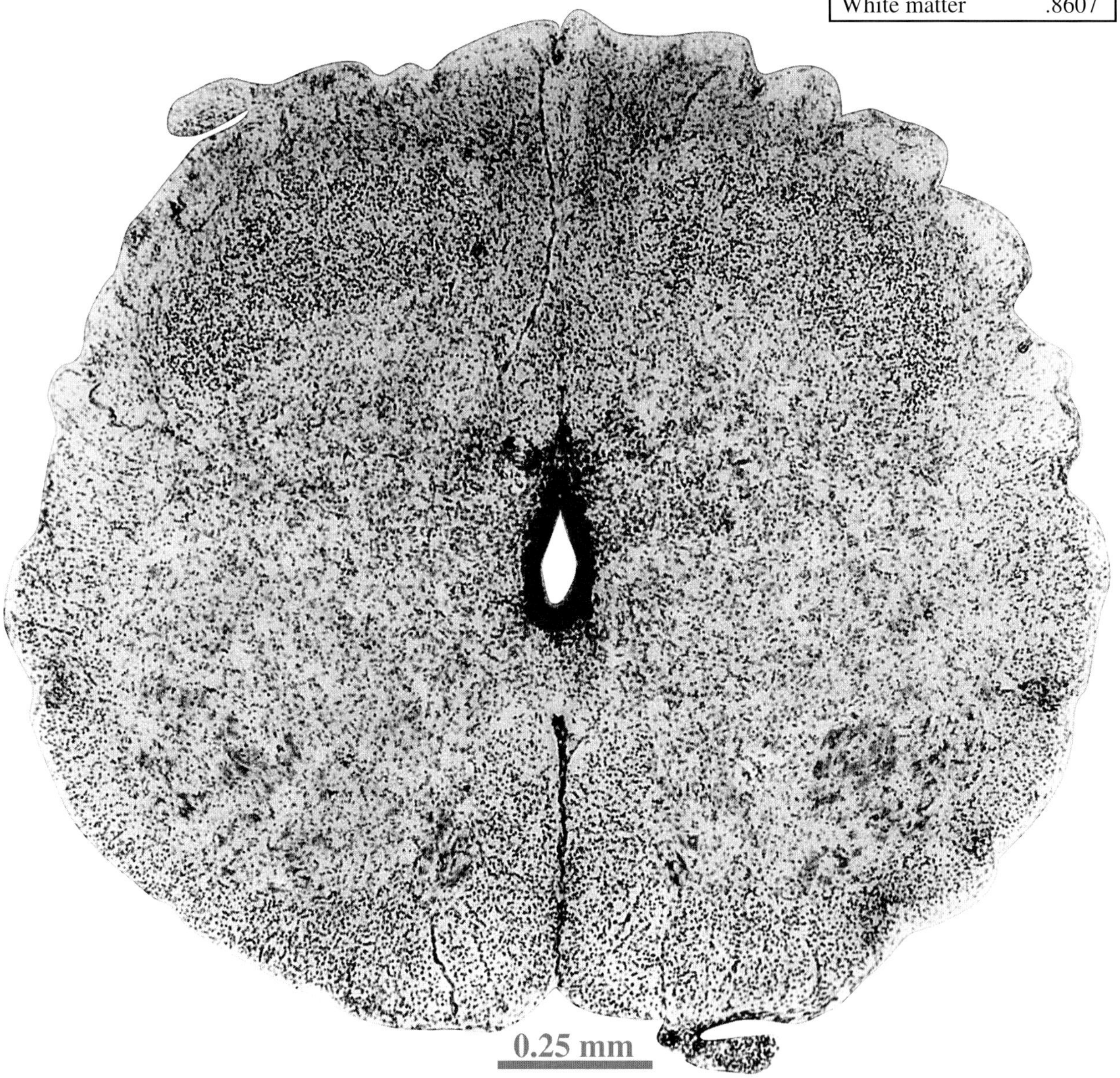

See Table III B-1 for the list of glial densities in each fiber tract. Examples of glial densities in each category are labeled in the lower left corner of Plate 39B.

PLATE 43B

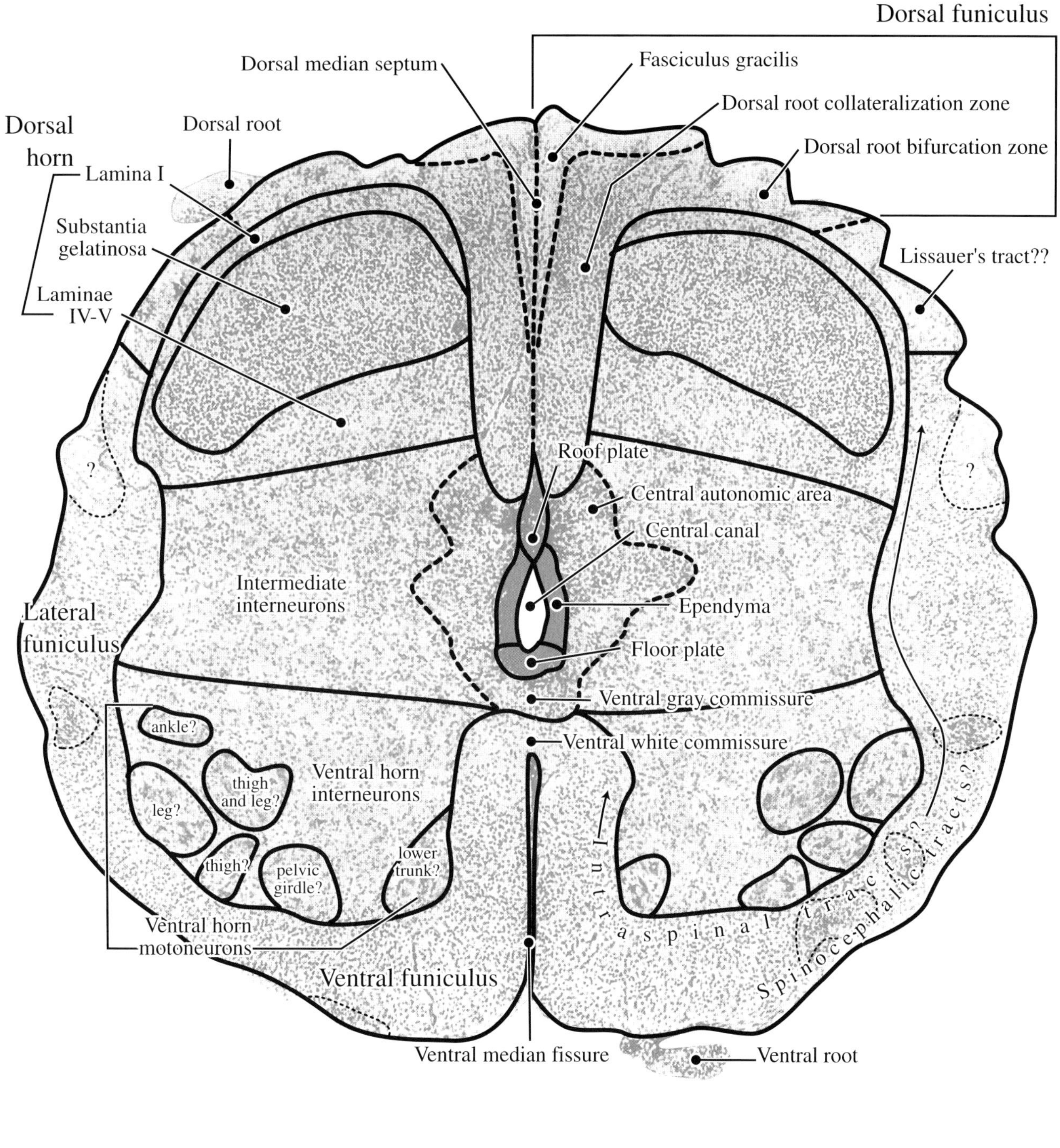

PLATE 44A

GW19
CR 130 mm
Y52-61
Lumbar Enlargement
Cell body stain

Areas (mm^2)

Central canal	.0072
Neuroepithelium	.0151
Roof plate	.0170
Floor plate	.0031
Gray matter	1.8235
White matter	.7822

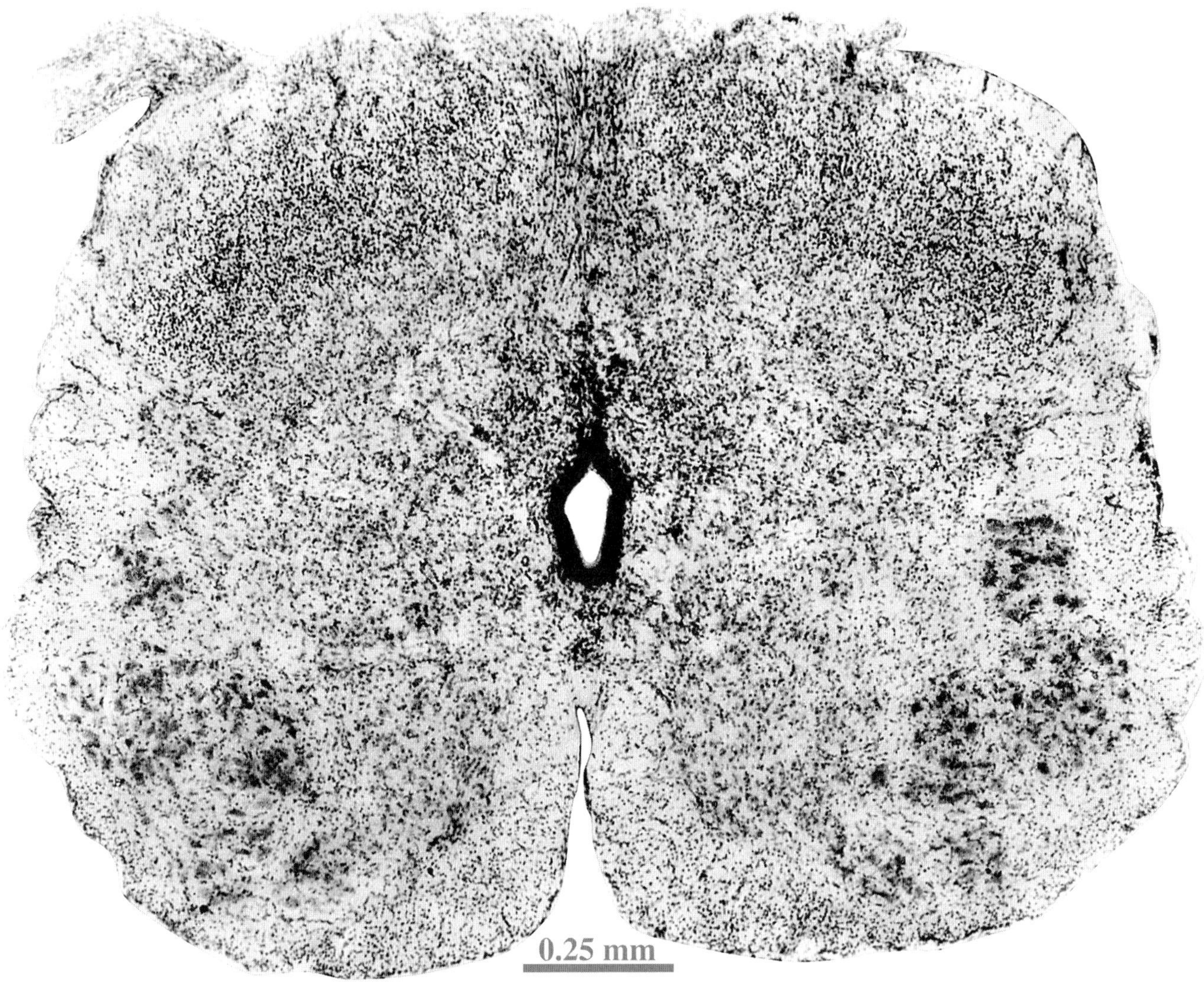

See Table III B-1 for the list of glial densities in each fiber tract. Examples of glial densities in each category are labeled in the lower left corner of Plate 39B.

PLATE 44B

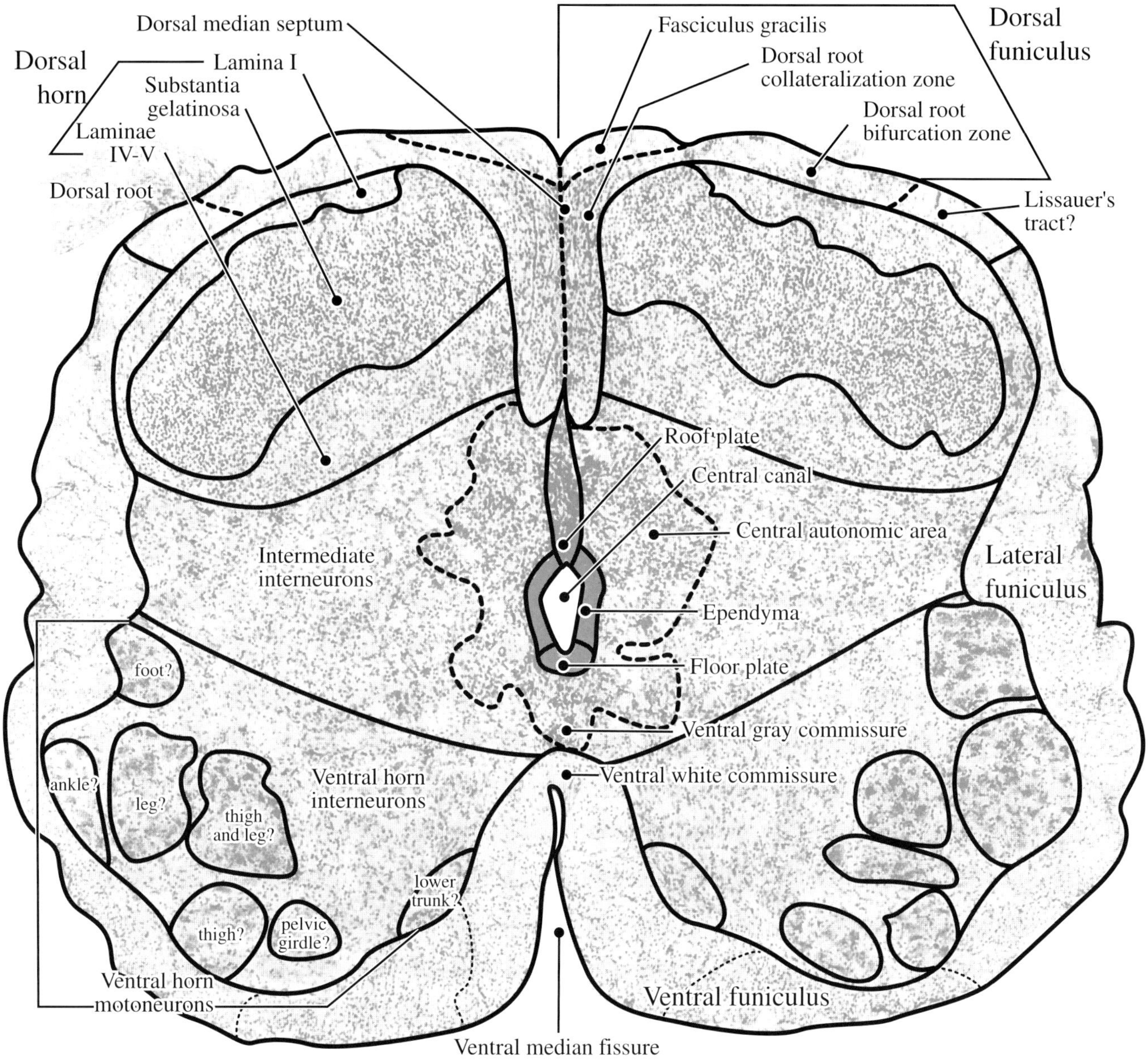

PLATE 45A

GW19
CR 130 mm
Y52-61
Lumbosacral
Cell body stain

Areas (mm^2)

Central canal	.0027
Neuroepithelium	.0187
Roof plate	.0201
Floor plate	.0019
Gray matter	1.4461
White matter	.5269

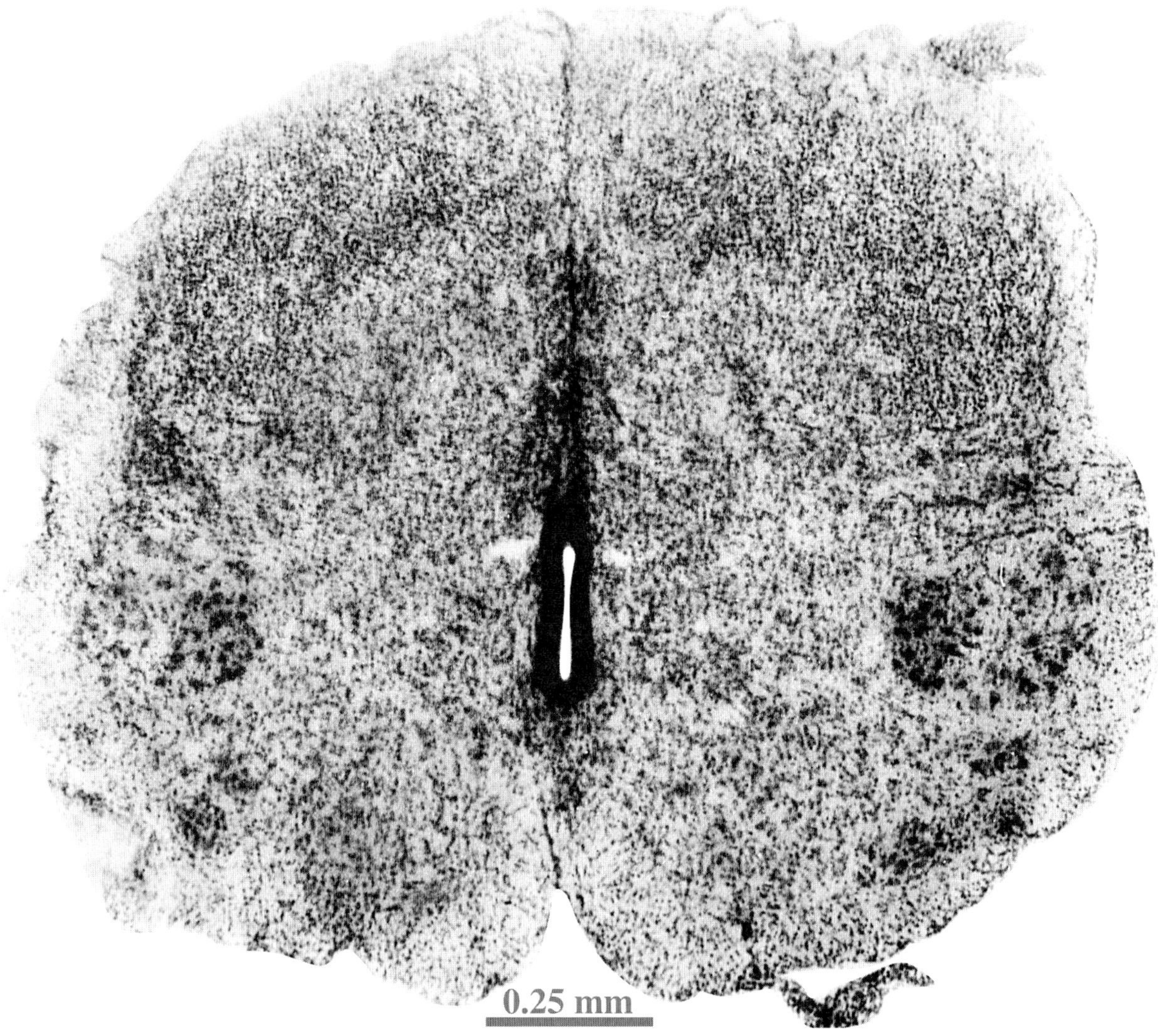

See Table III B-1 for the list of glial densities in each fiber tract. Examples of glial densities in each category are labeled in the lower left corner of Plate 39B.

PLATE 45B

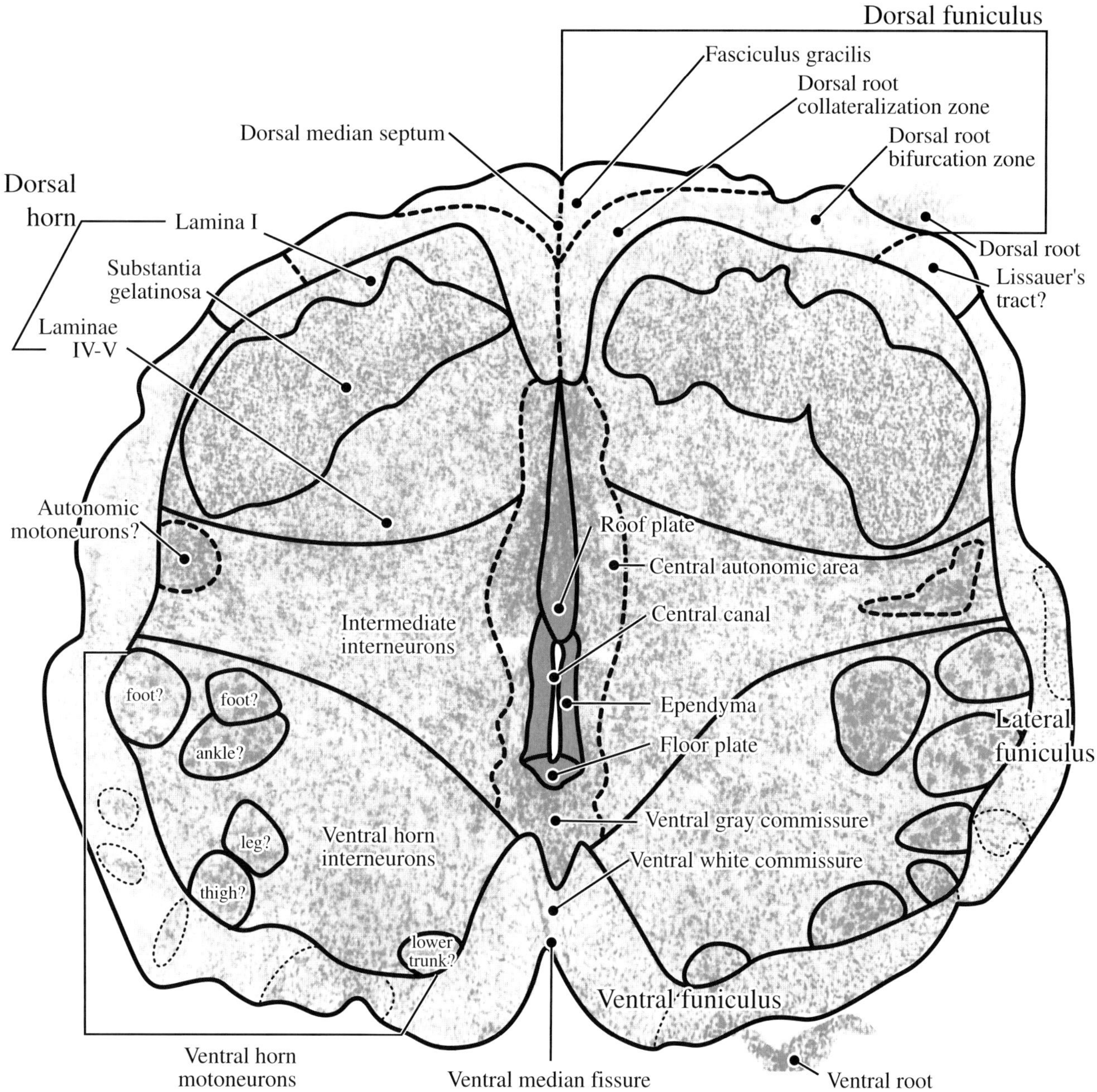

Part III: The Second Trimester (concluded)

C. Matched myelin and cell body stained sections in the spinal cord of a GW26 fetus

Plate 46 is a survey of matched myelin stained and cell body stained sections from Y60-61, a specimen in the Yakovlev Collection with a crown rump length of 210 mm. All sections are shown at the same scale. The boxes enclosing each section list the approximate level and the total area (post-fixation) of the section in square millimeters (mm^2). Full-page normal contrast photographs of each specimen are in **Plates 47A-52A.** Low contrast photographs with superimposed labels and outlines of structural details are in **Plates 47B-52B.** In this specimen, the myelin stained and cell body stained sections were preserved on separate large glass plates without any section numbers. Eighteen myelin stained and 17 cell body stained sections were photographed ranging from upper cervical to thoracic levels. There were no sections preserved at lumbar, sacral, and coccygeal levels. The 35 photographic prints were intuitively arranged in order from upper cervical to thoracic levels, using internal features such as the size of the corticospinal tracts, and the width of the ventral horn. Then, myelin and cell body stained sections were matched. The upper cervical level sections and several middle to lower thoracic level sections were either damaged or had no matches. There were good matches in the region of the cervical enlargement (upper and lower levels are illustrated) and in the upper level of the thoracic cord.

The cross- sectional area of a myelin stained section is smaller than the matching cell body stained section in all cases. Evidently, the myelin staining procedure produces greater tissue shrinkage than the cell body staining procedure. The upper cervical enlargement section has a larger ventral corticospinal tract but is nearly the same size as the lower section in the cervical enlargement; the thoracic section is 47% smaller than the cervical sections.

Y60-61 is one of the youngest specimens to show any myelin stained areas. Actually, staining for myelin products begins around GW20 (Y27-60, CR 160 mm, Fig. 6-2 in Altman and Bayer, 2001), but that specimen is very incomplete, and is not shown in this Atlas. Three structures contain solid black stain indicative of myelinating axons (see *myelination column* in **Table III C-1**). Throughout the rest of the white matter in the myelin stained sections, there is either punctate staining or no staining. Dense punctate stained areas are assumed to have high concentrations of glia that *react* with the stain prior to production of the myelin sheath, sparse punctate stained areas have low concentrations, and unstained areas have none (*reactive glia column*, **Table III C-1**). In the cell body stained sections, there are various densities of what is assumed to be proliferating interfascicular glia in the dorsal, lateral, and ventral funiculi (*proliferating glia column,* **Table III C-1**), but generally, the concentration of proliferating glia is less pronounced than in the GW19 specimen.

The densities of reactive and proliferating glia within a fiber tract vary independently of each other (compare rows in **Table III C-1**). Some tracts have dense reactive glia and dense proliferating glia (for example, deep gracile and cuneate fasciculi). Others have dense reactive glia but sparse proliferating glia (for example, intraspinal tracts), sparse reactive glia and dense proliferating glia (for example, superficial gracile and cuneate fasciculi), sparse reactive glia and sparse proliferating glia (for example, the spinocephalic tracts at the upper cervical enlargement level). These variations are probably due to the different rates of myelination in each fiber tract. In general, glial proliferation precedes reactive gliosis (see the differing concentrations of proliferating glia in the GW19 specimen, **Plates 38-45**), and that precedes myelination in different fiber tracts (see Chapter 6 in Altman and Bayer, 2001).

Table III C-1: Glia types and concentration in the white matter at GW26

Name	Myelination	Reactive glia	Proliferating glia
DORSAL ROOT	---	Dense	Dense
VENTRAL ROOT	Advanced	---	sparse
DORSAL FUNICULUS:			
dorsal root bif. zone	Some fibers	---	Dense
dorsal root col. zone	---	Dense	Dense
deep fas. gracilis	---	Dense	Dense
superficial fas. gracilis	---	Sparse	Dense
deep fas. cuneatus	---	Dense	Dense
superficial fas. cuneatus	---	Sparse	Dense
Lissauer's tract	---	None	Very sparse
LATERAL and VENTRAL FUNICULI:			
lat. corticospinal tract	---	None	Very sparse
ven. corticospinal tract	---	None	Very sparse
spinocerebellar tracts	---	Dense	Dense
ven. commissure	Some fibers	Dense	Dense
**intraspinal tracts	---	Gradient*	Sparse
††spinocephalic tracts	---	Gradient†	Sparse
med. long. fasiculus	---	Dense	Sparse
tectospinal tract	---	Dense	Sparse
vestibulospinal tract	---	Sparse	Sparse

** Overlaps with the medial longitudinal fasciculus, the tectospinal tract, and the lateral reticulospinal tract.

* The central part of the tract (adjacent to the ventral and lateral parts of the ventral horn) stains more intensely than medial parts or dorsolateral parts.

†† Contains ventral and lateral parts. Overlaps with the vestibulospinal, spinotectal, and spino-olivary tracts.

† Deep parts (adjacent to intraspinal tracts) are more dense than superficial parts (adjacent to the ventrolateral pial membrane).

PLATE 46

GW26, CR 210 mm, Y60-61

MYELIN STAIN | **CELL BODY STAIN**

Plates 47A, 47B
Upper Cervical Enlargement
Total area:
11.455 mm^2

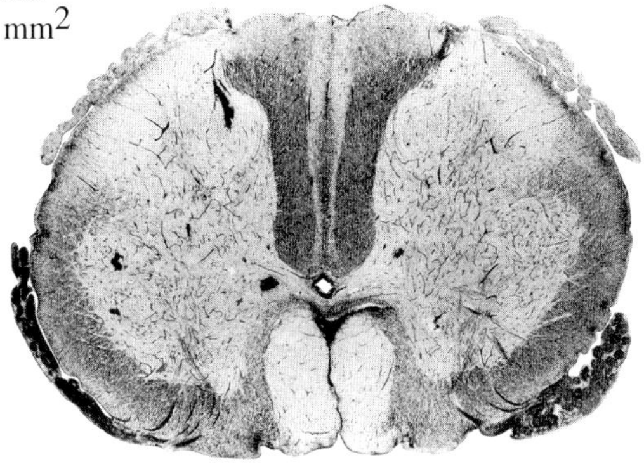

Plates 48A, 48B
Upper Cervical Enlargement
Total area:
12.070 mm^2

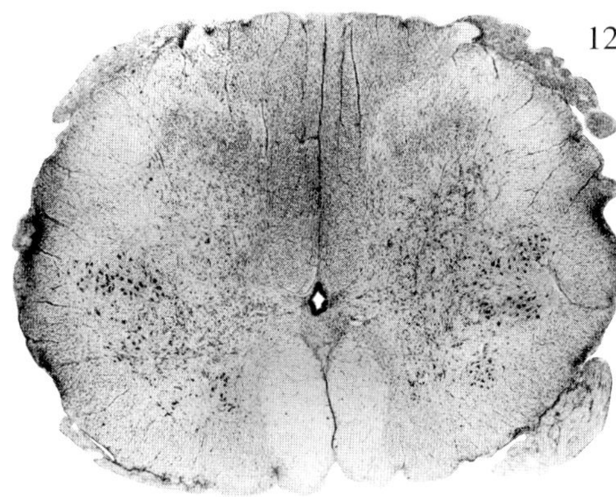

Plates 49A, 49B
Lower Cervical Enlargement
Total area:
11.380 mm^2

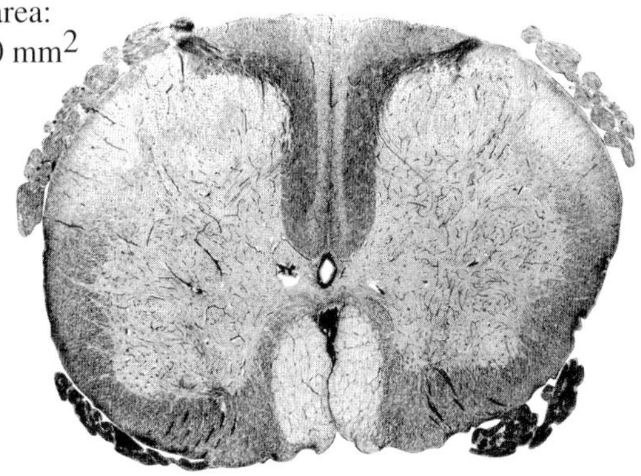

Plates 50A, 50B
Lower Cervical Enlargement
Total area:
12.589 mm^2

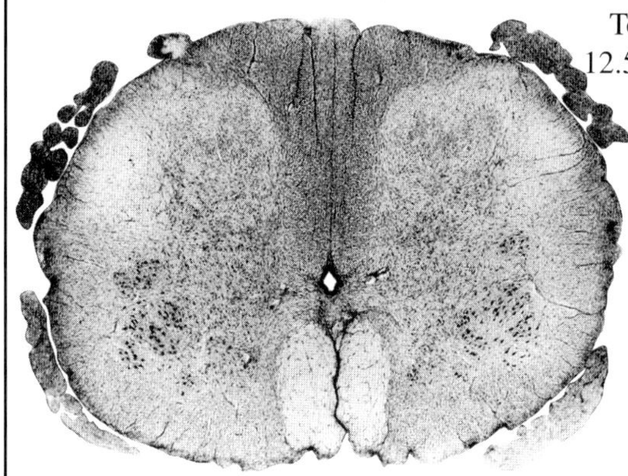

Plates 51A, 51B
Upper Thoracic
Total area:
6.029 mm^2

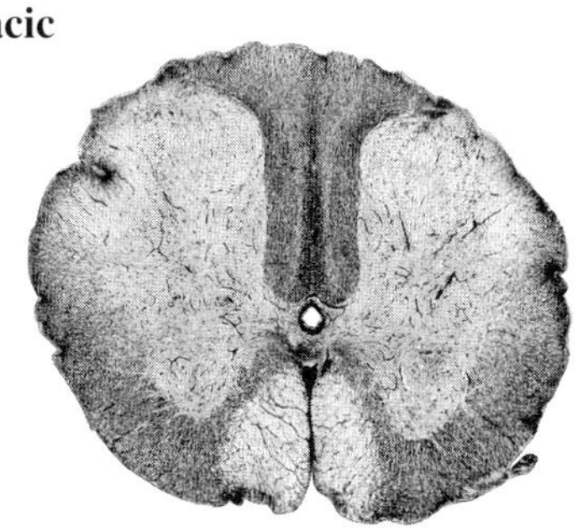

0.5 mm

Plates 52A, 52B
Upper Thoracic
Total area:
6.982 mm^2

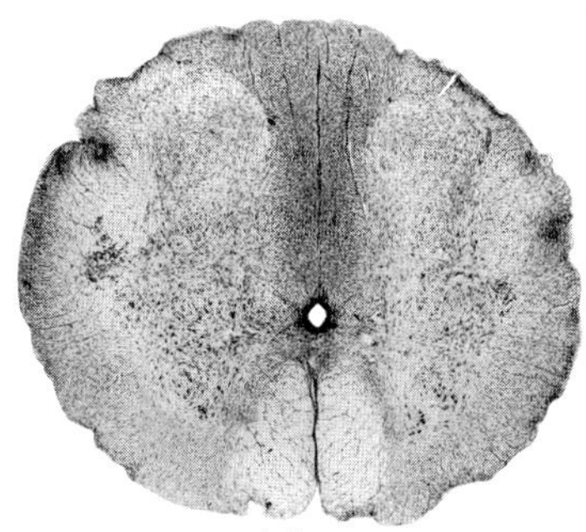

0.5 mm

PLATE 47A

GW26
CR 210 mm
Y60-61
Upper Cervical Enlargement
Myelin stain

Areas (mm^2)

Central canal	.0069
Ependyma	.0163
Gray matter	4.7293
White matter	6.7021

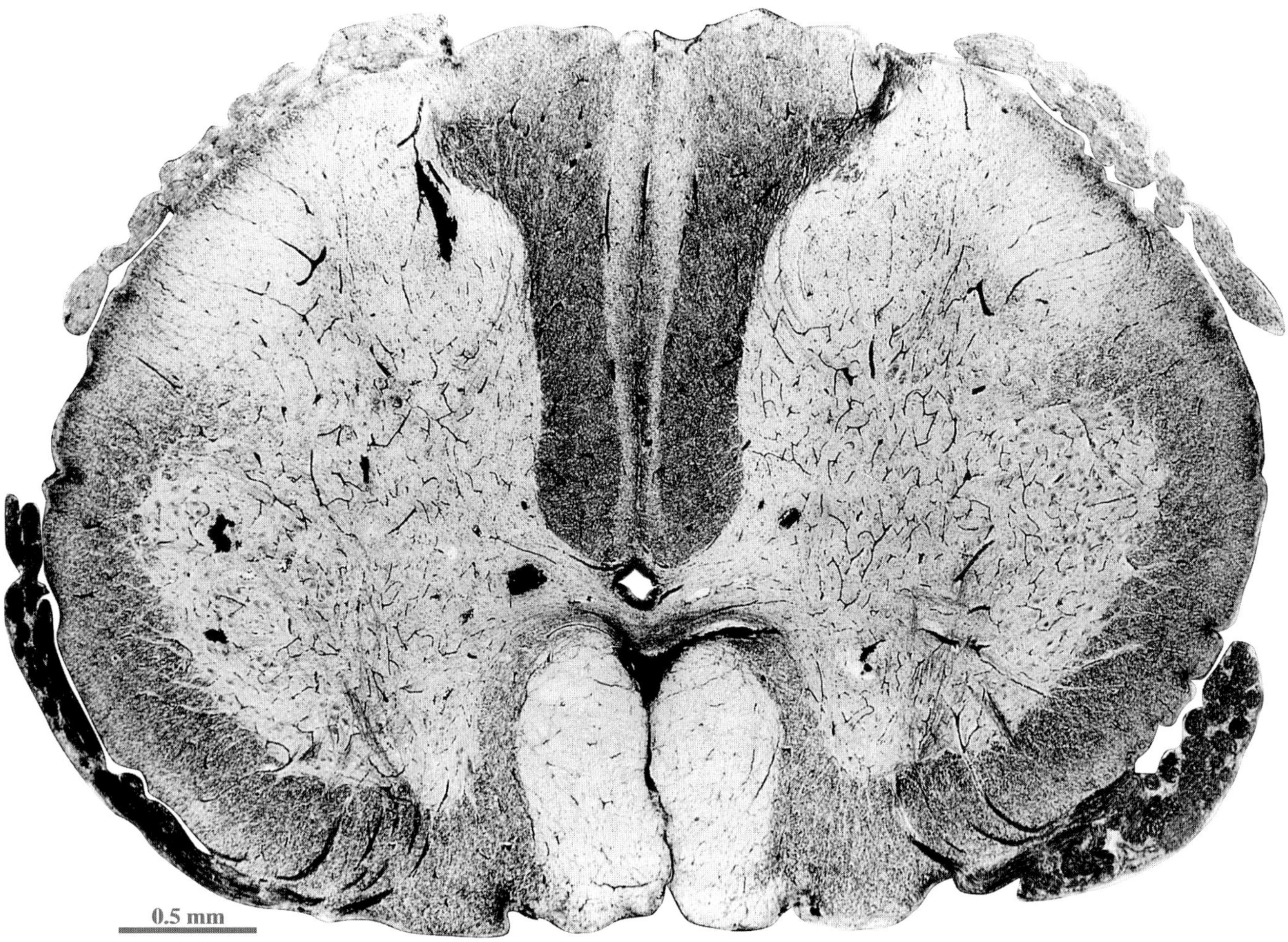

PLATE 47B

See the matched cell body stained section in Plates 48A and B

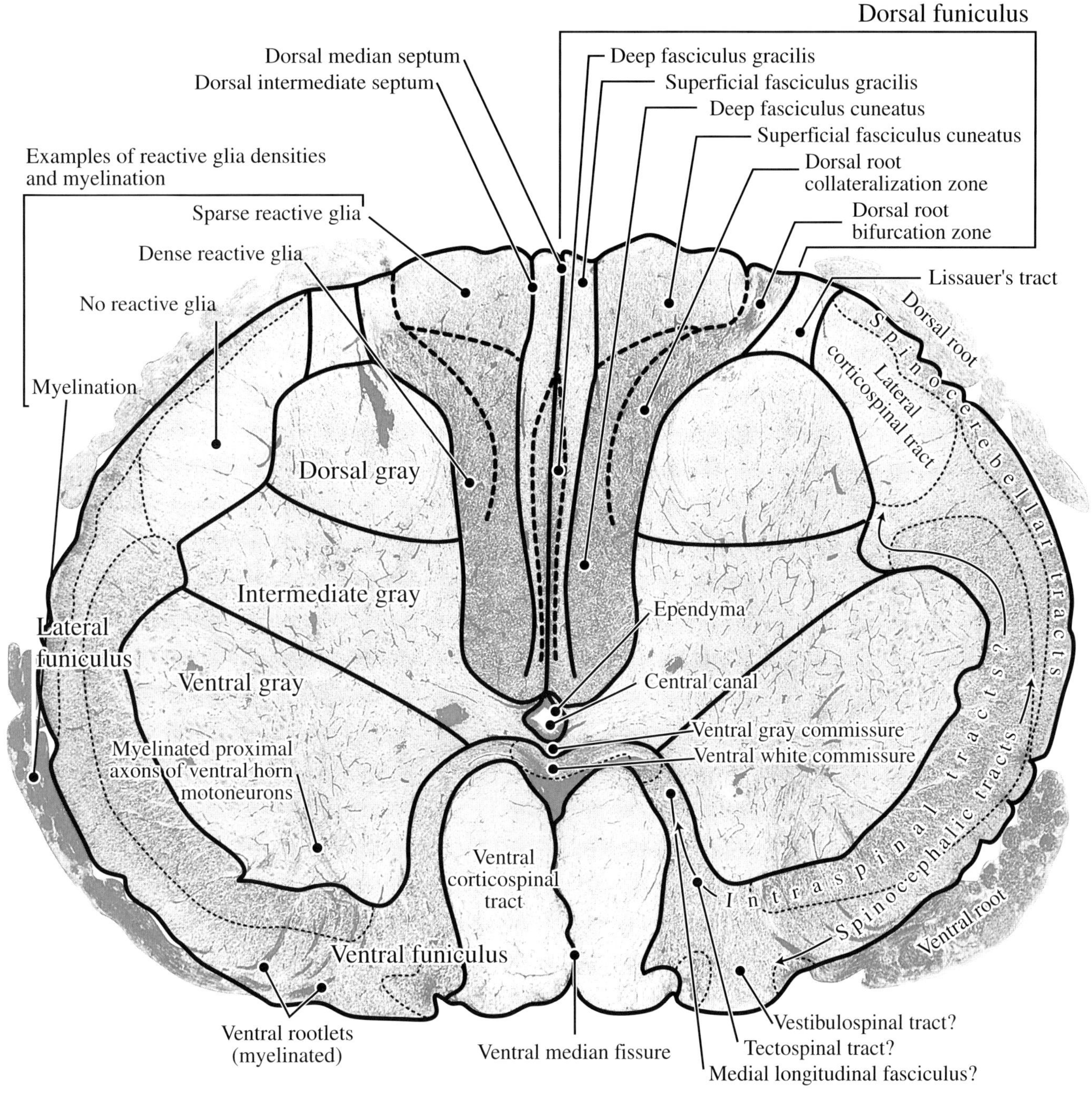

The fine lines in the lateral and ventral funiculi segregate regions of differing densities of reactive glia, not the borders of fiber tracts. These lines may correspond to the borders of some fiber tracts, such as the spinocerebellar and the lateral corticospinal tracts. Other lines run across fiber tracts, such as the dense region adjacent to the ventral and lateral part of the ventral horn that includes the central part of the intraspinal tracts and the deep parts of the spinocephalic tracts. Many of the fiber tracts in the ventral and lateral funiculi overlap and do not have distinct borders. The spinotectal, spino-olivary, spinoreticular, and ventrolateral reticulospinal tracts (all unlabeled) intermingle with the spinocephalic fibers. The medial longitudinal fasciculus and tectospinal tract are interspersed with the medial fibers of the intraspinal tracts; dorsal and lateral fibers of the intraspinal tract are infiltrated by the lateral reticulospinal tract (unlabeled).

PLATE 48A

GW26
CR 210 mm
Y60-61
Upper Cervical Enlargement
Cell body stain

Areas (mm²)

Central canal	.0055
Ependyma	.0206
Gray matter	5.0505
White matter	6.9934

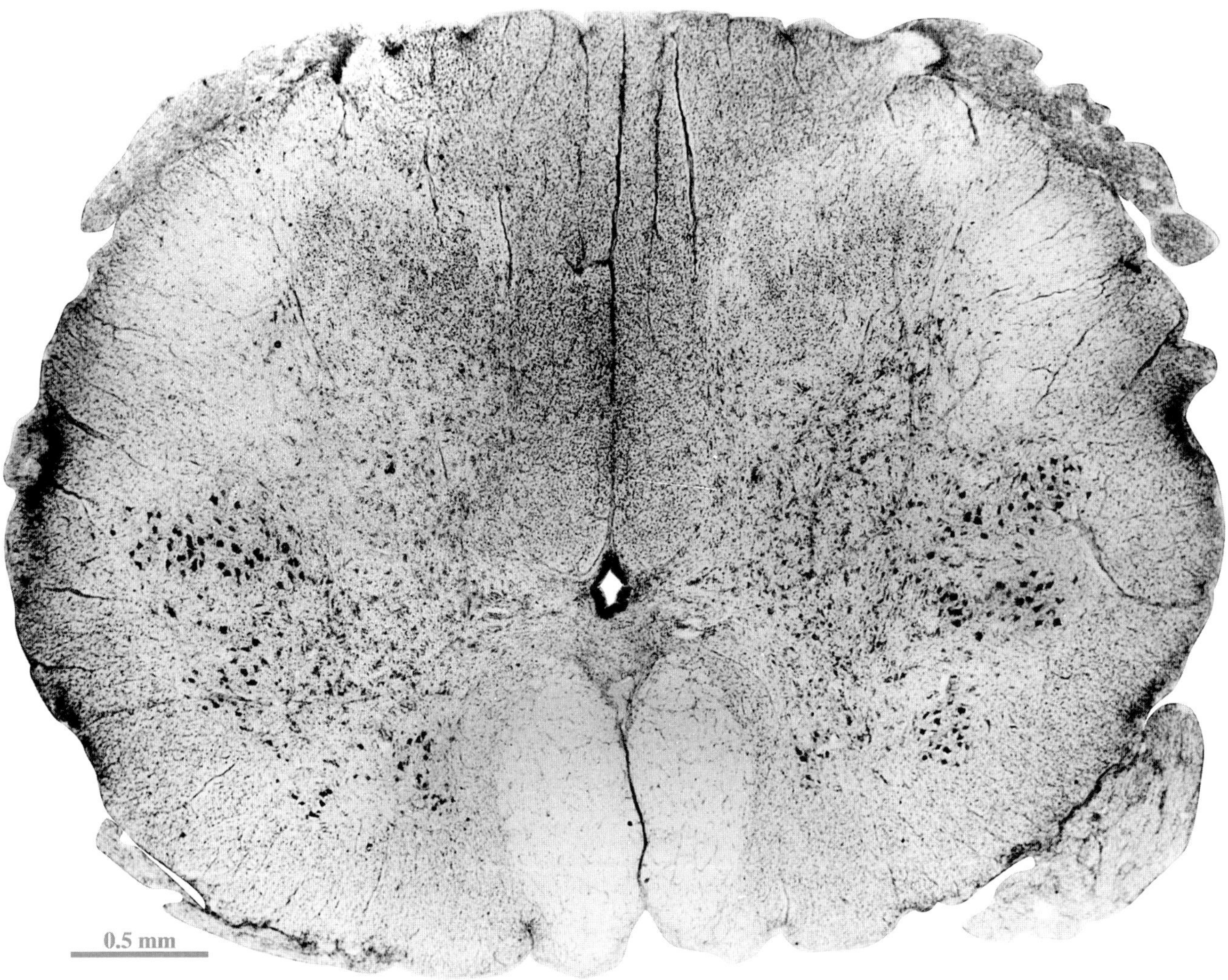

PLATE 48B

See the matched myelin stained section in Plates 47A and B

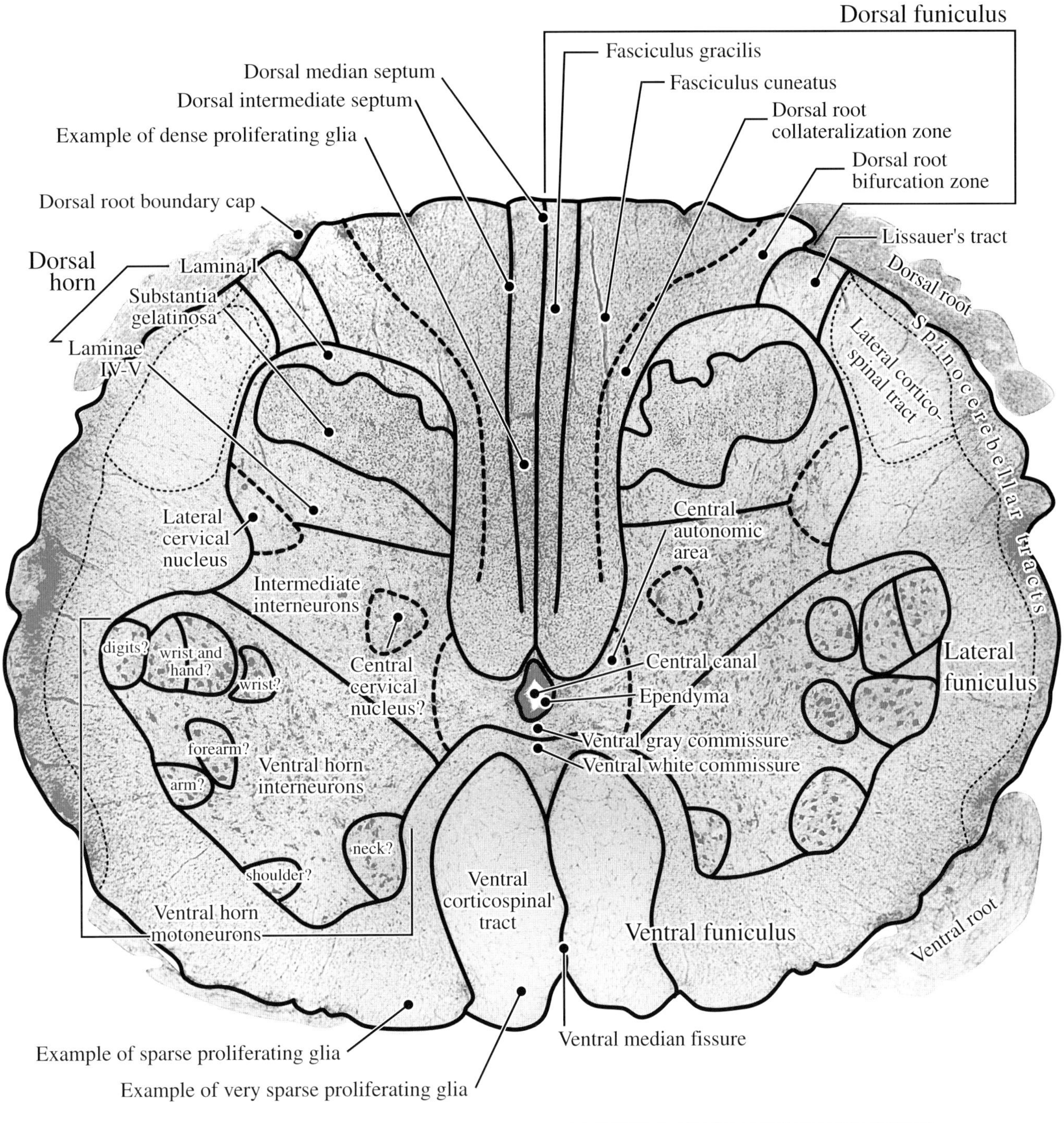

Only the corticospinal and spinocerebellar tracts can be clearly delineated in the ventral and lateral funiculi. The corticospinal tracts stand out as clear areas with very sparse proliferating glia. The spinocerebellar tracts have a more dense concentration of proliferating glia. A sparse population of proliferating glia fills the remaining ventral and lateral funiculi and contains several fiber tracts (unlabeled in this section). Refer to the matching myelin stained section for the approximate locations of the medial longitudinal fasciculus, tectospinal, vestibulospinal, intraspinal, and spinocephalic tracts.

PLATE 49A

GW26
CR 210 mm
Y60-61
Lower Cervical Enlargement
Myelin stain

Areas (mm^2)

Central canal	.0090
Ependyma	.0196
Gray matter	5.1514
White matter	6.2003

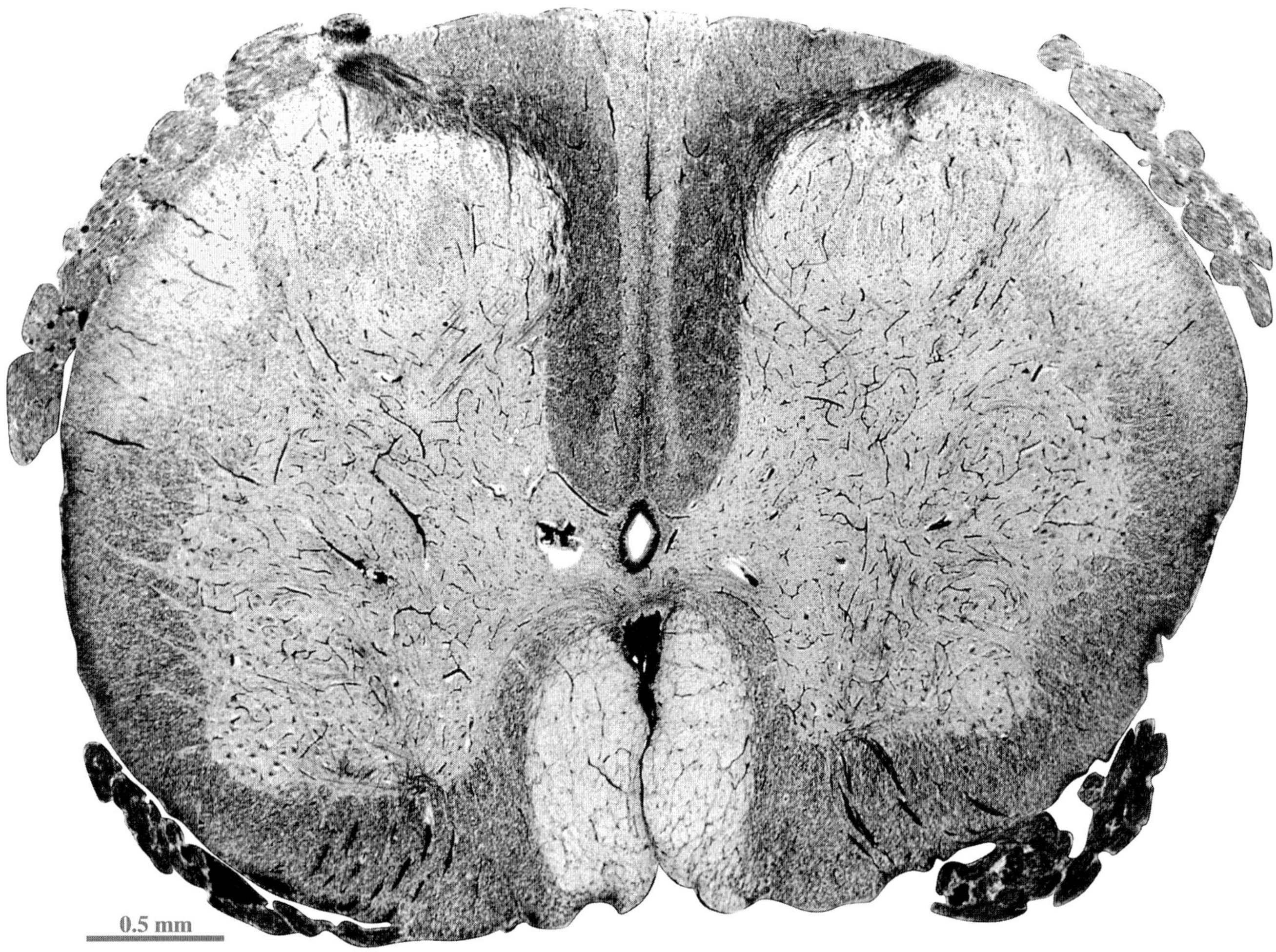

PLATE 49B

See the matched cell body stained section in Plates 50A and B

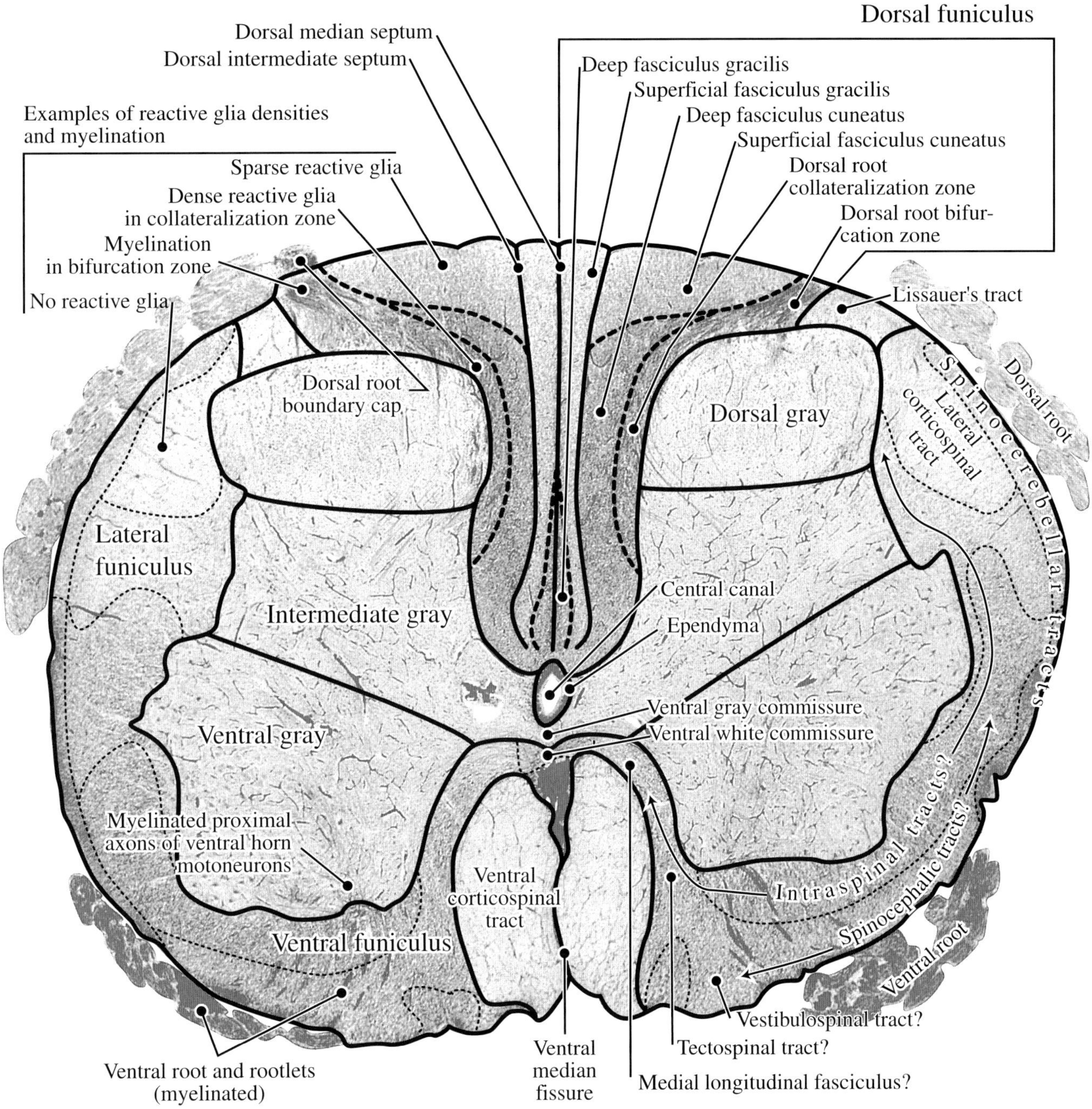

PLATE 50A

GW26
CR 210 mm
Y60-61
Lower Cervical Enlargement
Cell body stain

Areas (mm^2)

Central canal	.0086
Ependyma	.0234
Gray matter	5.3014
White matter	7.2555

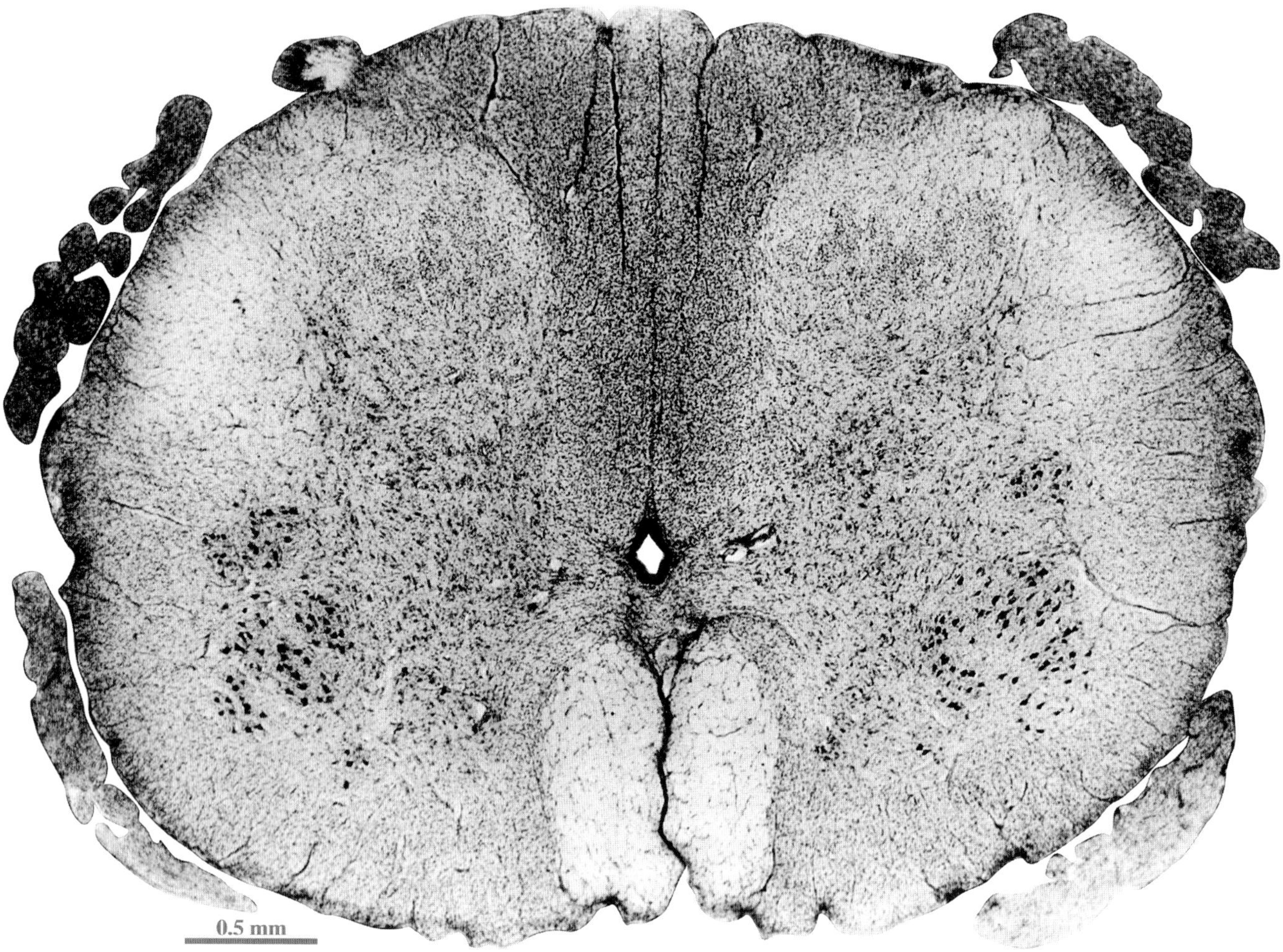

PLATE 50B

See the matched myelin stained section in Plates 49A and B

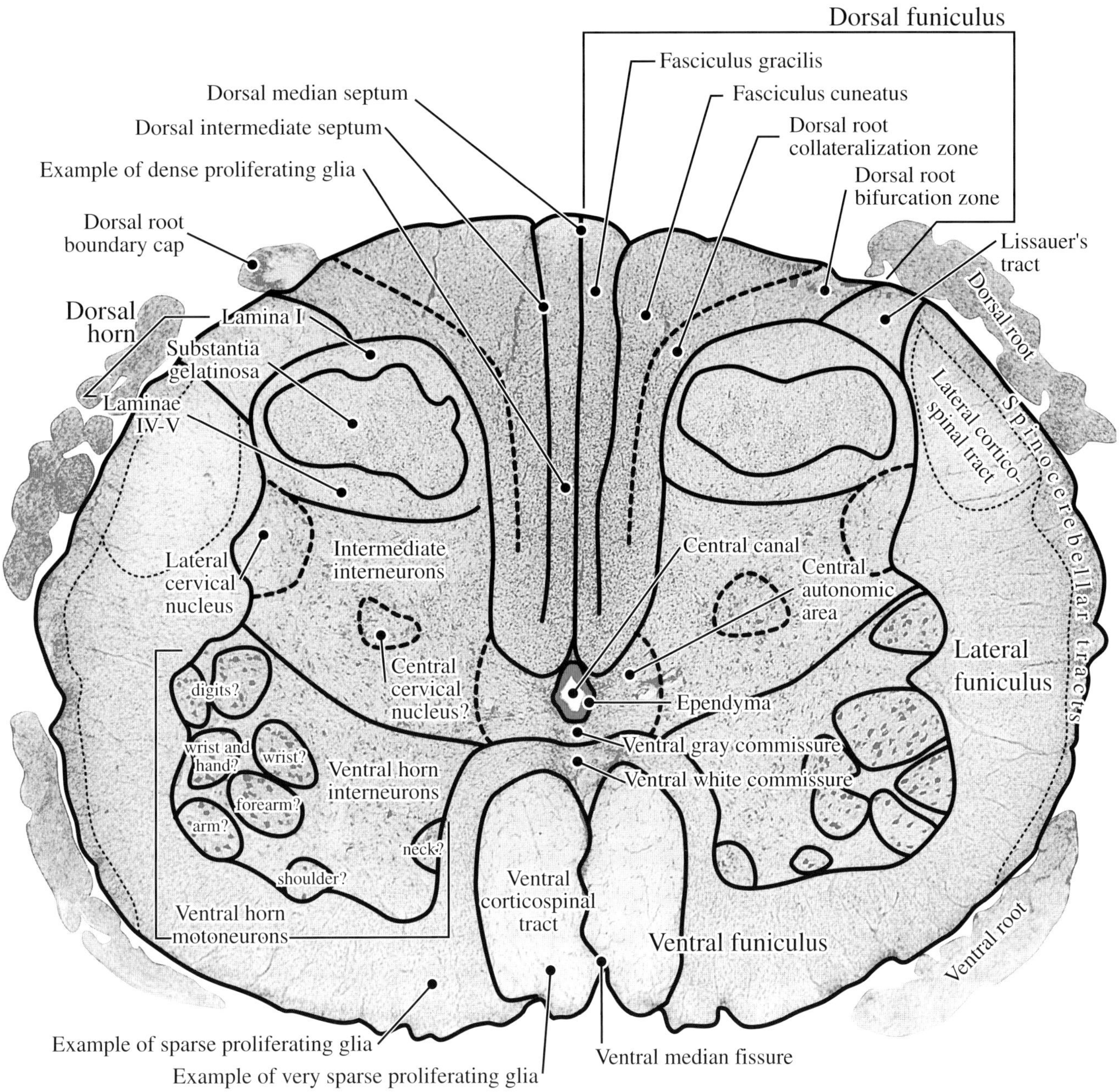

Refer to the matching myelin stained section for the approximate locations of the medial longitudinal fasciculus, tectospinal, vestibulospinal, intraspinal, and spinocephalic tracts.

PLATE 51A

GW26
CR 210 mm
Y60-61
Upper Thoracic
Myelin stain

Areas (mm^2)

Central canal	.0071
Ependyma	.0144
Gray matter	2.3030
White matter	3.7045

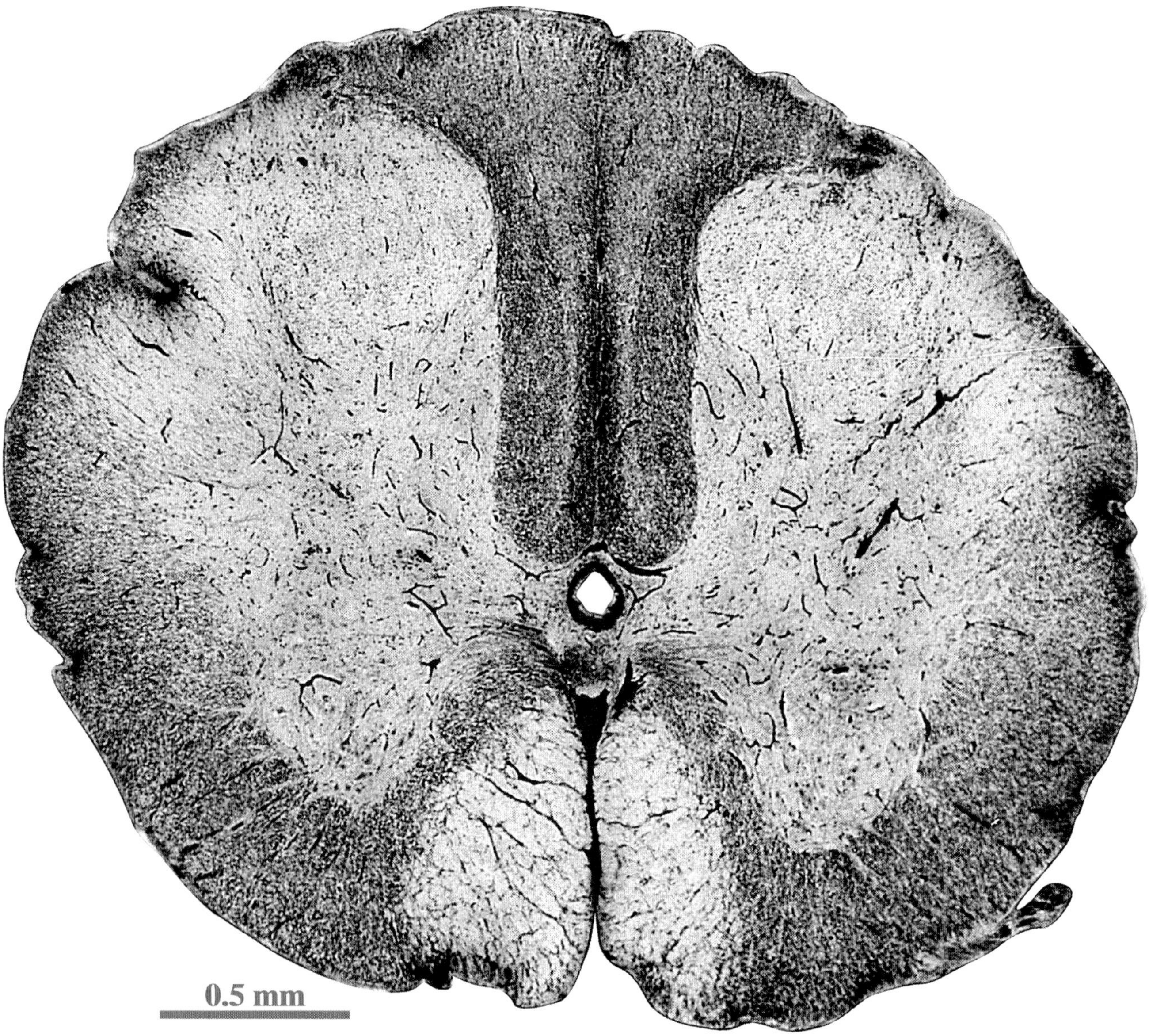

PLATE 51B

See the matched cell body stained section in Plates50A and B

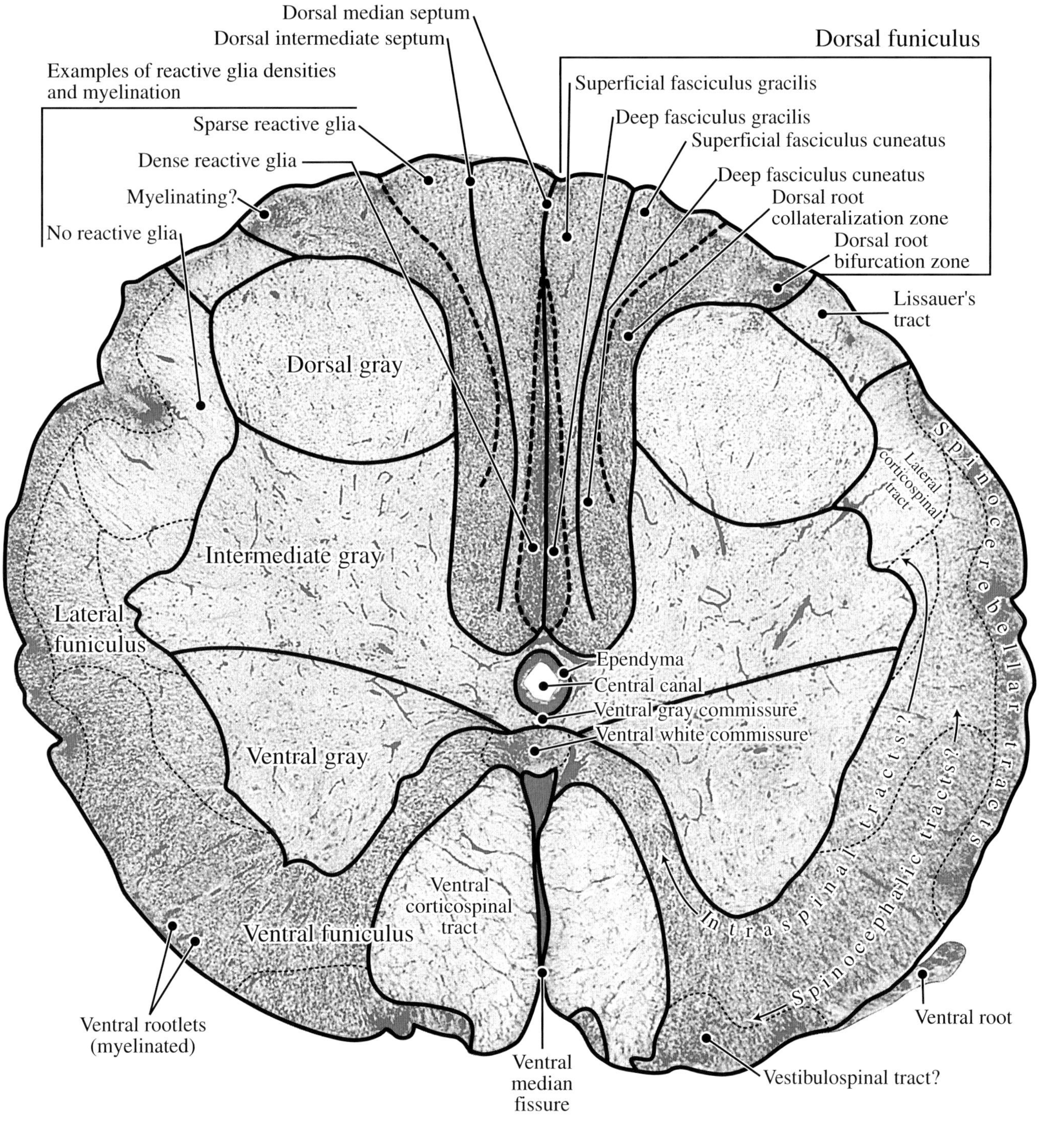

PLATE 52A

GW26
CR 210 mm
Y60-61
Upper Thoracic
Cell body stain

Areas (mm^2)

Central canal	.0083
Ependyma	.0204
Gray matter	2.8518
White matter	4.1012

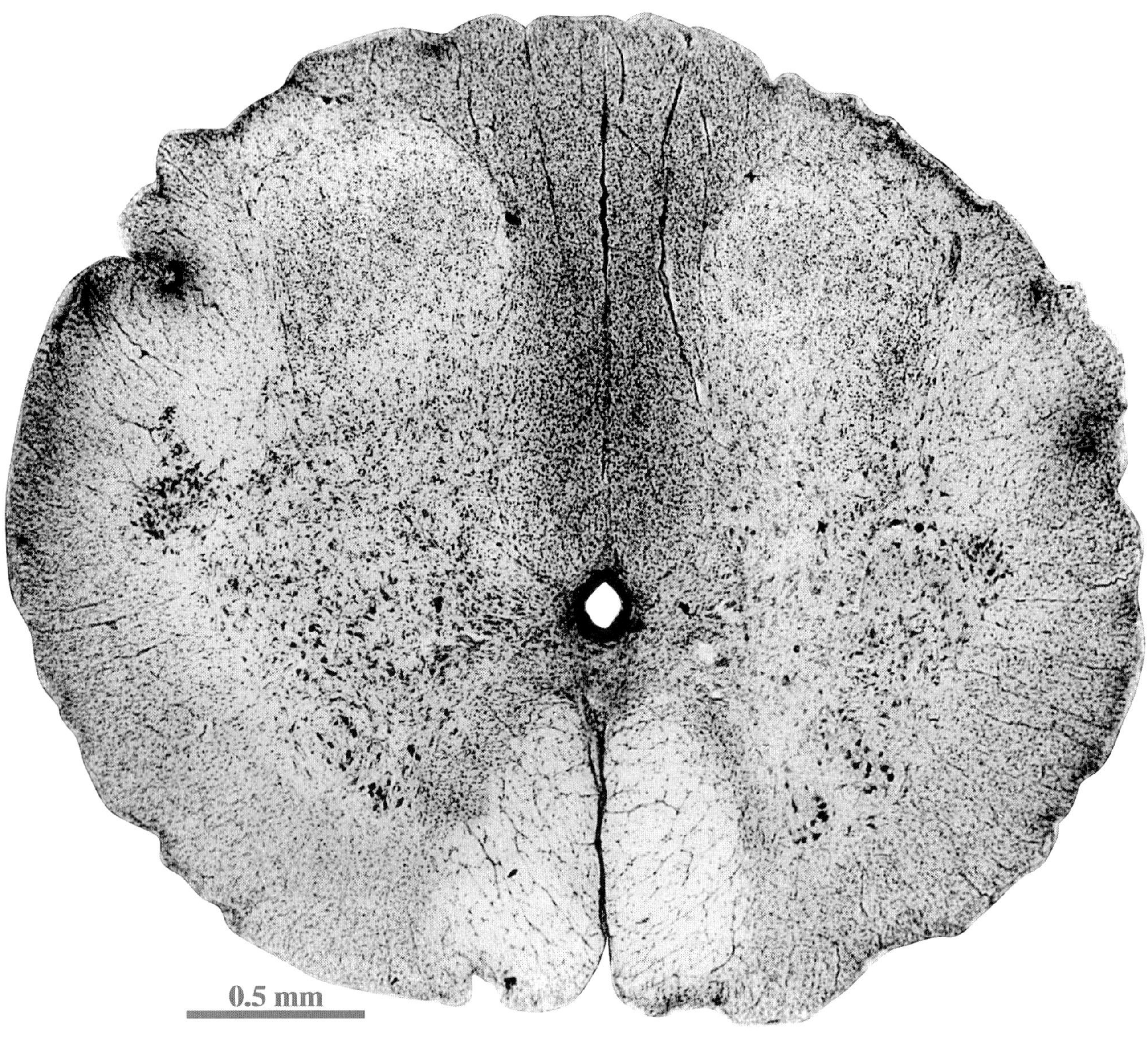

PLATE 52B

See the matchedmyelin stained section in Plates 48A and B

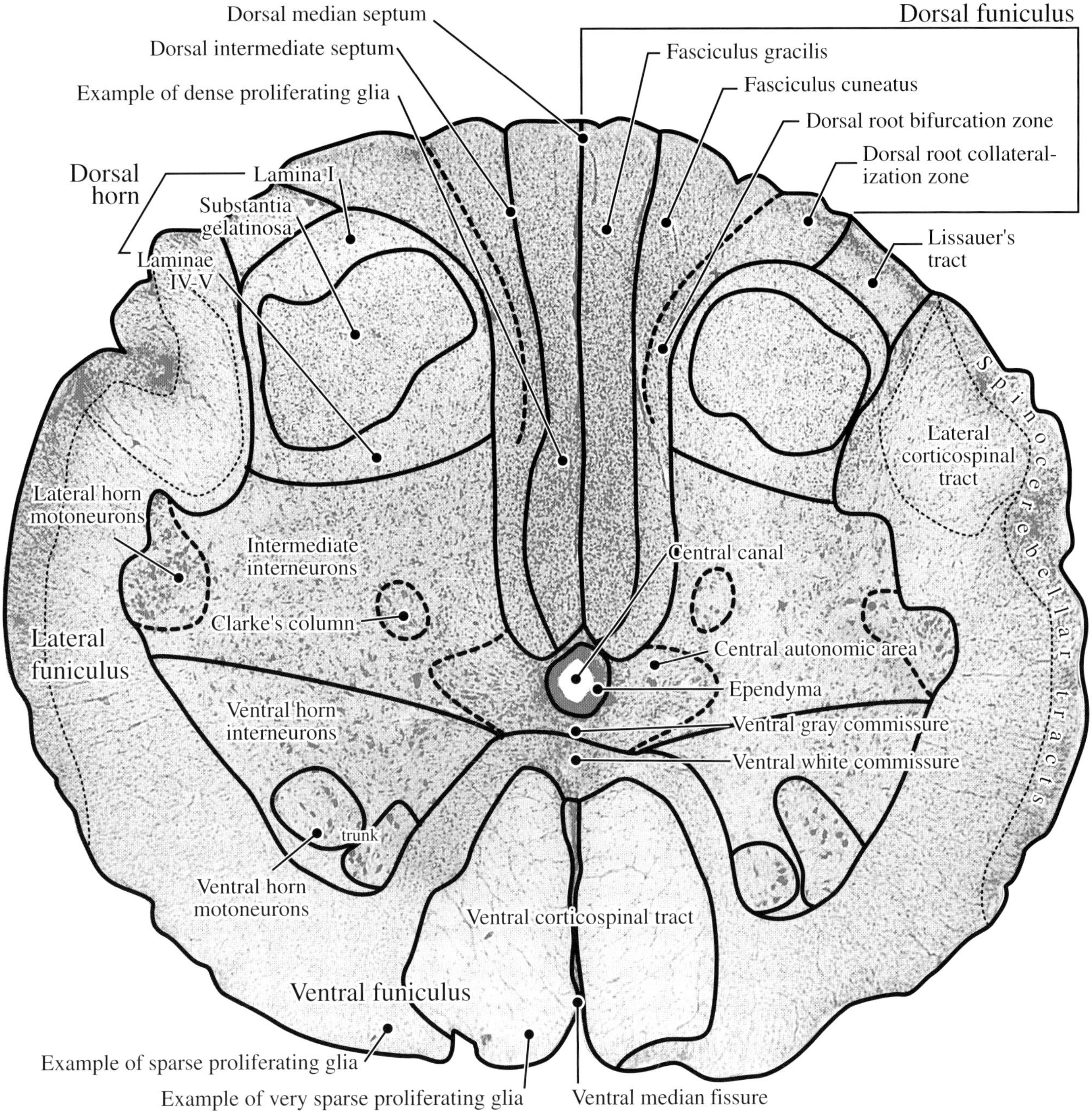

Refer to the matching myelin stained section for the approximate locations of the vestibulo-spinal, intraspinal, and spinocephalic tracts.

IV. The Third Trimester

A. Matched myelin and cell body stained sections in the spinal cord of a GW31 fetus

Plate 53 is a survey of matched myelin stained and cell body stained sections from Y162-61, a specimen in the Yakovlev Collection with a crown rump length of 270 mm (see Chapter 6 in Altman and Bayer, 2001). All sections are shown at the same scale. The boxes enclosing each section list the approximate level and the total area (post-fixation) of the section in square millimeters (mm^2). Full-page normal contrast photographs of each specimen are in **Plates 54A-61A.** Low contrast photographs with superimposed labels and outlines of structural details are in **Plates 54B-61B.** In this specimen, the myelin stained and cell body stained sections were preserved on separate large glass plates without any numbers. Thirteen myelin stained and 13 cell body stained sections were photographed ranging from lower thoracic to sacral/coccygeal levels. There were no sections preserved at cervical, upper thoracic, and middle thoracic levels. The 26 photographic prints were intuitively arranged in order from thoracic to sacral/coccygeal levels, using internal features such as the diminishing size of the corticospinal tracts from rostral to caudal levels, and the width of the ventral horn. Then, the myelin stained and the cell body stained sections were matched.

As in the previous specimen, the cross-sectional area of a myelin stained section is smaller than the matching cell body stained section in all cases. Evidently, the myelin staining procedure produces greater tissue shrinkage than the cell body staining procedure. The lumbar enlargement expands relative to regions above and below. Using the myelin-stained section areas for comparison, the lumbar enlargement is 90% larger than the lower thoracic level, 41% larger than the upper lumbar level, and 278% larger than the sacral/coccygeal level.

Myelination in this specimen is more advanced than in the previous specimen, even though the two are only 5 weeks apart, and we are dealing with a more immature (caudal) region of the spinal cord. Dense staining indicative of true myelination is seen in the ventral commissure, the ventral rootlets, the dorsal root bifurcation zone, the dorsal root collateralization zone, and deep regions of the fasciculus gracilis (see *column 2* in **Table IV A-1**). Myelinated fibers from the dorsal root collateralization zone penetrate the gray matter, and there is a light dusting of reactive glia in the subgelatinosal plexus in the dorsal gray as finer collateral axons prepare for later myelination (see Fig. 6-39 in Altman and Bayer, 2001). Clumps of myelinated axons are in the lateral part of the intermediate gray and the lateral neck region of the dorsal gray (the reticulated area, labeled in Figs. 9-25 through 9-31 in Altman and Bayer, 2001). The lateral corticospinal tract and Lissauer's tract contain very sparse reactive glia. The dorsal root bifurcation zone is a complex area that not only contains heavily myelinated fibers, but also unmyelinated fibers. The heavily myelinated axons are presumably from large sensory neurons in the dorsal root ganglia, while the unmyelinated axons are from small nociceptive ganglion cells. The remaining fiber tracts contain varying densities of reactive glia (see *column 3* in **Table IV A-1**). In the cell body stained sections, there is a different density of proliferating glia in various fiber tracts. It is lowest in the lateral and ventral corticospinal tracts and highest in the dorsal funiculus (except Lissauer's tract) and spinocerebellar tracts (see *column 4* in **Table IV A-1**).

In the cell body stained sections, columns of motoneurons continue to be prominent and show greater degrees of segregation in the ventral horn. The dorsal horn has fairly well-defined clusters of small neurons in the substantia gelatinosa. The accumulation of lateral horn motoneurons is very obvious at the thoracic level. Clarke's column is also prominent in this specimen.

Table IV A-1: Glia types and concentration in the white matter at GW31

Name	Myelination	Reactive glia	Proliferating glia
DORSAL FUNICULUS:			
dorsal root bif. zone	Many fibers*	---	Sparse
dorsal root col. zone	Many fibers	---	Very dense
deep fas. gracilis	Many fibers	---	Very dense
superficial fas. gracilis	Some fibers	Dense	Dense
Lissauer's tract	---	None	Very sparse
LATERAL and VENTRAL FUNICULI:			
lat. cortricospinal tract	---	Very sparse	Very sparse
ven. cortricospinal tract	---	Sparse	Very sparse
rubrospinal tract	Some fibers	Dense	Sparse
spinocerebellar tracts	Some fibers	Dense	Dense
lat. reticulospinal tract	Some fibers	Dense	Sparse
intraspinal tracts	Some fibers	Dense	Sparse
spinocephalic tracts	Some fibers	Dense	Sparse
vestibulospinal tract	Some fibers	Dense	Sparse
ven. commissure	Many fibers	---	Sparse

* intermingled in a bed of nonreactive glia (associated with Lissauer's tract fibers?)

PLATE 53

GW31, CR 270 mm, Y162-61

MYELIN STAIN | **CELL BODY STAIN**

Plates 54A, 54B
Lower Thoracic
Total area:
5.1272 mm^2

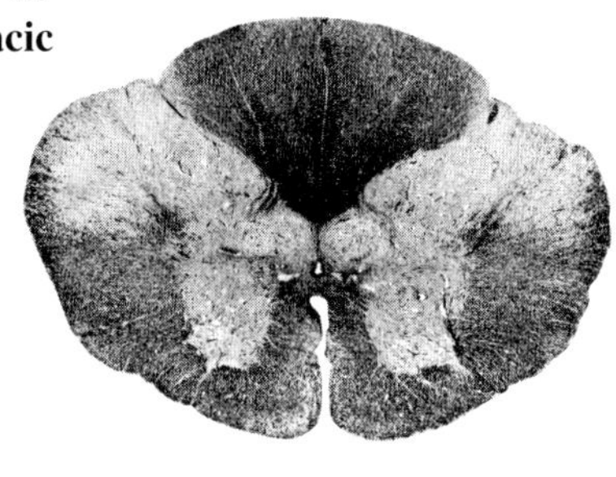

Plates 55A, 55B
Lower Thoracic
Total area:
5.2380 mm^2

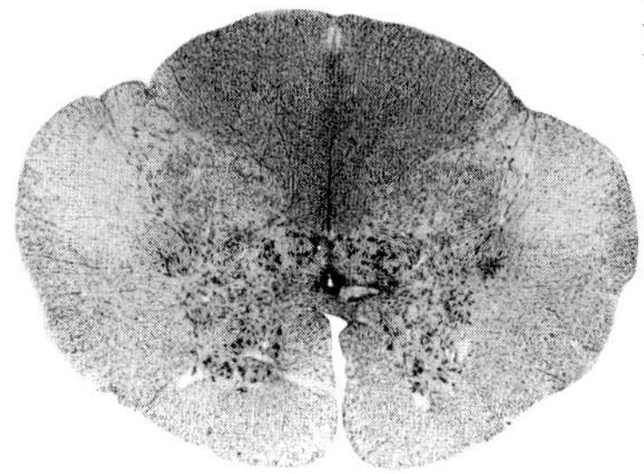

Plates 56A, 56B
Upper Lumbar
Total area:
6.9121 mm^2

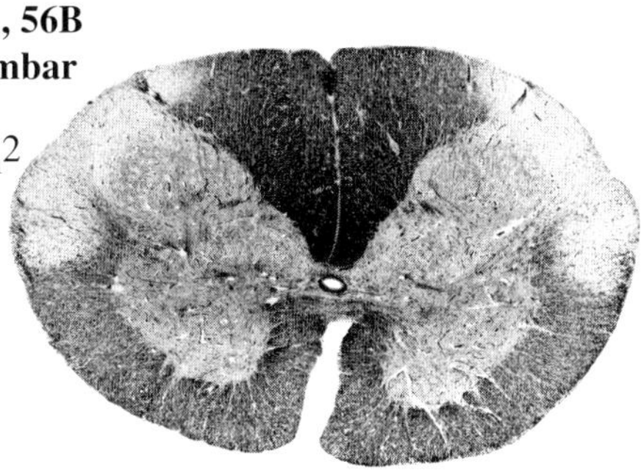

Plates 57A, 57B
Upper Lumbar
Total area:
7.1380 mm^2

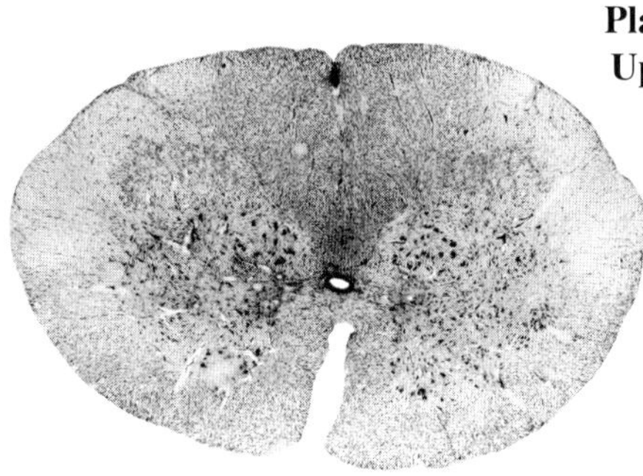

Plates 58A, 58B
Lumbar Enlargement
Total area:
9.7553 mm^2

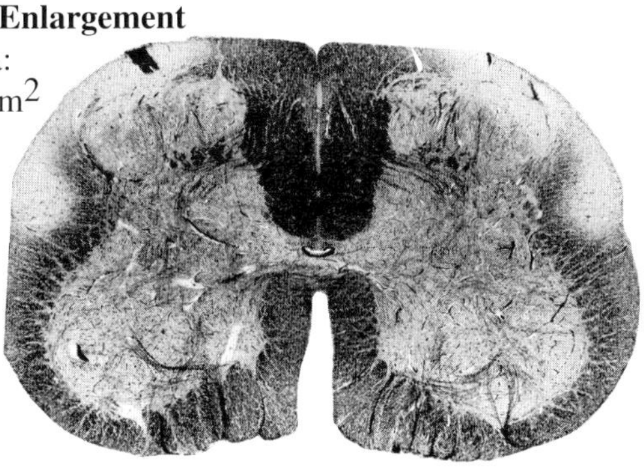

Plates 59A, 59B
Lumbar Enlargement
Total area:
10.918 mm^2

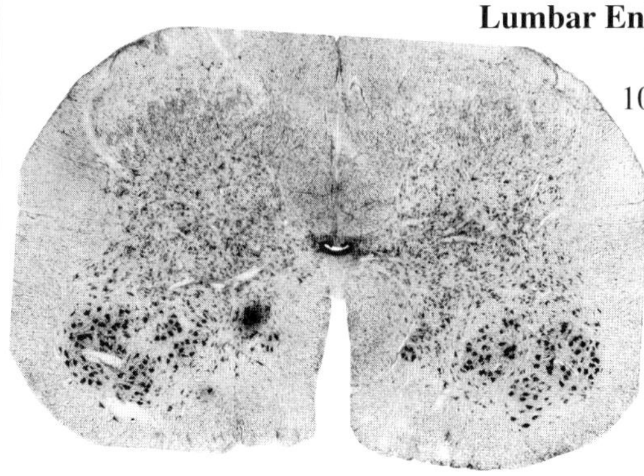

Plates 60A, 60B
Sacral/Coccygeal
Total area:
3.5153 mm^2

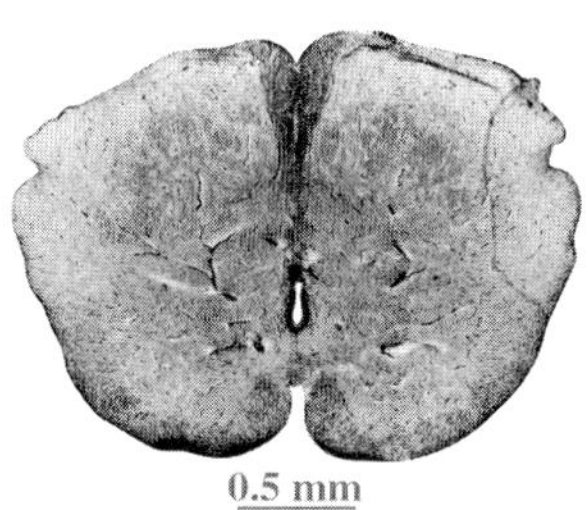

Plates 61A, 61B
Sacral/Coccygeal
Total area:
3.5751 mm^2

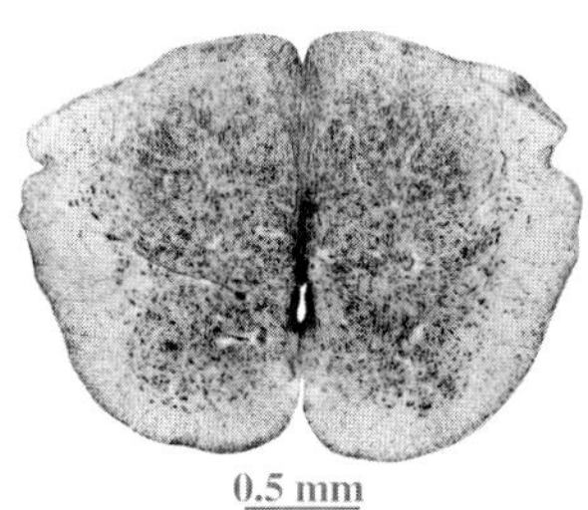

PLATE 54A

GW31
CR 270 mm
Y162-61
Lower Thoracic
Myelin stain

Areas (mm^2)

Central canal	.0012
Ependyma	.0057
Gray matter	1.4171
White matter	3.7032

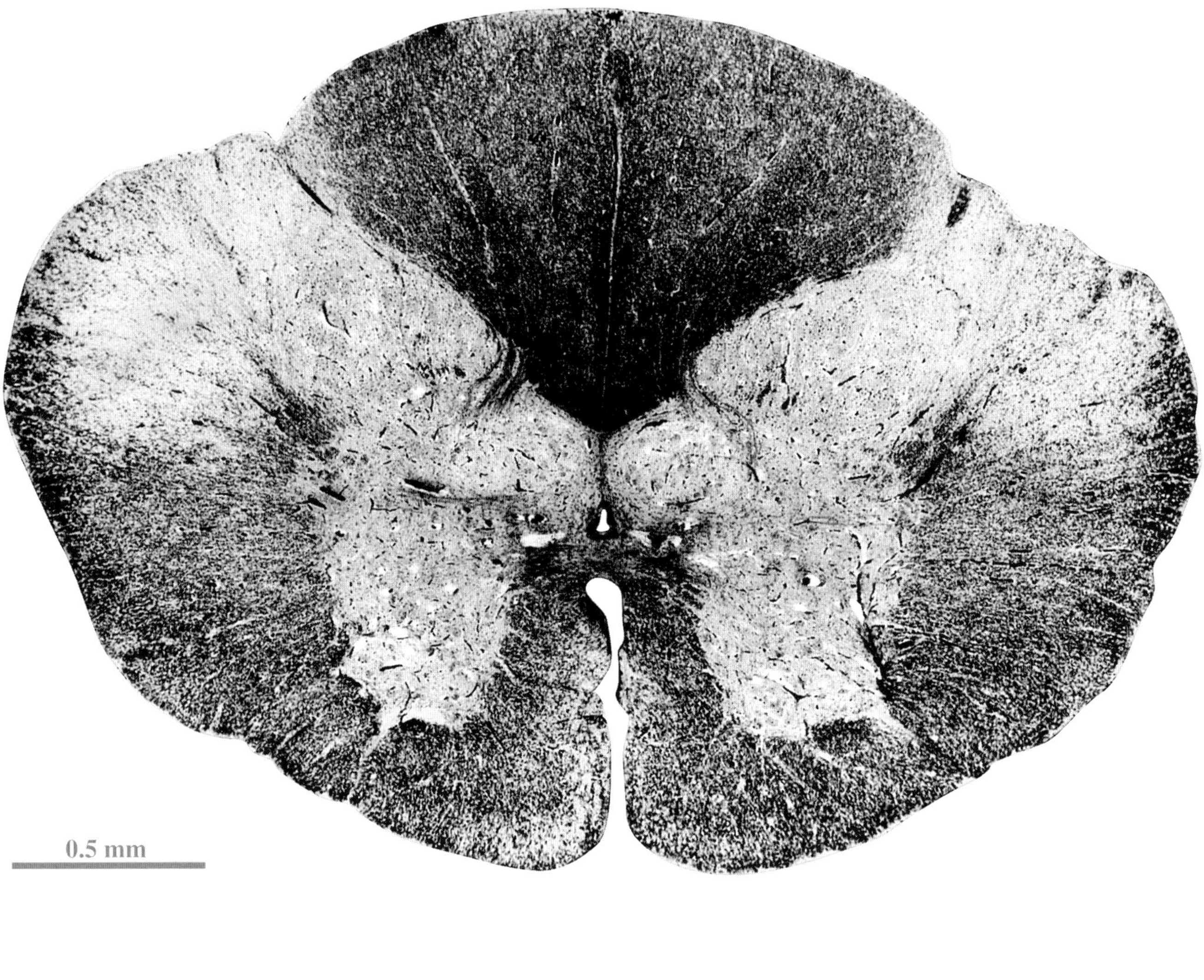

PLATE 54B

See the matched cell body stained section in Plates 55A and B

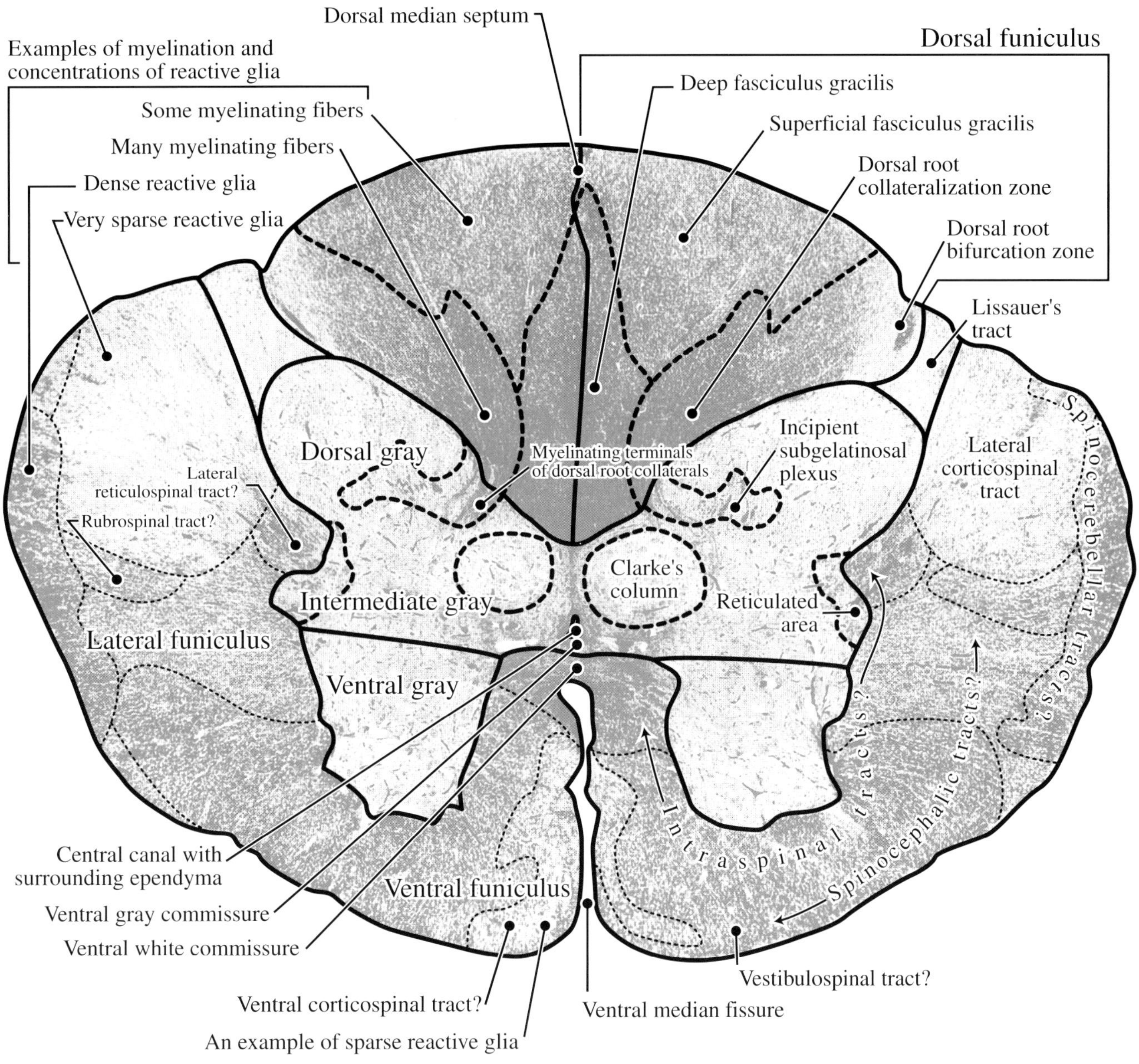

The fine lines in the lateral and ventral funiculi segregate regions of differing densities of the myelin stain, not the borders of fiber tracts. These lines may correspond to the borders of some fiber tracts, such as the spinocerebellar and the lateral corticospinal tracts. Other lines run across fiber tracts, such as the area surrounding the ventral horn that includes the central part of the intraspinal tracts, the vestibulospinal tract, and the ventral and central fibers of the spinocephalic tracts. Most tracts in the ventral and lateral funiculi overlap and do not have distinct borders. The ventral part of the spinocephalic tracts overlaps with the vestibulospinal tract; its central part includes fibers in the ventrolateral reticulospinal tract (not labeled), the spinotectal tract (not labeled), and the spino-olivary tract (not labeled). The intraspinal tracts overlap with the lateral reticulospinal tract. Some fibers of the rubrospinal tract may intermingle with ventral fibers of the lateral corticospinal tract.

PLATE 55A

GW31
CR 270 mm
Y162-61
Lower Thoracic
Cell body stain

Areas (mm^2)

Central canal	.0011
Ependyma	.0087
Gray matter	1.5228
White matter	3.7054

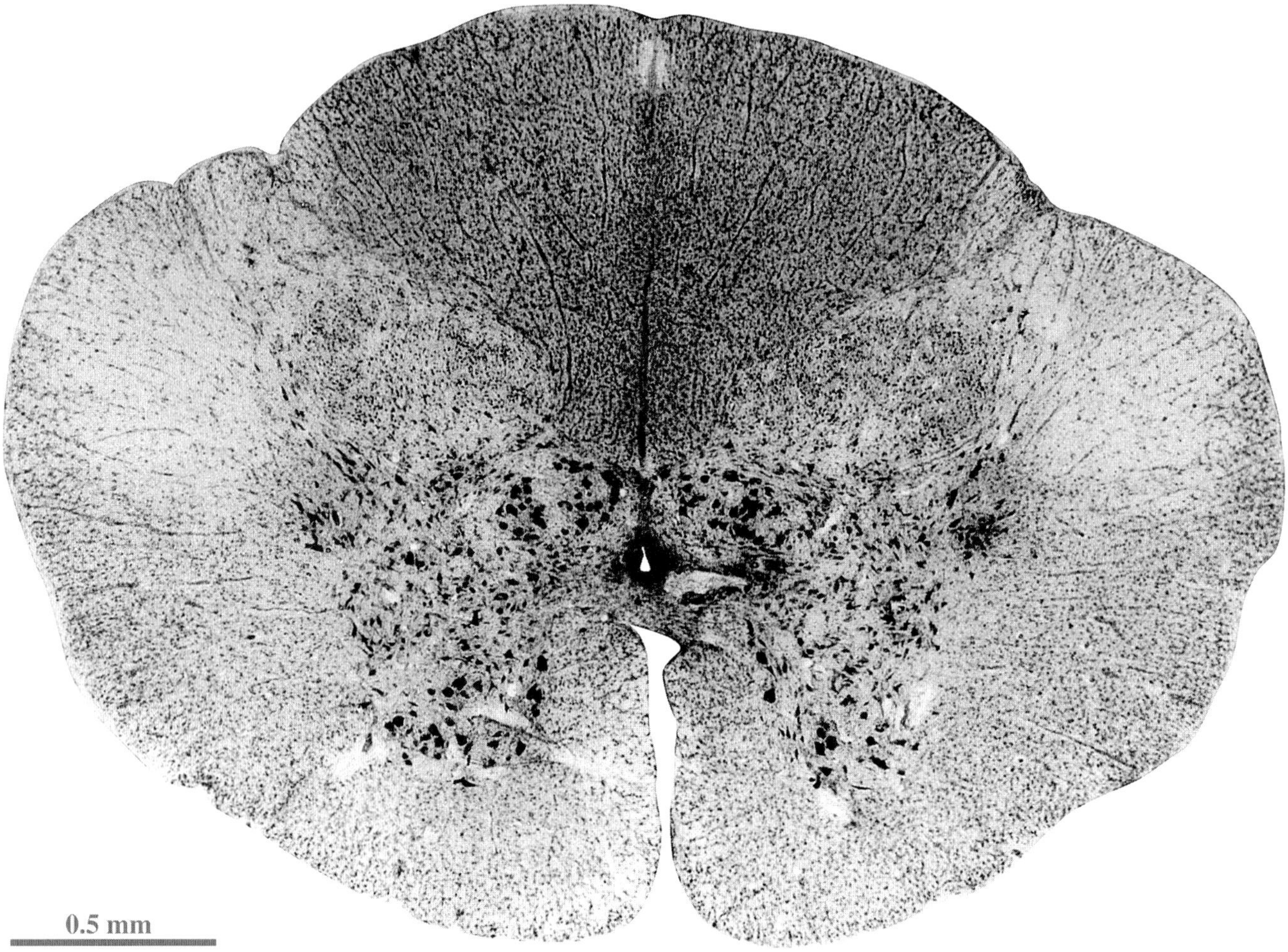

PLATE 55B

See the matched myelin stained section in Plates 54A and B

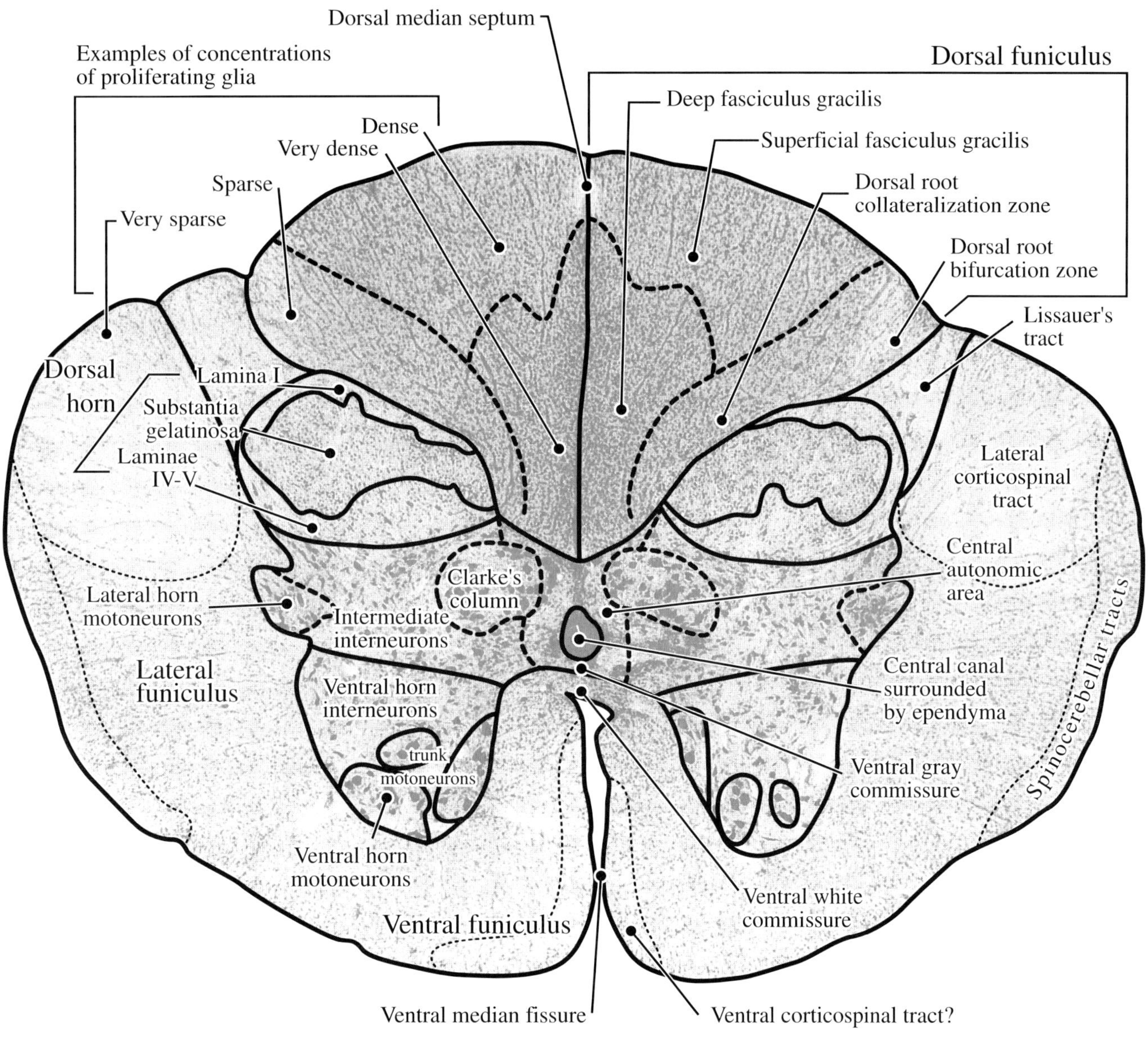

Only the corticospinal and spinocerebellar tracts can be clearly delineated in the ventral and lateral funiculi. The corticospinal tracts stand out as clear areas with very sparse proliferating glia. The spinocerebellar tracts have a slightly more dense concentration of proliferating glia. A sparse population of proliferating glia fills the remaining ventral and lateral funiculi and contains several fiber tracts (unlabeled in this section). Refer to the matching myelin stained section for the approximate locations of the rubrospinal, lateral reticulospinal, vestibulospinal, intraspinal, and spinocephalic tracts.

PLATE 56A

GW31
CR 270 mm
Y162-61
Upper Lumbar
Myelin stain

Areas (mm^2)

Central canal	.0038
Ependyma	.0110
Gray matter	2.5641
White matter	4.3333

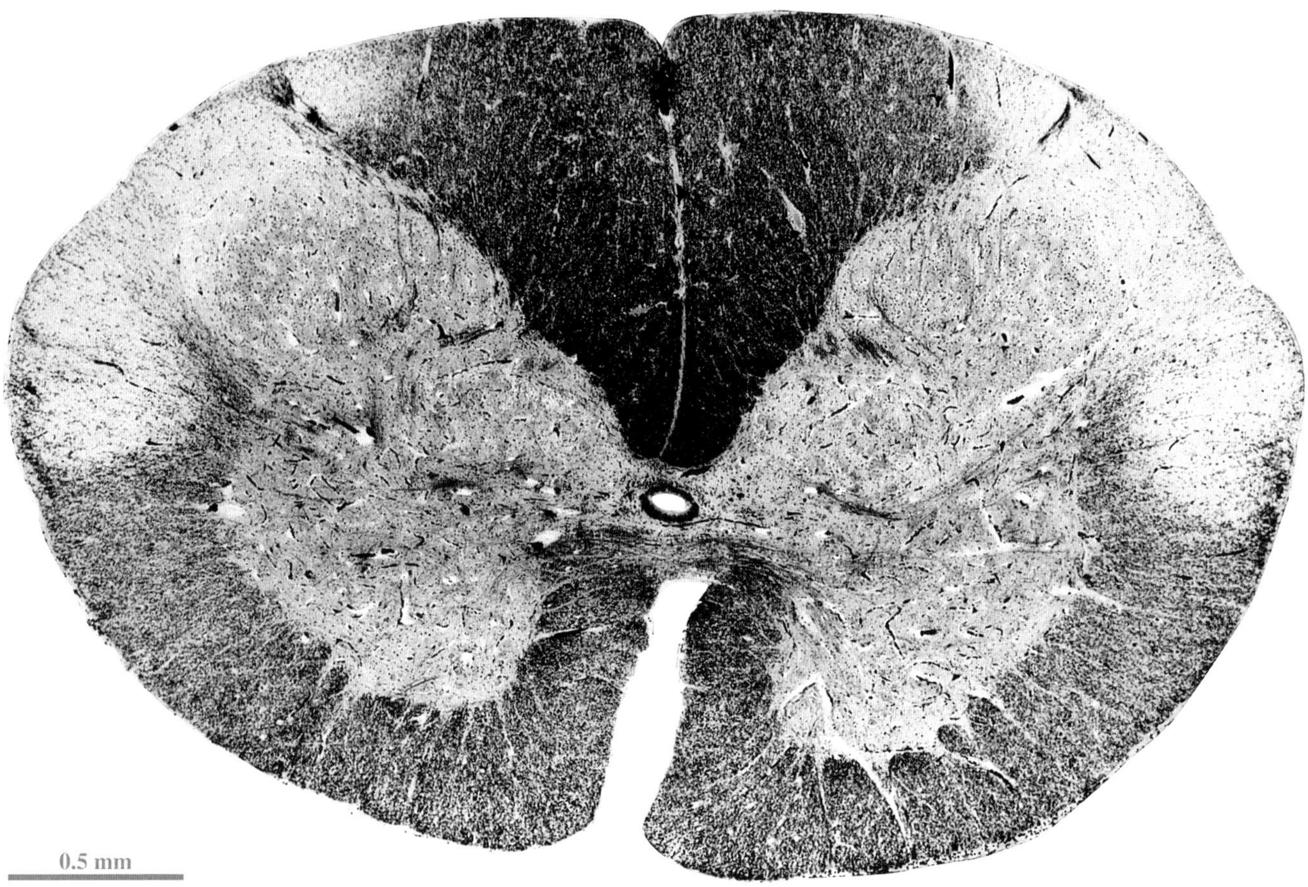

PLATE 56B

See the matched cell body stained section in Plates 57A and B

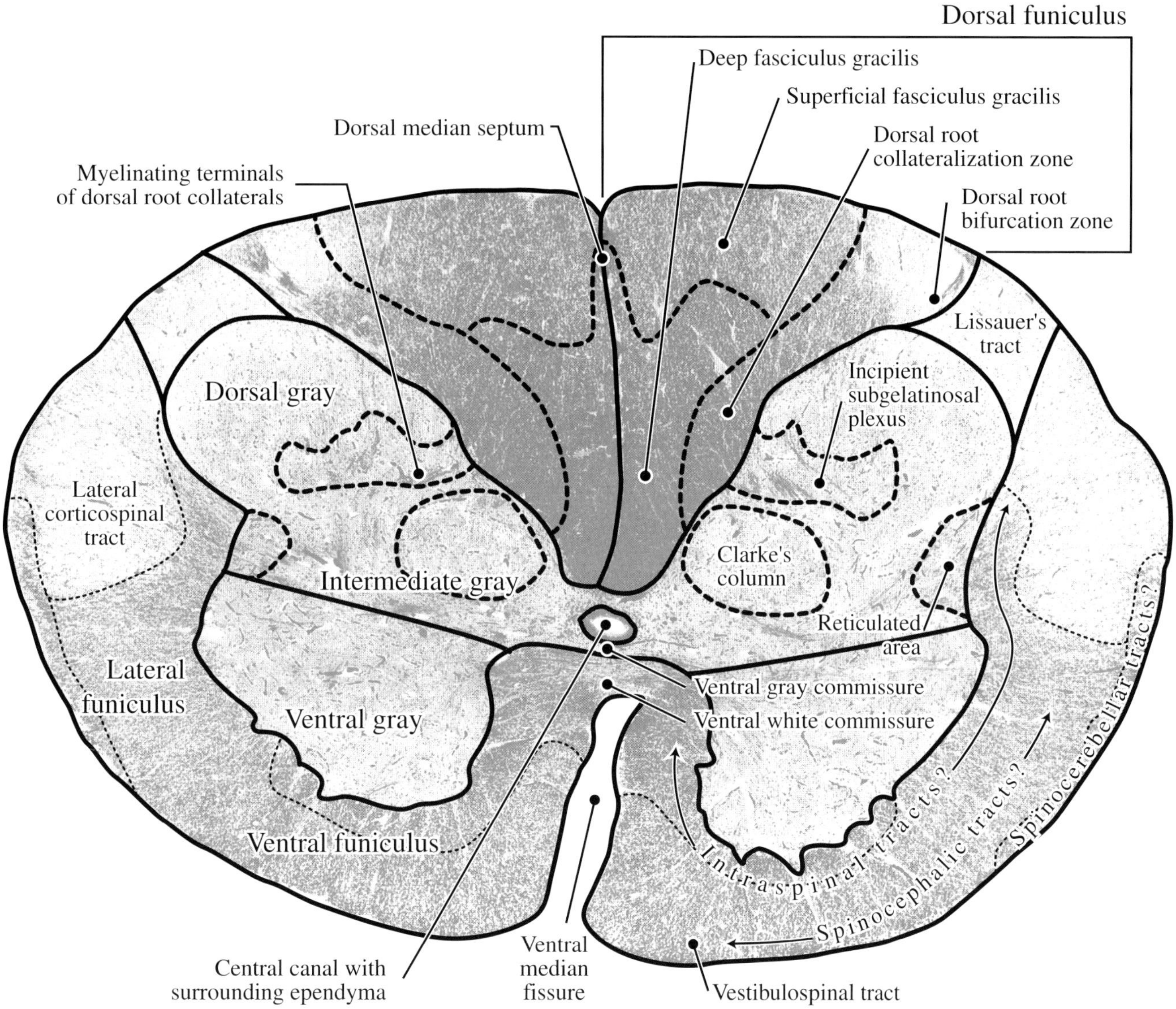

The increased density of the myelin stain that was interpreted to be the rubrospinal and lateral reticulospinal tracts in **Plate 54** cannot be detected in this section, so these fiber tracts are not labeled. However, the medial and central fibers in the intraspinal tracts have a greater staining density than the dorsolateral parts. That same gradient is seen in the thoracic cord of the GW26 specimen (**Plate 51**).

PLATE 57A

GW31
CR 270 mm
Y162-61
Upper Lumbar
Cell body stain

Areas (mm^2)

Central canal	.0050
Ependyma	.0135
Gray matter	2.7017
White matter	4.4178

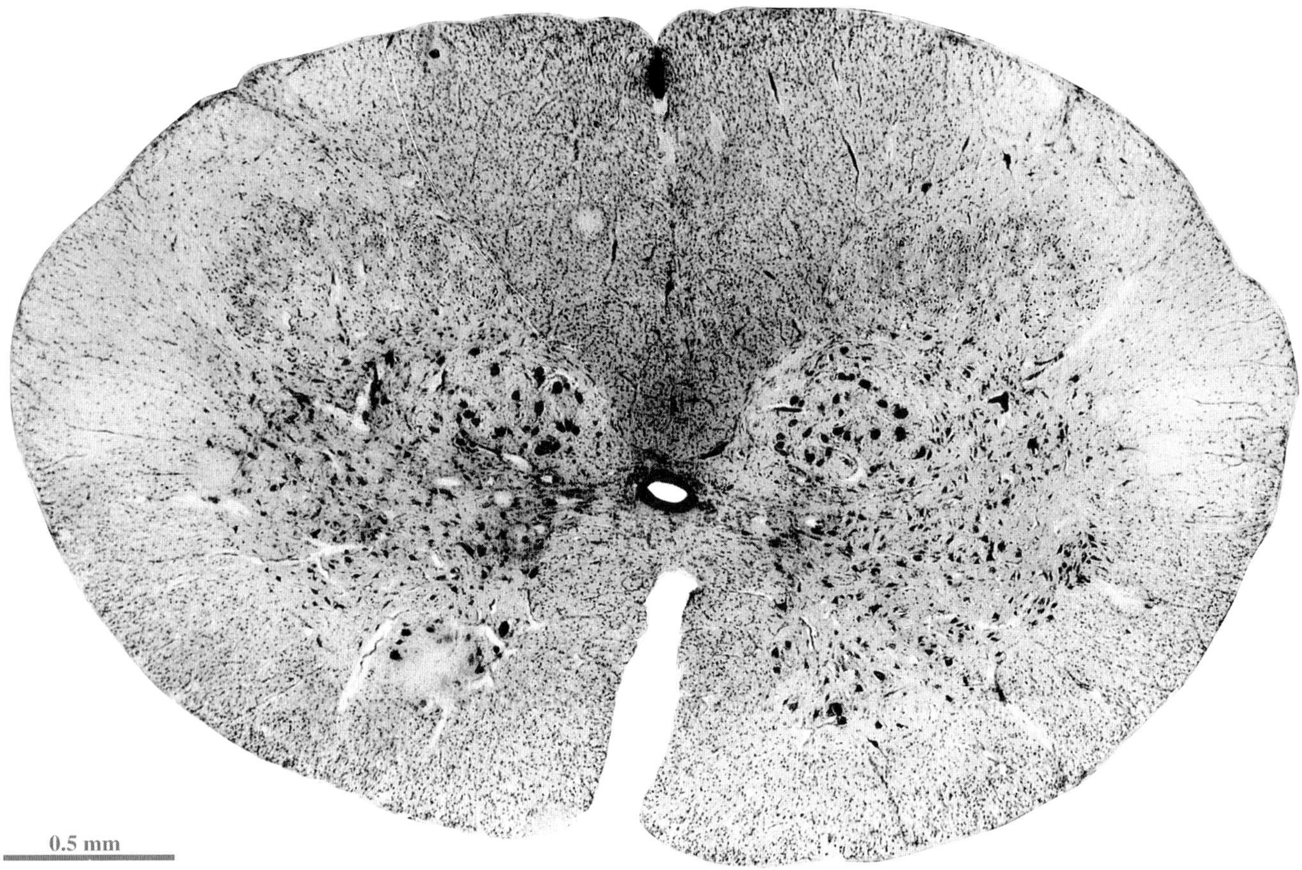

0.5 mm

PLATE 57B

See the matched myelin stained section in Plates 56A and B

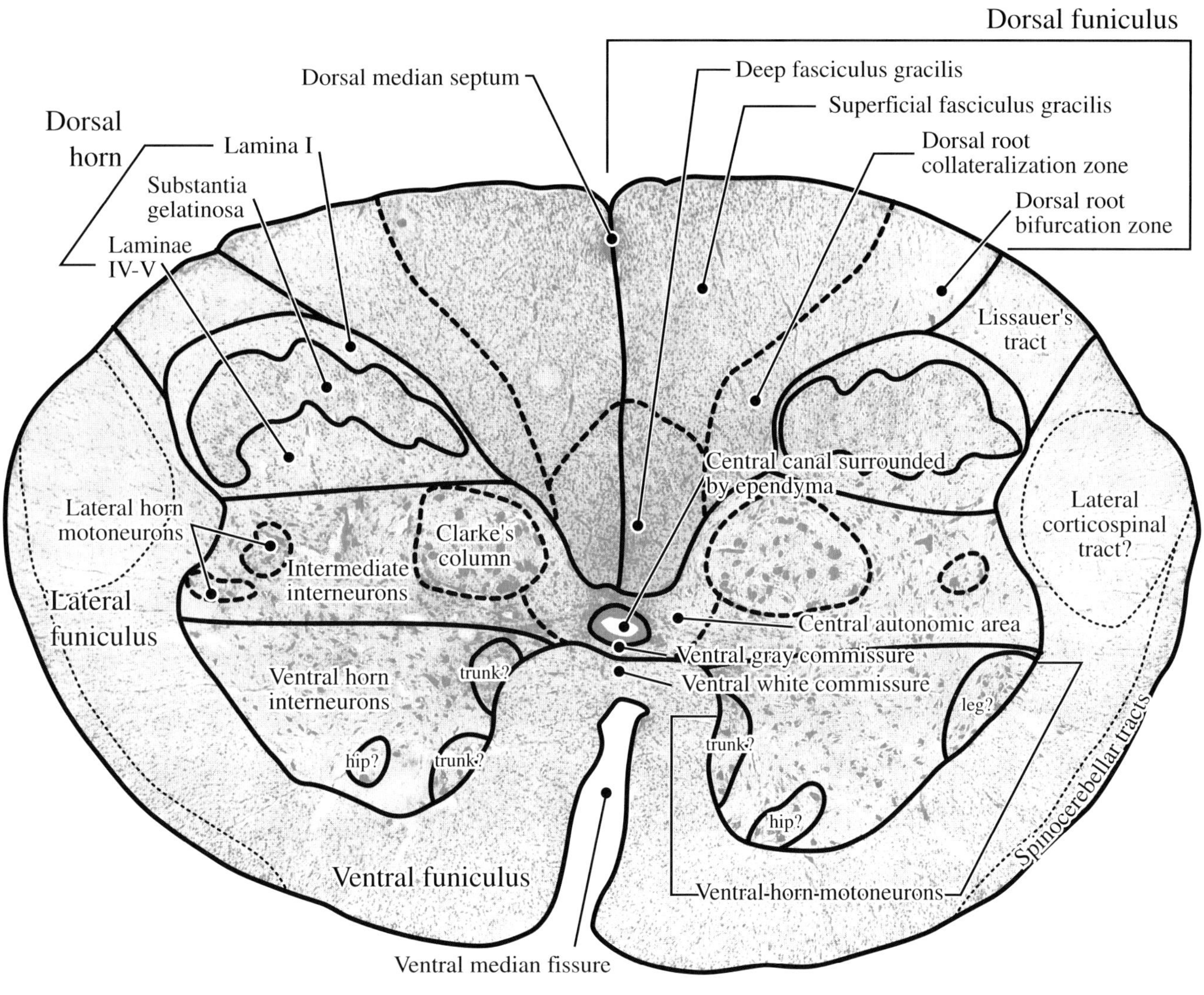

Refer to the matching myelin stained section for the approximate locations of the vestibulospinal, intraspinal, and spinocephalic tracts.

PLATE 58A

GW31
CR 270 mm
Y162-61
Lumbar Enlargement
Myelin stain

Areas (mm^2)

Central canal	.0030
Ependyma	.0085
Gray matter	4.9128
White matter	4.8309

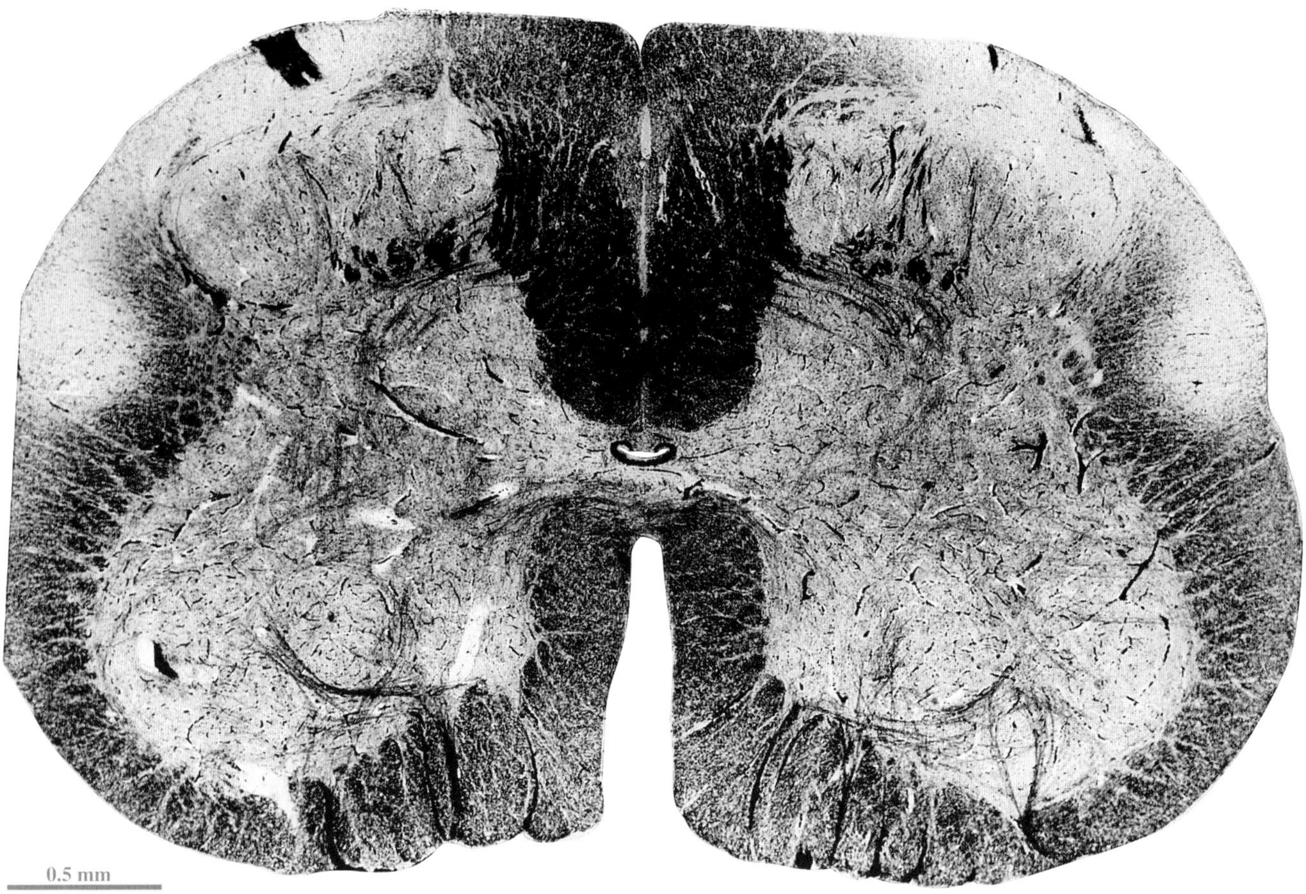

PLATE 58B

See the matched cell body stained section in Plates 59A and B

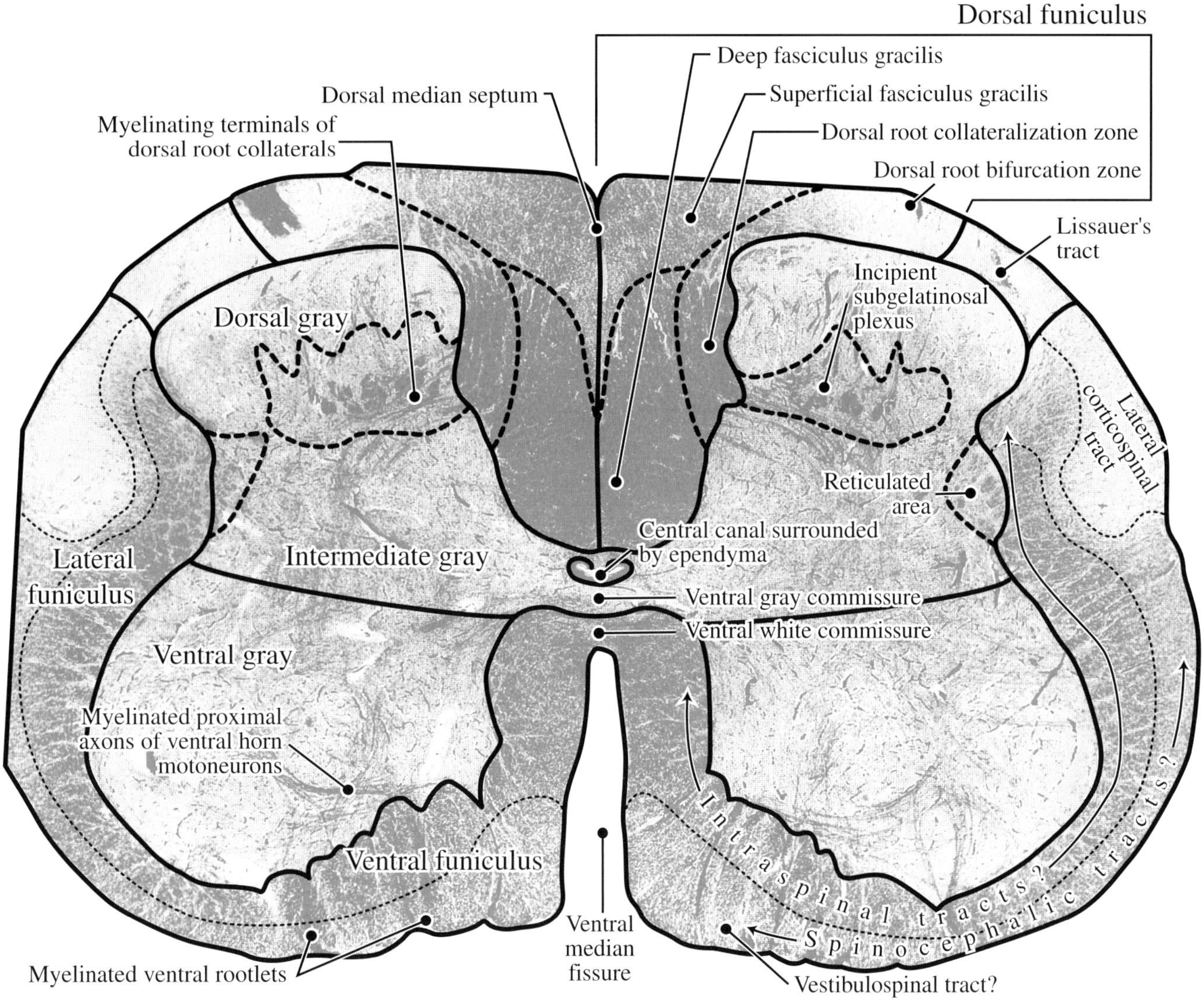

PLATE 59A

GW31
CR 270 mm
Y162-61
Lumbar Enlargement
Cell body stain

Areas (mm^2)

Central canal	.0028
Ependyma	.0154
Gray matter	5.7639
White matter	5.1354

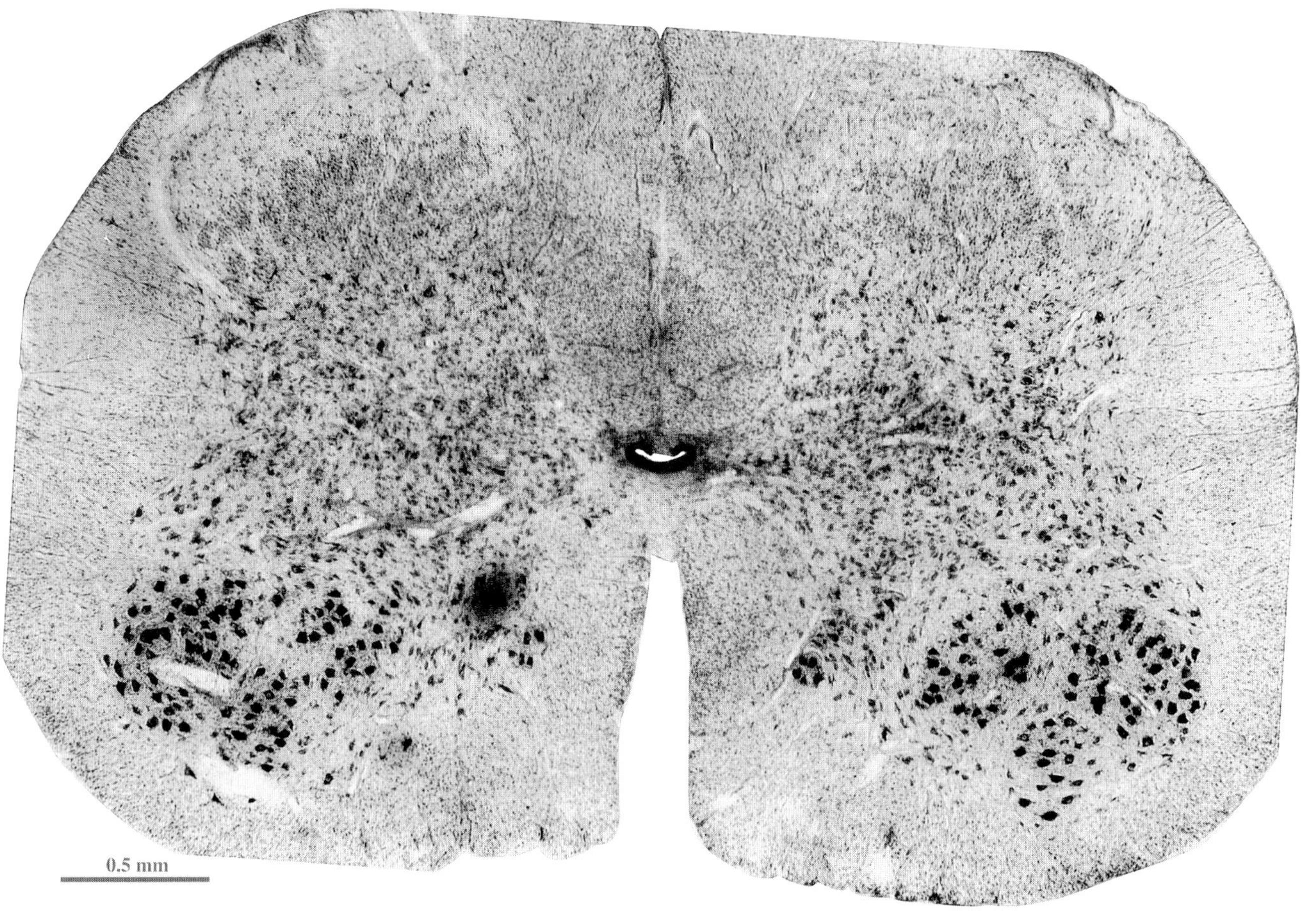

PLATE 59B

See the matched myelin stained section in Plates 58A and B

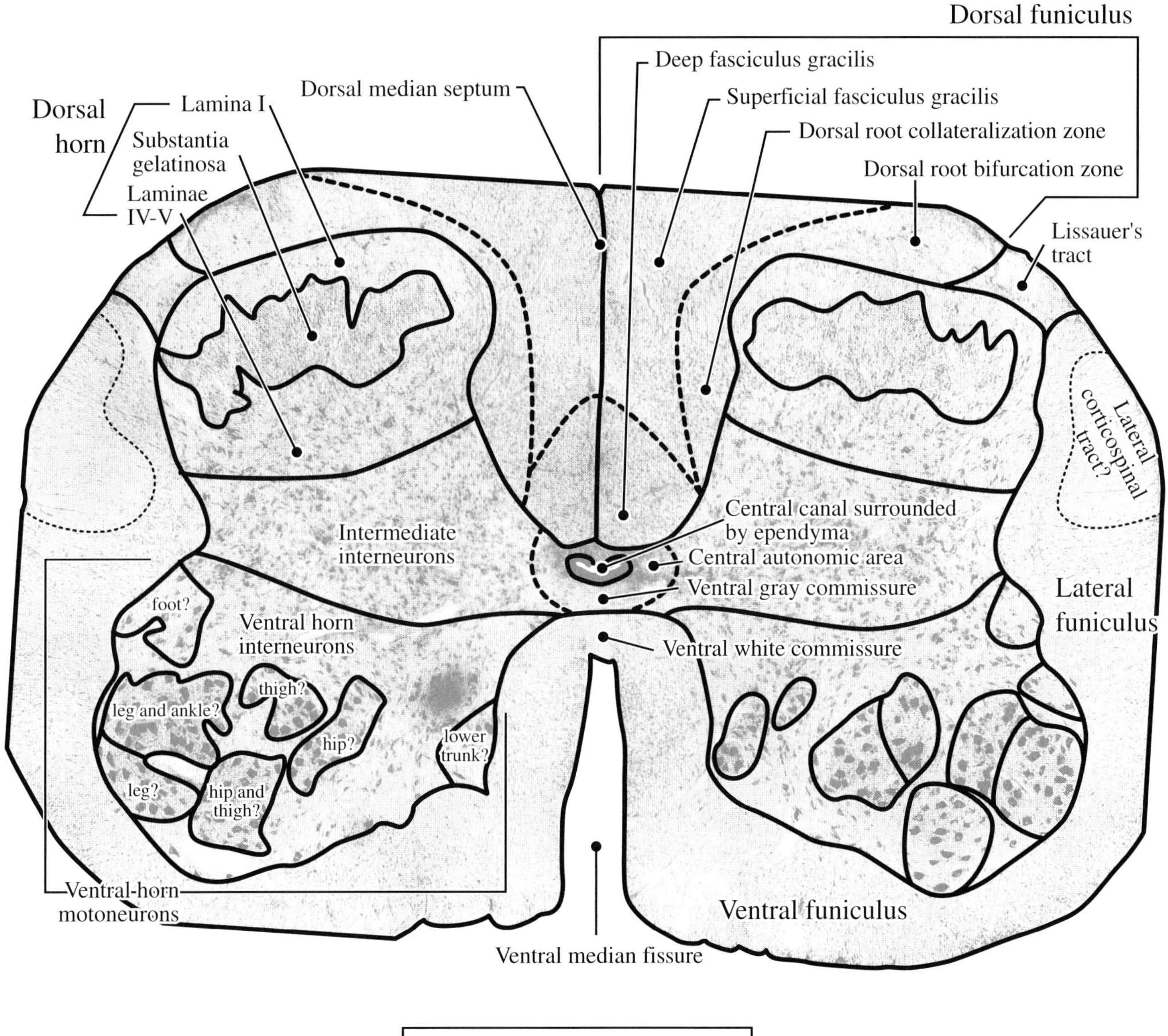

Refer to the matching myelin stained section for the approximate locations of the vestibulo-spinal, intraspinal, and spinocephalic tracts.

PLATE 60A

GW31
CR 270 mm
Y162-61
Sacral/Coccygeal
Myelin stain

Areas (mm^2)

Central canal	.0053
Ependyma	.0184
Gray matter	2.1433
White matter	1.3482

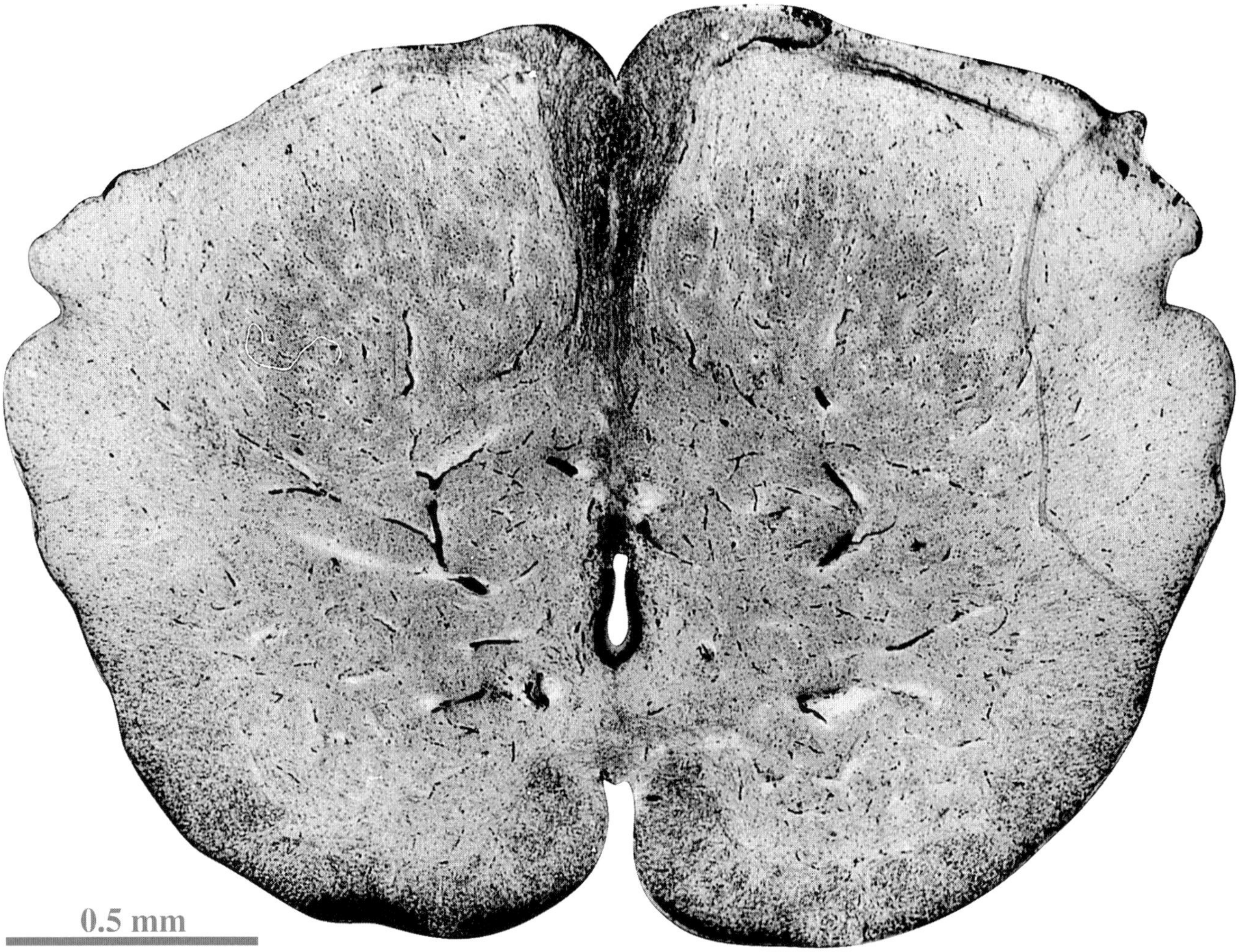

PLATE 60B

See the matched cell body stained section in Plates 61A and B

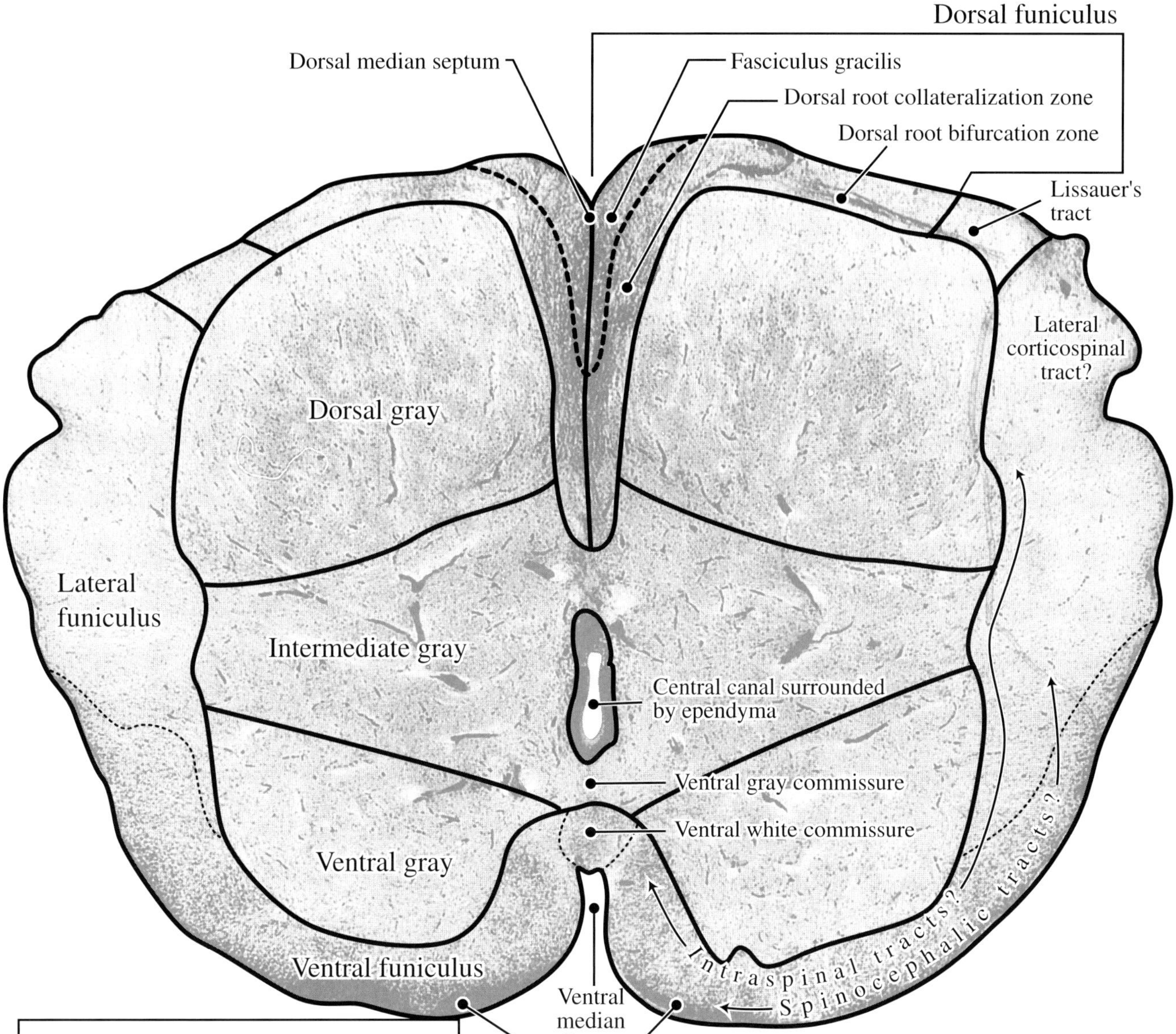

The border of the lateral corticospinal tract is no longer detectable in this section because the dorsal half of the lateral funiculus has very sparse reactive glia. The fiber tracts at this level of the spinal cord are more immature than the three rostral levels shown in this specimen. Thus, the overall rostral (first) to caudal (last) gradient of maturation is evident in this third trimester specimen.

PLATE 61A

GW31
CR 270 mm
Y162-61
Sacral/Coccygeal
Cell body stain

Areas (mm²)

Central canal	.0046
Ependyma	.0253
Gray matter	2.2183
White matter	1.3268

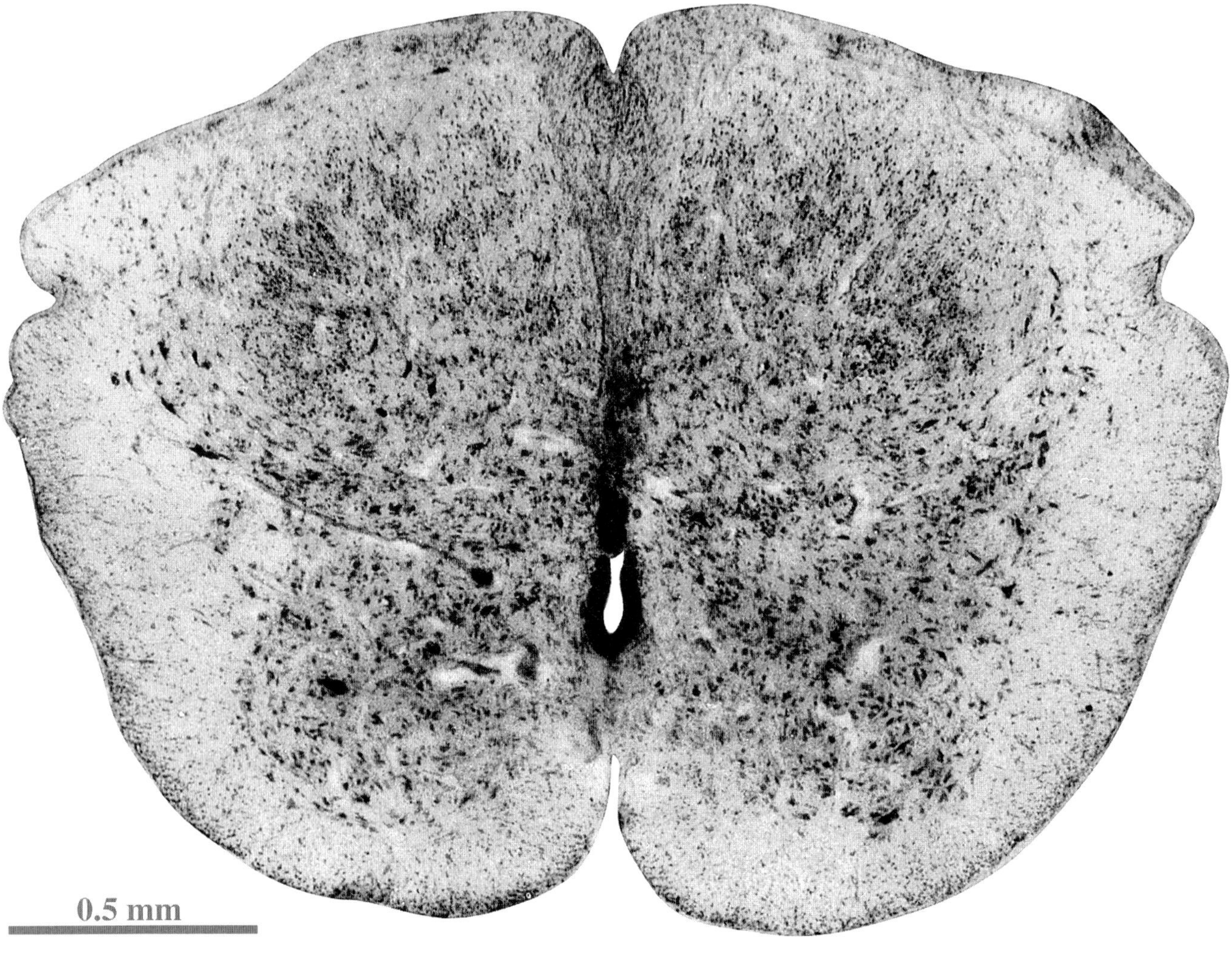

PLATE 61B

See the matched myelin stained section in Plates 60A and B

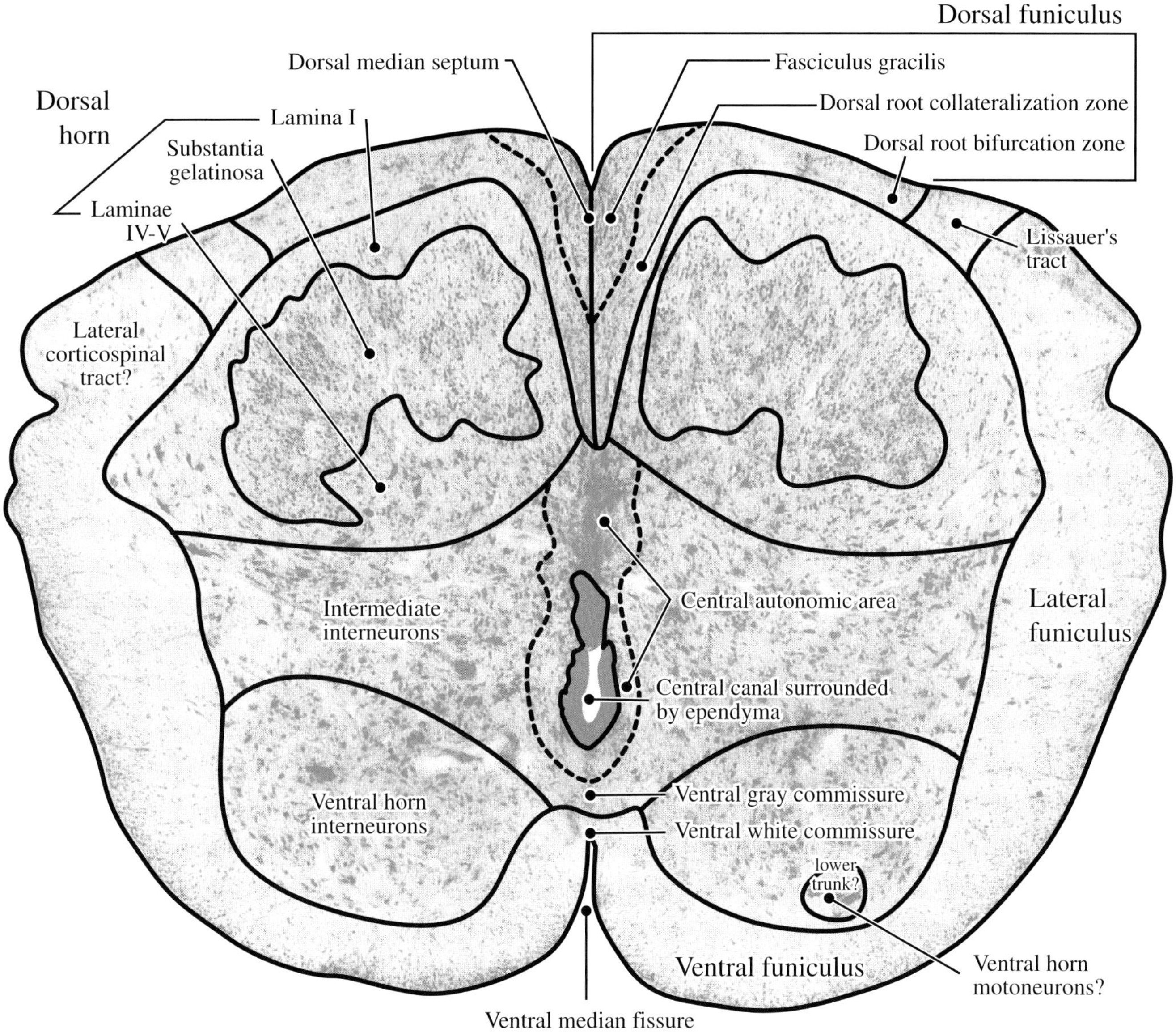

Refer to the matching myelin stained section for the approximate locations of the intraspinal and spinocephalic tracts.

B. Matched myelin and cell body stained sections in the spinal cord of a GW37 fetus

Plate 62 is a survey of matched myelin stained and cell body stained sections from Y117-61, a specimen in the Yakovlev Collection with a crown rump length of 310 mm (see Chapter 6 in Altman and Bayer, 2001). All sections are shown at the same scale. The boxes enclosing each section list the approximate level and the total area (post fixation) of the section in square millimeters (mm^2). Full-page normal contrast photographs of each specimen are in **Plates 63A-78A.** Low contrast photographs with superimposed labels and outlines of structural details are in **Plates 63B-78B.** In this specimen, the myelin stained and cell body stained sections were preserved on separate large glass plates without any section numbers. This specimen has the most complete set of spinal cord sections in the Yakovlev Collection with all levels preserved. Twenty myelin stained sections and 19 cell body stained sections were photographed ranging from upper cervical to sacral/coccygeal levels. The 39 photographic prints were intuitively arranged in order from upper cervical to sacral/coccygeal levels, using internal features such as the size of the corticospinal tracts and the width of the ventral horn. Then, myelin and cell body stained sections were matched and eight different levels were analyzed.

As in the previous specimens, the cross-sectional area of a myelin stained section is smaller than the matching cell body stained section in all cases, except at the upper thoracic level, where the areas are the same. Using the total areas of the myelin stained sections, the overall size differences between levels indicate the following: the cervical enlargement has the largest area, being larger than the lumbar enlargement by 30%. The middle thoracic level has the smallest area and is 6% smaller than the sacral/coccygeal level.

Myelination continues to advance in this specimen (see *columns 2 and 3* in **Table IV B-1**). Dense staining indicative of either advanced or beginning myelination is seen throughout the ventral funiculus and the lateral funiculus except the ventral and lateral corticospinal tracts, the sacral/coccygeal level excepted. All parts of the dorsal funiculus are either myelinated or are myelinating, except Lissauer's tract and the areas in the dorsal root bifurcation zone which most likely contains incoming axons that will join Lissauer's tract. The reticulated area is more prominently myelinated in the lateral intermediate gray and lateral neck region of the dorsal gray. The reactive glia in the subgelatinosal plexus is more concentrated, indicating that terminal parts of the dorsal root collaterals are approaching myelination. It is interesting to note that the white matter in the cell body stained sections (except the sacral/coccygeal level, **Plate 78**) show a uniformly reduced density of glia, suggesting the decline or cessation of myelination gliosis (see *column 4* in **Table IV B-1**). But again, the lateral and ventral corticospinal tracts generally stand out as having slightly lower concentrations of glia; in these tracts, myelination gliosis has not yet begun.

Table IV B-1: Glia types and concentration in the white matter at GW37

Name	Myelination	Reactive glia	Proliferating glia
DORSAL ROOT	Myelinated	---	Dense
VENTRAL ROOT	Myelinated	---	---
DORSAL FUNICULUS:			
dorsal root bif. zone	Many fibers*	---	Sparse
dorsal root col. zone	Myelinated	---	Sparse†
deep fas. gracilis	Myelinated	---	Sparse
superficial fas. gracilis	Many fibers	---	Sparse
deep fas. cuneatus	Myelinated	---	Sparse
superficial fas. cuneatus	Many fibers	---	Sparse
Lissauer's tract	---	None	Sparse
LATERAL and VENTRAL FUNICULI:			
lat. cortricospinal tract	---	Very sparse	Very sparse
ven. cortricospinal tract	---	Very sparse	Very sparse
lat. reticulospinal tract	Many fibers	---	Sparse
spinocerebellar tracts	Many fibers	---	Sparse
ven. commissure	Myelinated	---	Sparse
intraspinal tract	Many fibers	---	Sparse
spinocephalic tract	Some fibers	Very dense	Sparse
med. long. fasiculus	Many fibers	---	Sparse
vestibulospinal tract	---	Dense	Sparse

* intermingled in a bed of nonreactive glia (associated with Lissauer's tract fibers?)
† dense at the sacral/coccygeal level

PLATE 62

GW37, CR 310 mm, Y117-61

MYELIN STAIN **CELL BODY STAIN**

Plates 63A, 63B
Upper Cervical
Total area:
15.138 mm^2

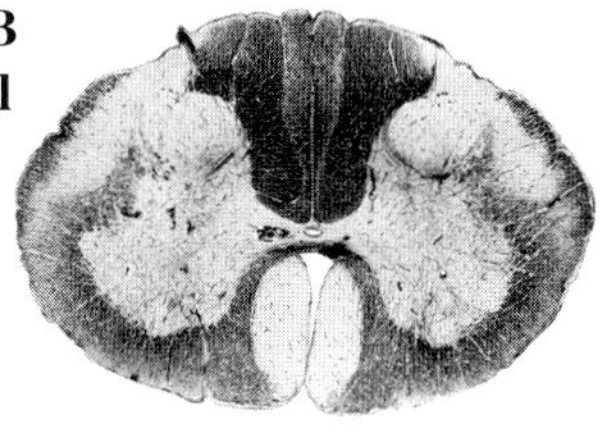

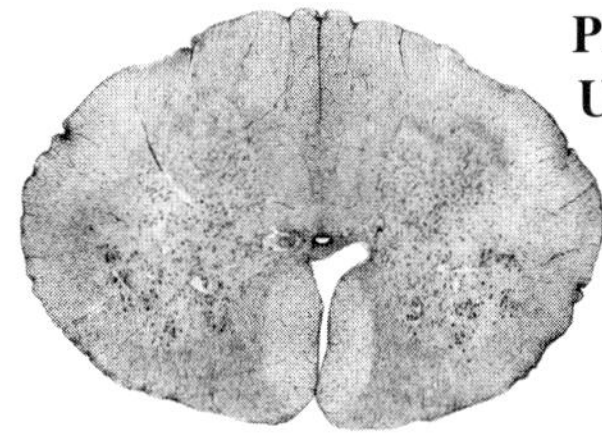

Plates 64A, 64B
Upper Cervical
Total area:
15.896 mm^2

Plates 65A, 65B
Cervical Enlargement
Total area:
16.046 mm^2

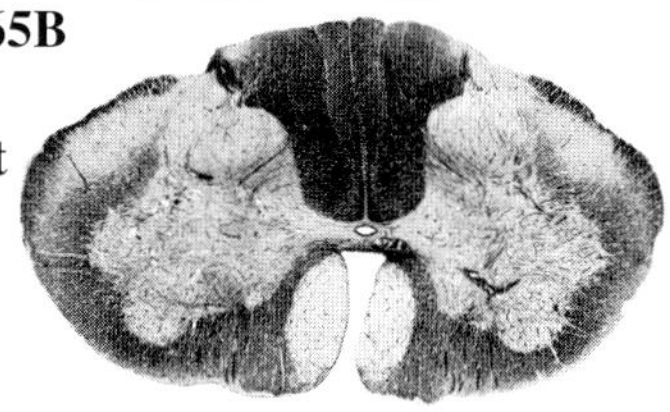

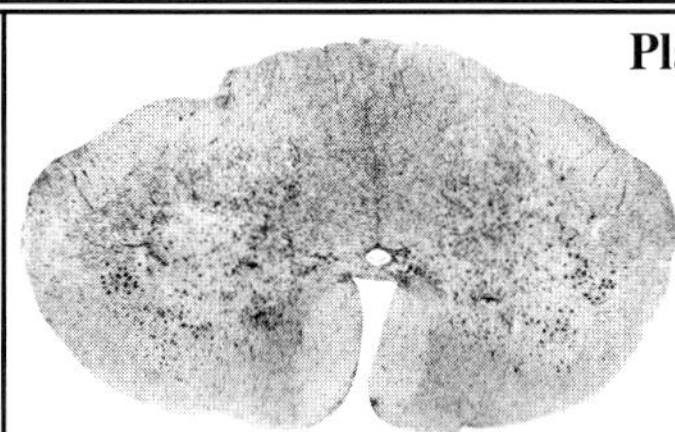

Plates 66A, 66B
Cervical Enlargement
Total area:
17.036 mm^2

Plates 67A, 67B
Upper Thoracic
Total area:
8.4267 mm^2

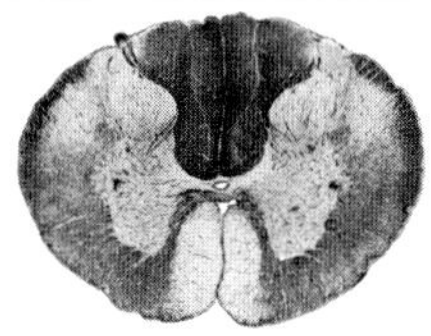

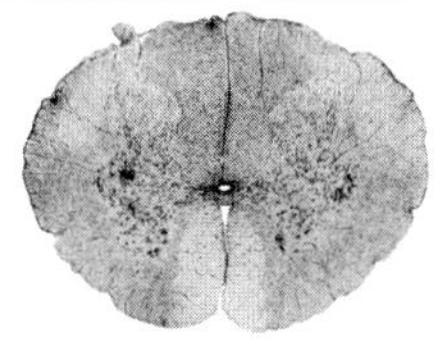

Plates 68A, 68B
Upper Thoracic
Total area:
8.4266 mm^2

Plates 69A, 69B
Middle Thoracic
Total area:
6.1539 mm^2

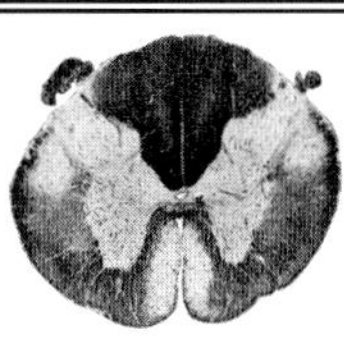

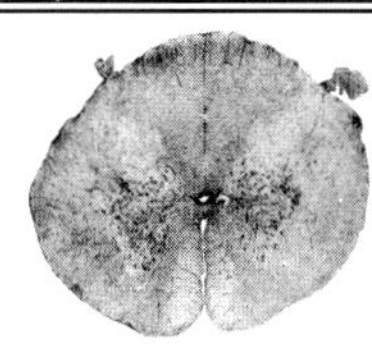

Plates 70A, 70B
Middle Thoracic
Total area:
6.7020 mm^2

Plates 71A, 71B
Lower Thoracic
Total area:
7.2005 mm^2

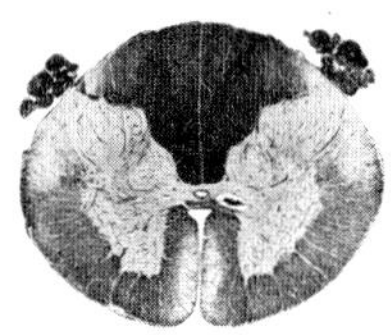

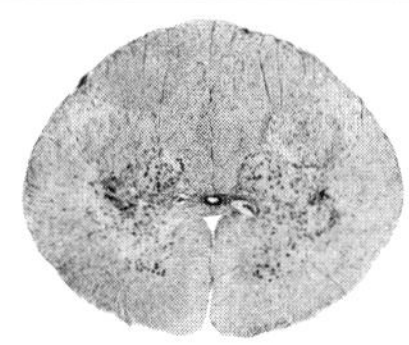

Plates 72A, 72B
Lower Thoracic
Total area:
7.9554 mm^2

Plates 73A, 73B
Upper Lumbar
Total area:
11.530 mm^2

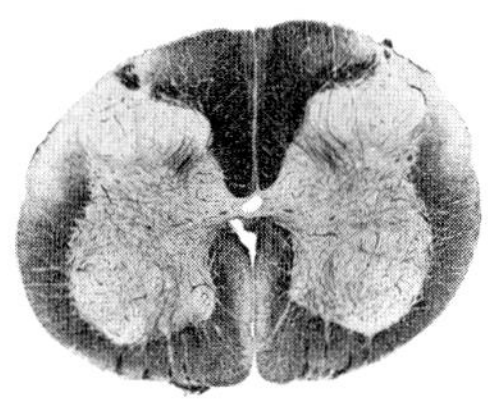

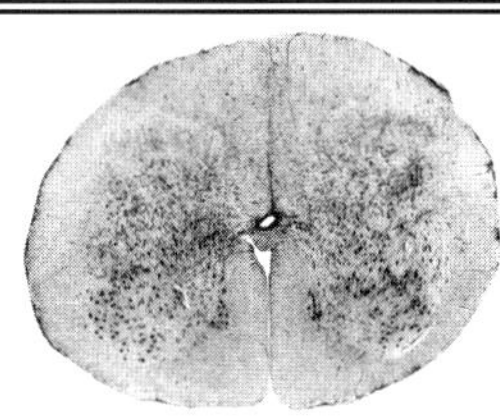

Plates 74A, 74B
Upper Lumbar
Total area:
12.297 mm^2

Plates 75A, 75B
Lumbar Enlargement
Total area:
12.297 mm^2

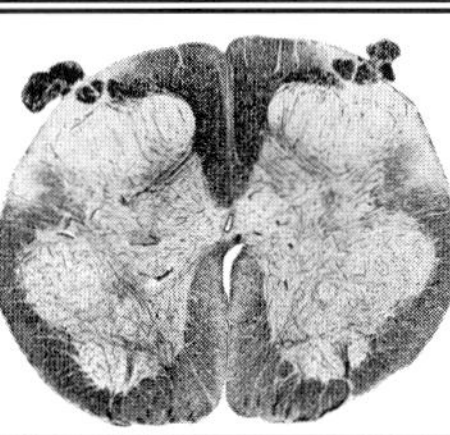

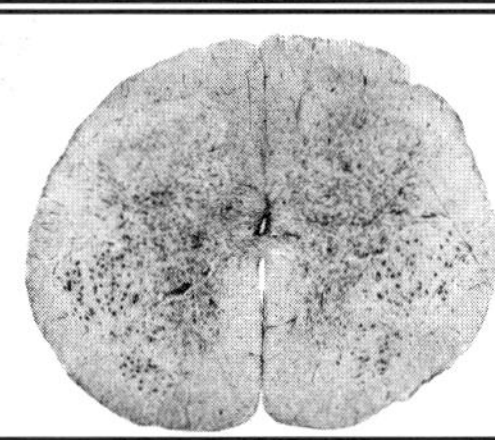

Plates 76A, 76B
Lumbar Enlargement
Total area:
12.872 mm^2

Plates 77A, 77B
Sacral/Coccygeal
Total area:
6.5068 mm^2

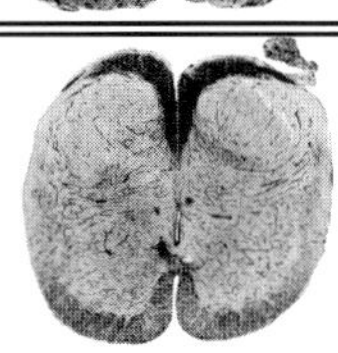

0.75 mm

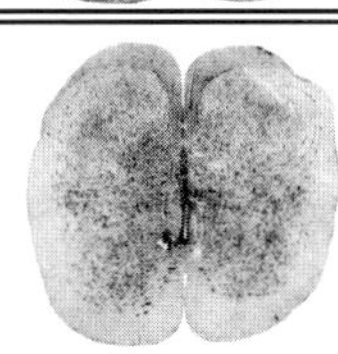

0.75 mm

Plates 78A, 78B
Sacral/Coccygeal
Total area:
6.7829 mm^2

PLATE 63A

GW37
CR 310 mm
Y117-61
Upper Cervical
Myelin stain

Areas (mm^2)

Central canal	.0044
Ependyma	.0102
Gray matter	5.2519
White matter	9.8713

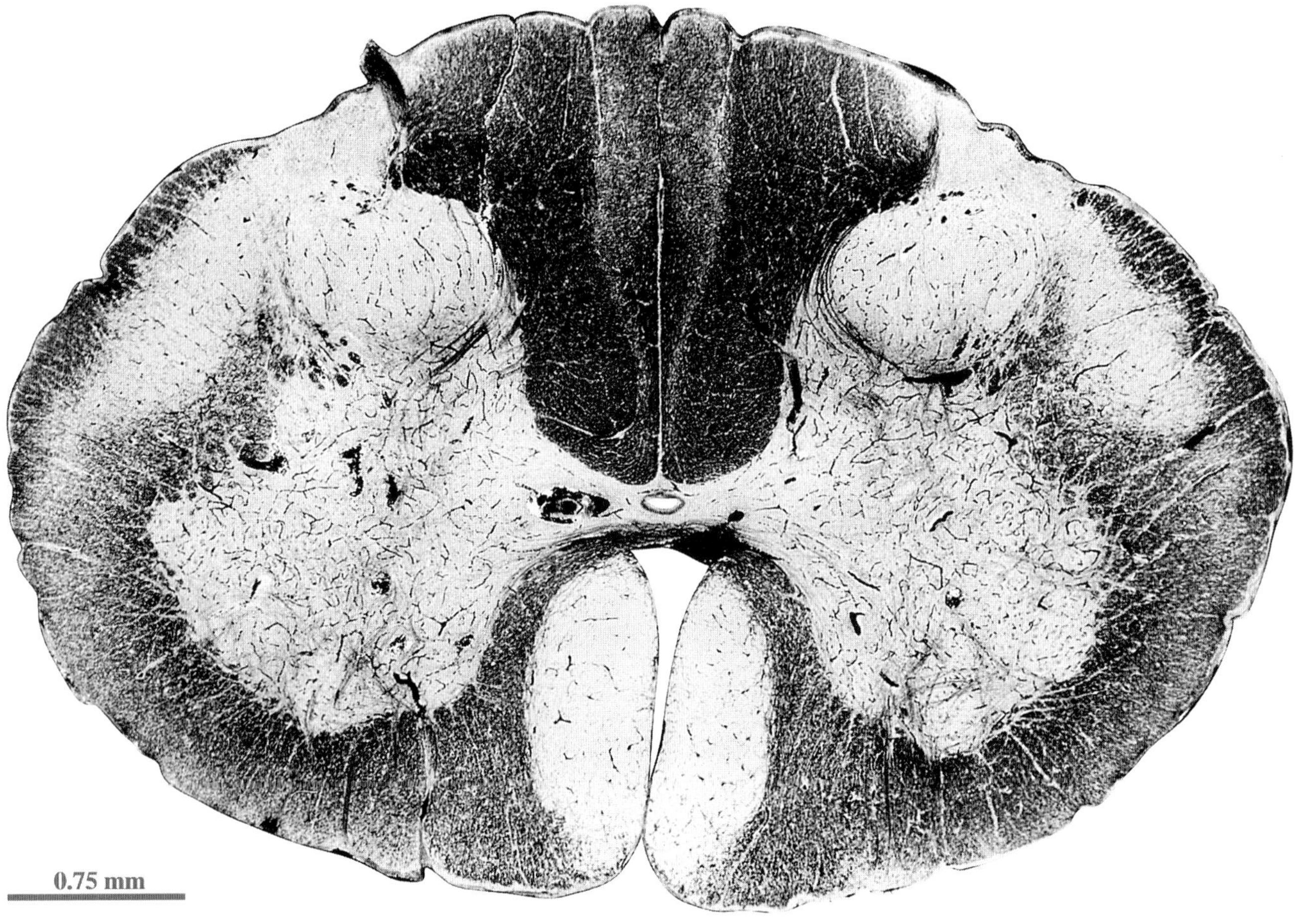

PLATE 63B

See the matched cell body stained section in Plates 64A and B

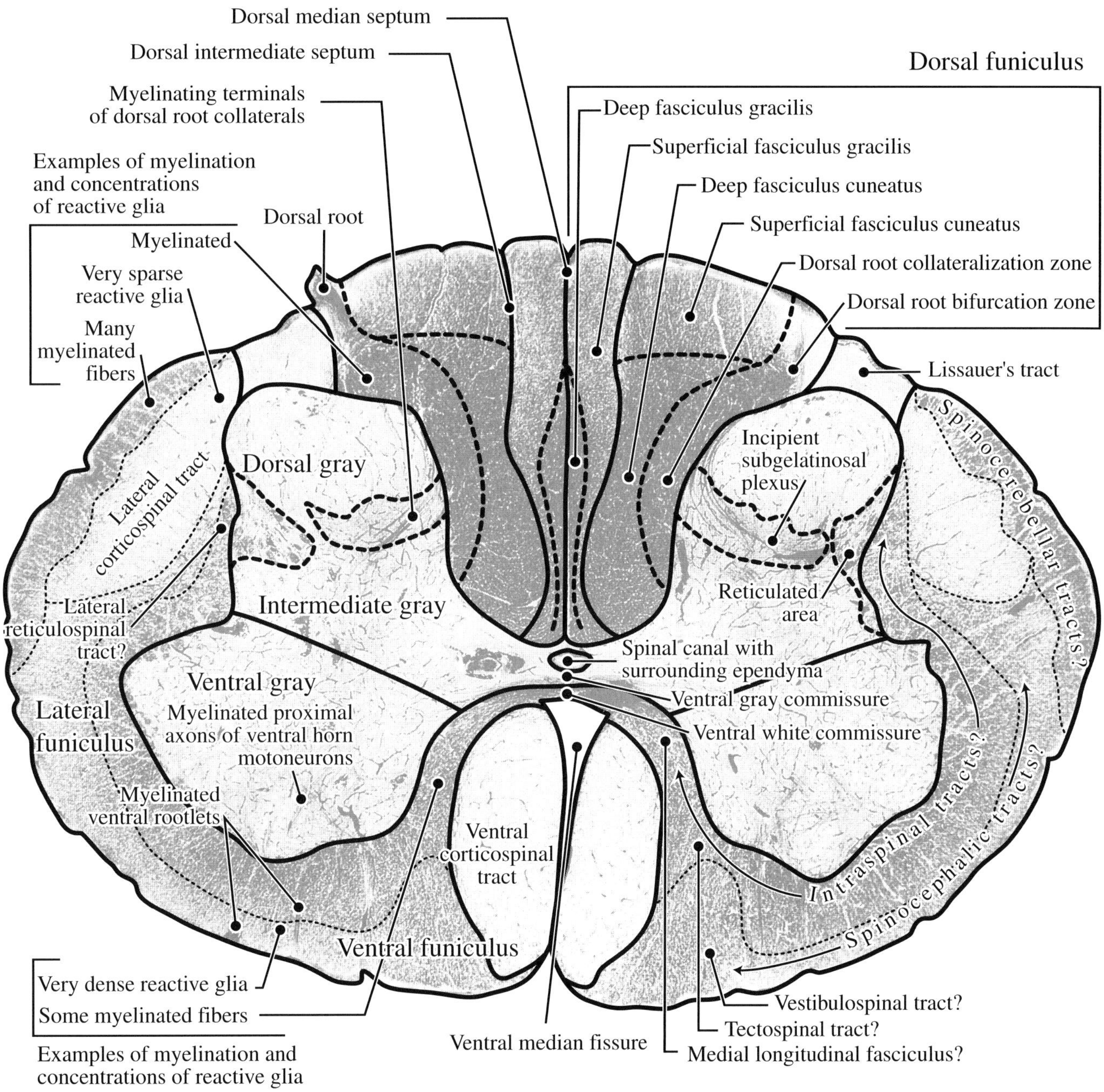

The fine lines in the lateral and ventral funiculi segregate regions of differing densities of the myelin stain, not the borders of fiber tracts. These lines may correspond to the borders of some fiber tracts, such as the spinocerebellar and the lateral corticospinal tracts. Other lines run across fiber tracts, such as the area surrounding the ventral horn that includes the medial and central parts of the intraspinal tracts, and the internal fibers of the spinocephalic tracts. Most tracts in the ventral and lateral funiculi overlap and do not have distinct borders. The ventral part of the spinocephalic tracts overlaps with the vestibulospinal tract; its central part includes fibers in the ventrolateral reticulospinal tract (not labeled), the spinotectal tract (not labeled), and the spino-olivary tract (not labeled). The intraspinal tracts overlap with the medial longitudinal fasciculus, the tectospinal tract, and the lateral reticulospinal tract.

PLATE 64A

GW37
CR 310 mm
Y117-61
Upper Cervical
Cell body stain

Areas (mm^2)

Central canal	.0038
Ependyma	.0175
Gray matter	5.8125
White matter	10.0620

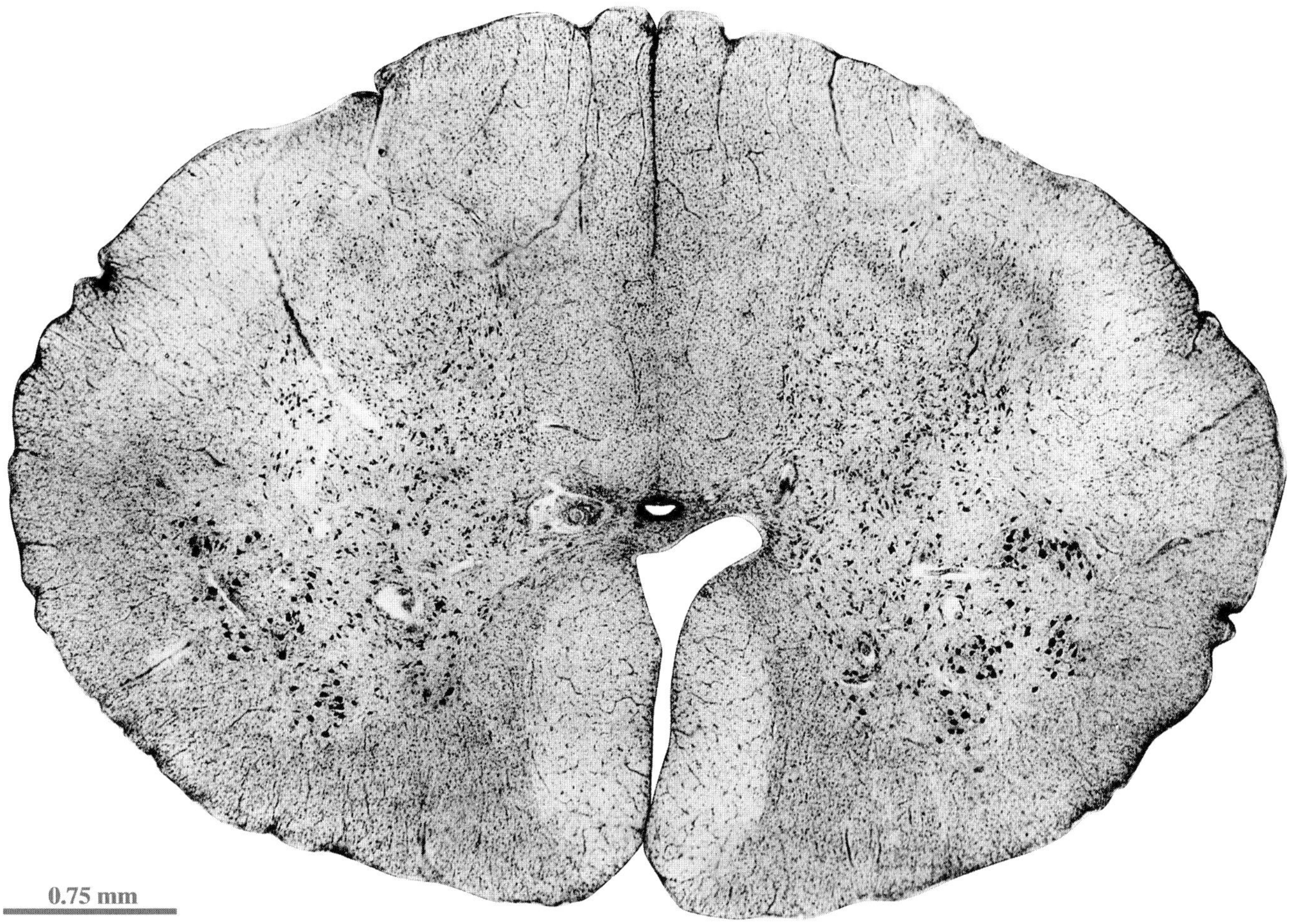

PLATE 64B

See the matched myelin stained section in Plates 63A and B

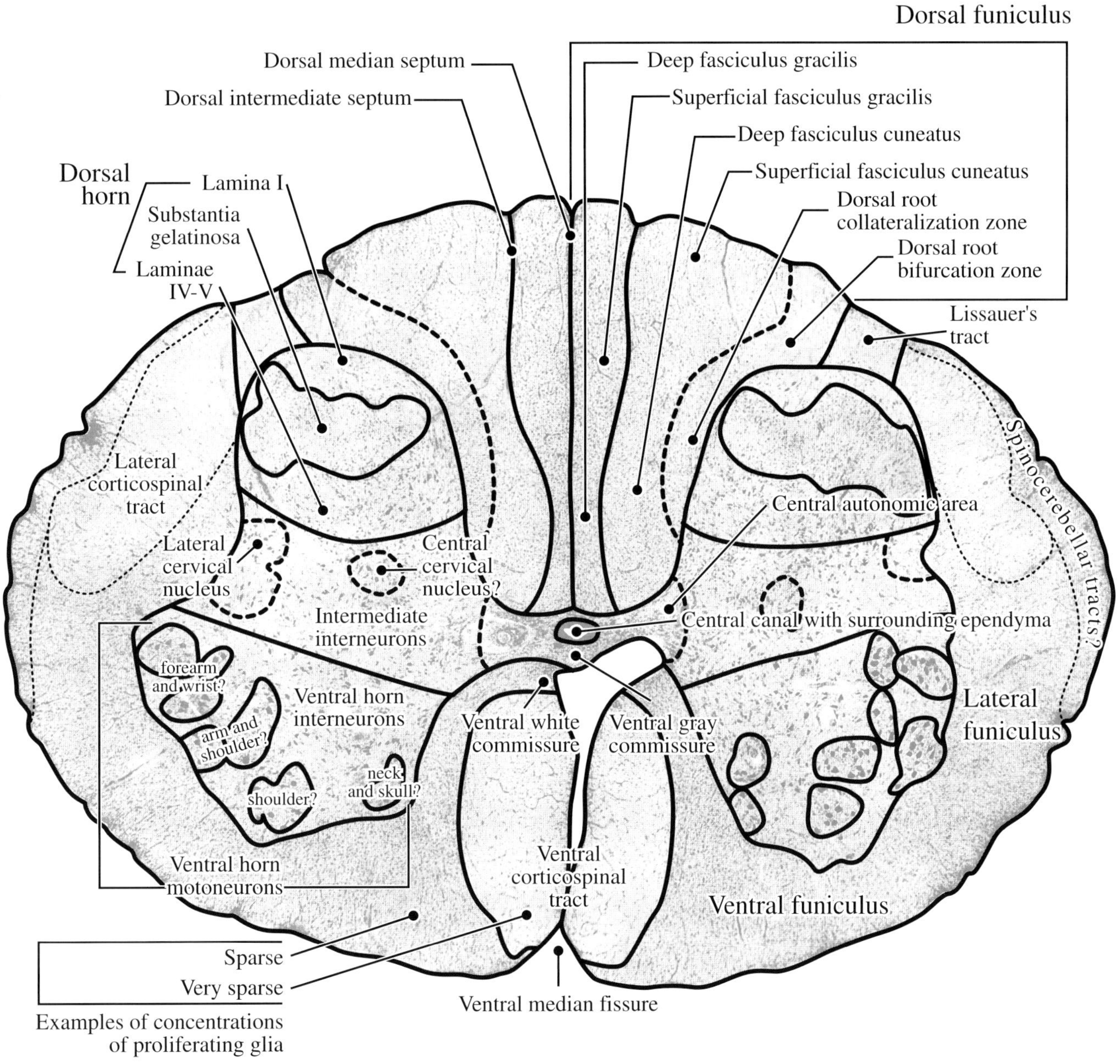

Only the corticospinal and spinocerebellar tracts can be clearly delineated in the ventral and lateral funiculi. The corticospinal tracts stand out as clear areas with very sparse proliferating glia. The spinocerebellar tracts have a slightly more dense concentration of proliferating glia. A sparse population of proliferating glia fills the remaining ventral and lateral funiculi and contains several fiber tracts (unlabeled in this section). Refer to the matching myelin stained section for the approximate locations of the lateral reticulospinal, vestibulospinal, intraspinal, and spinocephalic tracts.

PLATE 65A

GW37
CR 310 mm
Y117-61
Cervical Enlargement
Myelin stain

Areas (mm^2)

Central canal	.0073
Ependyma	.0137
Gray matter	6.0566
White matter	9.9683

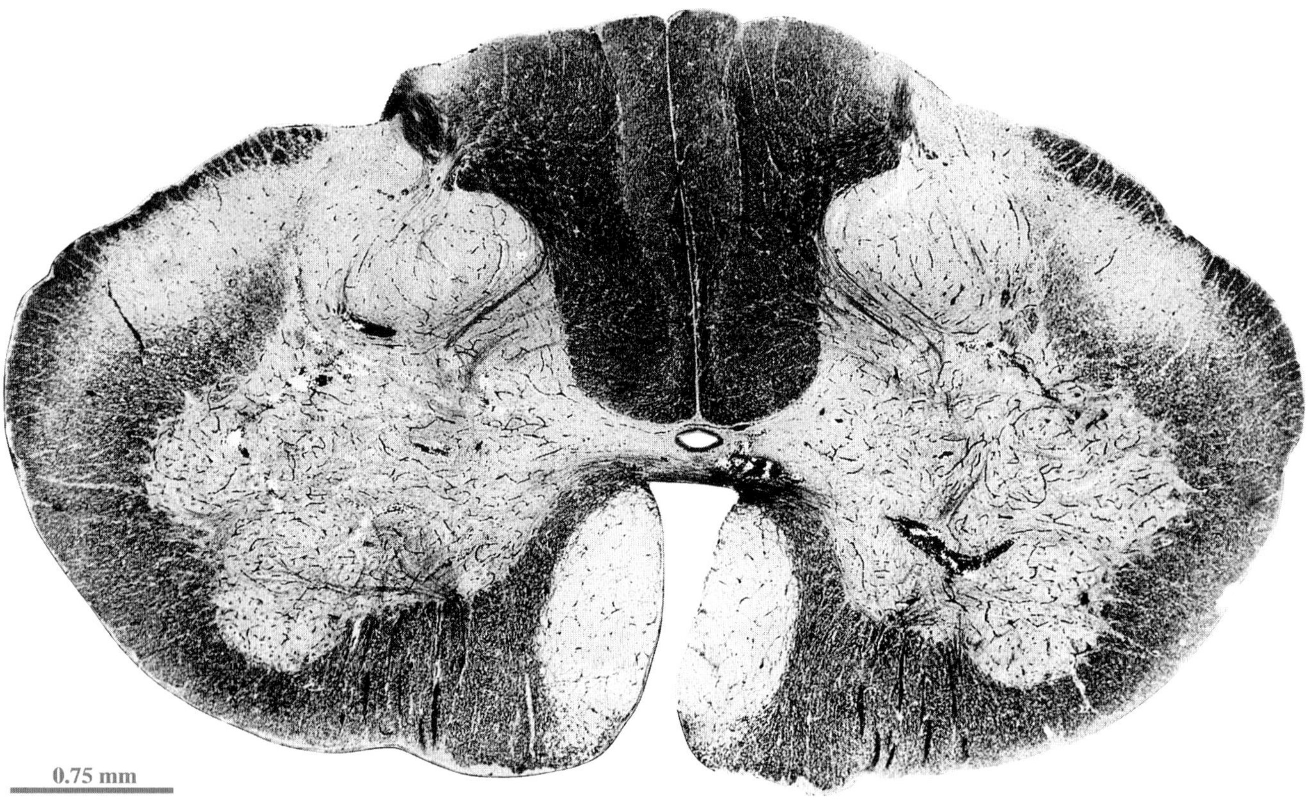

PLATE 65B

See the matched cell body stained section in Plates 66A and B

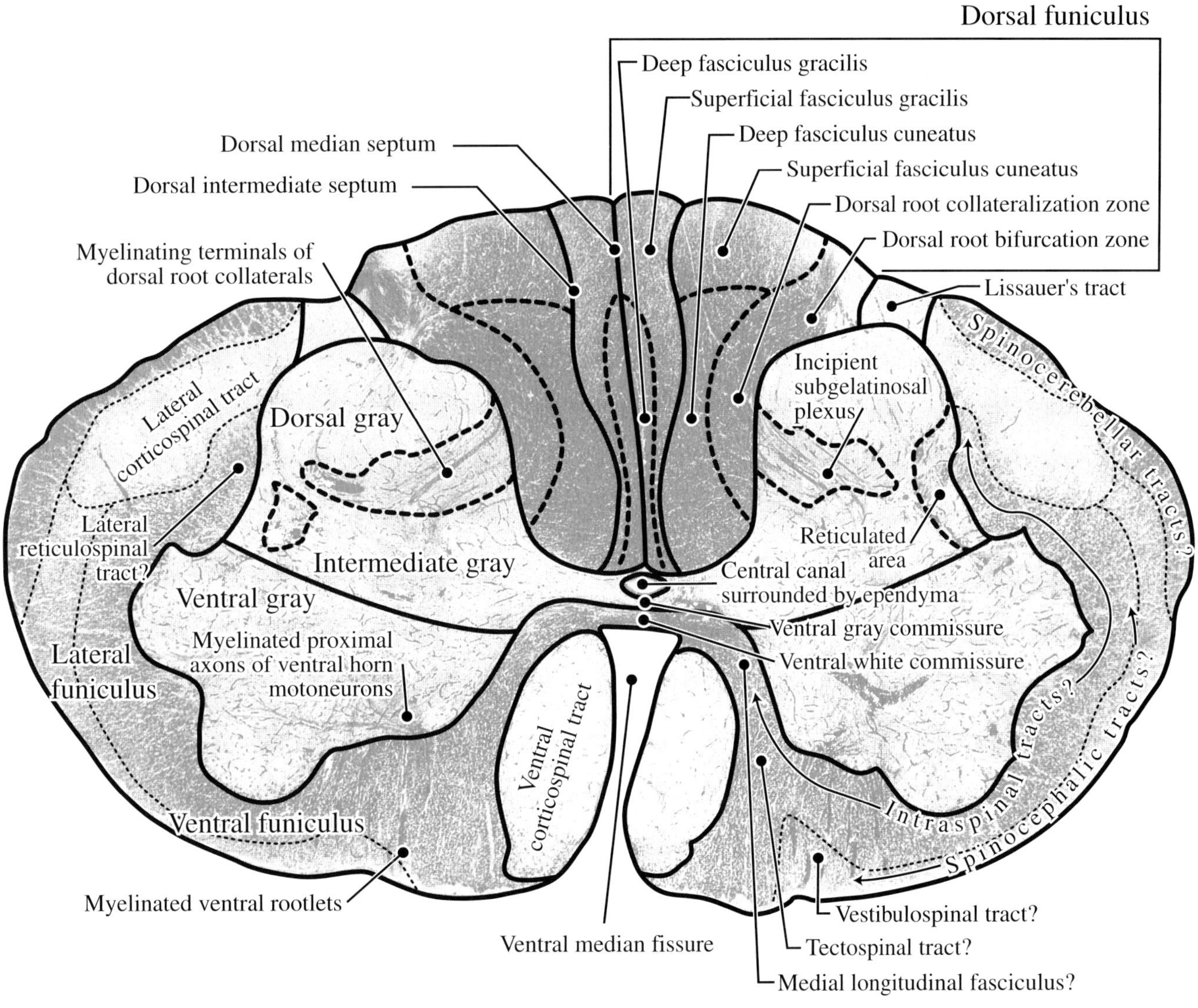

PLATE 66A

GW37
CR 310 mm
Y117-61
Cervical Enlargement
Cell body stain

Areas (mm^2)

Central canal	.0079
Ependyma	.0230
Gray matter	6.6710
White matter	10.3340

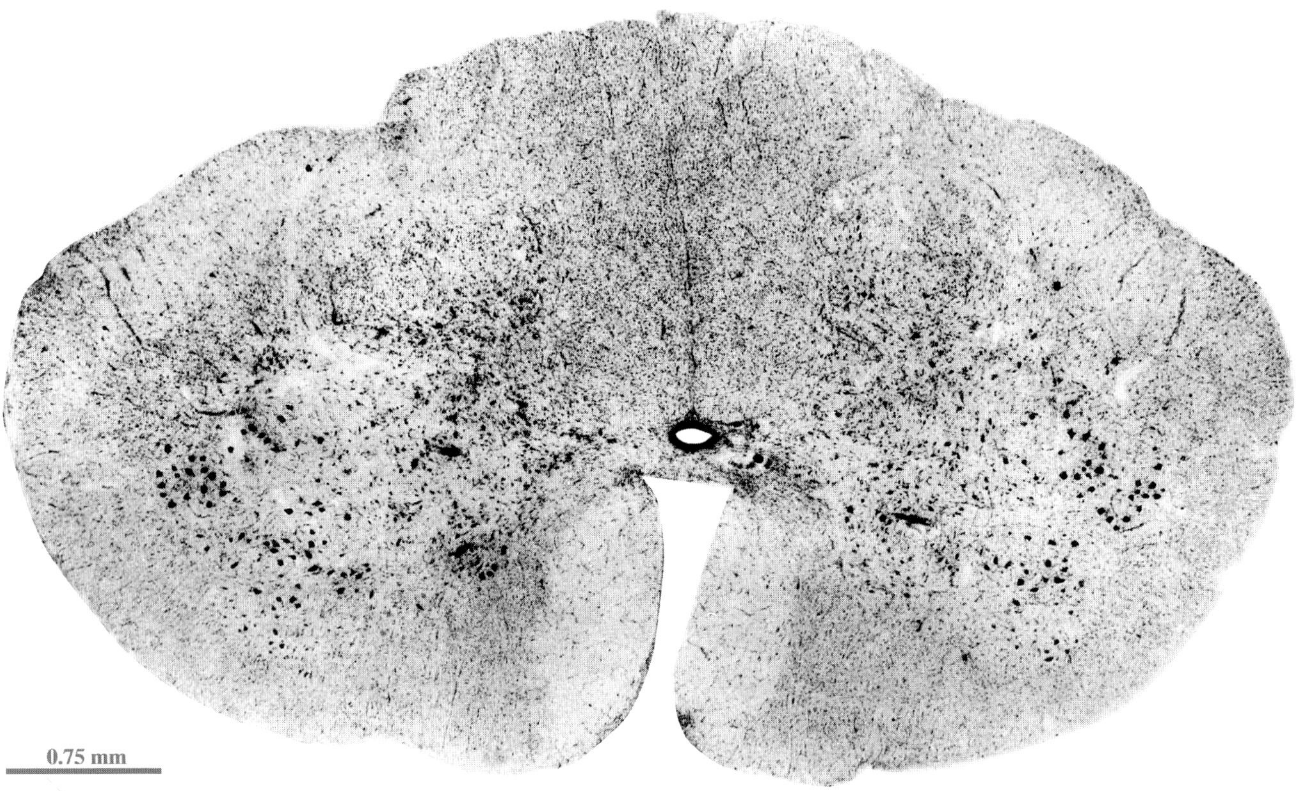

PLATE 66B

See the matched myelin stained section in Plates 65A and B

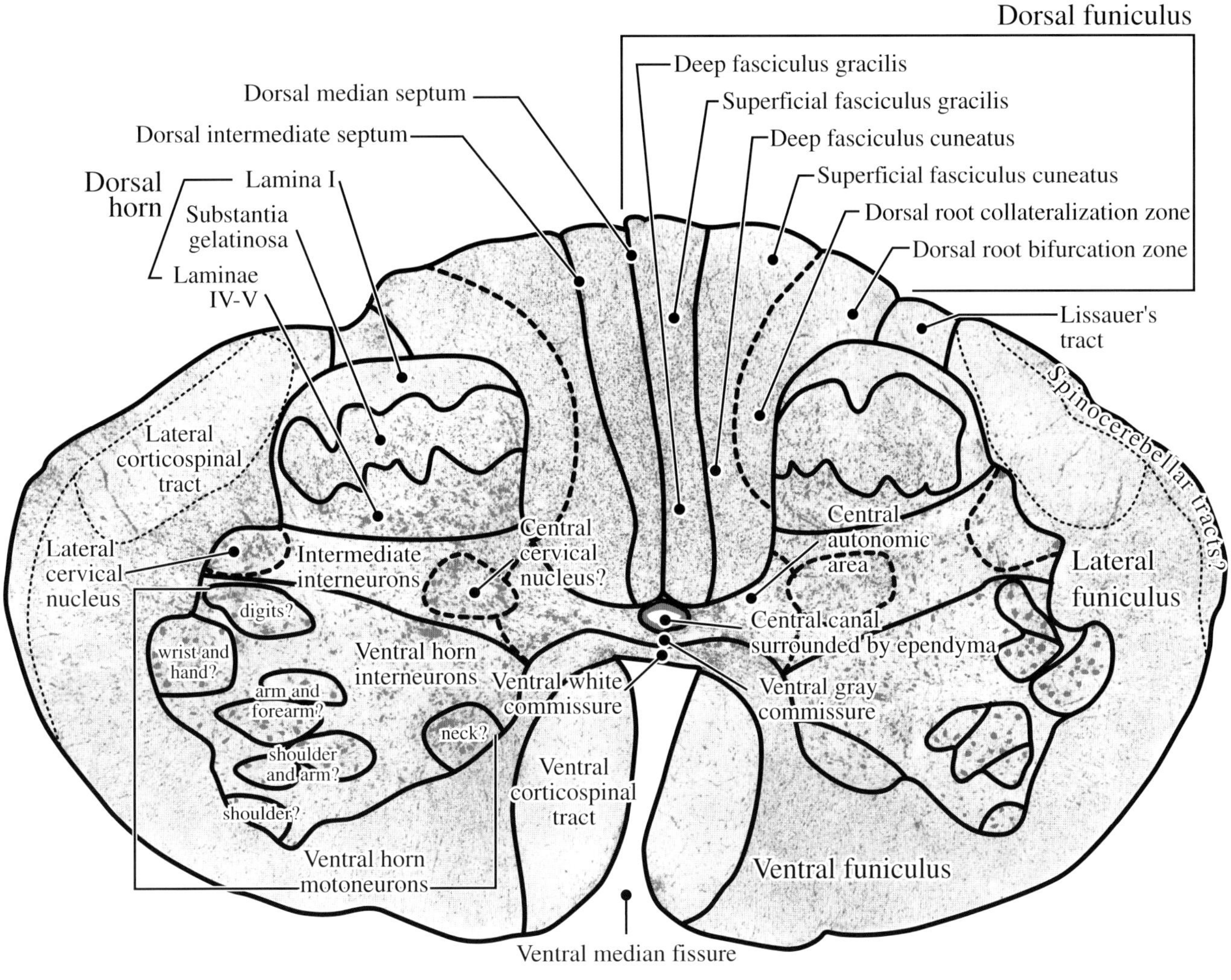

Refer to the matching myelin stained section for the approximate locations of the lateral reticulospinal, vestibulospinal, intraspinal, and spinocephalic tracts.

PLATE 67A

GW37
CR 310 mm
Y117-61
Upper Thoracic
Myelin stain

Areas (mm^2)

Central canal	.0037
Ependyma	.0086
Gray matter	2.1599
White matter	6.2545

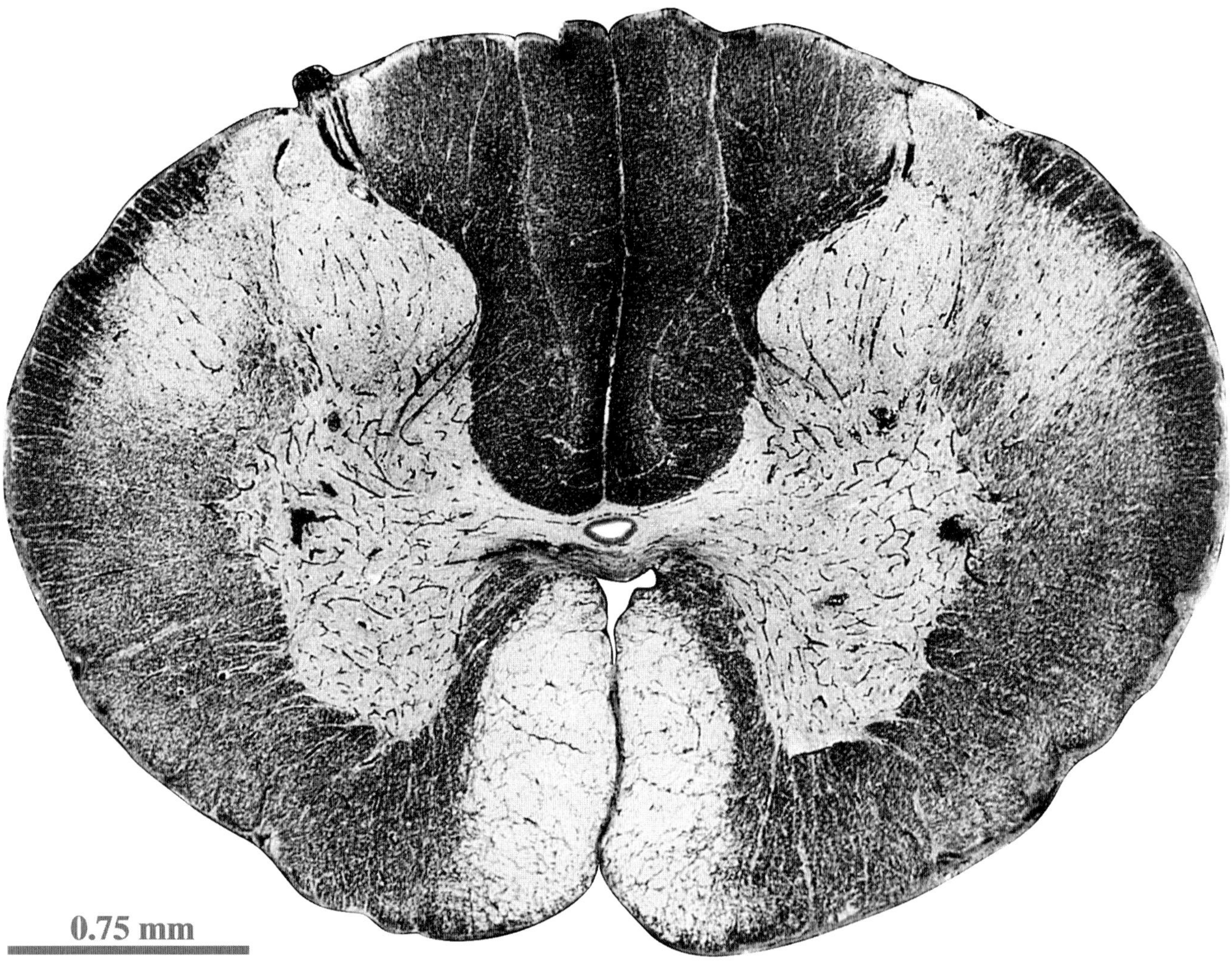

0.75 mm

PLATE 67B

See the matched cell body stained section in Plates 68A and B

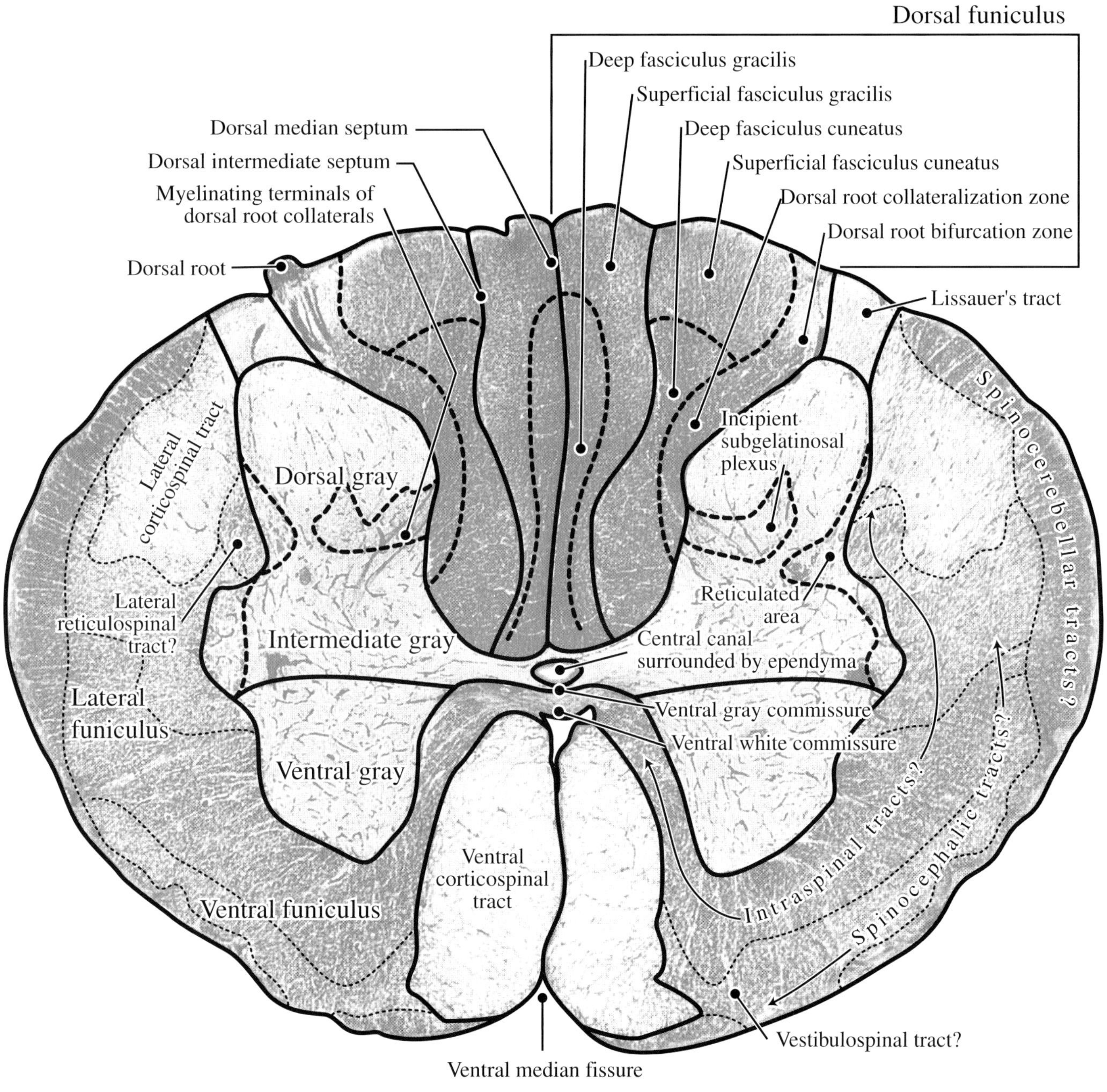

The medial and central fibers in the intraspinal tracts have a greater staining density than the dorsolateral parts. That same gradient is seen at the thoracic level of the GW26 (**Plate 51**) and GW 31 (**Plate 56**) specimens.

PLATE 68A

GW37
CR 310 mm
Y117-61
Upper Thoracic
Cell body stain

Areas (mm^2)

Central canal	.0030
Ependyma	.0152
Gray matter	2.3167
White matter	6.0916

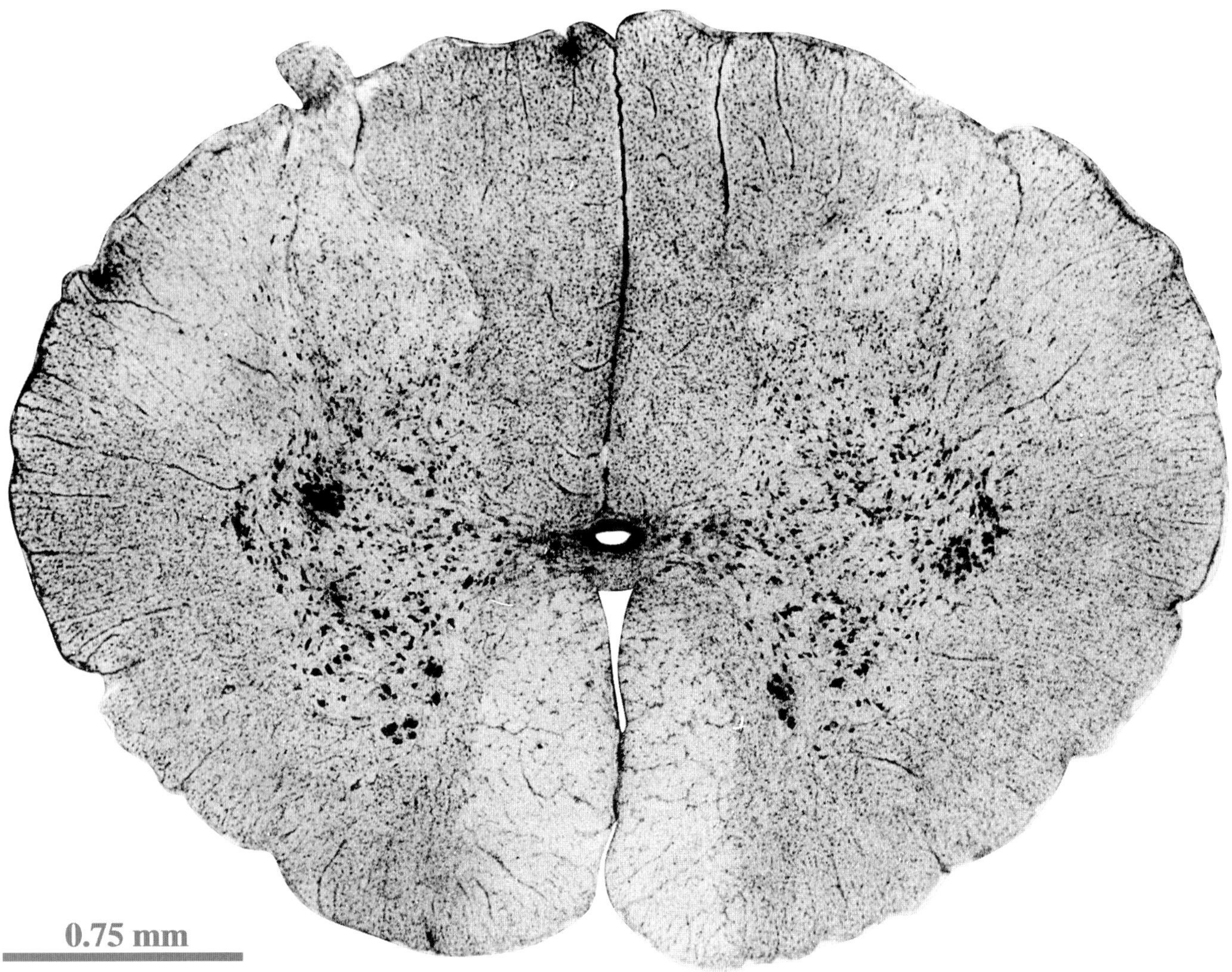

PLATE 68B

See the matched myelin stained section in Plates 67A and B

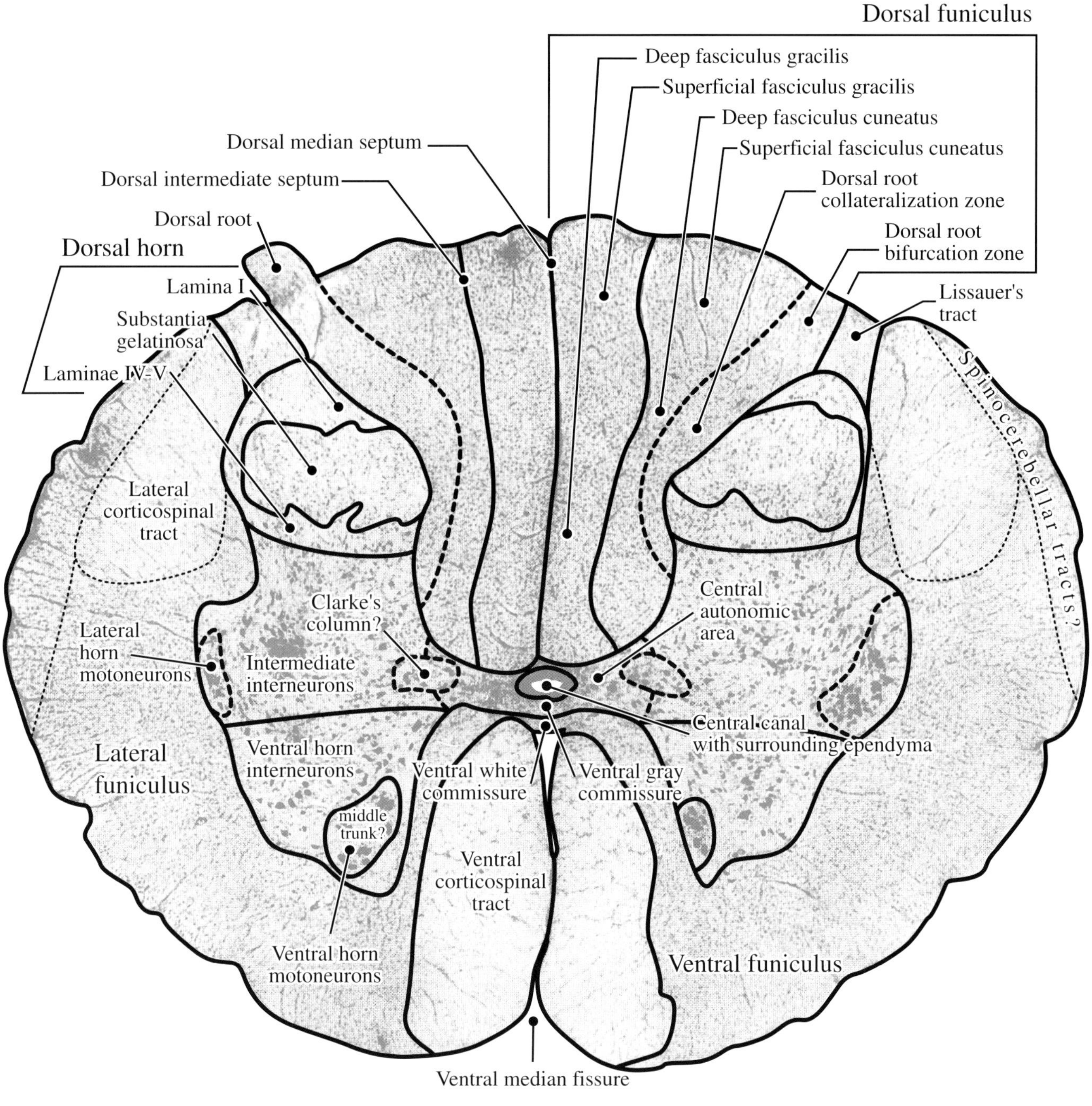

Refer to the matching myelin stained section for the approximate locations of the lateral reticulospinal, vestibulospinal, intraspinal, and spinocephalic tracts.

PLATE 69A

GW37
CR 310 mm
Y117-61
Middle Thoracic
Myelin stain

Areas (mm²)

Central canal	.0029
Ependyma	.0069
Gray matter	1.5695
White matter	4.5747

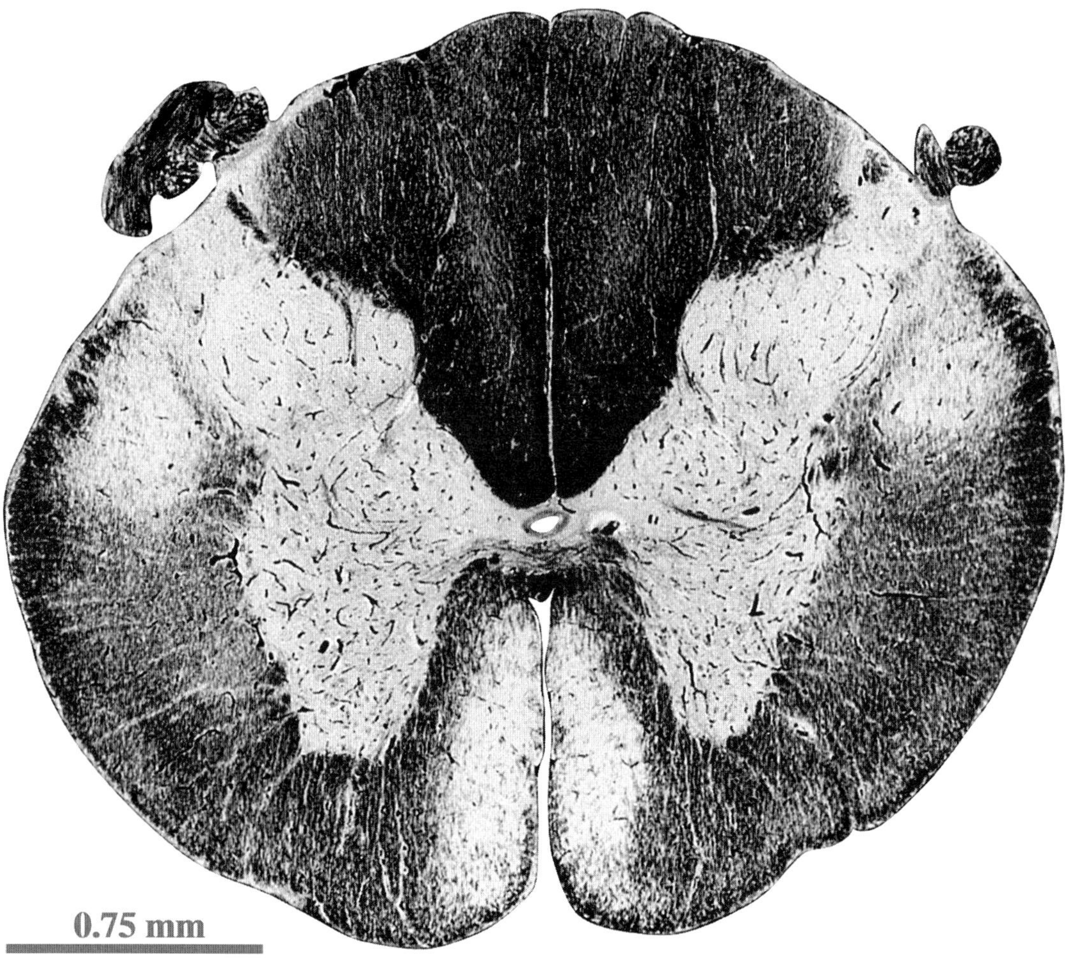

PLATE 69B

See the matched cell body stained section in Plates 70A and B

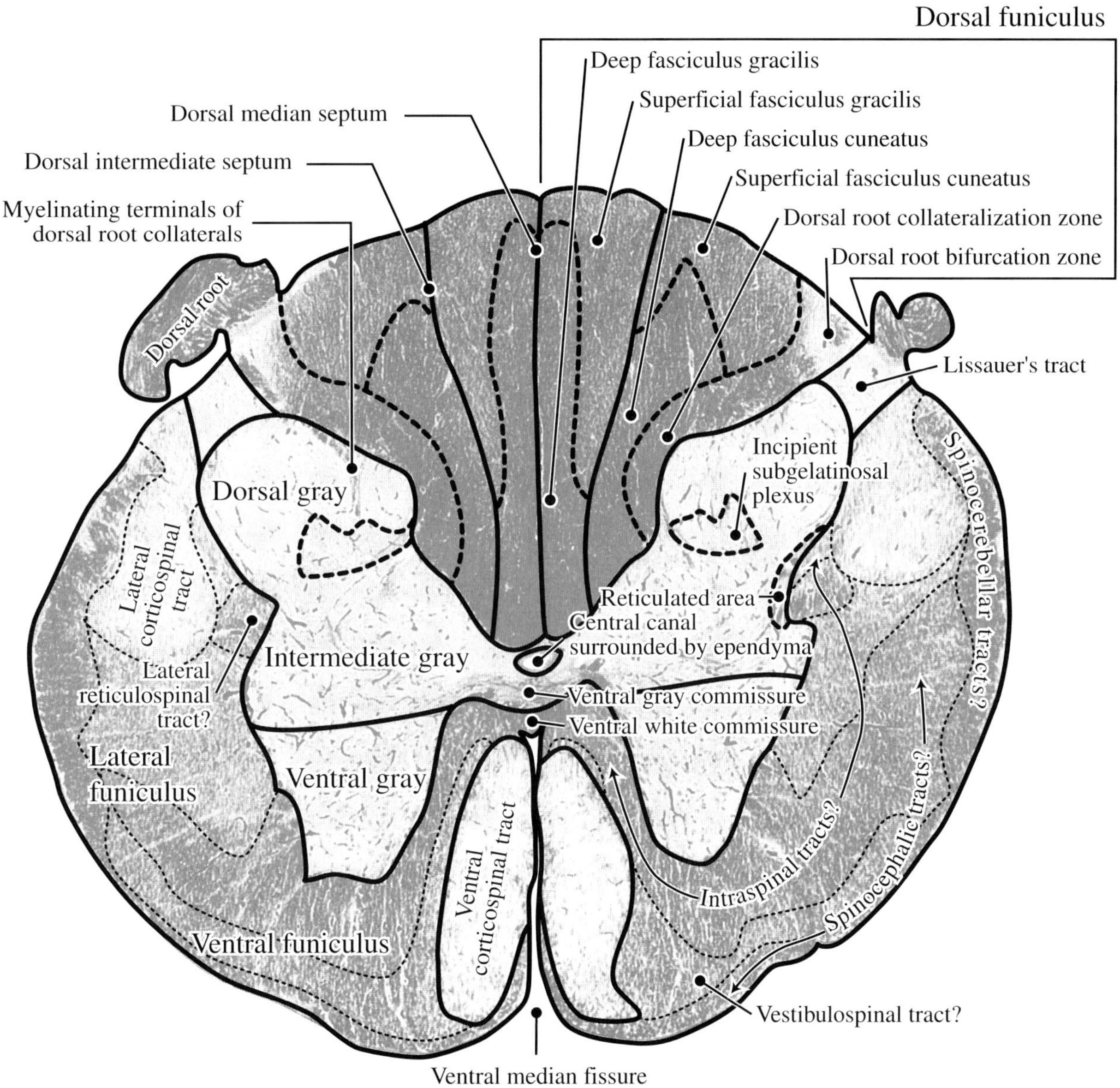

PLATE 70A

GW37
CR 310 mm
Y117-61
Middle Thoracic
Cell body stain

Areas (mm^2)

Central canal	.0045
Ependyma	.0132
Gray matter	1.6976
White matter	4.9867

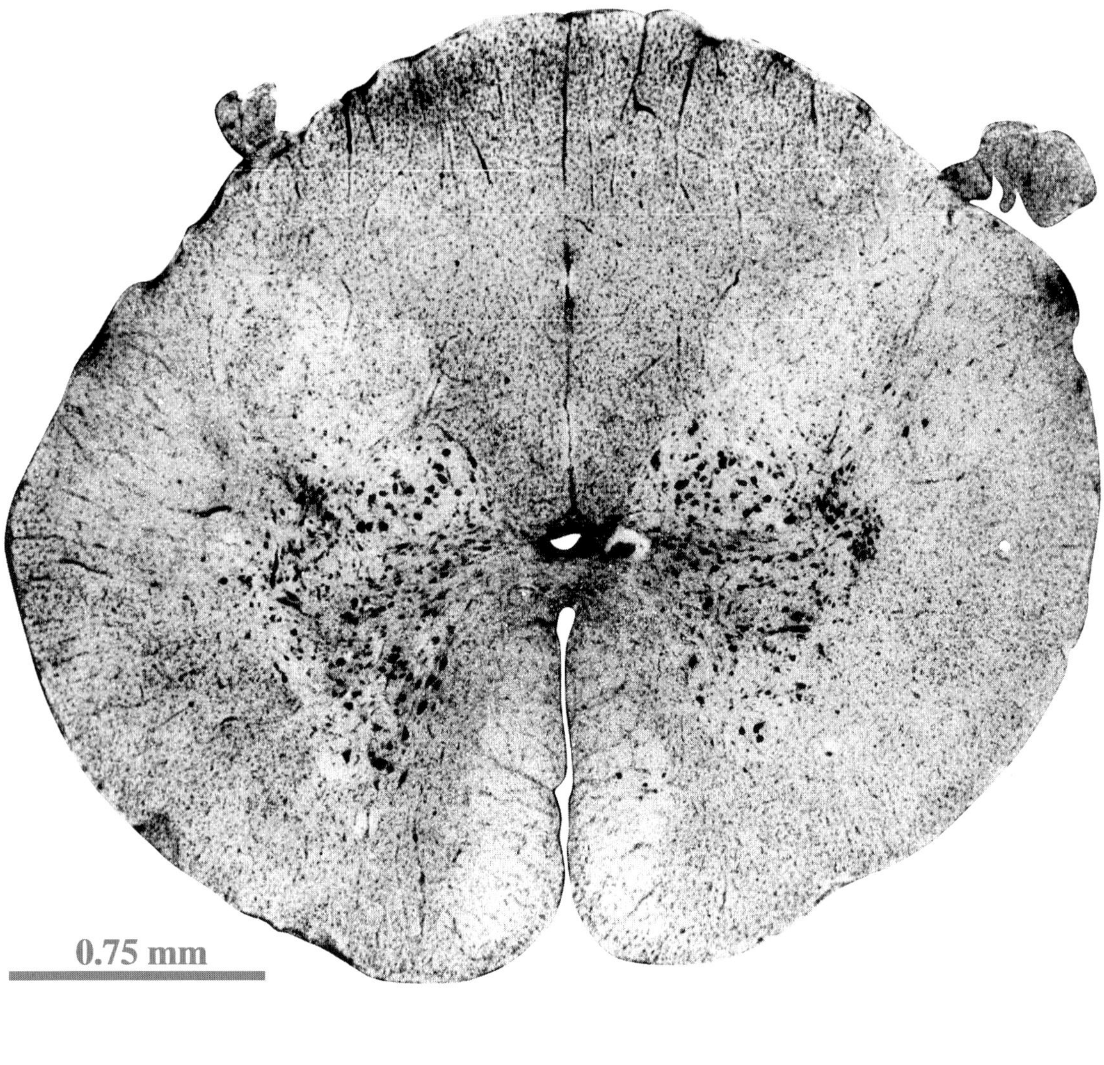

PLATE 70B

See the matched myelin stained section in Plates 69A and B

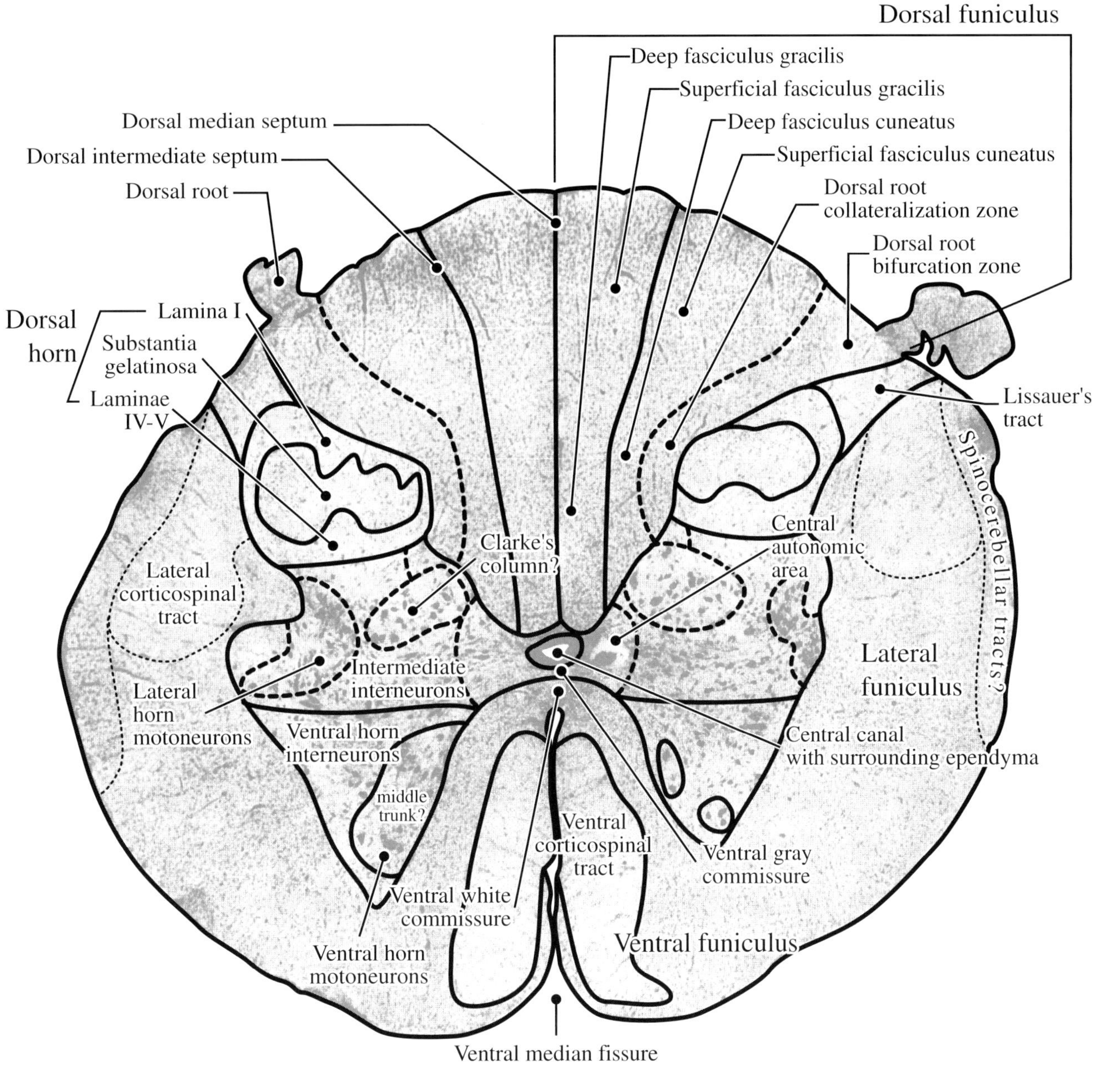

Refer to the matching myelin stained section for the approximate locations of the lateral reticulospinal, vestibulospinal, intraspinal, and spinocephalic tracts.

PLATE 71A

GW37
CR 310 mm
Y117-61
Lower Thoracic
Myelin stain

Areas (mm²)

Central canal	.0057
Ependyma	.0070
Gray matter	2.3538
White matter	4.8340

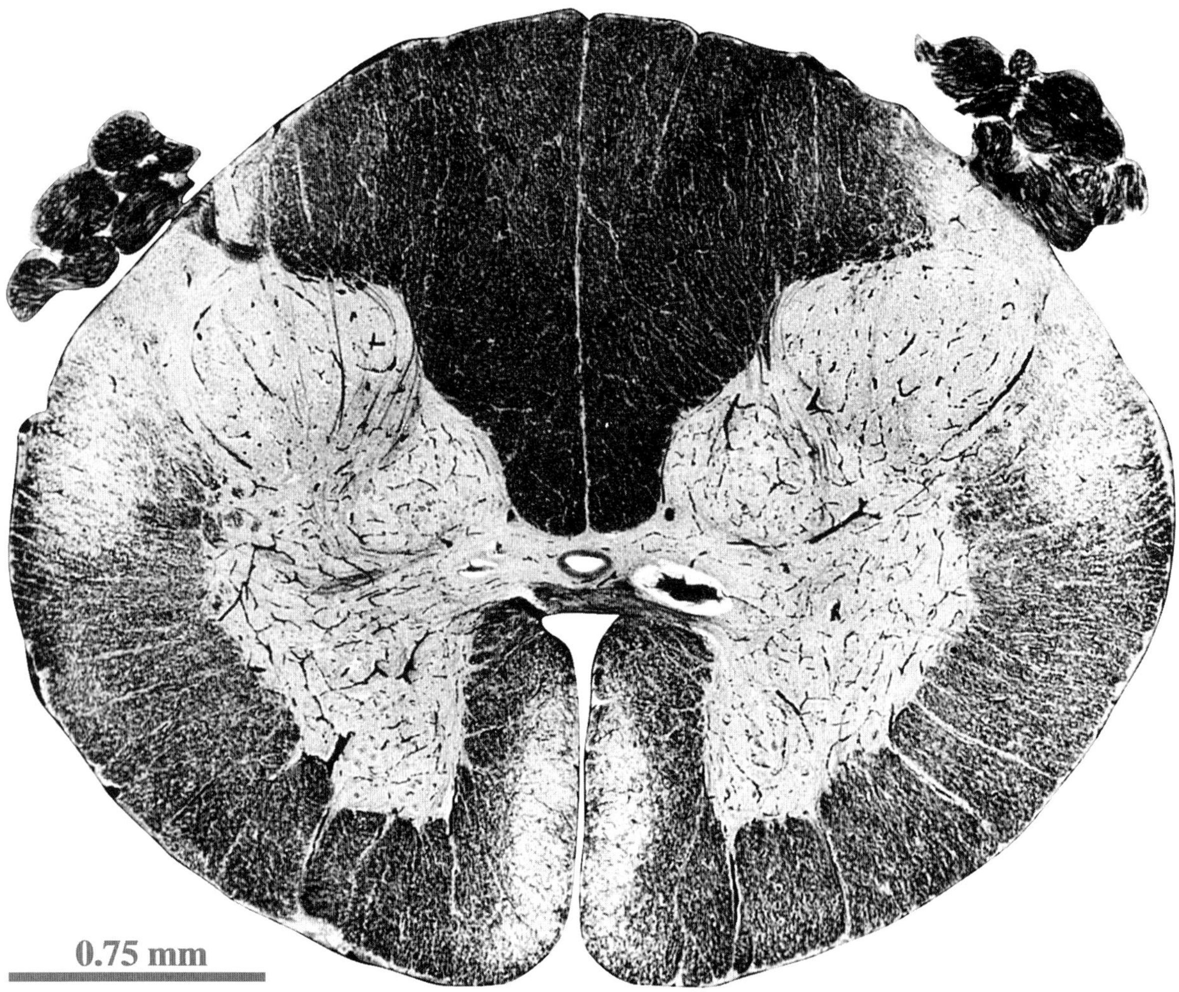

PLATE 71B

See the matched cell body stained section in Plates 72A and B

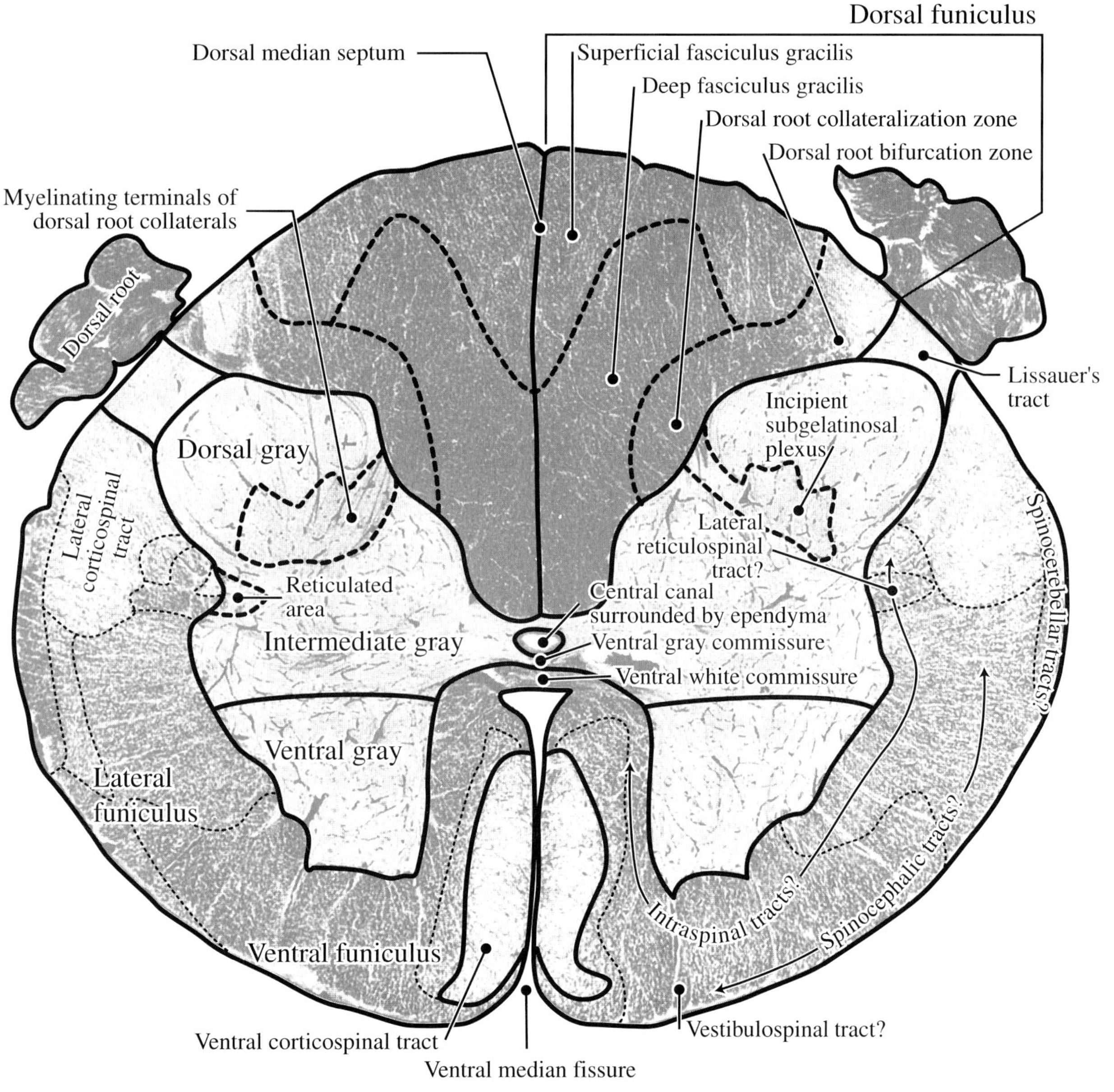

PLATE 72A

GW37
CR 310 mm
Y117-61
Lower Thoracic
Cell body stain

Areas (mm^2)

Central canal	.0031
Ependyma	.0130
Gray matter	2.5442
White matter	5.3951

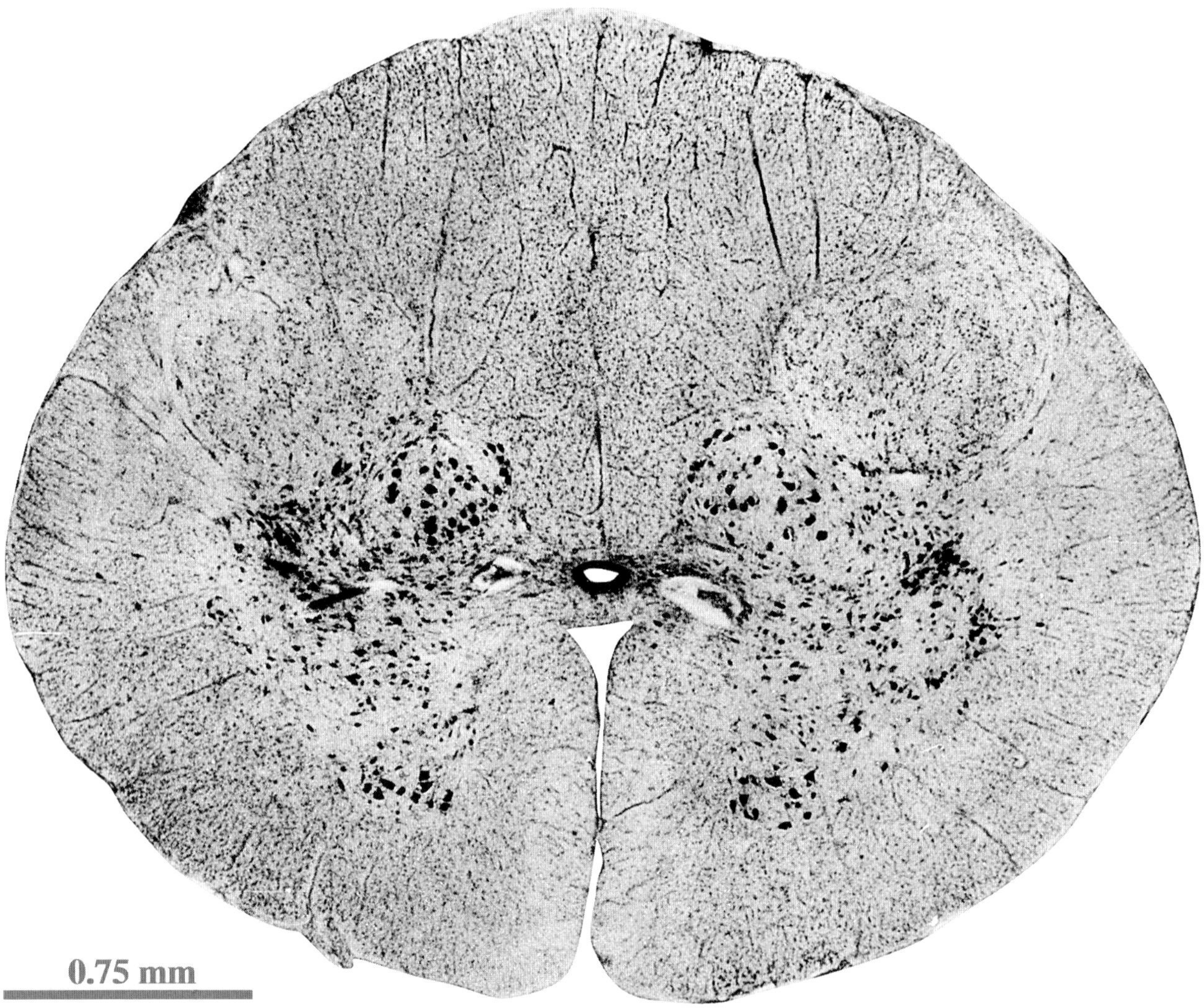

PLATE 72B

See the matched myelin stained section in Plates 71A and B

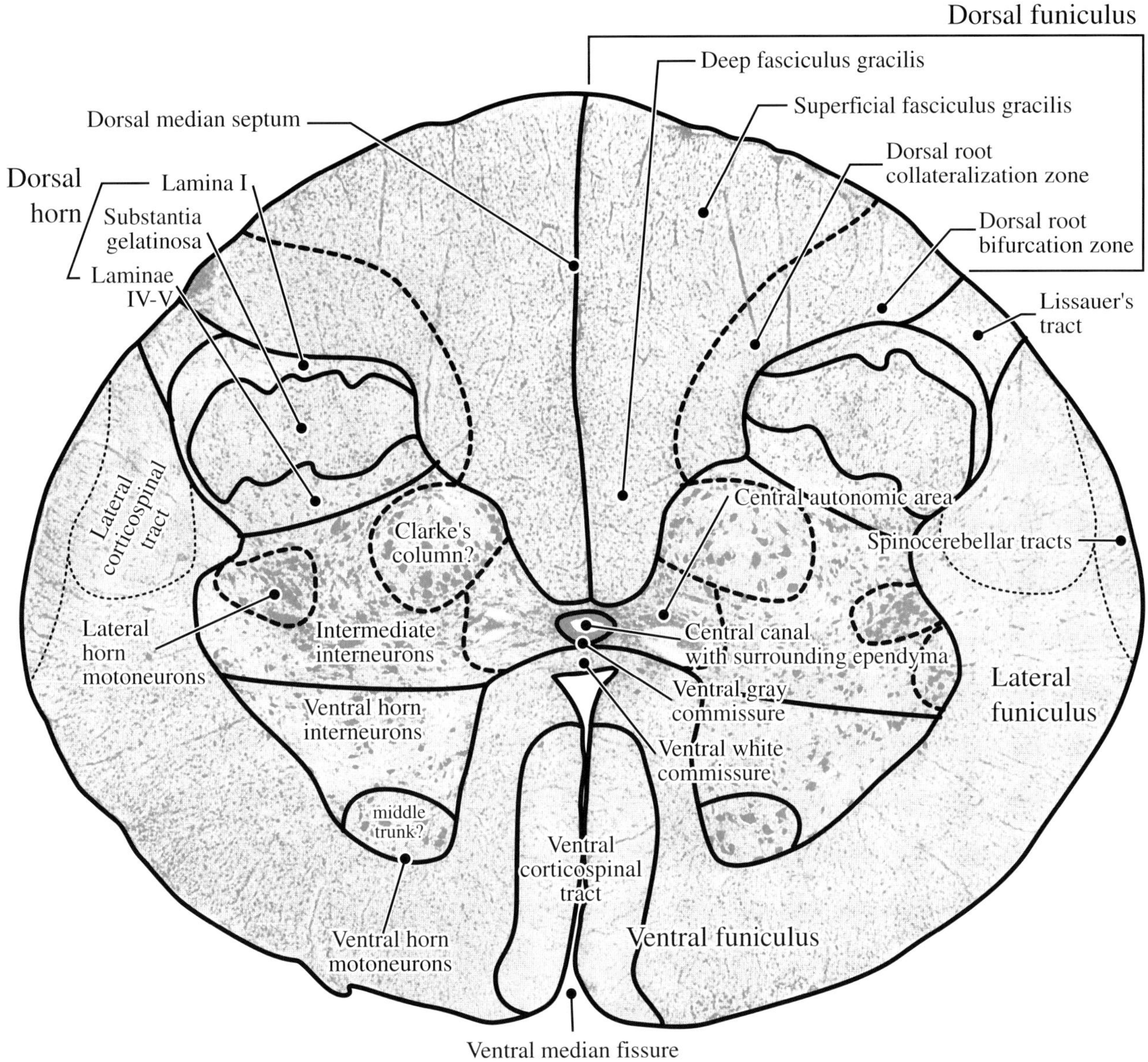

Refer to the matching myelin stained section for the approximate locations of the lateral reticulospinal, vestibulospinal, intraspinal, and spinocephalic tracts.

PLATE 73A

GW37
CR 310 mm
Y117-61
Upper Lumbar
Myelin stain

Areas (mm^2)

Central canal	.0109
Ependyma	.0113
Gray matter	5.4305
White matter	6.0778

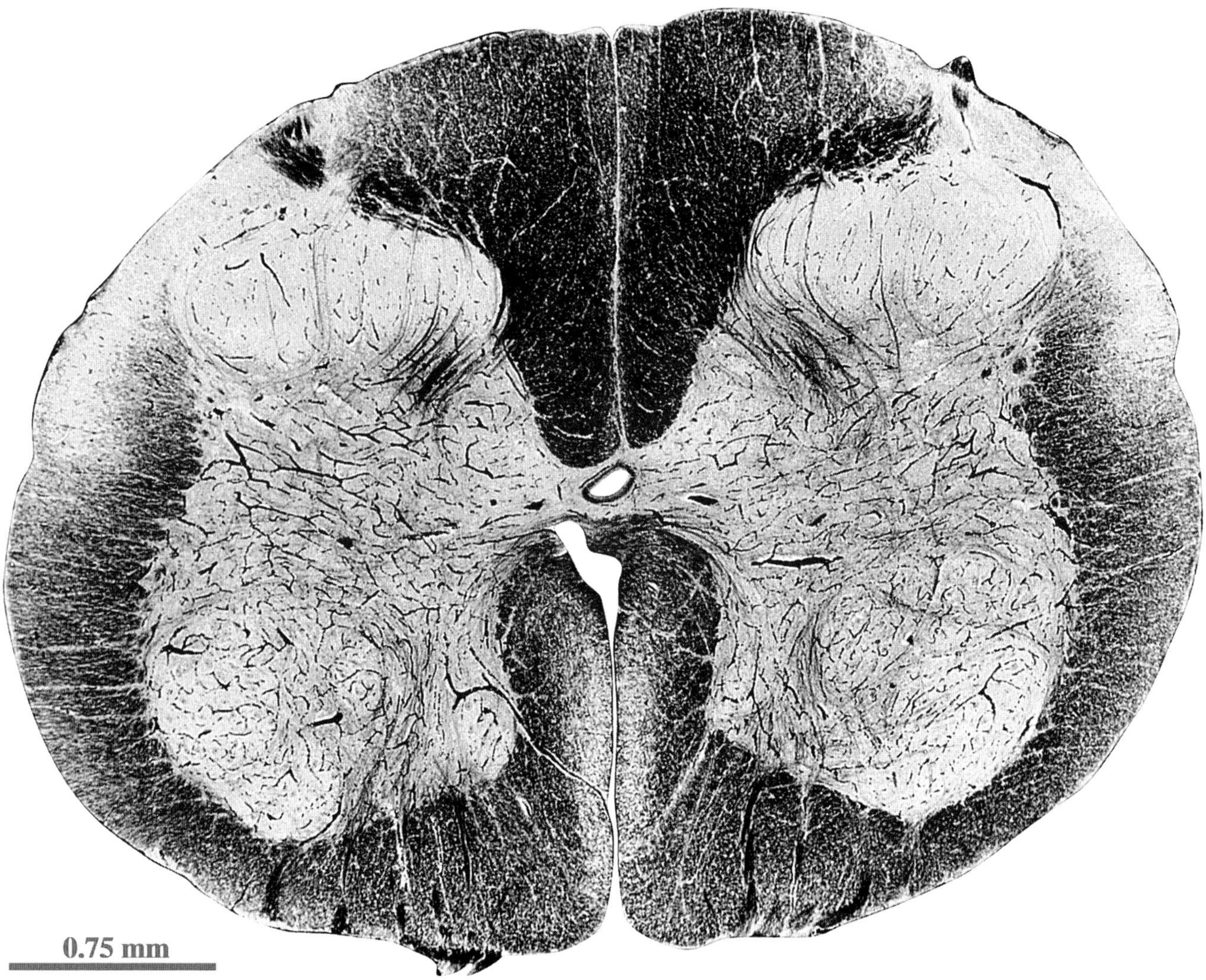

PLATE 73B

See the matched cell body stained section in Plates 74A and B

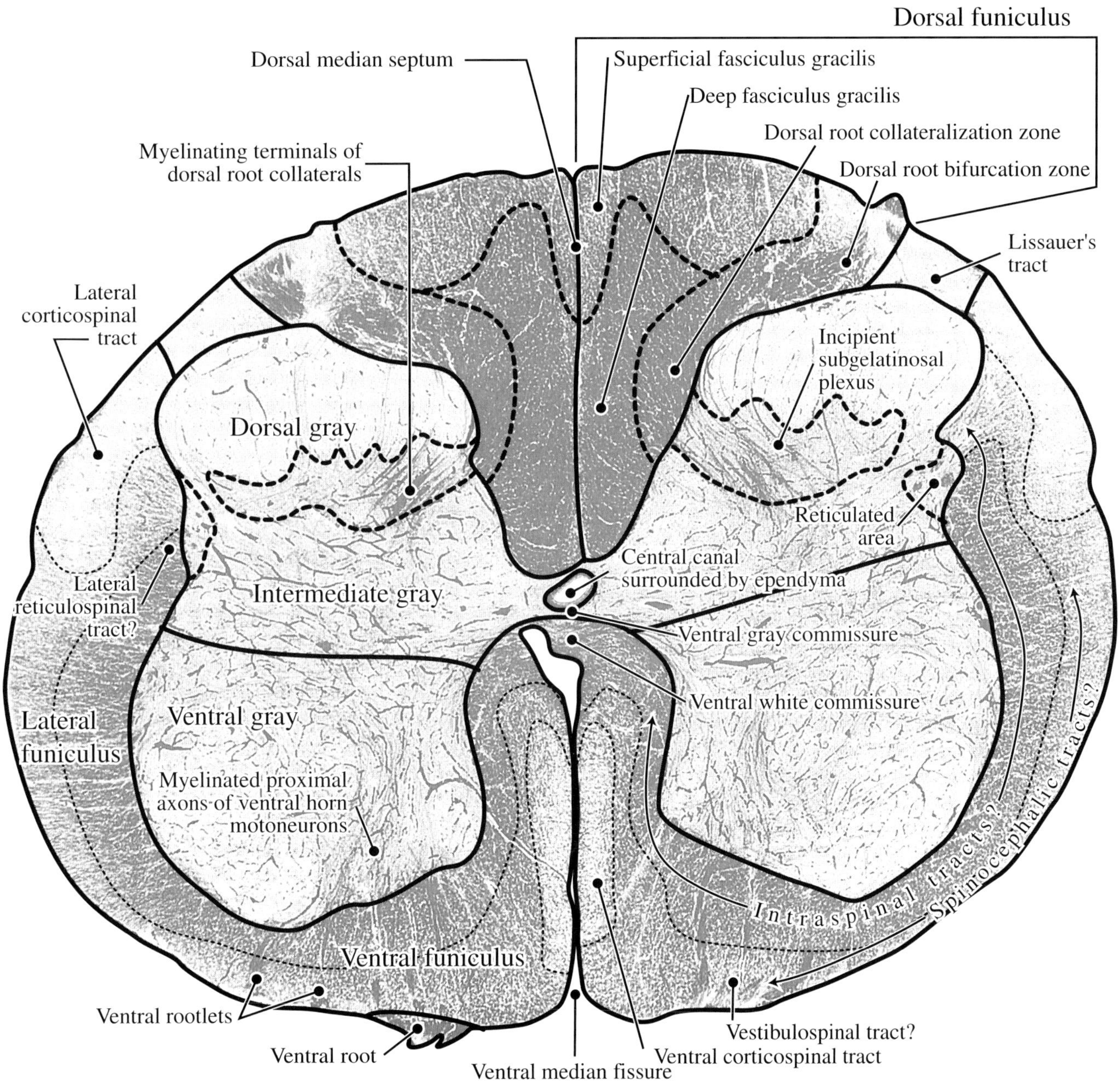

PLATE 74A

GW37
CR 310 mm
Y117-61
Upper Lumbar
Cell body stain

Areas (mm^2)

Central canal	.0087
Ependyma	.0178
Gray matter	5.7810
White matter	6.4894

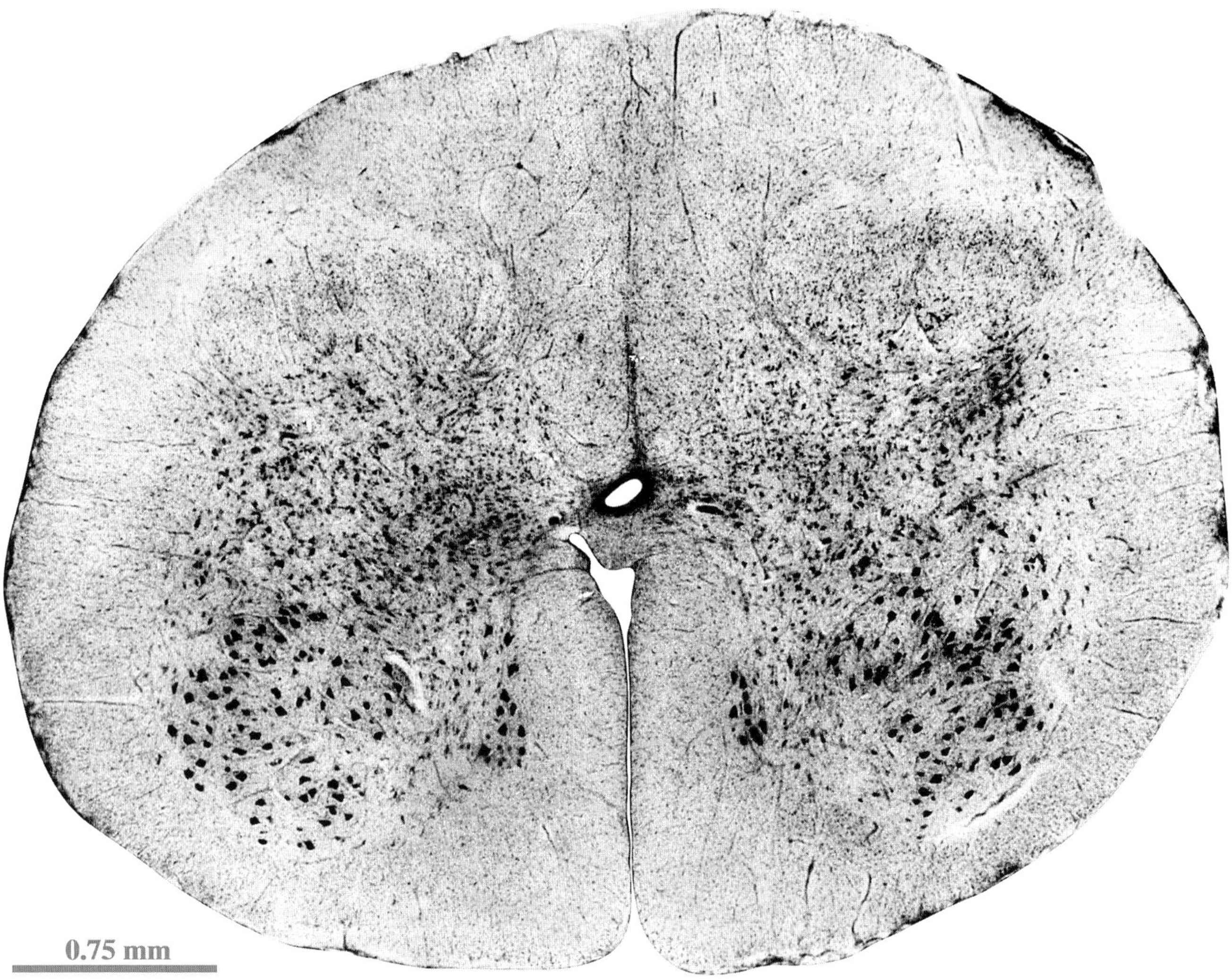

PLATE 74B

See the matched myelin stained section in Plates 73A and B

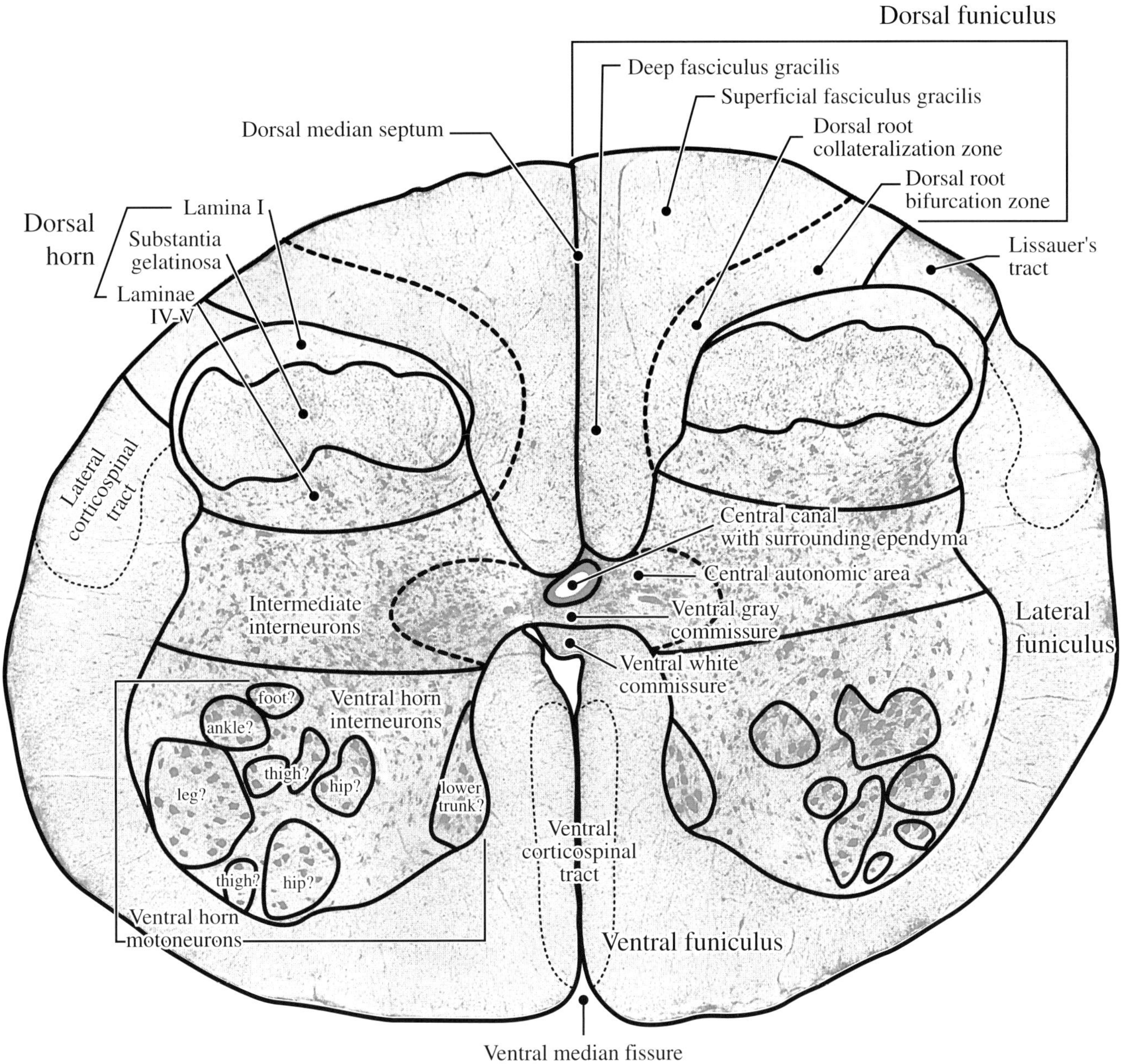

Refer to the matching myelin stained section for the approximate locations of the lateral reticulospinal, vestibulospinal, intraspinal, and spinocephalic tracts.

PLATE 75A

GW37
CR 310 mm
Y117-61
Lumbar Enlargement
Myelin stain

Areas (mm^2)

Central canal	.0043
Ependyma	.0087
Gray matter	6.6891
White matter	5.5948

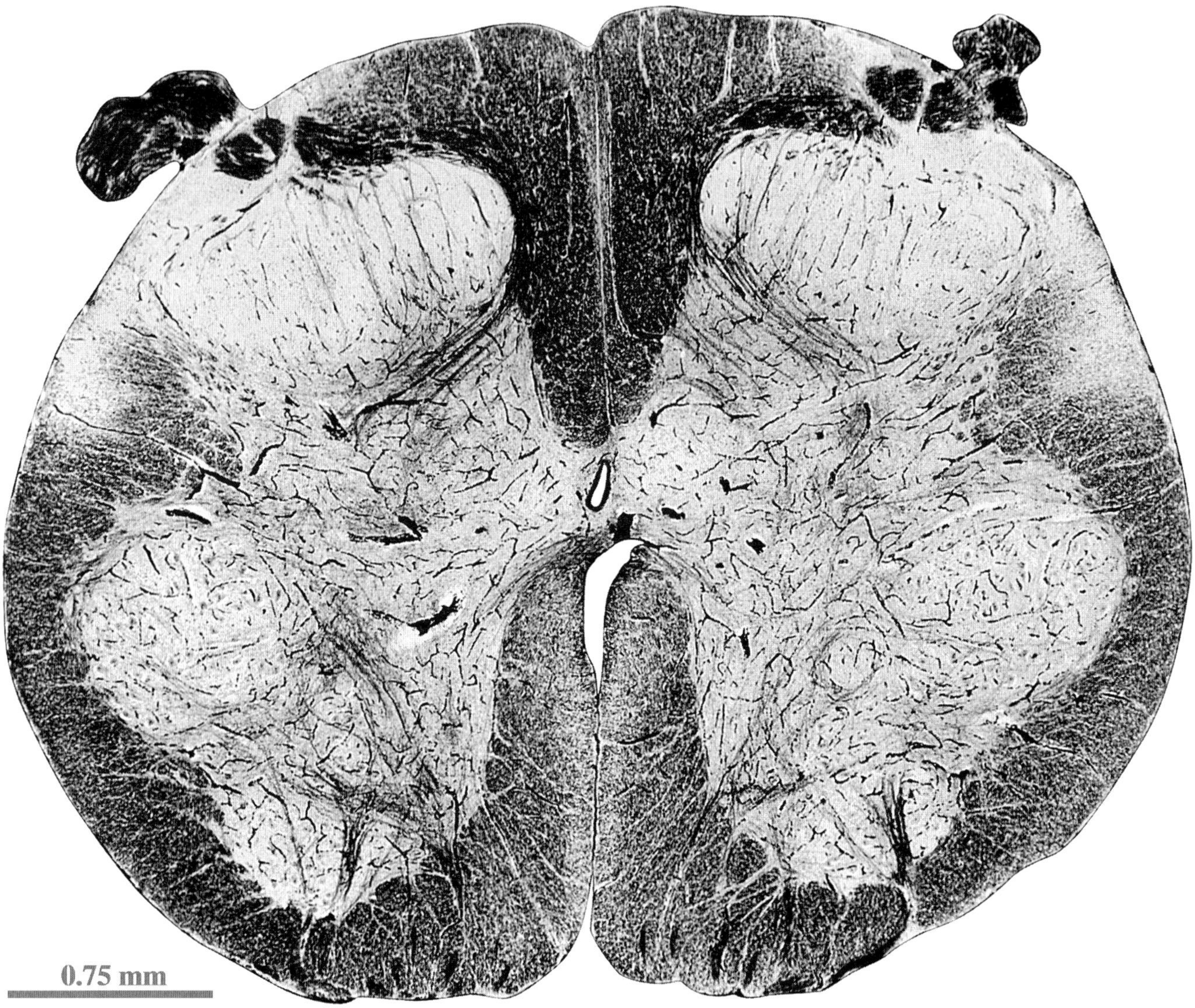

PLATE 75B

See the matched cell body stained section in Plates 76A and B

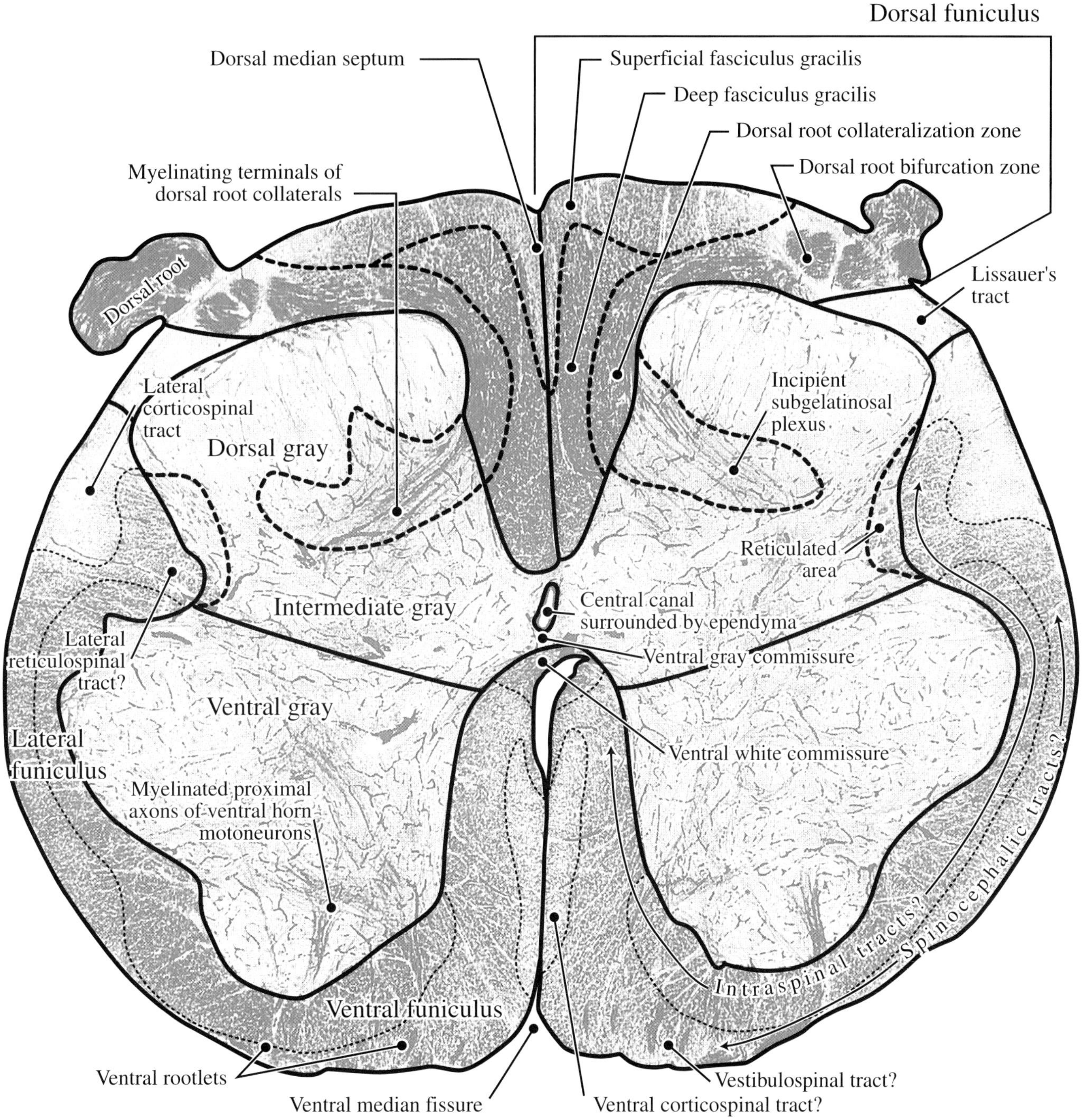

PLATE 76A

GW37
CR 310 mm
Y117-61
Lumbar Enlargement
Cell body stain

Areas (mm²)

Central canal	.0036
Ependyma	.0140
Gray matter	6.7877
White matter	6.0663

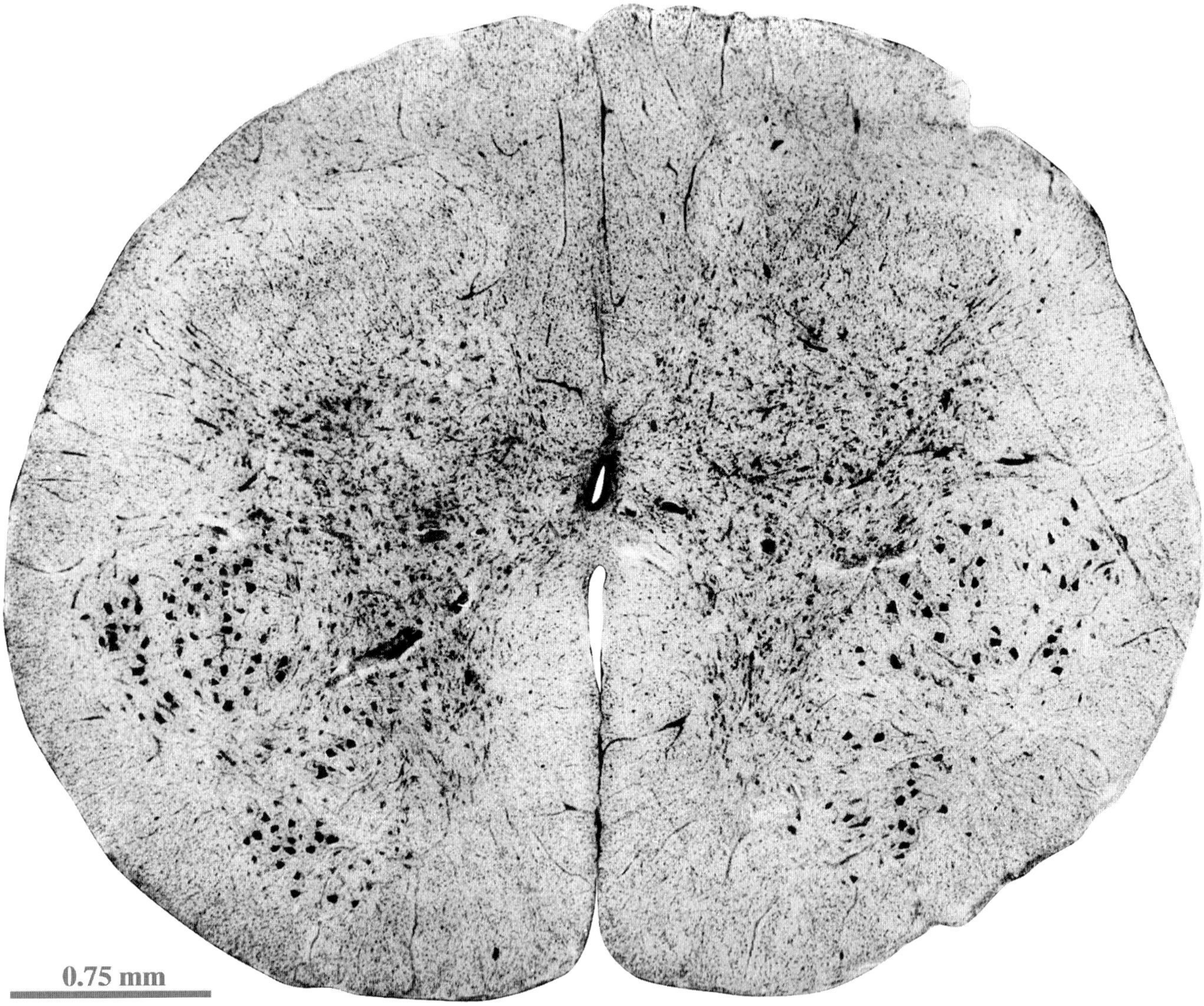

PLATE 76B

See the matched myelin stained section in Plates 75A and B

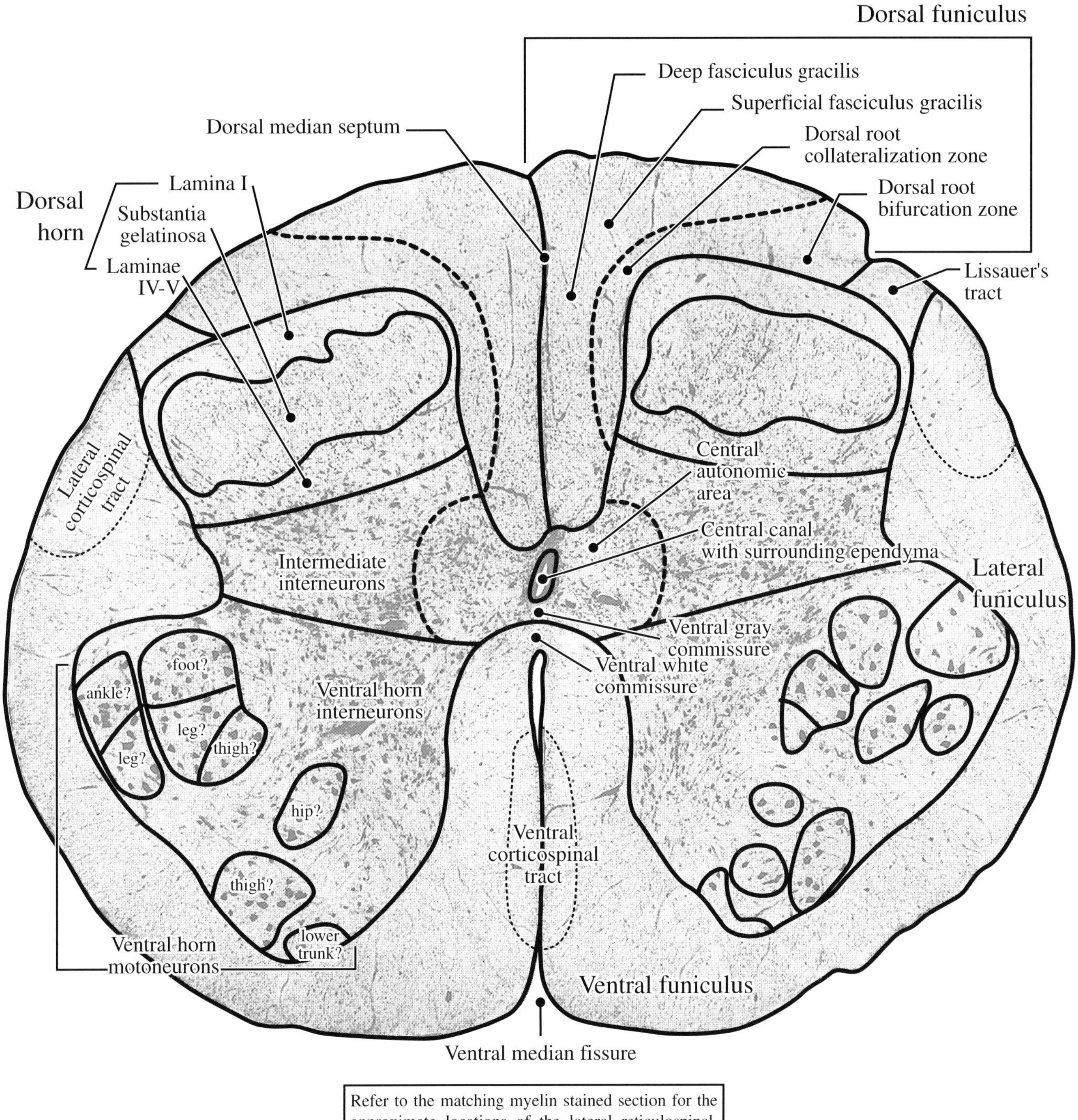

Refer to the matching myelin stained section for the approximate locations of the lateral reticulospinal, vestibulospinal, intraspinal, and spinocephalic tracts.

PLATE 77A

GW37
CR 310 mm
Y117-61
Sacral/Coccygeal
Myelin stain

Areas (mm²)

Central canal	.0050
Ependyma	.0160
Gray matter	4.3017
White matter	2.1841

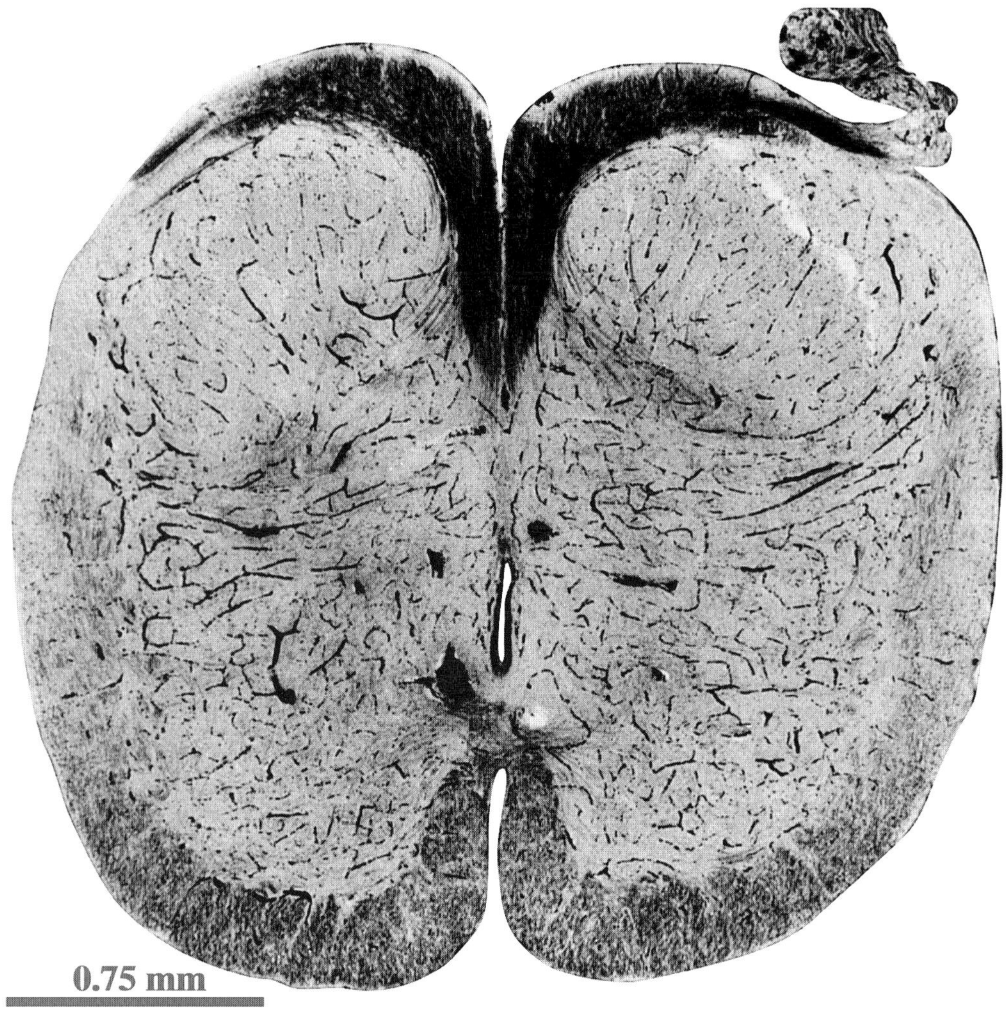

PLATE 77B

See the matched cell body stained section in Plates 78A and B

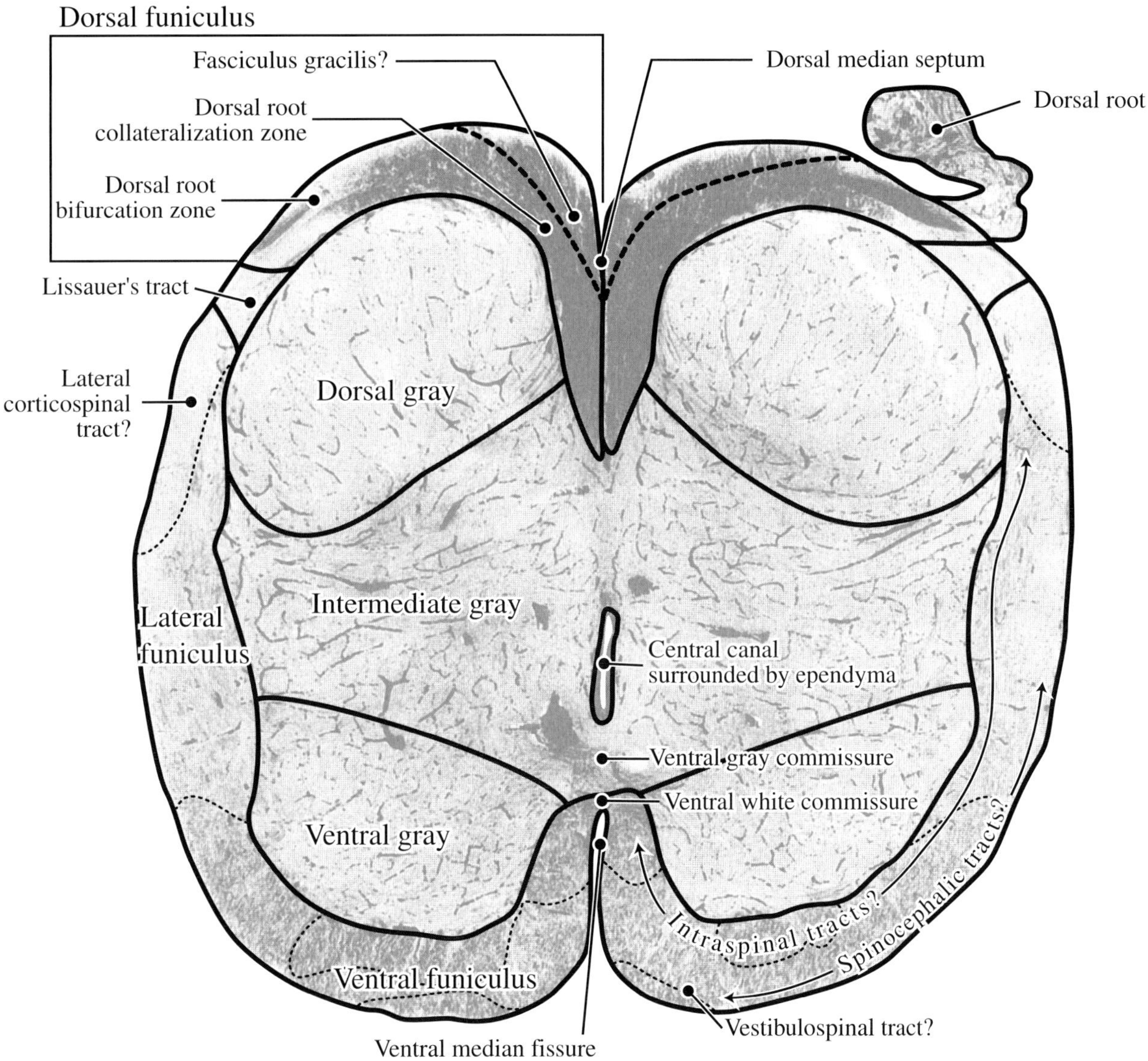

PLATE 78A

GW37
CR 310 mm
Y117-61
Sacral/Coccygeal
Cell body stain

Areas (mm^2)

Central canal	.0044
Ependyma	.0271
Gray matter	4.4850
White matter	2.2665

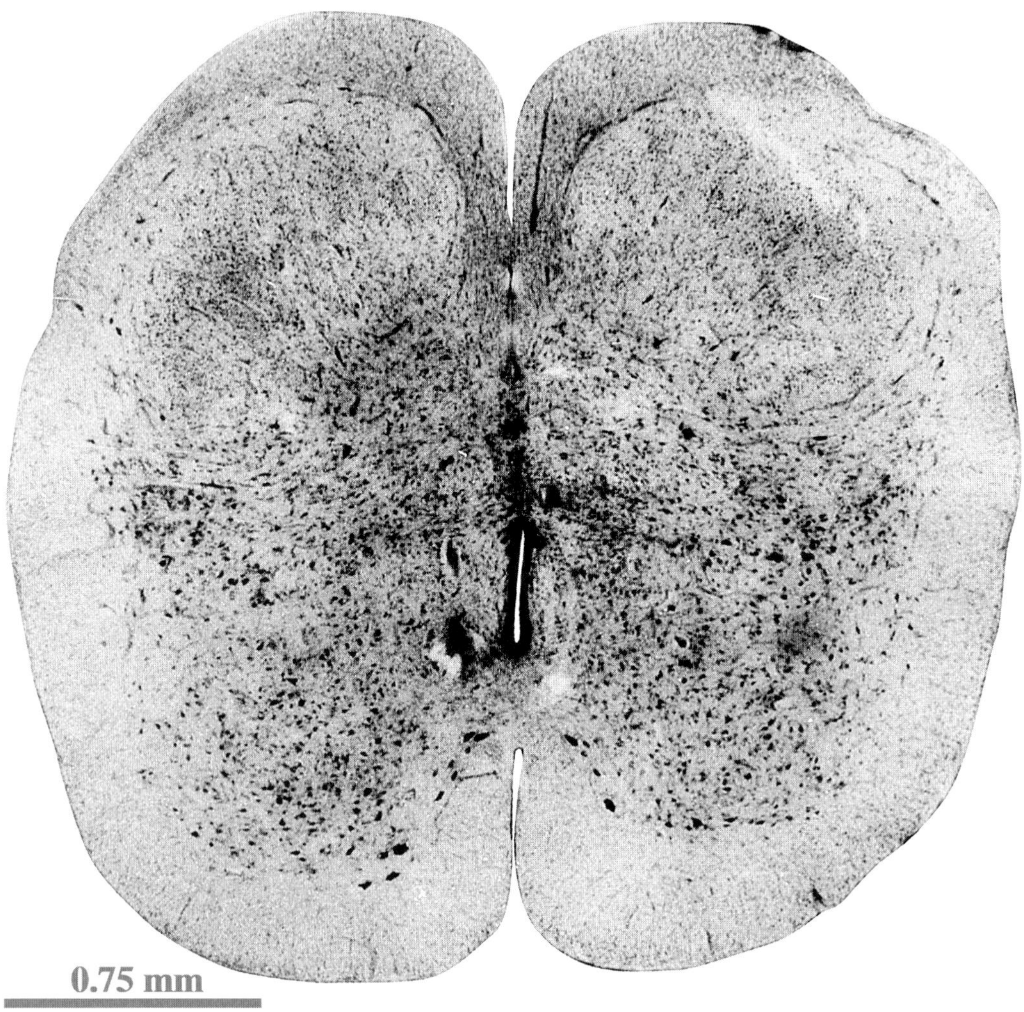

PLATE 78B

See the matched myelin stained section in Plates 77A and B

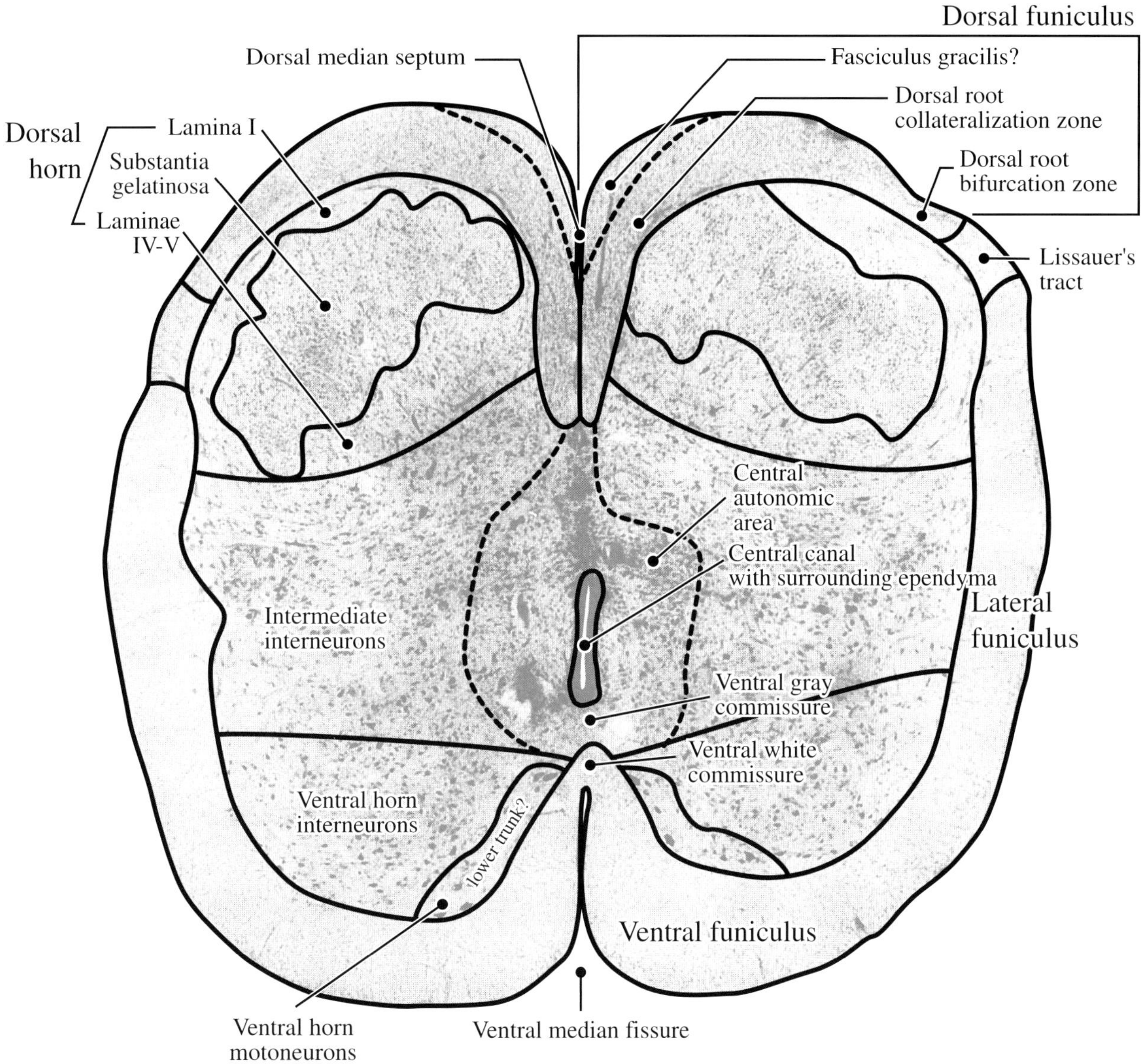

In this section, the dorsal funiculus still shows some myelination gliosis, indicating that this level of the spinal cord is more immature than rostral levels. The lateral funiculus has a very sparse population of proliferating glia throughout, and the border of the lateral corticospinal tract is not visible. Refer to the matching myelin stained section for the approximate locations of the lateral reticulospinal, vestibulospinal, intraspinal, and spinocephalic tracts.

V. The Early Postnatal Period: Infants at 4 days, 4 weeks, and 4 months

A. Matched myelin and cell body stained sections in the spinal cord of a 4-day-old infant

Plate 79 is a survey of matched myelin stained and cell body stained sections from Y299-62, a specimen in the Yakovlev Collection with a crown rump length of 350 mm. All sections are shown at the same scale. The boxes enclosing each section list the approximate level and the total area of the section in square millimeters (mm^2). Full-page normal contrast photographs of each specimen are in **Plates 80A-91A.** Low contrast photographs with superimposed labels and outlines of structural details are in **Plates 80B-91B.** In this specimen, the myelin stained and cell body stained sections were preserved on separate large glass plates without any section numbers. Twenty-two myelin stained sections and 22 cell body stained sections were photographed ranging from the cervical enlargement to sacral/coccygeal levels. The 44 photographic prints were intuitively arranged in order from upper cervical to sacral/coccygeal levels, using internal features such as the size of the corticospinal tracts and the width of the ventral horn. Then myelin and cell body stained sections were matched and six different levels were analyzed. Since many sections in the middle thoracic region are damaged, that level is not illustrated.

As in the previous specimens, the cross-sectional area of a myelin stained section is smaller than the matching cell body stained section in all cases. Using the total areas of the myelin stained sections, the overall size differences between levels indicate the following comparisons: the cervical enlargement is the level with the largest cross-sectional area, being larger than the lumbar enlargement by 65%. The sacral/coccygeal level has the smallest cross-sectional area and is 37% smaller than the lower thoracic level.

Myelination in this specimen is slightly advanced compared to the GW37 specimen (*columns 2 and 3*, **Table V A-1**) because there is no longer a gradient between superficial and deep parts of the gracile and cuneate fasciculi, and more myelinated axons are in the subgelatinosal plexus. The ventral corticospinal tract is smaller in this specimen than the one at GW37. For a discussion of individual differences between specimens regarding the corticospinal tracts, see Chapter 7, Section 7.2.2 in Altman and Bayer, 2001). The corticospinal tracts in the 4-day-old infant indicate more advanced myelination than in the GW37 fetus by having at least a sparse concentration of reactive glia and a high concentration of proliferating glia (*columns 3 and 4*, **Table V A-1**). Another indication of maturation is the sparse (medial) to very sparse (lateral) gradient of reactive glia in the lateral corticospinal tract. Myelination follows a medial to lateral gradient in that tract (Altman and Bayer, 2001). There is an overall gradient of sparse reactive glia (superficial fibers in the vestibulospinal and spinocephalic tracts) to the beginnings of myelination (deep fibers in the intraspinal tracts) in the ventral and lateral funiculi that cuts across the borders of the fiber tracts.

Table V A-1: Glia types and concentration in the white matter in a 4-day-old Infant

Name	Myelination	Reactive glia	Proliferating glia
DORSAL ROOT	Myelinated	---	Sparse**
VENTRAL ROOT	Myelinated	---	---
DORSAL FUNICULUS:			
dorsal root bif. zone	Many fibers*	---	Sparse
dorsal root col. zone	Myelinated	---	Sparse
fas. gracilis	Myelinated	---	Sparse
fas. cuneatus	Myelinated	---	Sparse
Lissauer's tract	---	None	Sparse
LATERAL and VENTRAL FUNICULI:			
lat. cortricospinal tract	---	Very sparse	Sparse
ven. cortricospinal tract	---	Very sparse	Sparse
spinocerebellar tracts	Myelinated	---	Sparse
ven. commissure	Myelinated	---	Dense
intraspinal tracts	Many fibers	---	Dense
lat. reticulospinal tract	Some fibers	Very sparse	Sparse
spinocephalic tracts	Some fibers†	Very sparse•	Dense
med. long. fasiculus	Some fibers	---	Sparse
vestibulospinal tract	Some fibers	---	Sparse

** dense band in a sparse field (part of the boundary cap?)
* intermingled in a bed of nonreactive glia
† few fibers at the cervical level
• peripheral fibers at the cervical level

PLATE 79

INFANT, 4 DAYS, CR 350 mm, Y299-62

MYELIN STAIN | **CELL BODY STAIN**

Plates 80A, 80B
Cervical Enlargement
Total Area:
20.063 mm^2

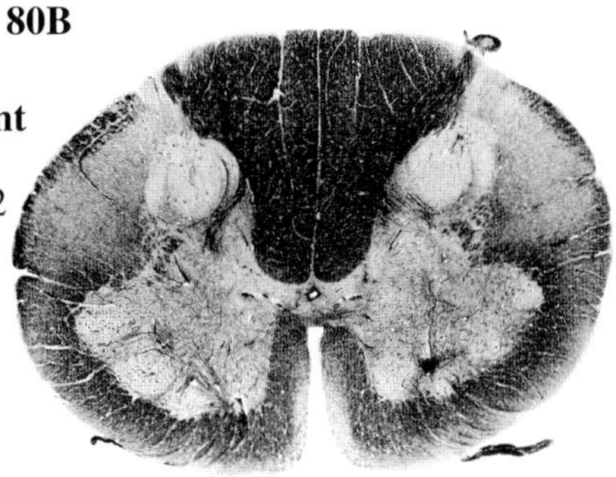

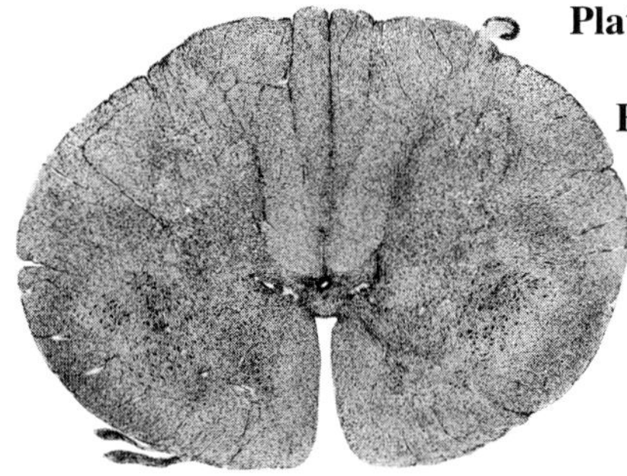

Plates 81A, 81B
Cervical Enlargement
Total Area:
21.370 mm^2

Plates 82A, 82B
Upper Thoracic
Total Area:
10.177 mm^2

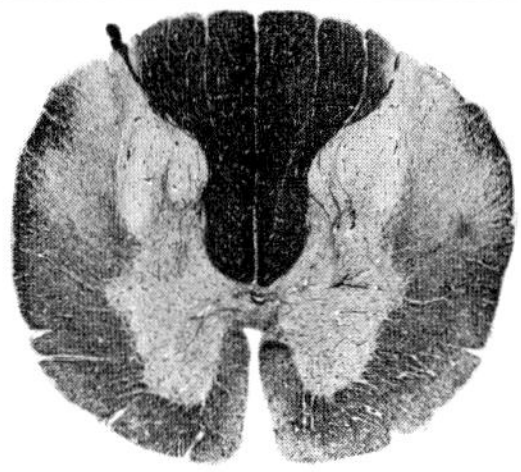

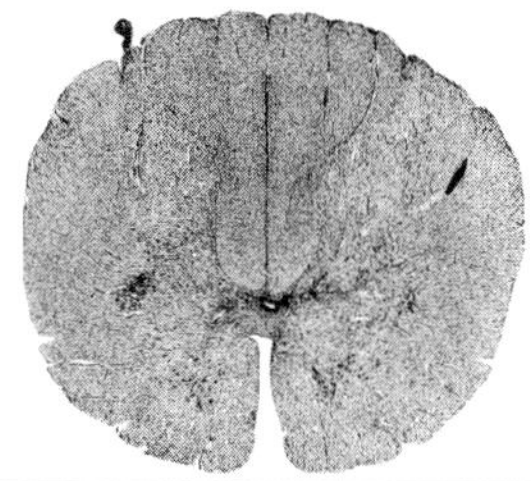

Plates 83A, 83B
Upper Thoracic
Total Area:
10.514 mm^2

Plates 84A, 84B
Lower Thoracic
Total Area:
7.7961 mm^2

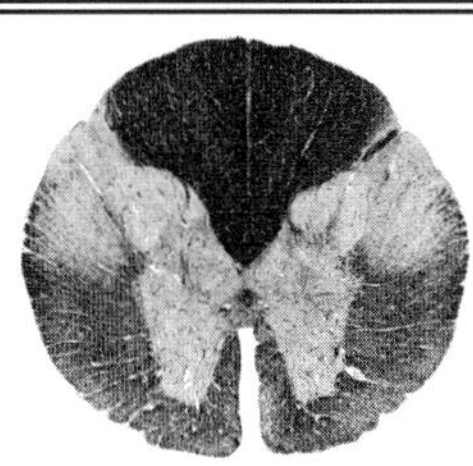

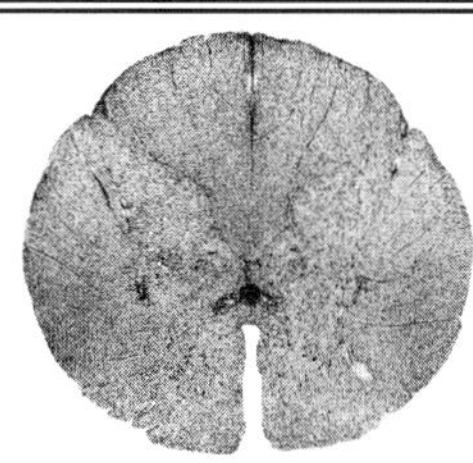

Plates 85A, 85B
Lower Thoracic
Total Area:
8.2168 mm^2

Plates 86A, 86B
Upper Lumbar
Total Area:
11.735 mm^2

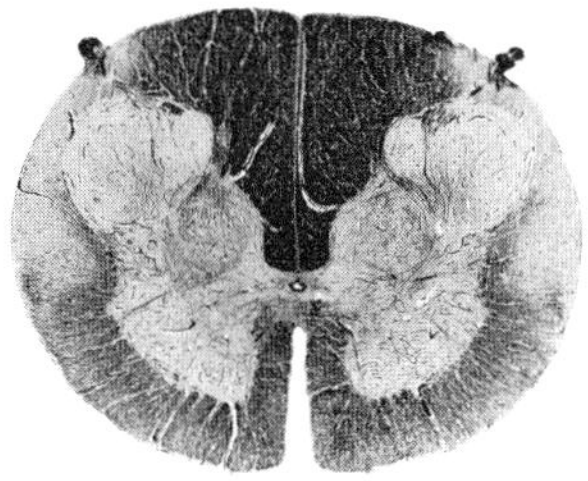

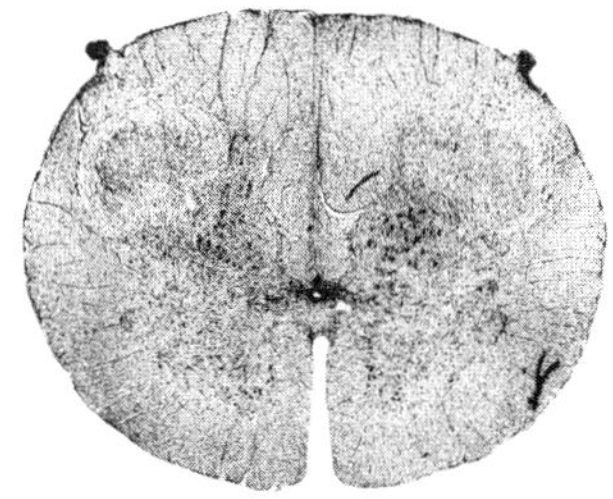

Plates 87A, 87B
Upper Lumbar
Total Area:
12.730 mm^2

Plates 88A, 88B
Lumbar Enlargement
Total Area:
12.152 mm^2

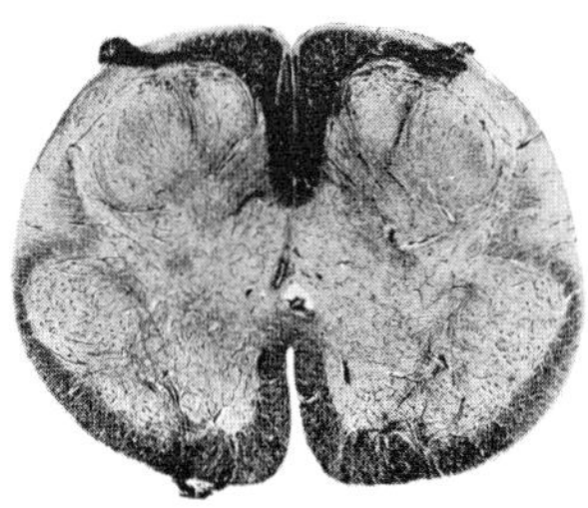

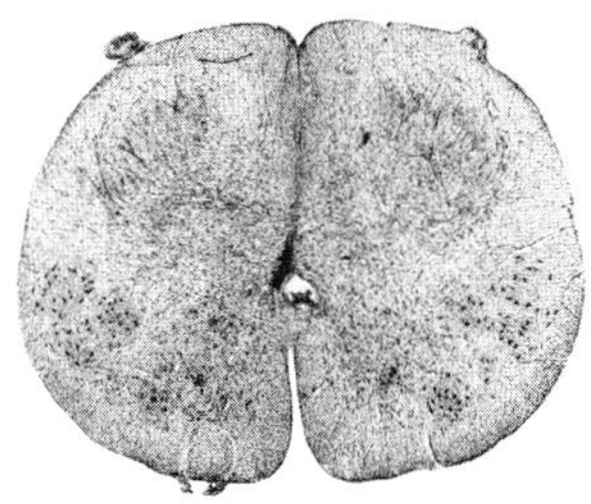

Plates 89A, 89B
Lumbar Enlargement
Total Area:
12.162 mm^2

Plates 90A, 90B
Sacral/Coccygeal
Total Area:
5.6792 mm^2

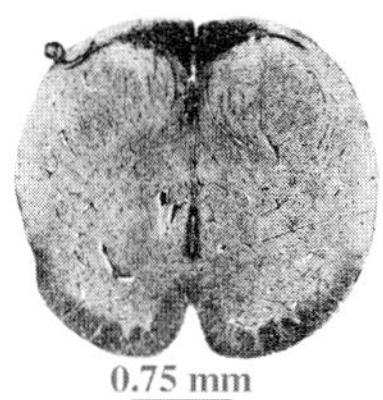

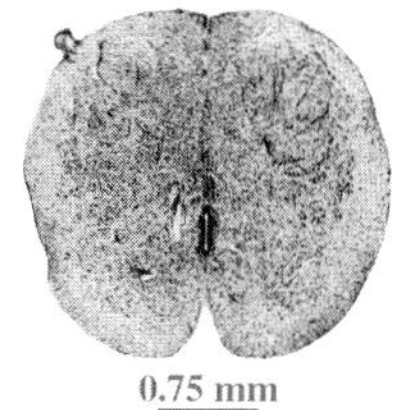

Plates 91A, 91B
Sacral/Coccygeal
Total Area:
5.9052 mm^2

PLATE 80A

Infant, 4 Days
CR 350 mm
Y299-62
Cervical Enlargement
Myelin stain

Areas (mm^2)

Central canal	.0029
Ependyma	.0119
Gray matter	6.9875
White matter	13.0610

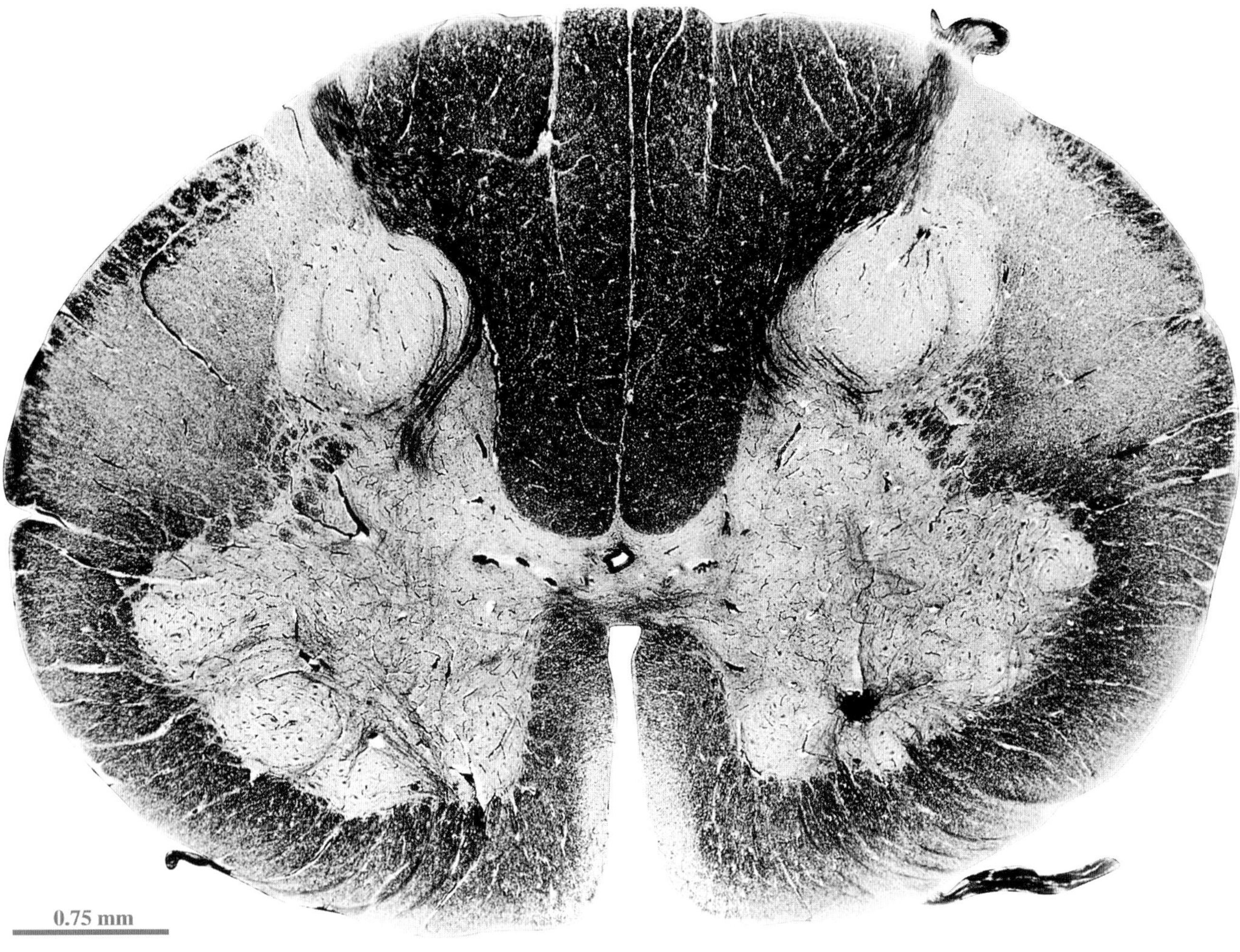

PLATE 80B

See the matched cell body stained section in Plates 81A and B

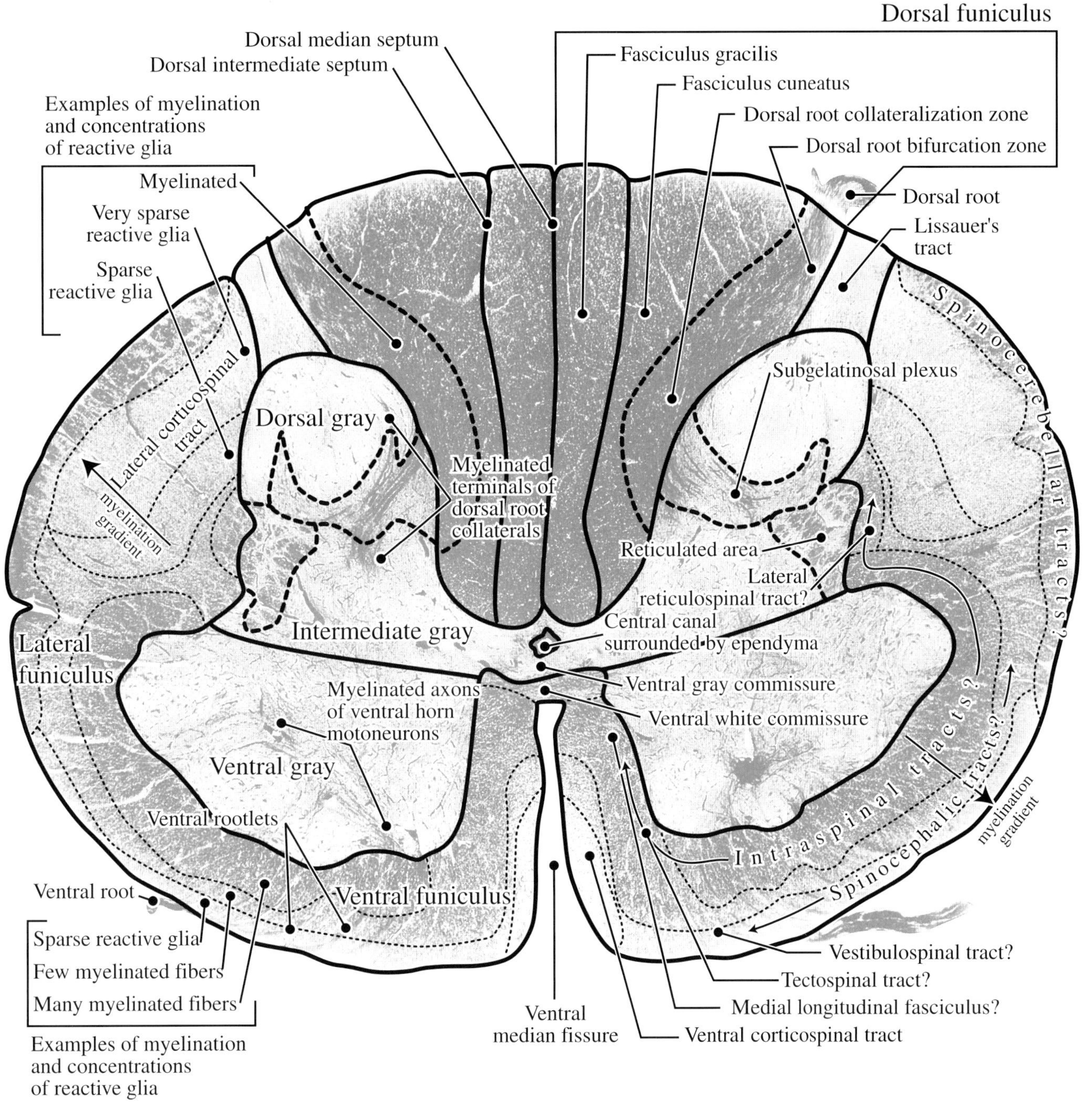

The fine lines in the lateral and ventral funiculi segregate regions of differing densities of the myelin stain, not the borders of fiber tracts. These lines may correspond to the borders of some fiber tracts, such as the spinocerebellar and the lateral corticospinal tracts. Other lines run across fiber tracts, such as the area surrounding the ventral horn that includes the medial and central parts of the intraspinal tracts, and the internal fibers of the spinocephalic tracts. Most tracts in the ventral and lateral funiculi overlap and do not have distinct borders. The ventral part of the spinocephalic tracts overlaps with the vestibulospinal tract; its central part includes fibers in the ventrolateral reticulospinal tract (not labeled), the spinotectal tract (not labeled), and the spino-olivary tract (not labeled). The intraspinal tracts overlap with the medial longitudinal fasciculus, the tectospinal tract, and the lateral reticulospinal tract.

PLATE 81A

Infant, 4 Days
CR 350 mm
Y299-62
Cervical Enlargement
Cell body stain

Areas (mm^2)

Central canal	.0019
Ependyma	.0167
Gray matter	7.8633
White matter	13.4880

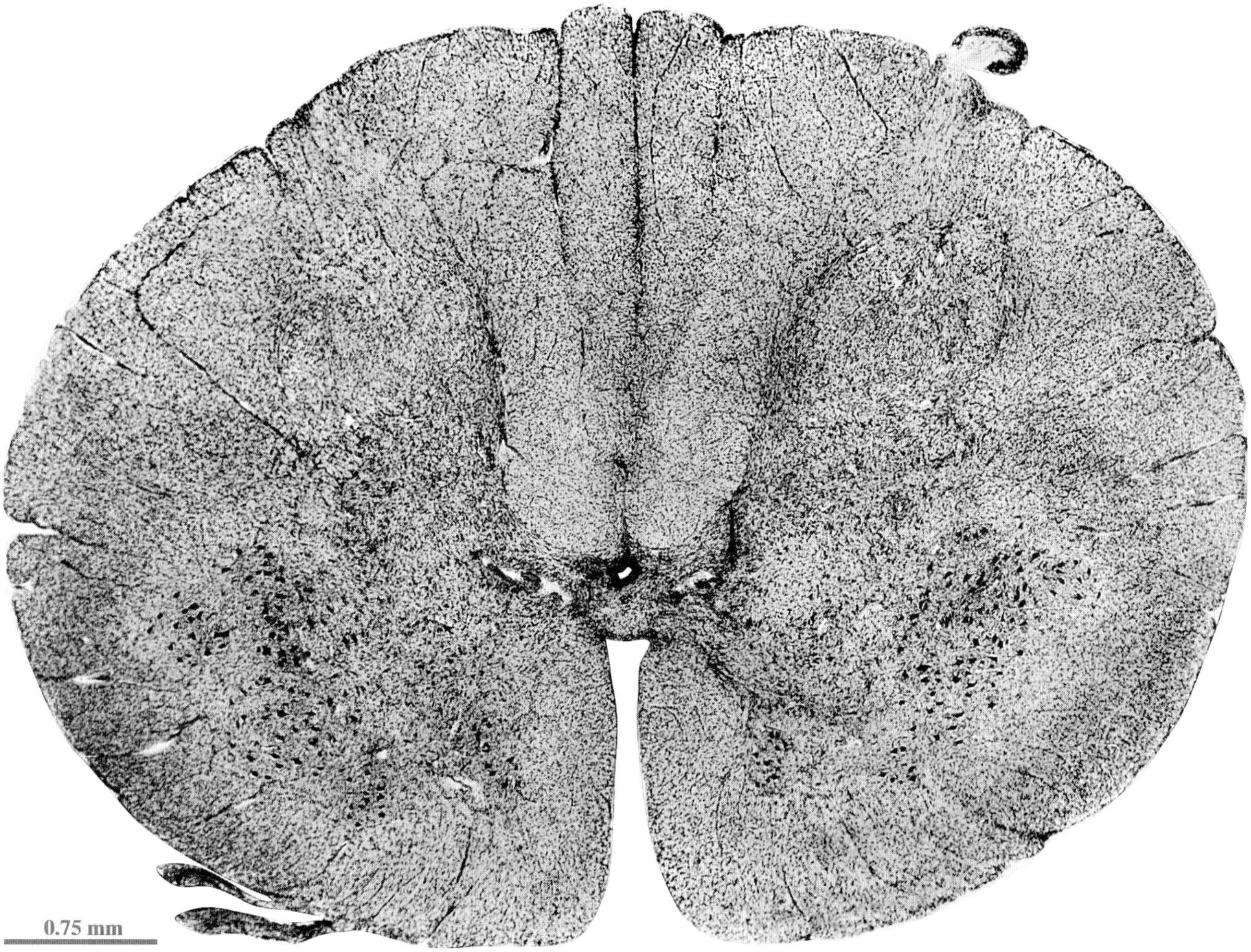

PLATE 81B

See the matched myelin stained section in Plates 80A and B

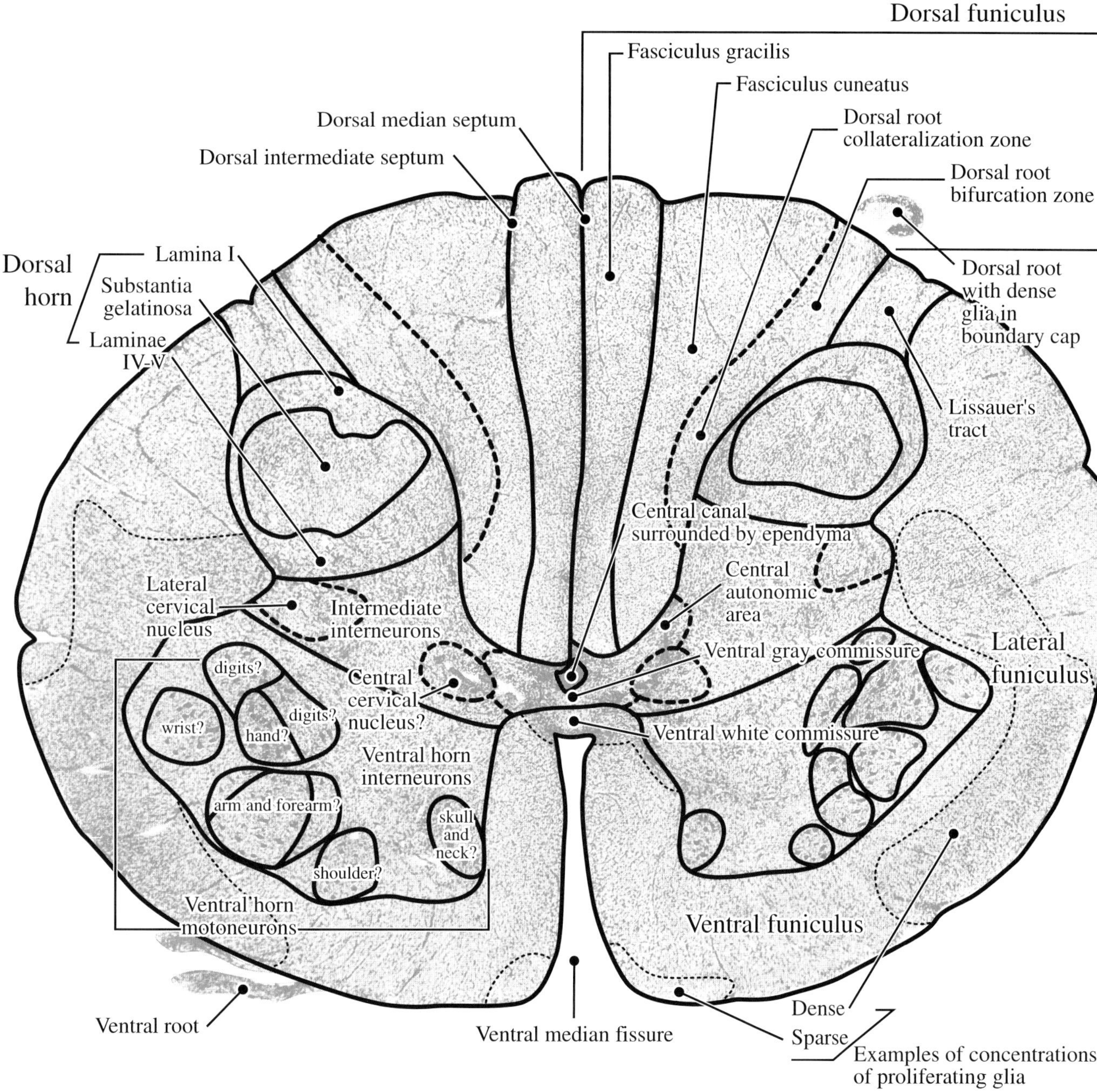

The corticospinal tracts no longer stand out as clear areas with very sparse proliferating glia in the white matter. There is a large area of slightly more dense proliferating glia in the ventral and lateral funiculi that corresponds partly to the regions that are still myelinating. Refer to the matching myelin stained section for the approximate locations of both corticospinal tracts, the lateral reticulospinal, vestibulospinal, intraspinal, and spinocephalic tracts.

PLATE 82A

Infant, 4 Days
CR 350 mm
Y299-62
Upper Thoracic
Myelin stain

Areas (mm^2)

Central canal	.0016
Ependyma	.0081
Gray matter	2.8495
White matter	7.3181

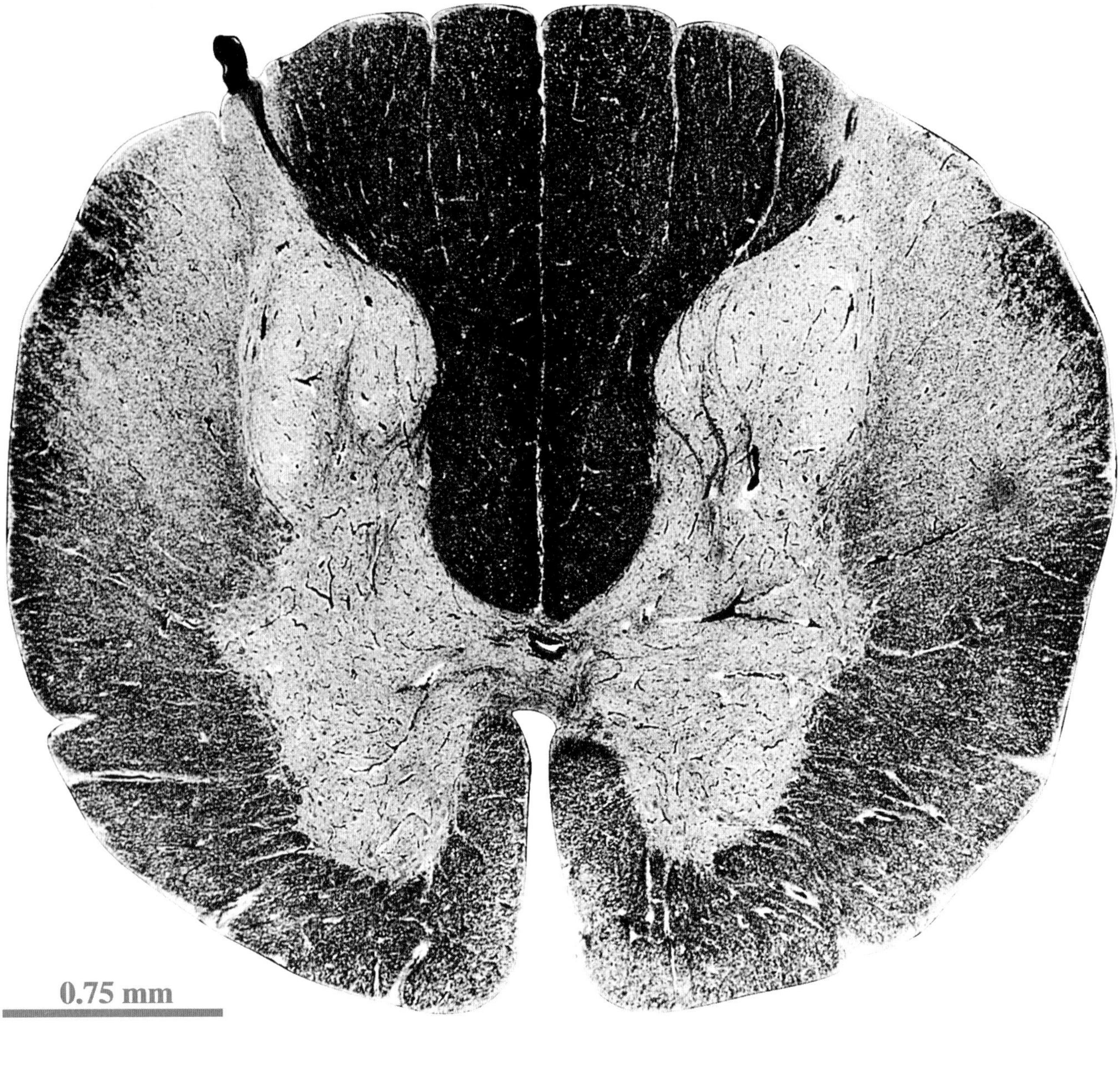

0.75 mm

PLATE 82B

See the matched cell body stained section in Plates 83A and B

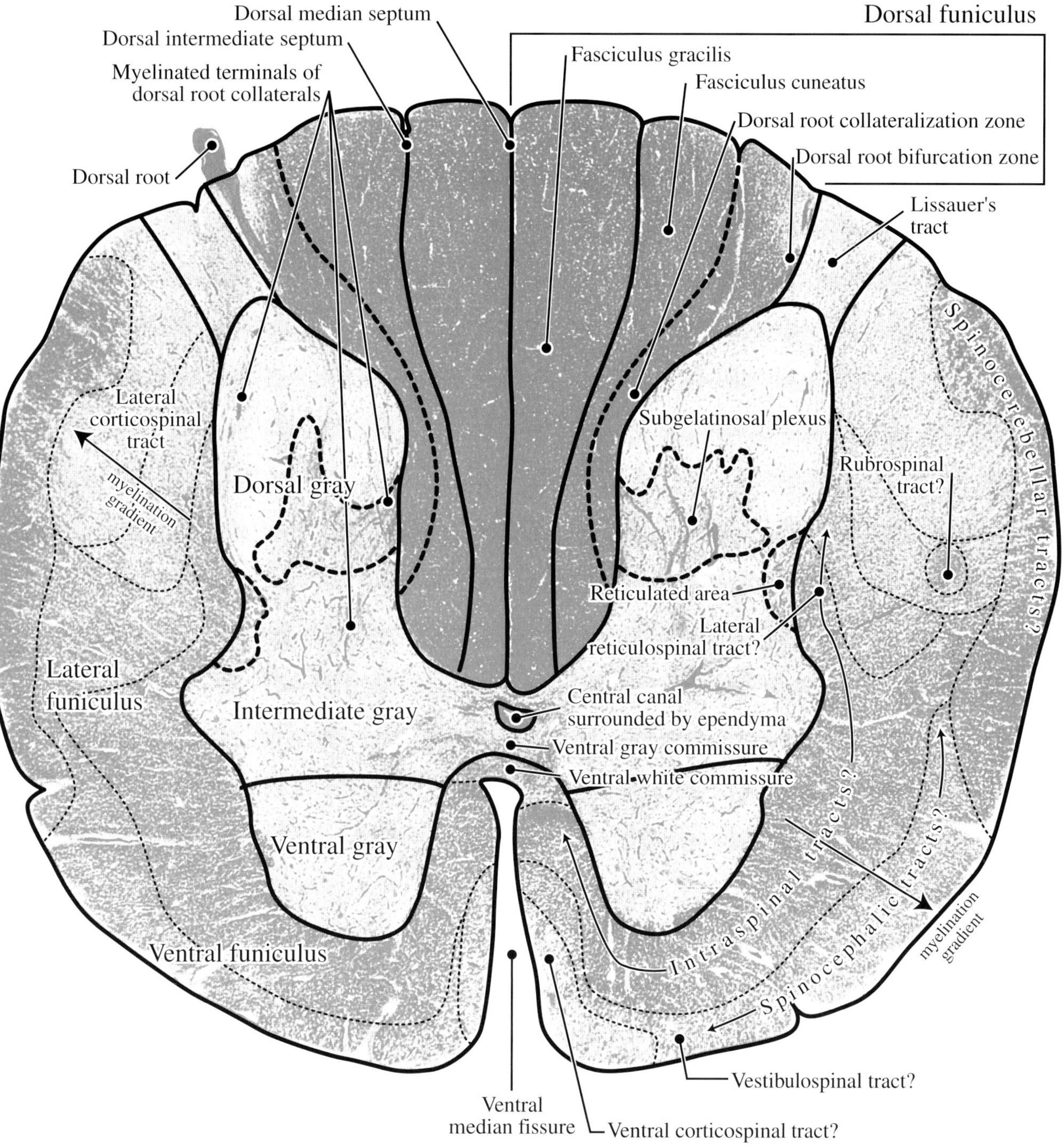

This section and **Plate 80** show the gradient of sparse to very sparse reactive glia in the lateral corticospinal tract (separated by a dashed line within the lateral corticospinal tract). On the right side, the circular area of more dense reactive glia may be the rubrospinal tract.

PLATE 83A

Infant, 4 Days
CR 350 mm
Y299-62
Upper Thoracic
Cell body stain

Areas (mm^2)

Central canal	.0019
Ependyma	.0141
Gray matter	2.9713
White matter	7.5265

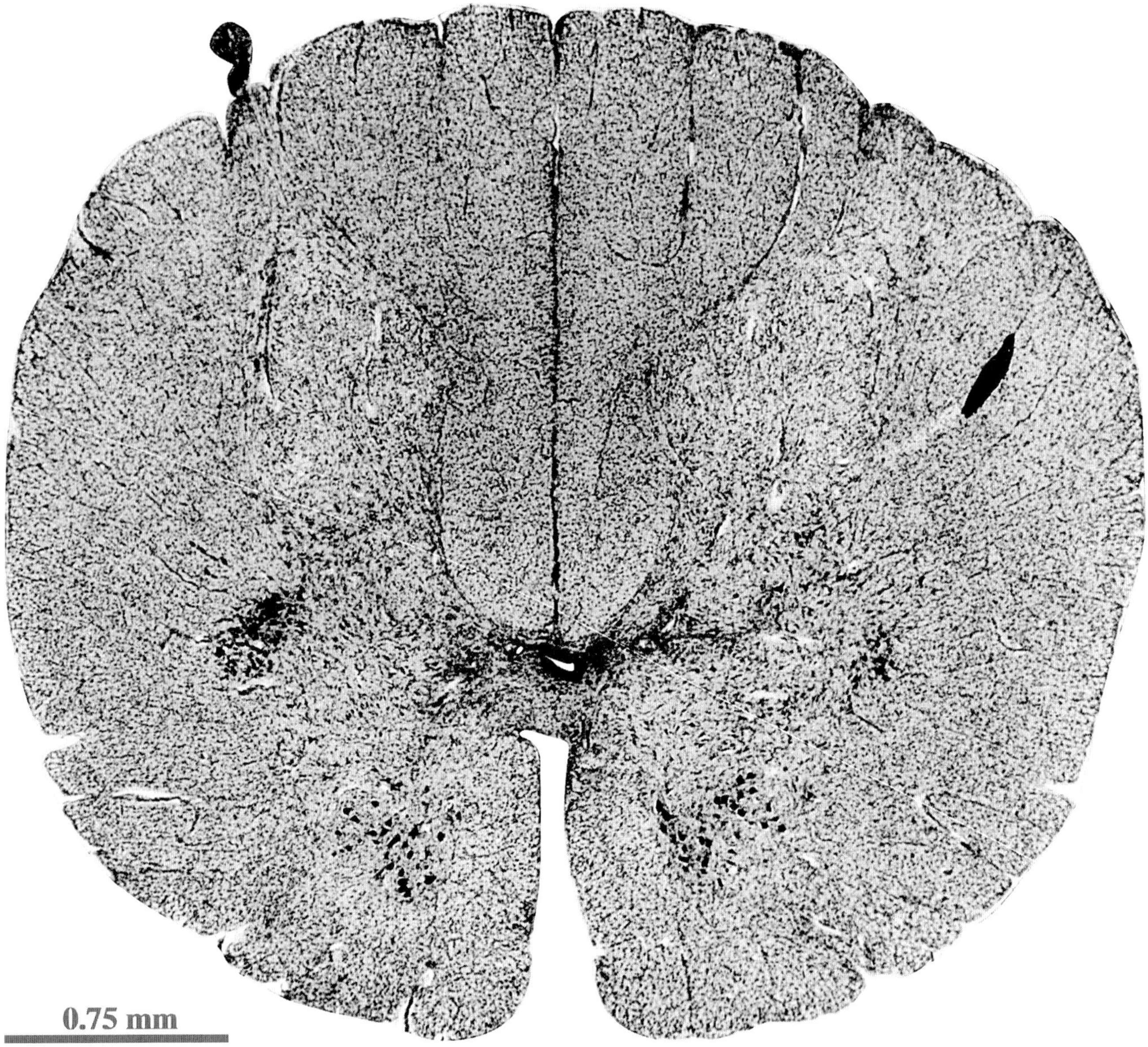

PLATE 83B

See the matched myelin stained section in Plates 82A and B

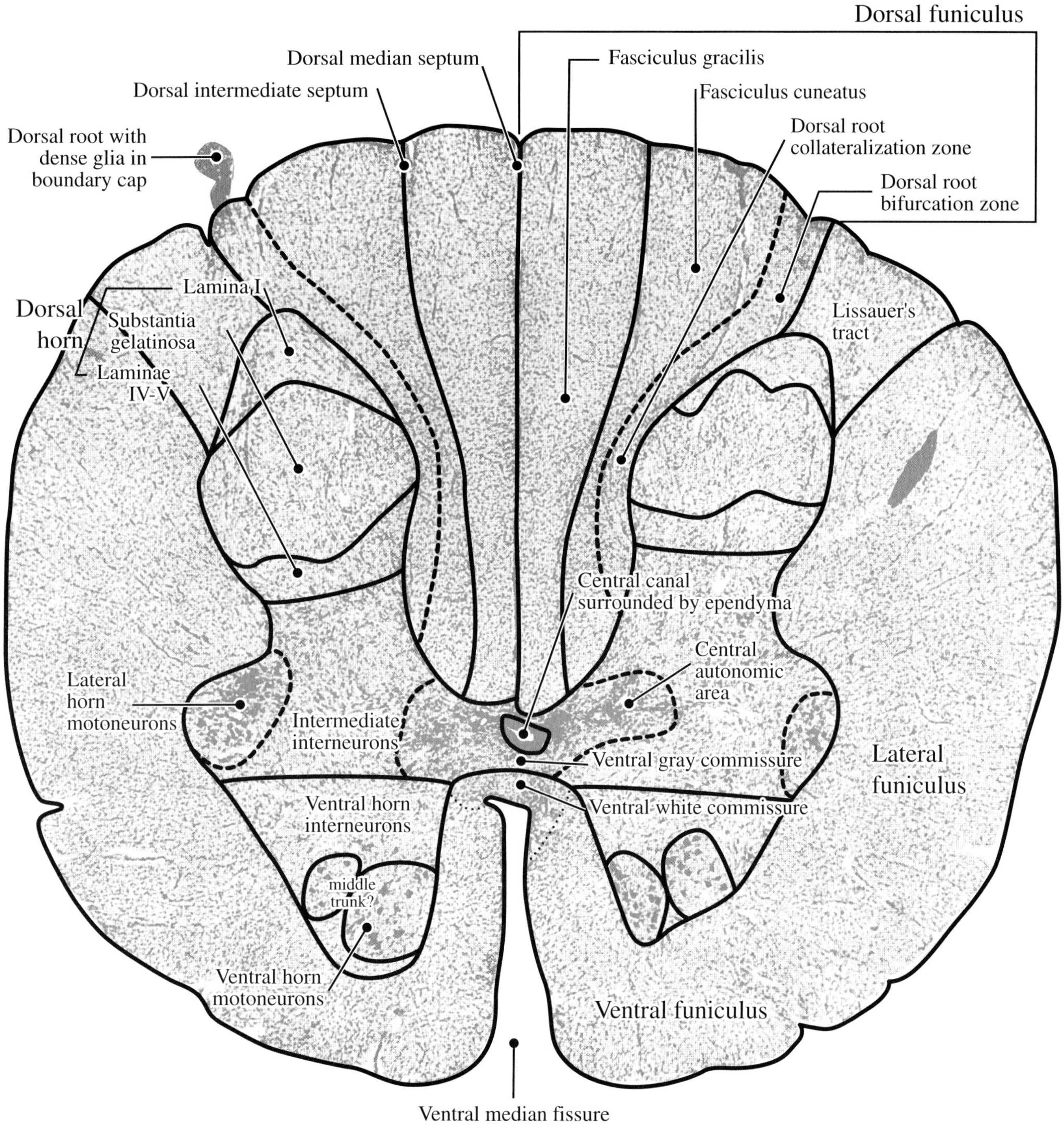

The white matter in this section has a uniformly sparse population of proliferating glia, and no fiber tracts in the ventral and lateral funiculi stand out with a different glial concentration. This suggests that myelination gliosis is finished. Refer to the matching myelin stained section for the approximate locations of the lateral corticospinal, ventral corticospinal, rubrospinal, lateral reticulospinal, vestibulospinal, intraspinal, and spinocephalic tracts.

PLATE 84A

Infant, 4 Days
CR 350 mm
Y299-62
Lower Thoracic
Myelin stain

Areas (mm²)

Central canal	.0005
Ependyma	.0047
Gray matter	1.9740
White matter	5.8169

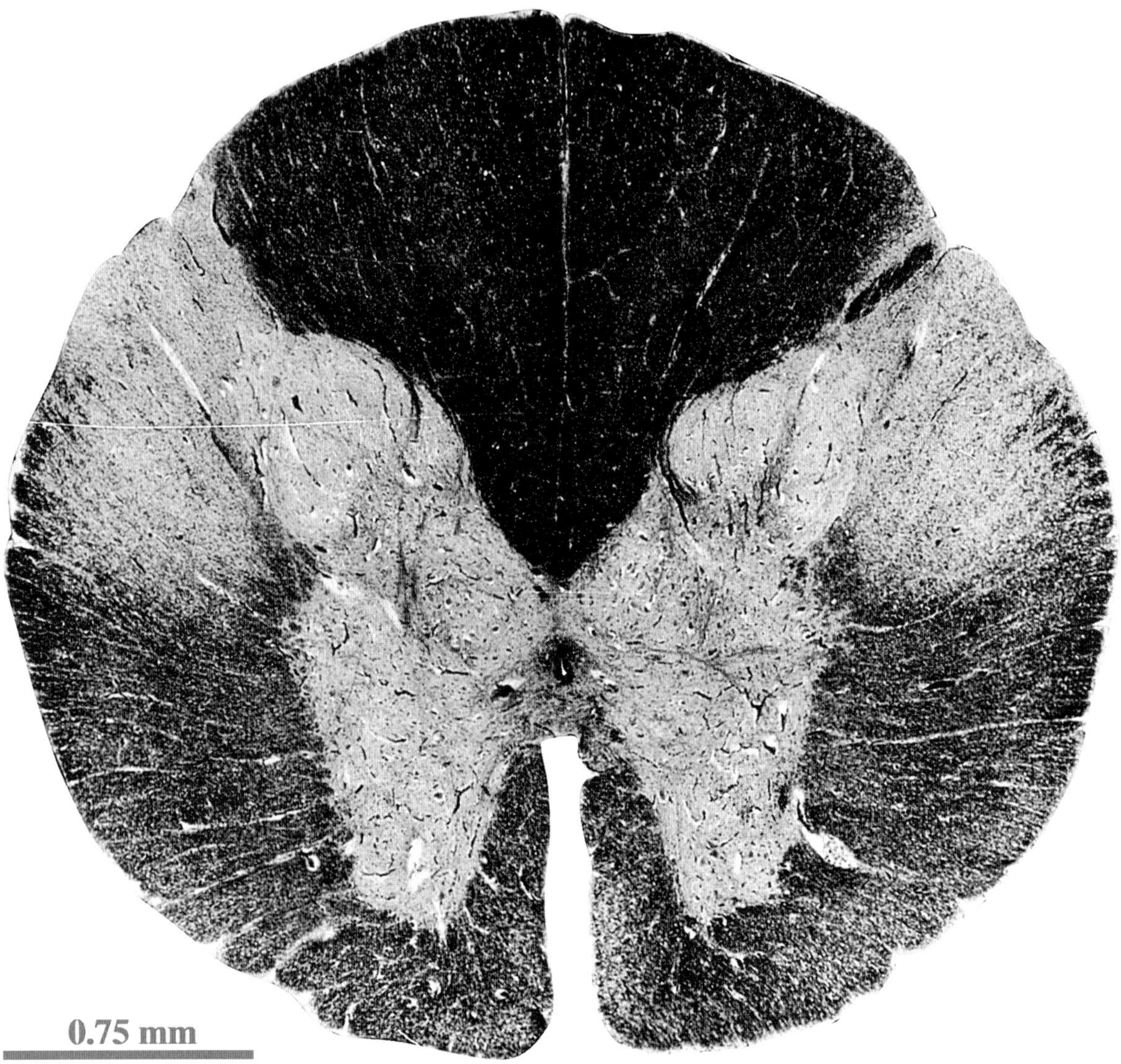

PLATE 84B

See the matched cell body stained section in Plates 85A and B

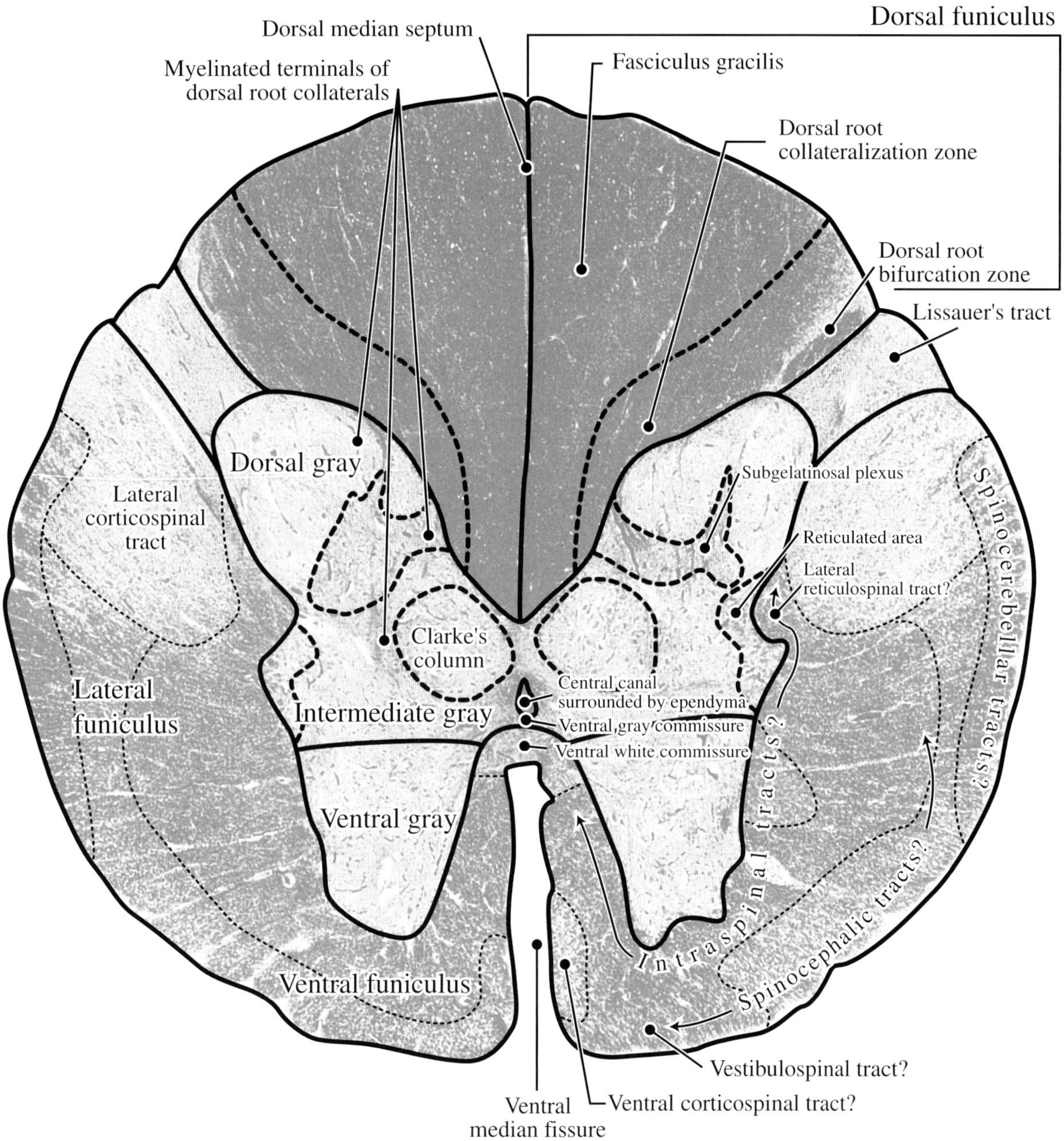

PLATE 85A

Infant, 4 Days
CR 350 mm
Y299-62
Lower Thoracic
Cell body stain

Areas (mm^2)

Central canal	.0015
Ependyma	.0148
Gray matter	2.1221
White matter	6.0785

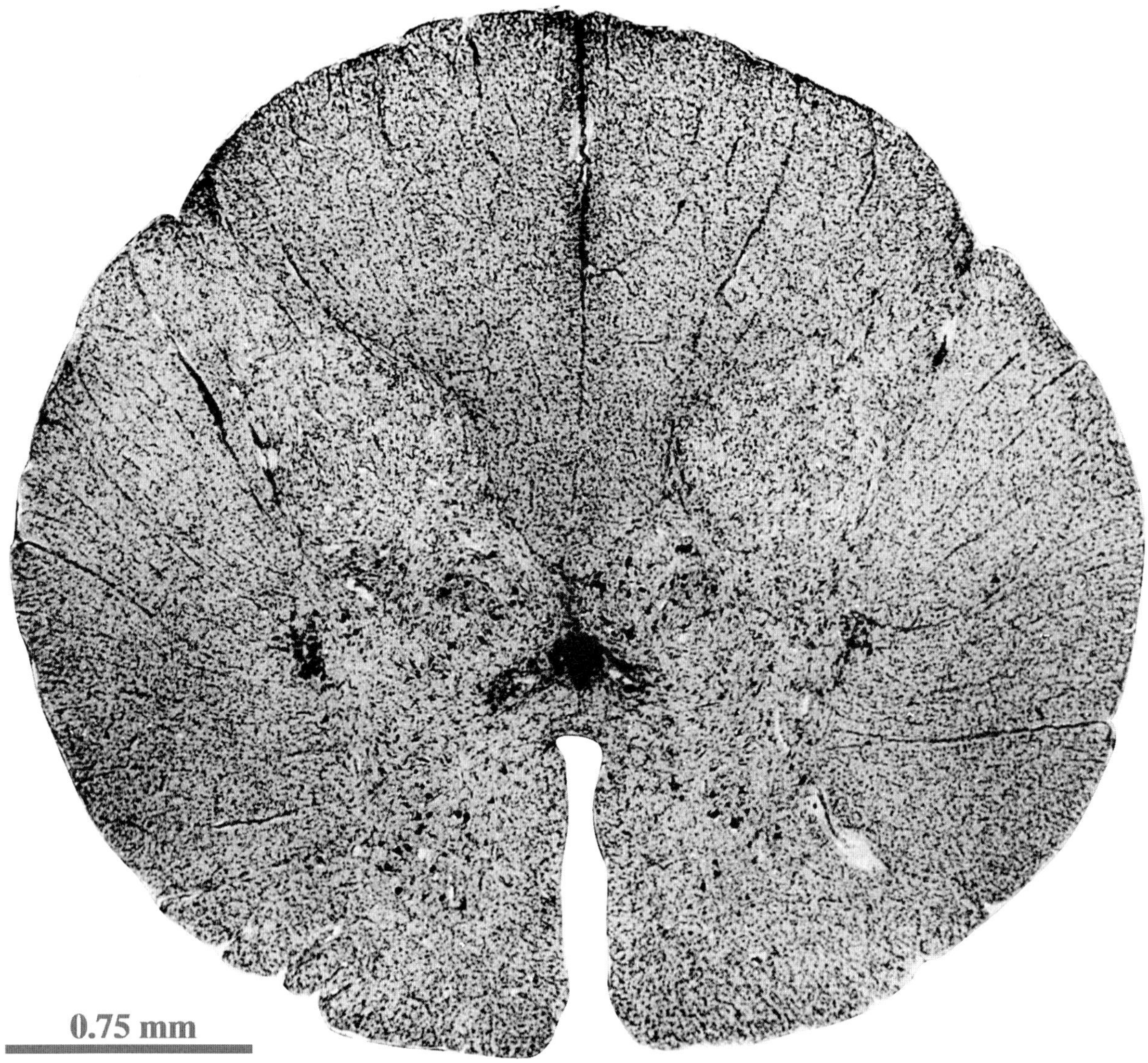

PLATE 85B

See the matched myelin stained section in Plates 84A and B

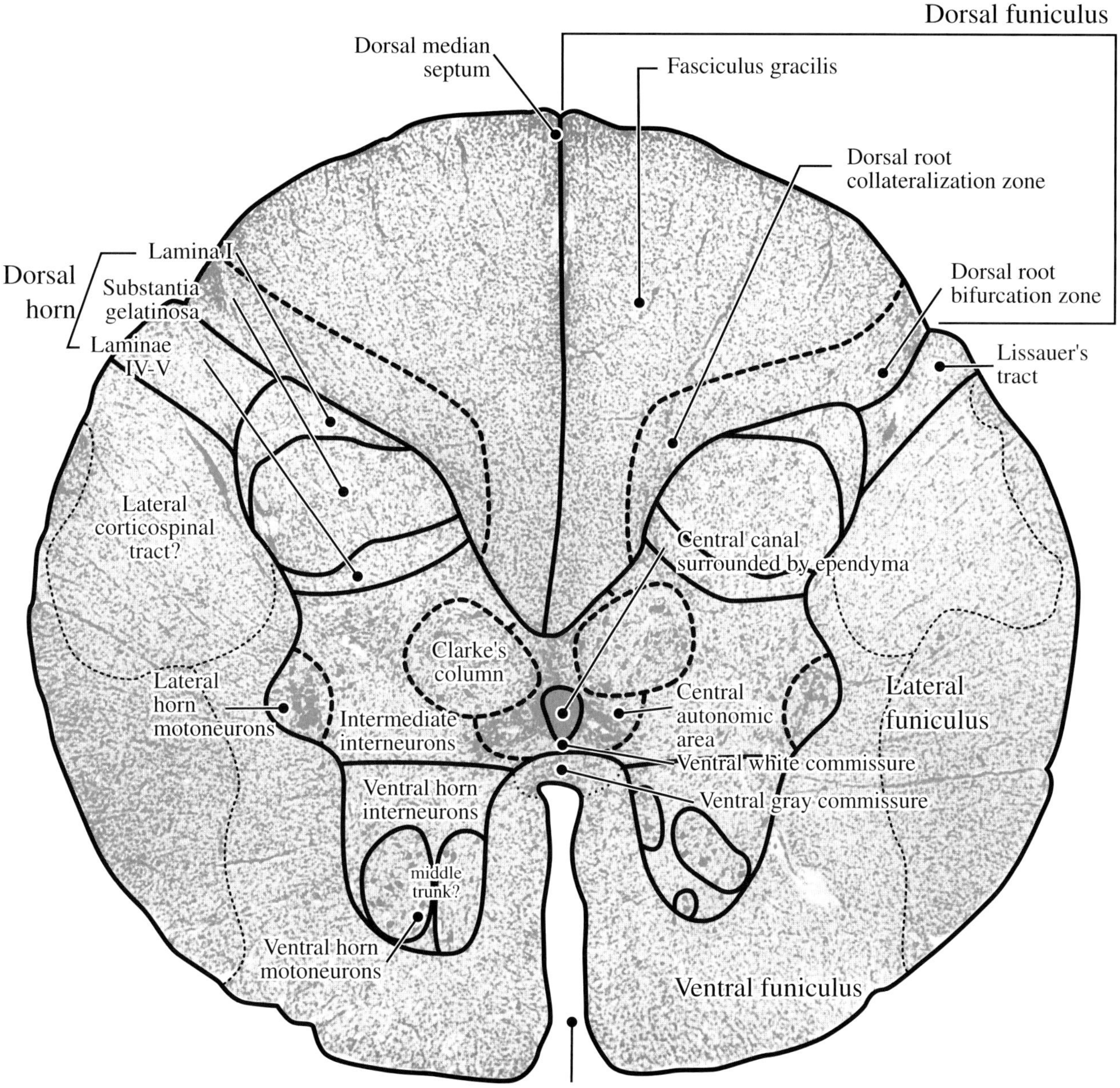

This caudal section shows a slight reduction in the concentration of proliferating glia in the lateral corticospinal tract; that suggests a rostral to caudal gradient in the myelination of the lateral corticospinal tract. There is also a region of slightly greater density in the ventral and lateral funiculi. Refer to the matching myelin stained section for the approximate locations of the lateral reticulospinal, vestibulospinal, intraspinal, and spinocephalic tracts.

PLATE 86A

Infant, 4 Days
CR 350 mm
Y299-62
Upper Lumbar
Myelin stain

Areas (mm^2)

Central canal	.0019
Ependyma	.0087
Gray matter	4.7867
White matter	6.9382

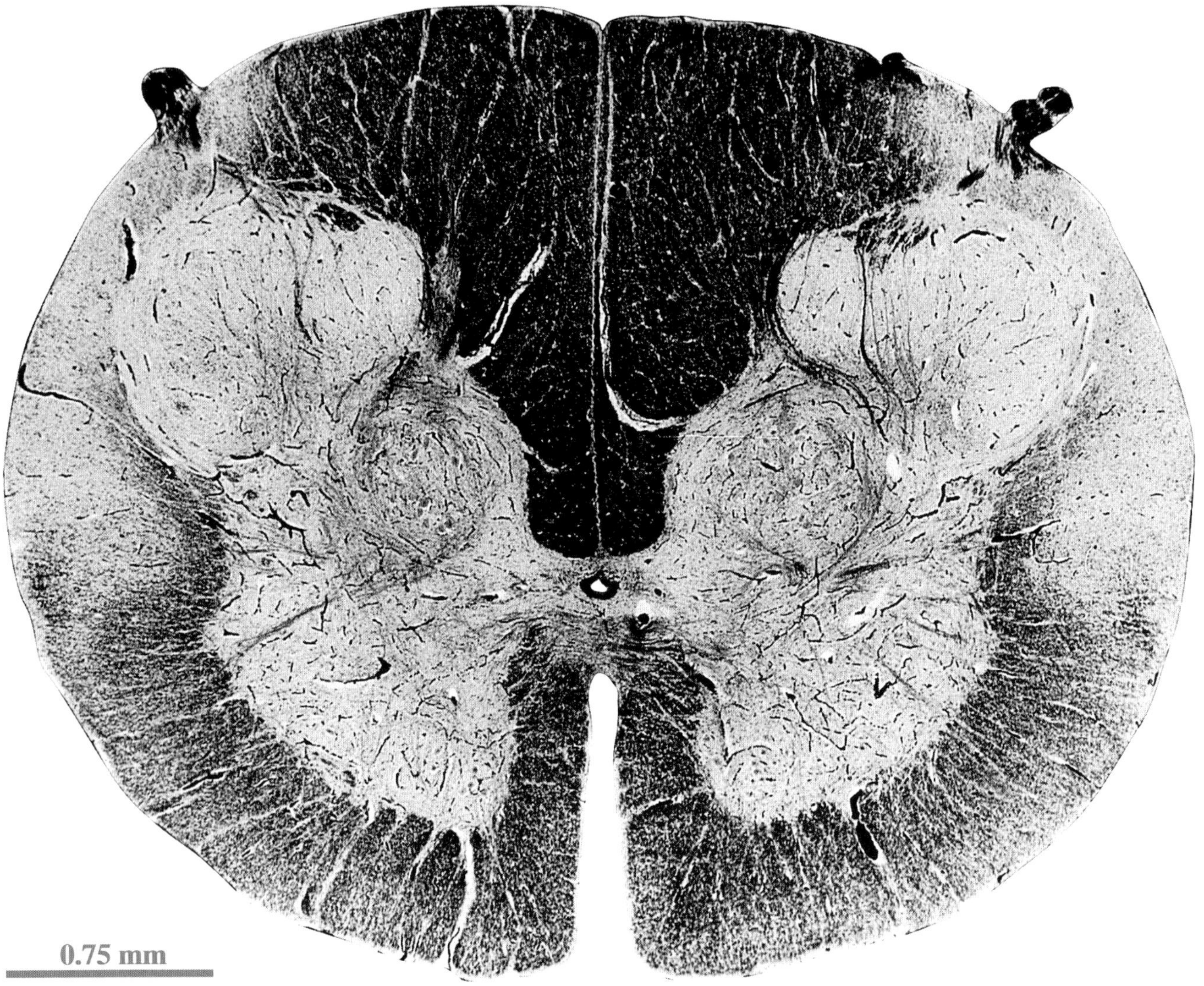

PLATE 86B

See the matched cell body stained section in Plates 87A and B

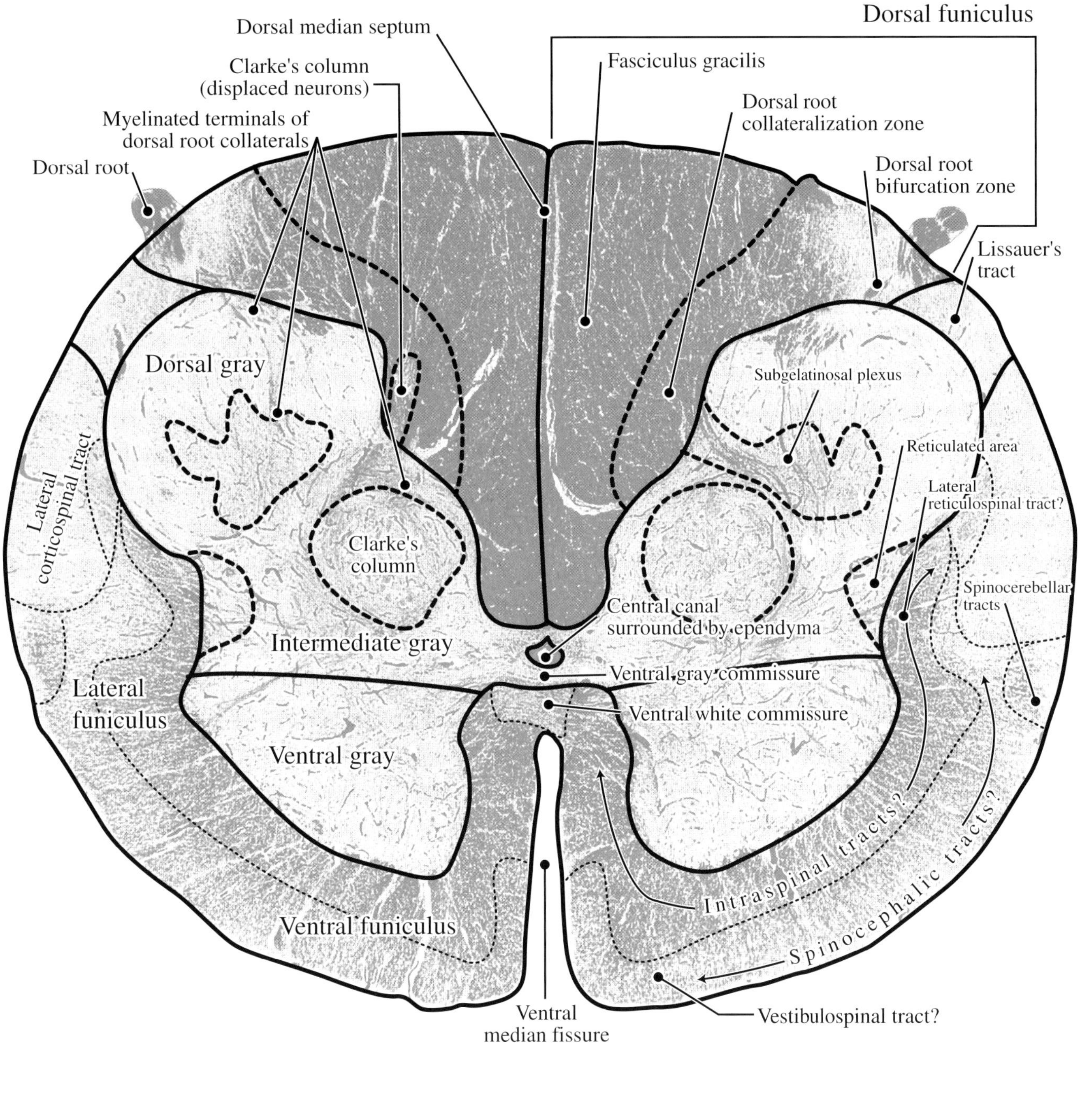

PLATE 87A

Infant, 4 Days
CR 350 mm
Y299-62
Upper Lumbar
Cell body stain

Areas (mm^2)

Central canal	.0014
Ependyma	.0147
Gray matter	5.2589
White matter	7.4545

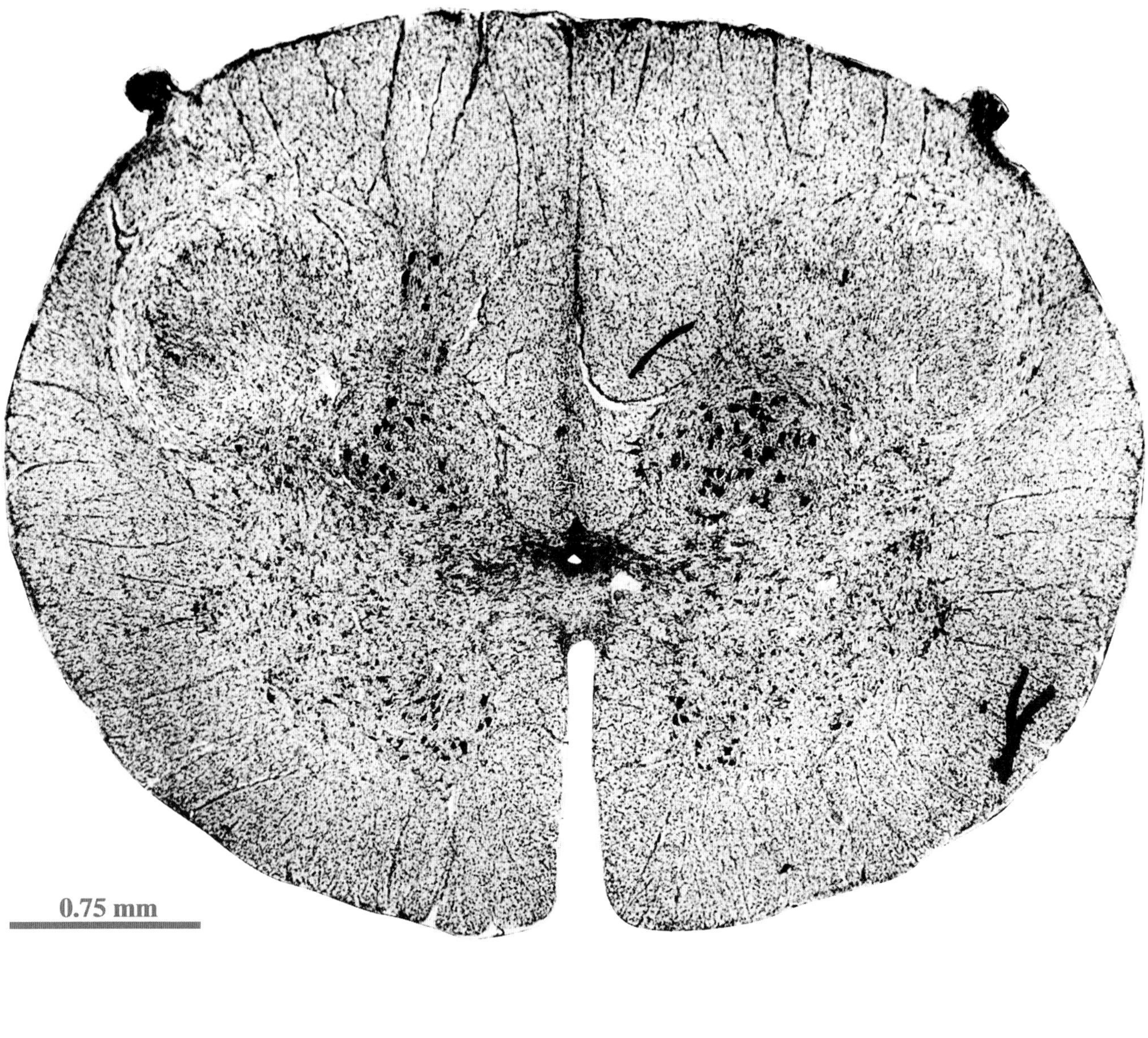

PLATE 87B

See the matched myelin stained section in Plates 86A and B

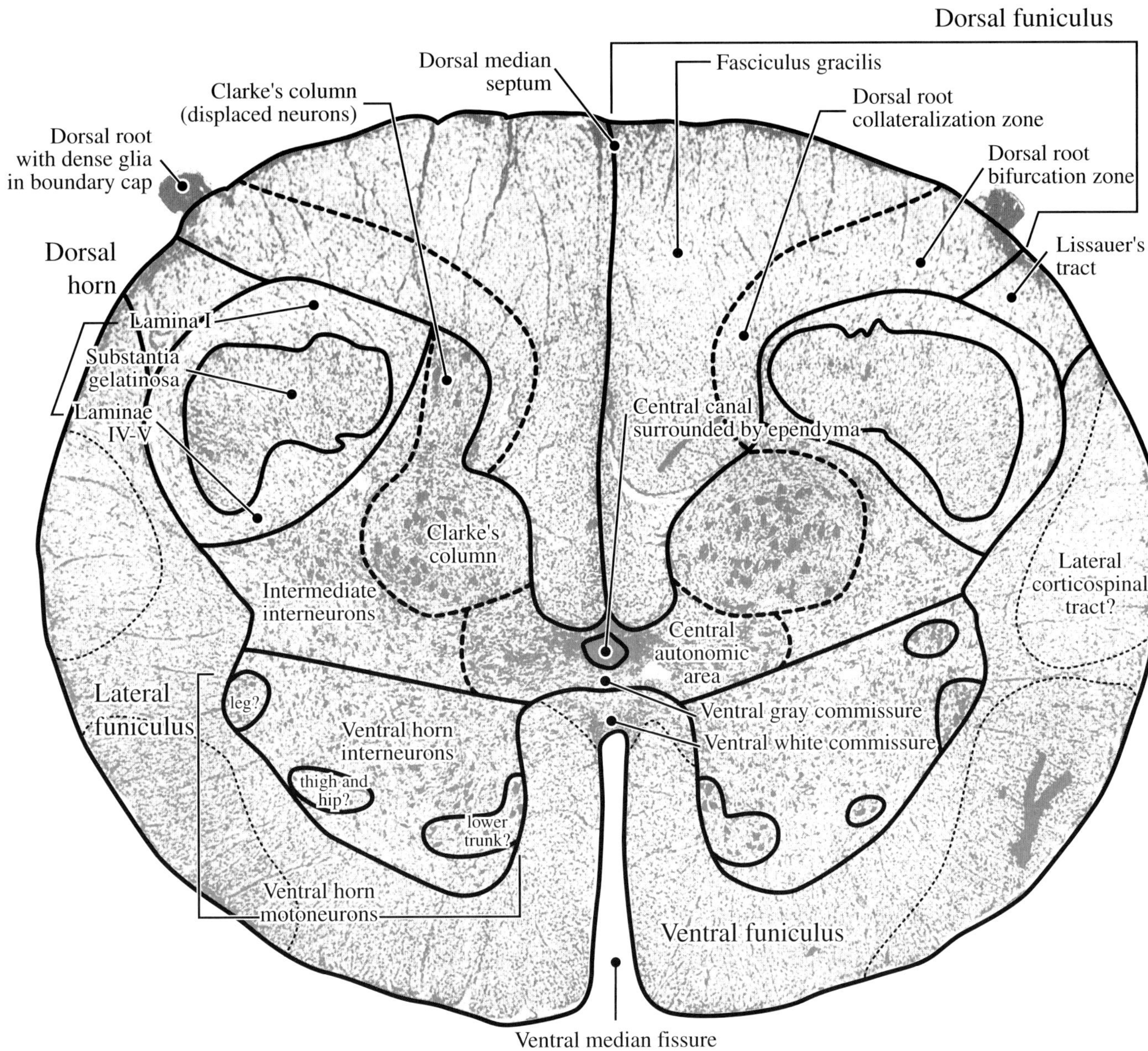

Refer to the matching myelin stained section for the approximate locations of the lateral reticulospinal, vestibulospinal, intraspinal, and spinocephalic tracts.

PLATE 88A

Infant, 4 Days
CR 350 mm
Y299-62
Lumbar Enlargement
Myelin stain

Areas (mm^2)

Central canal	.0020
Ependyma	.0104
Gray matter	8.0120
White matter	4.1273

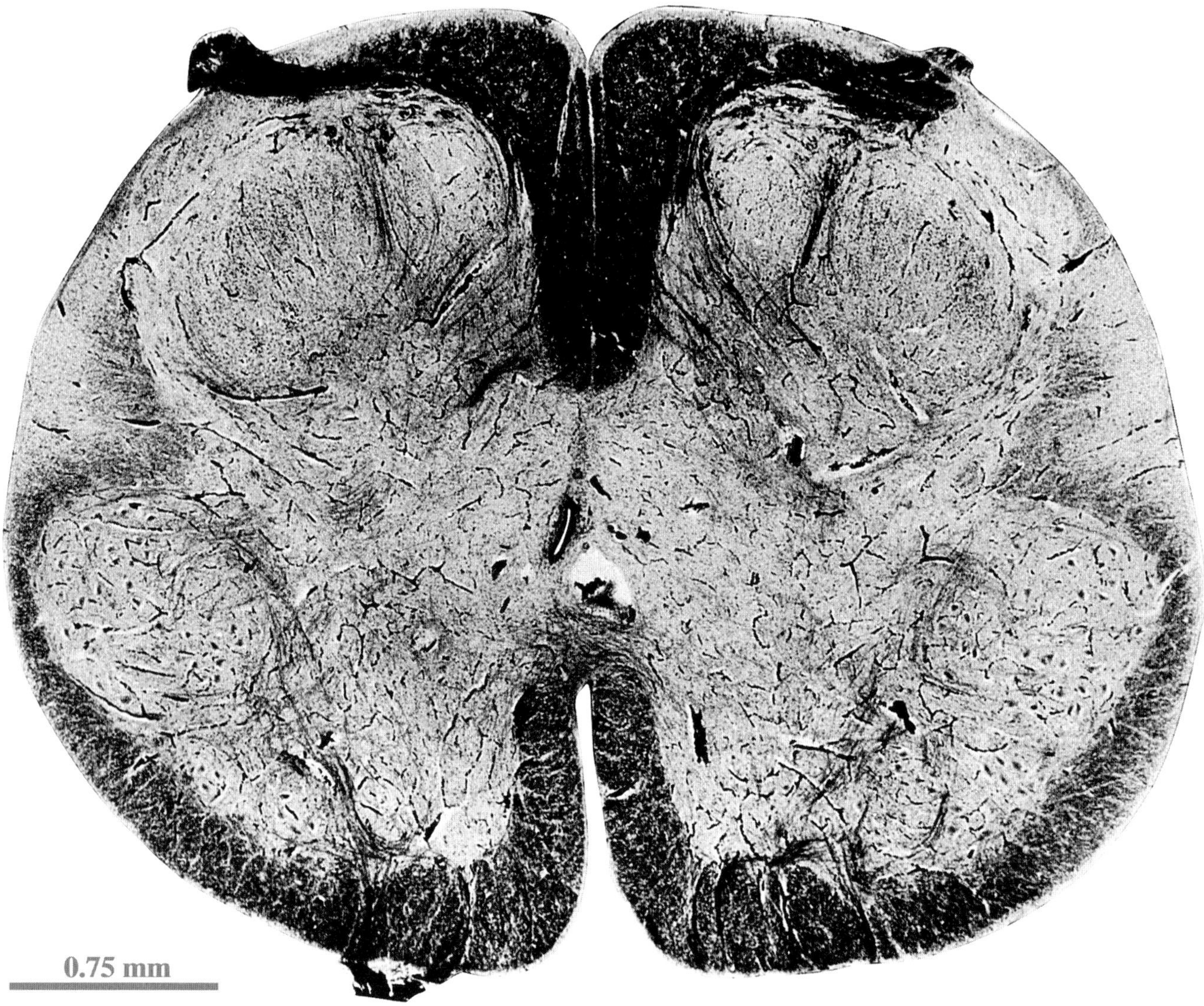

PLATE 88B

See the matched cell body stained section in Plates 89A and B

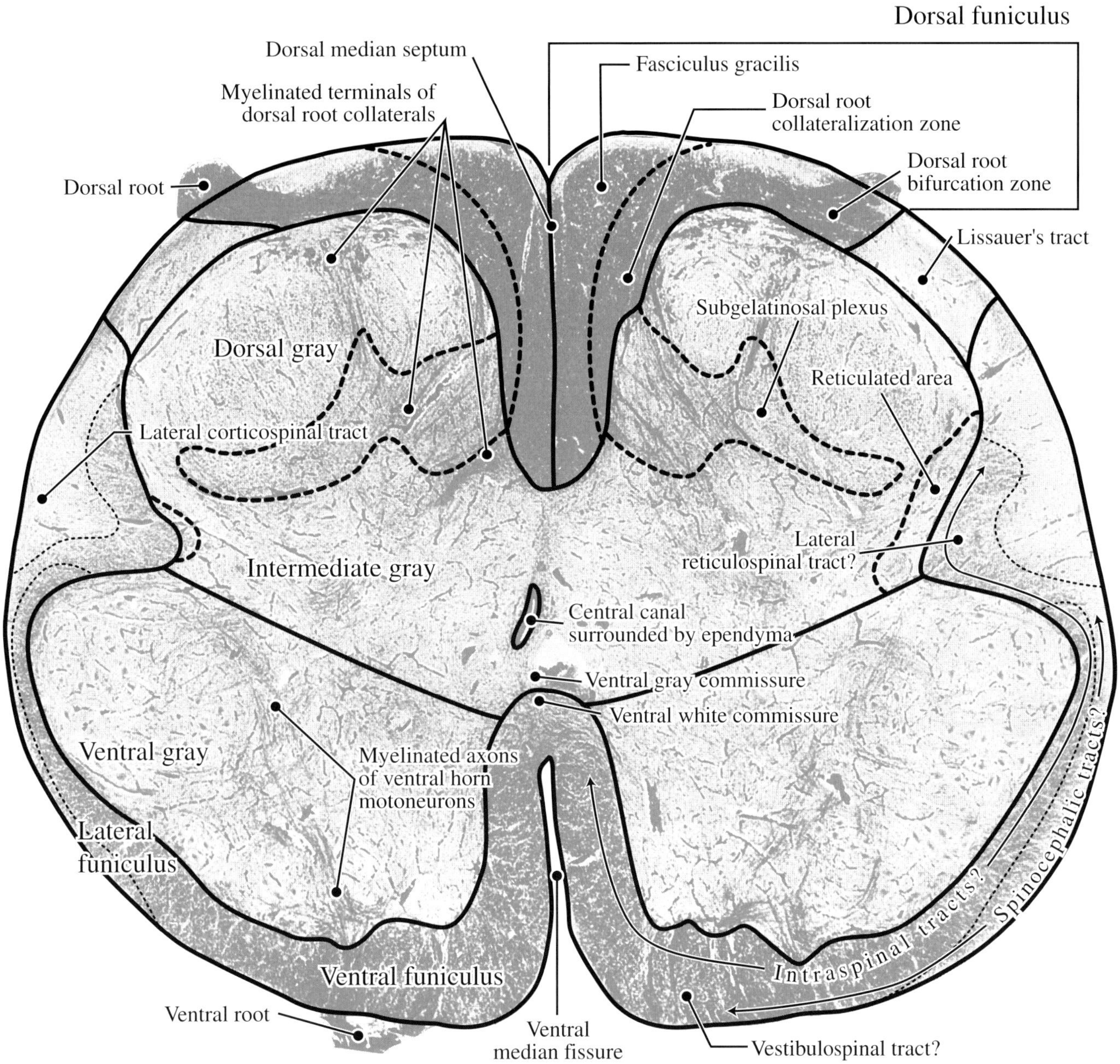

PLATE 89A

Infant, 4 Days
CR 350 mm
Y299-62
Lumbar Enlargement
Cell body stain

Areas (mm^2)

Central canal	.0046
Ependyma	.0302
Gray matter	7.9475
White matter	4.1796

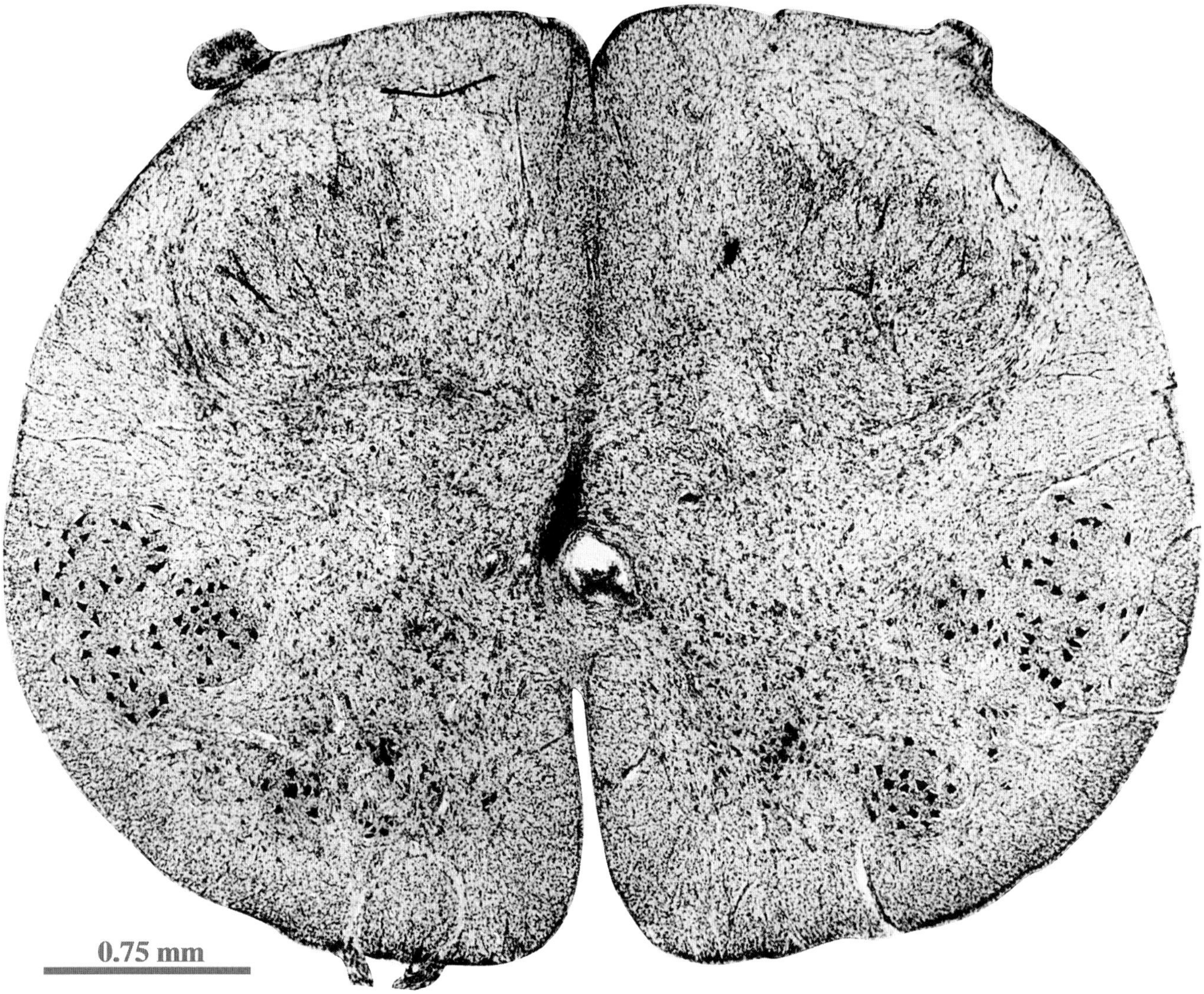

PLATE 89B

See the matched myelin stained section in Plates 88A and B

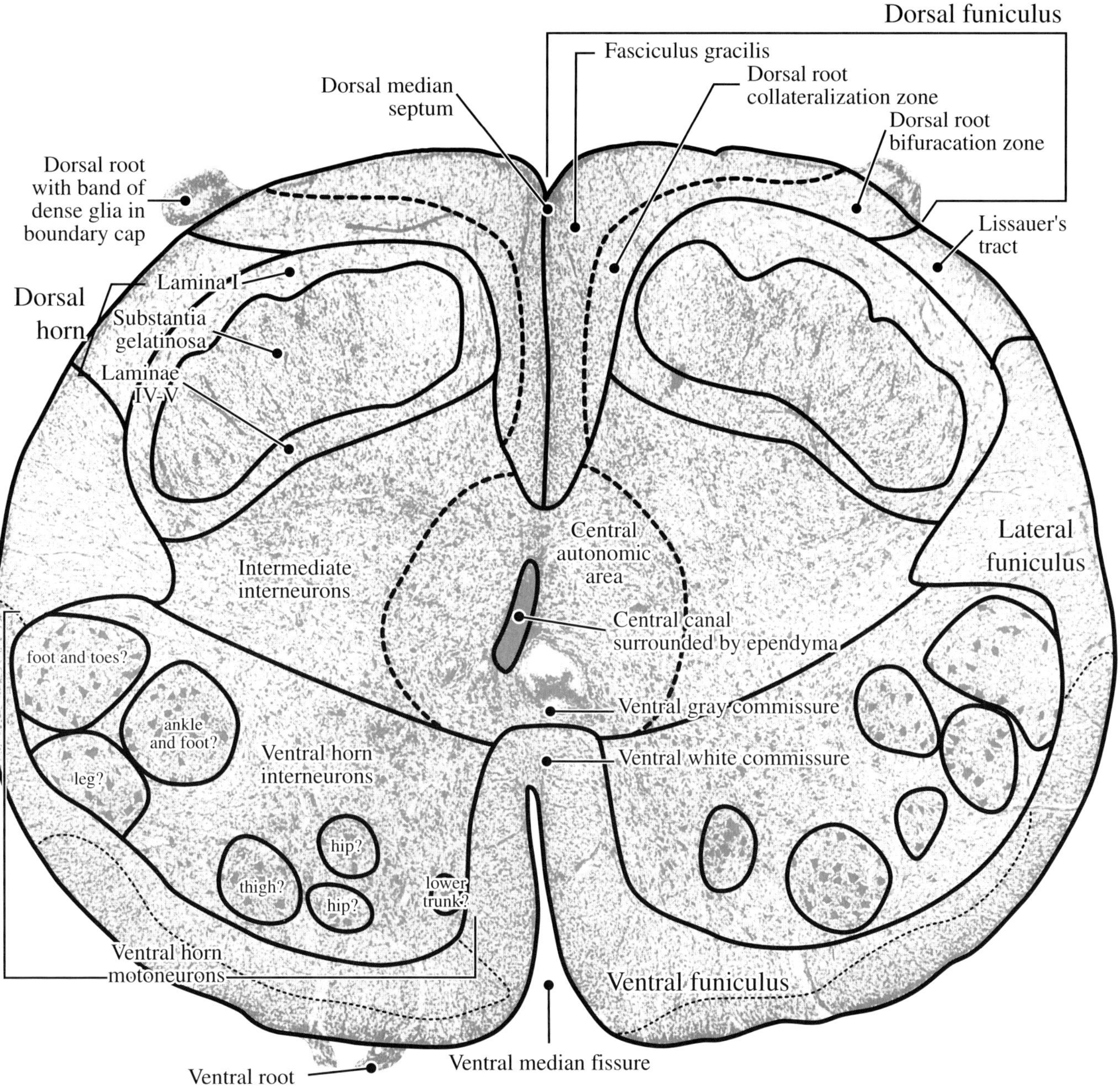

Refer to the matching myelin stained section for the approximate locations of the lateral corticospinal, lateral reticulospinal, vestibulospinal, intraspinal, and spinocephalic tracts.

PLATE 90A

Infant, 4 Days
CR 350 mm
Y299-62
Sacral/Coccygeal
Myelin stain

Areas (mm²)

Central canal	.0036
Ependyma	.0228
Gray matter	3.8397
White matter	1.8131

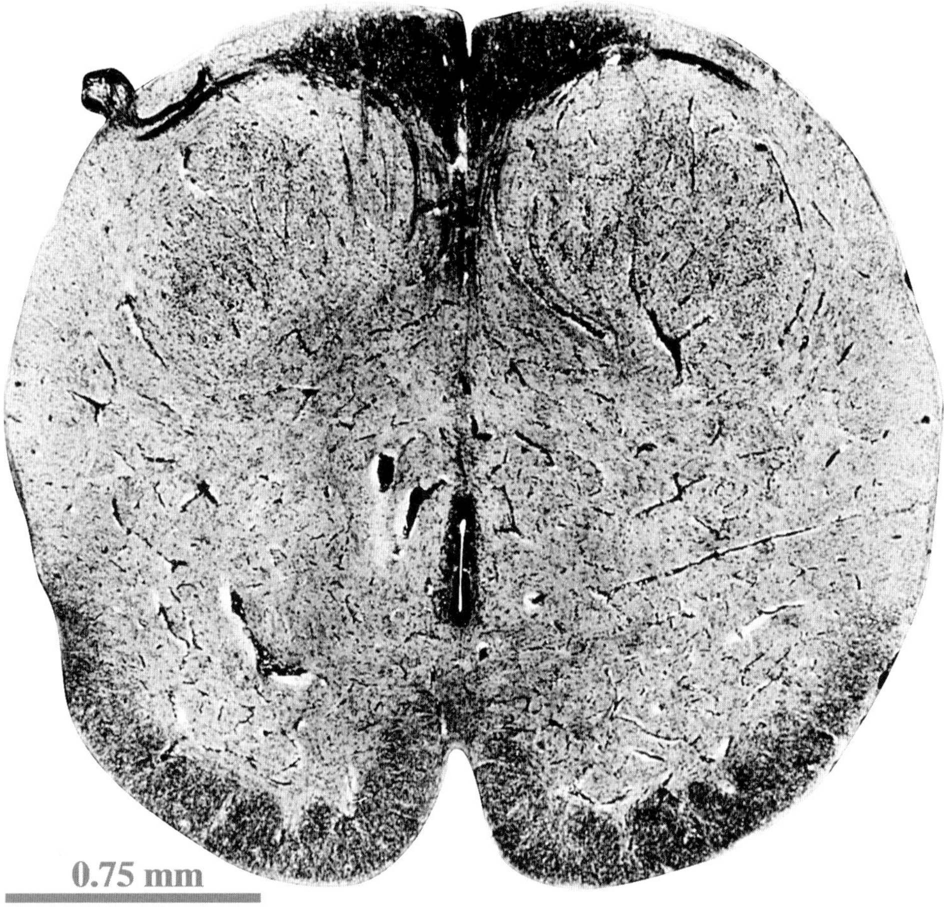

PLATE 90B

See the matched cell body stained section in Plates 91A and B

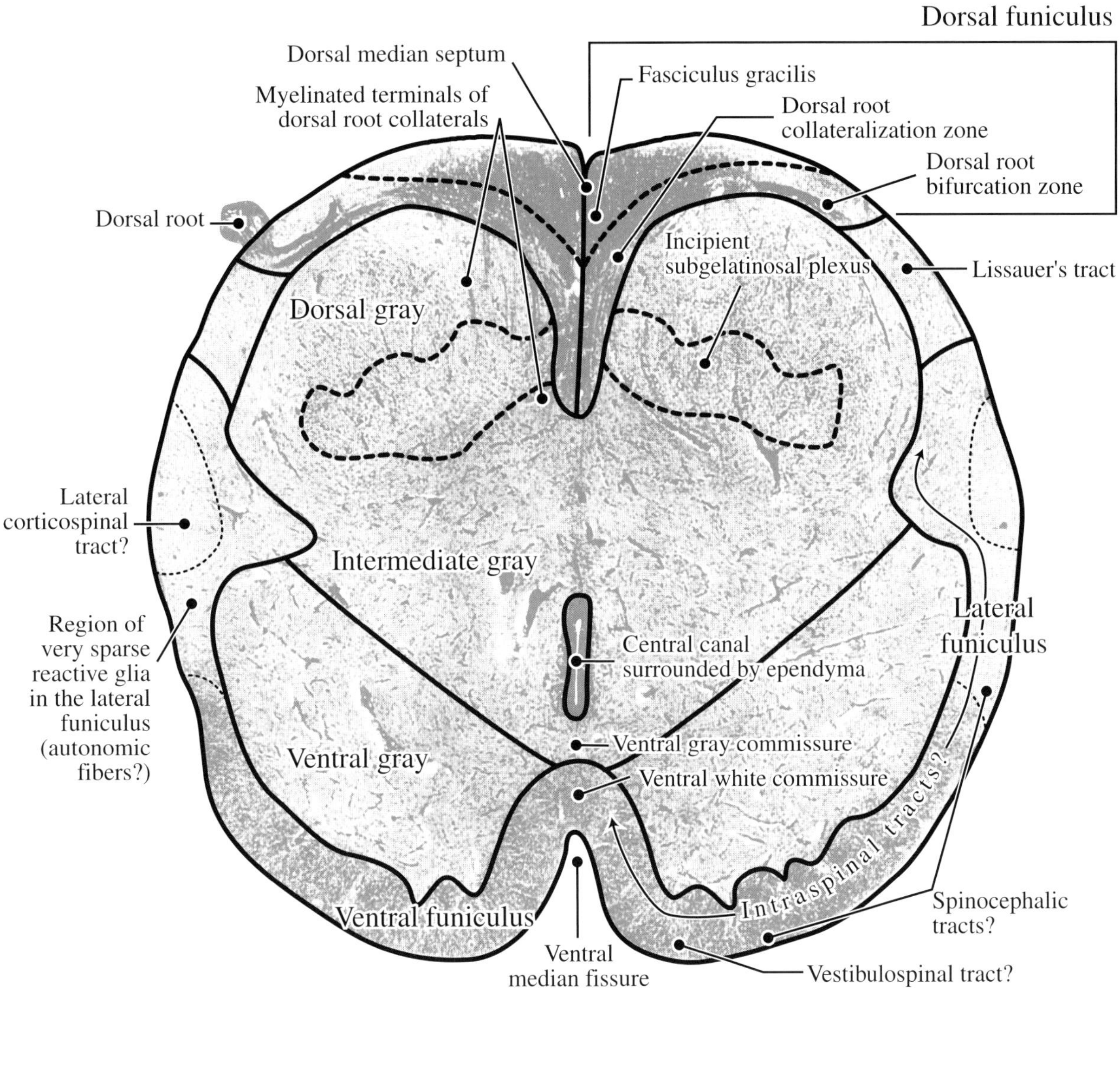

PLATE 91A

Infant, 4 Days
CR 350 mm
Y299-62
Sacral/Coccygeal
Cell body stain

Areas (mm^2)

Central canal	.0037
Ependyma	.0281
Gray matter	4.0451
White matter	1.8283

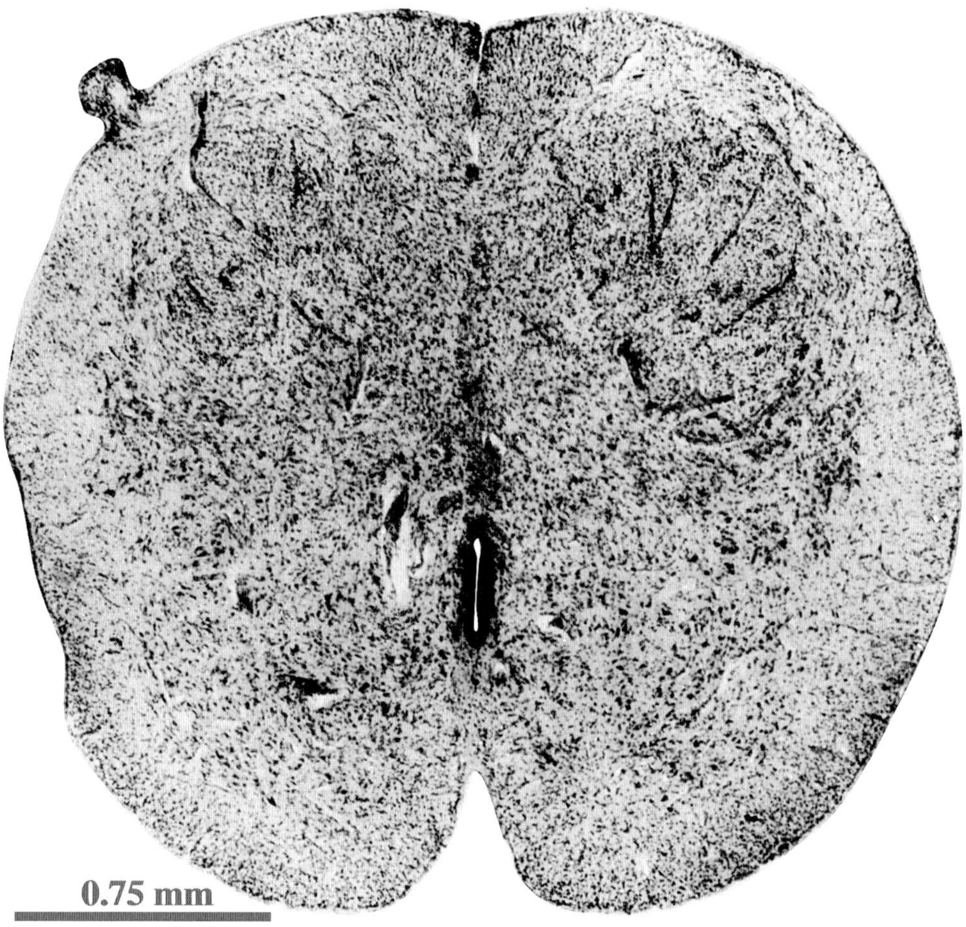

PLATE 91B

See the matched myelin stained section in Plates 90A and B

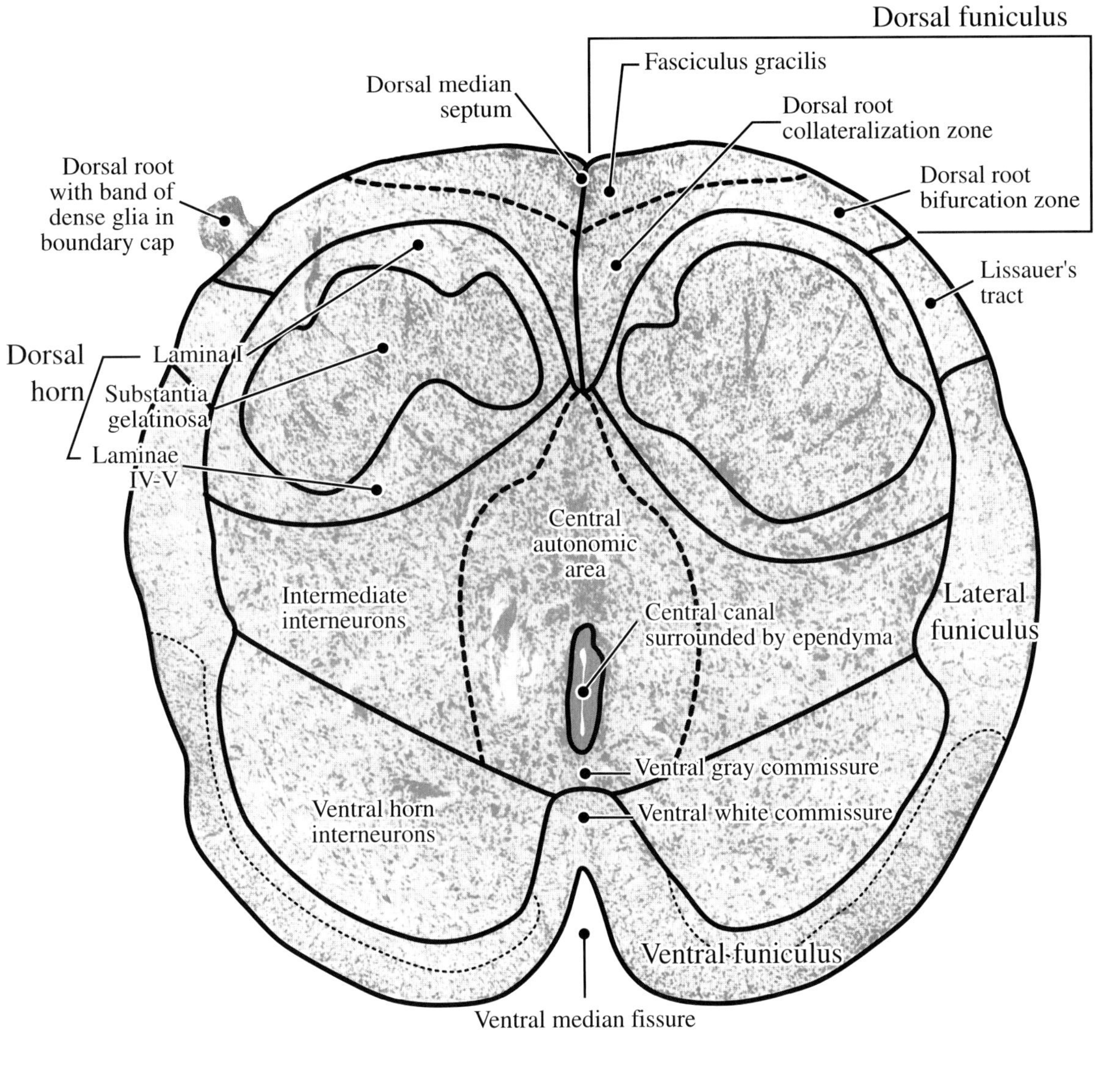

Refer to the matching myelin stained section for the approximate locations of the lateral corticospinal, vestibulospinal, intraspinal, and spinocephalic tracts.

Part V: The Early Postnatal Period (continued)

B. Matched myelin and cell body stained sections in the spinal cord of a 4-week-old infant

Plate 92 is a survey of matched myelin stained and cell body stained sections from Y23-60, a specimen in the Yakovlev Collection with a crown rump length of 410 mm. All sections are shown at the same scale. The boxes enclosing each section list the approximate level and the total area (post-fixation) of the section in square millimeters (mm^2). Full-page normal contrast photographs of each specimen are in **Plates 93A-102A.** Low contrast photographs with superimposed labels and outlines of structural details are in **Plates 93B-102B.** In this specimen, the myelin stained and cell body stained sections were preserved on separate large glass plates without consecutive section numbers. In addition, some sections were flipped over before they were placed on the plates. Sixteen myelin stained sections and 16 cell body stained sections were photographed ranging from upper cervical to lumbar enlargement levels. The 32 photographic prints were intuitively arranged in order from upper cervical to lumbar levels, using internal features such as the reduced size of the corticospinal tracts from rostral to caudal levels. Then, myelin and cell body stained sections were matched and 6 different levels are analyzed. Only myelin stained sections are illustrated at the cervical enlargement and upper lumbar levels because there are no matching cell body stained sections.

As in the previous specimens, the cross-sectional area of a myelin stained section is smaller than the matching cell body stained section with the exception of the upper thoracic level. At that level, the cell body and myelin stained sections are not perfectly matched; the cell body stained section is probably 3 to 4 sections below the myelin stained section. Using the total areas of the myelin-stained sections, the cervical enlargement has the largest cross-sectional area, being larger than the lumbar enlargement by 20%; the two thoracic levels have smaller cross-sectional areas, averaging 74% smaller than the cervical enlargement and 45% smaller than the lumbar enlargement.

Myelination in this specimen has two different characteristics (**Table V B-1**) than in the 4-day-old infant. *First*, there are myelination gradients in two fiber tracts. (1) The lateral corticospinal tract has a gradient of medial (dense) to lateral (sparse) reactive glia at cervical levels; myelinated fibers will first appear medially and last laterally (see Chapter 9, Figures 9-25 through 9-32 in Altman and Bayer, 2001). (2) There is some indication of a myelination gradient in the spinocephalic tracts at both cervical levels, because the outer fibers appear to myelinate later than the inner fibers (those closer to the intraspinal tracts). *Second*, the ventral commissure, which is either myelinating or is myelinated at GW37 and the 4-day-old infant, appears to be less advanced in this specimen, especially at cervical levels. That is probably due to the crossing of unmyelinated axons from the ventral corticospinal tract (see p. 198 in Ranson and Clark, 1959). Prior to this time, axons in the ventral corticospinal tract are present, but they may not have grown decussating collaterals.

The gray matter in the myelin stained sections shows heavy fascicles of myelinated fibers penetrating the dorsal horn from above and medially where it joins the intermediate gray. There is a large subgelatinosal plexus of myelinated fibers in the myelin stained sections. Fine fascicles of myelinated fibers are also scattered throughout the ventral horn, only labeled in the lumbar enlargement where there is a high ratio of gray matter to white matter. Many of these myelinated axons are from the large neurons in the dorsal root ganglion penetrating the dorsal horn from above and below (in the subgelatinosal plexus), arborizing in the intermediate gray (especially around Clarke's column), and in the ventral horn. Other myelinated axons are from the motoneurons in the ventral horn that later organize into fascicles to form the ventral rootlets of spinal nerves. The intraspinal tract is one of the earliest myelinating; see the dense reactive glia in this tract at GW26 (**Plates 47, 49,** and **51**).

Table V B-1: Glia types and concentration in the white matter in a 4-week-old infant

Name	Myelination	Reactive glia	Proliferating glia
DORSAL ROOT	Myelinated	---	Sparse•
VENTRAL ROOT	Myelinated	---	---
DORSAL FUNICULUS:			
dorsal root bif. zone	Many fibers*	---	Sparse
dorsal root col. zone	Myelinated	---	Sparse
fas. gracilis	Myelinated	---	Sparse
fas. cuneatus	Myelinated	---	Sparse
Lissauer's tract	---	None	Sparse
LATERAL and VENTRAL FUNICULI:			
lat. cortricospinal tract	---	Gradient†	Sparse
ven. cortricospinal tract	---	Sparse	Sparse
spinocerebellar tracts	Myelinated	---	Sparse
ven. commissure	Myelinated**	---	Sparse
intraspinal tract	Myelinated	---	Sparse
spinocephalic tract	Gradient††	Gradient††	Sparse
med. long. fasiculus	Many fibers	---	Sparse
vestibulospinal tract	Some fibers	---	Sparse

• dense band in a sparse field in **Plate 102**
* intermingled in a bed of nonreactive glia
† dense to sparse at cervical and thoracic levels, sparse at remaining levels
** some unmyelinated fibers at cervical levels
†† at cervical levels: myelinating internally, reactive glia externally; at other levels: myelinated or myelinating

PLATE 92

INFANT, 4 WEEKS, CR 410 mm, Y23-60

MYELIN STAIN	CELL BODY STAIN
Plates 93A, 93B **Upper Cervical** Total Area: 15.186 mm^2 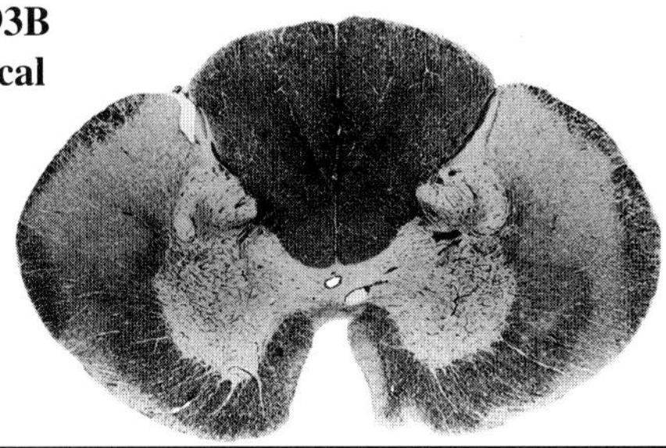	**Plates 94A, 94B** **Upper Cervical** Total Area: 15.947 mm^2 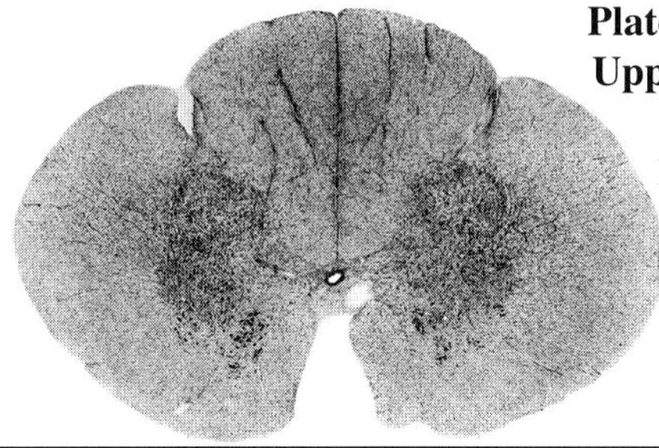
Plates 95A, 95B **Cervical Enlargement** Total Area: 17.136 mm^2 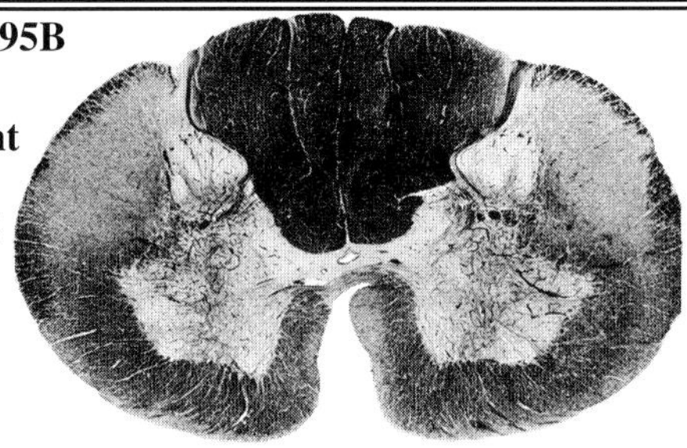	MATCHING SECTION NOT AVAILABLE
Plates 96A, 96B **Upper Thoracic** Total Area: 10.017 mm^2 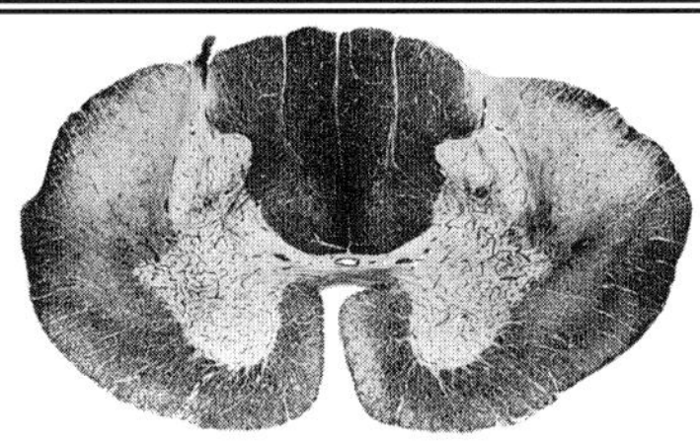	**Plates 97A. 97B** **Upper Thoracic** Total Area: 9.8042 mm^2 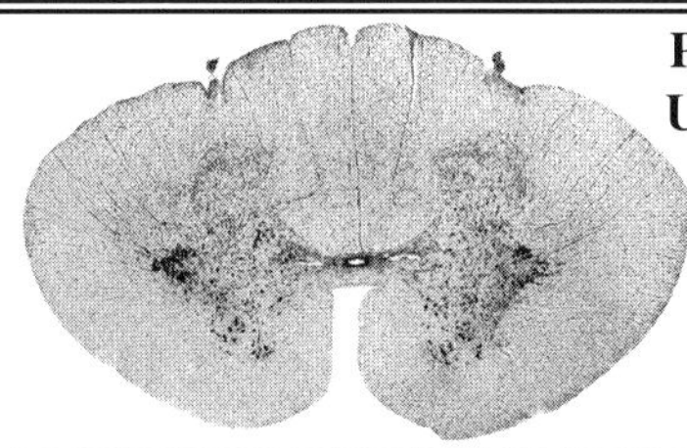
Plates 98A, 98B **Lower Thoracic** Total Area: 9.7045 mm^2 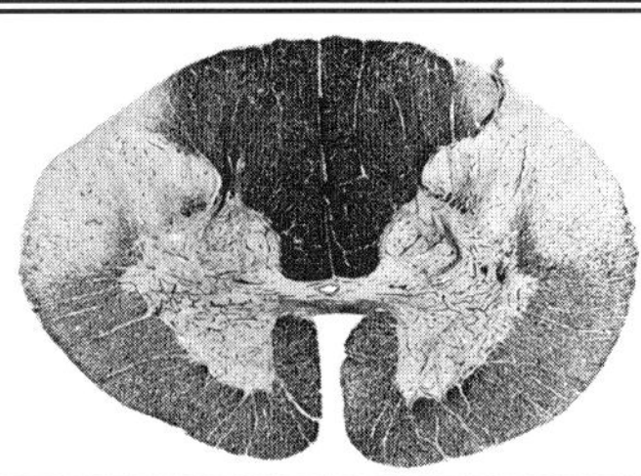	**Plates 99A, 99B** **Lower Thoracic** Total Area: 10.856 mm^2 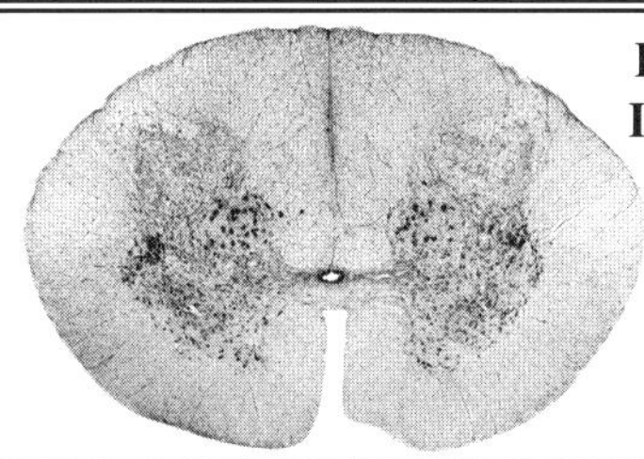
Plates 100A, 100B **Upper Lumbar** Total Area: 11.213 mm^2 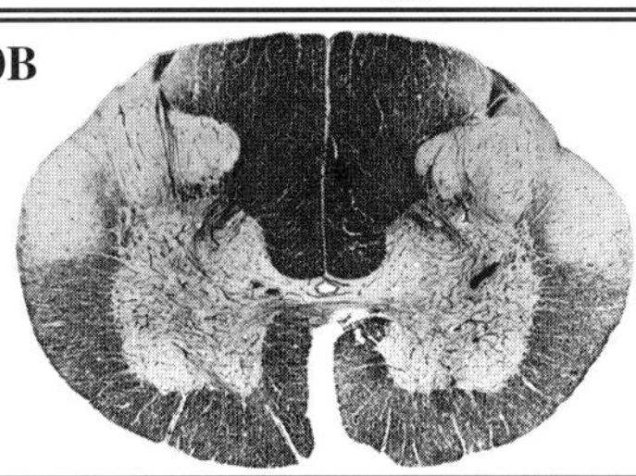	MATCHING SECTION NOT AVAILABLE
Plates 101A, 101B **Lumbar Enlargement** Total Area: 14.299 mm^2 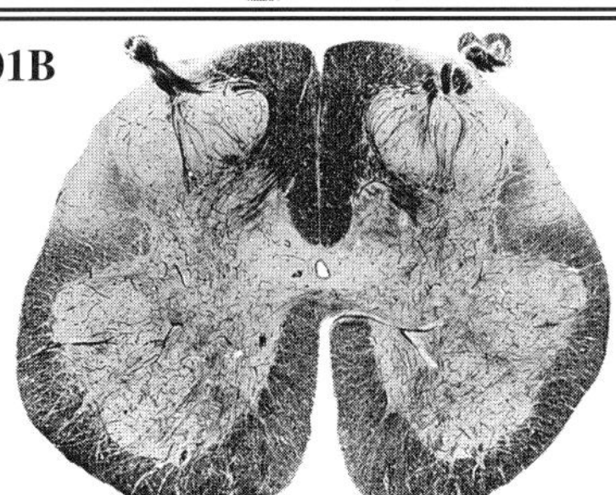 0.75 mm	**Plates 102A, 102B** **Lumbar Enlargement** Total Area: 14.478 mm^2 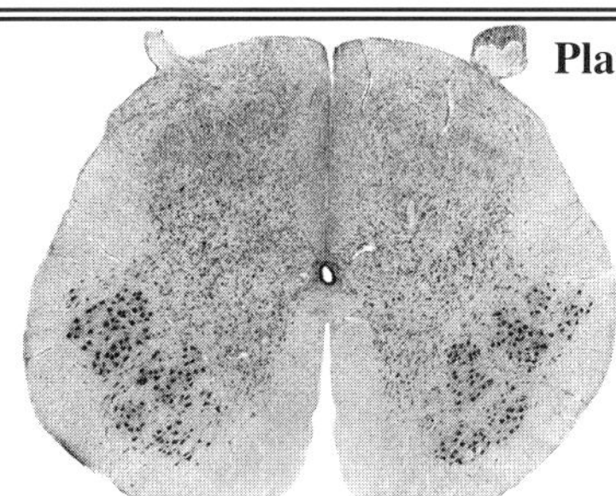 0.75 mm

PLATE 93A

Infant, 4 Weeks
CR 410 mm
Y23-60
Upper Cervical
Myelin stain

Areas (mm^2)

Central canal	.0054
Ependyma	.0115
Gray matter	3.5600
White matter	11.6090

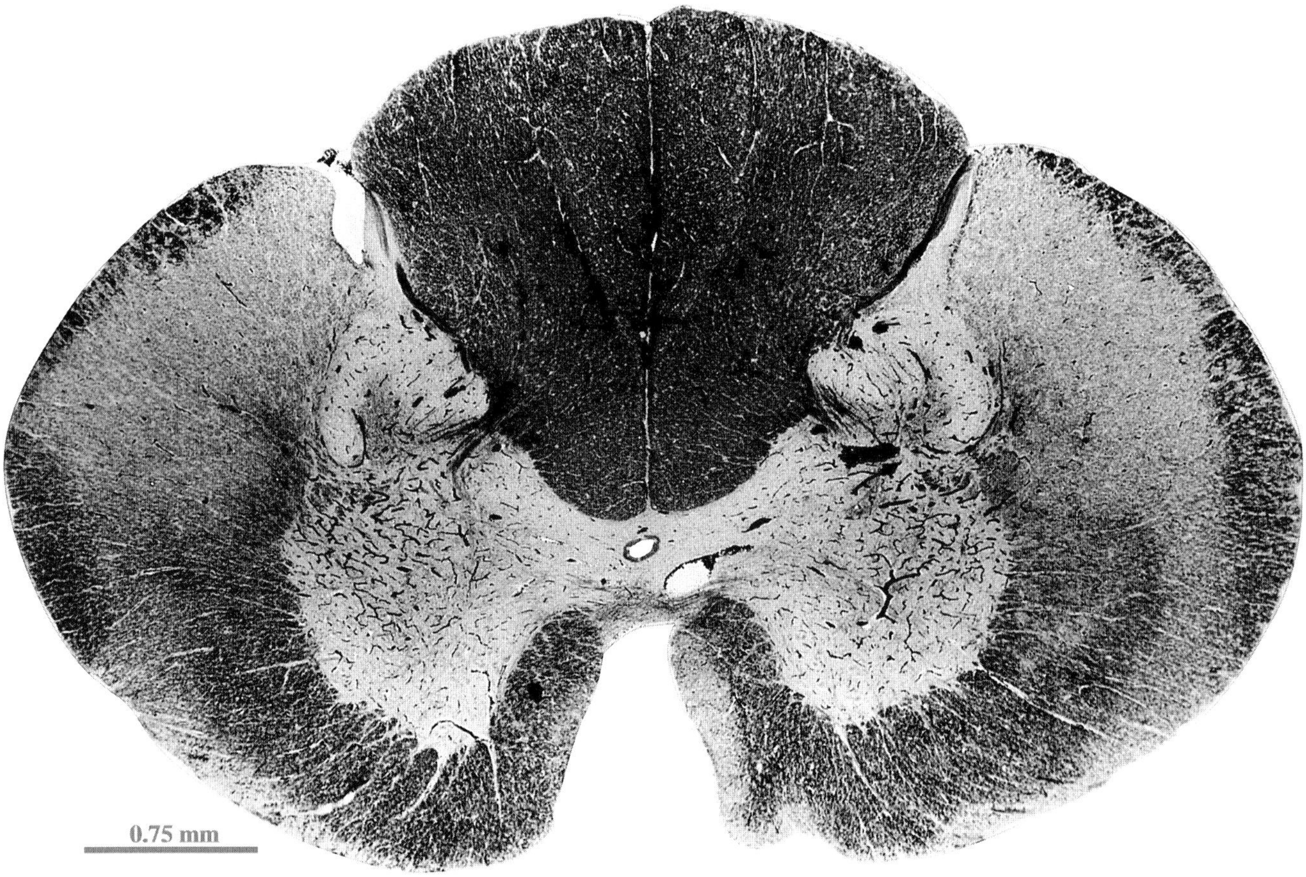

PLATE 93B

See the matched cell body stained section in Plates 94A and B

Note the lighter staining of the ventral commissure in this section. This commissure is myelinated in the GW37 fetus and 4-day-old infant. It is possible that the *unmyelinated* ventral corticospinal tract (present in each of the specimens at the cervical level) is growing decussating collaterals in the 4-week-old infant that dilute the number of *myelinated* fibers in the commissure.

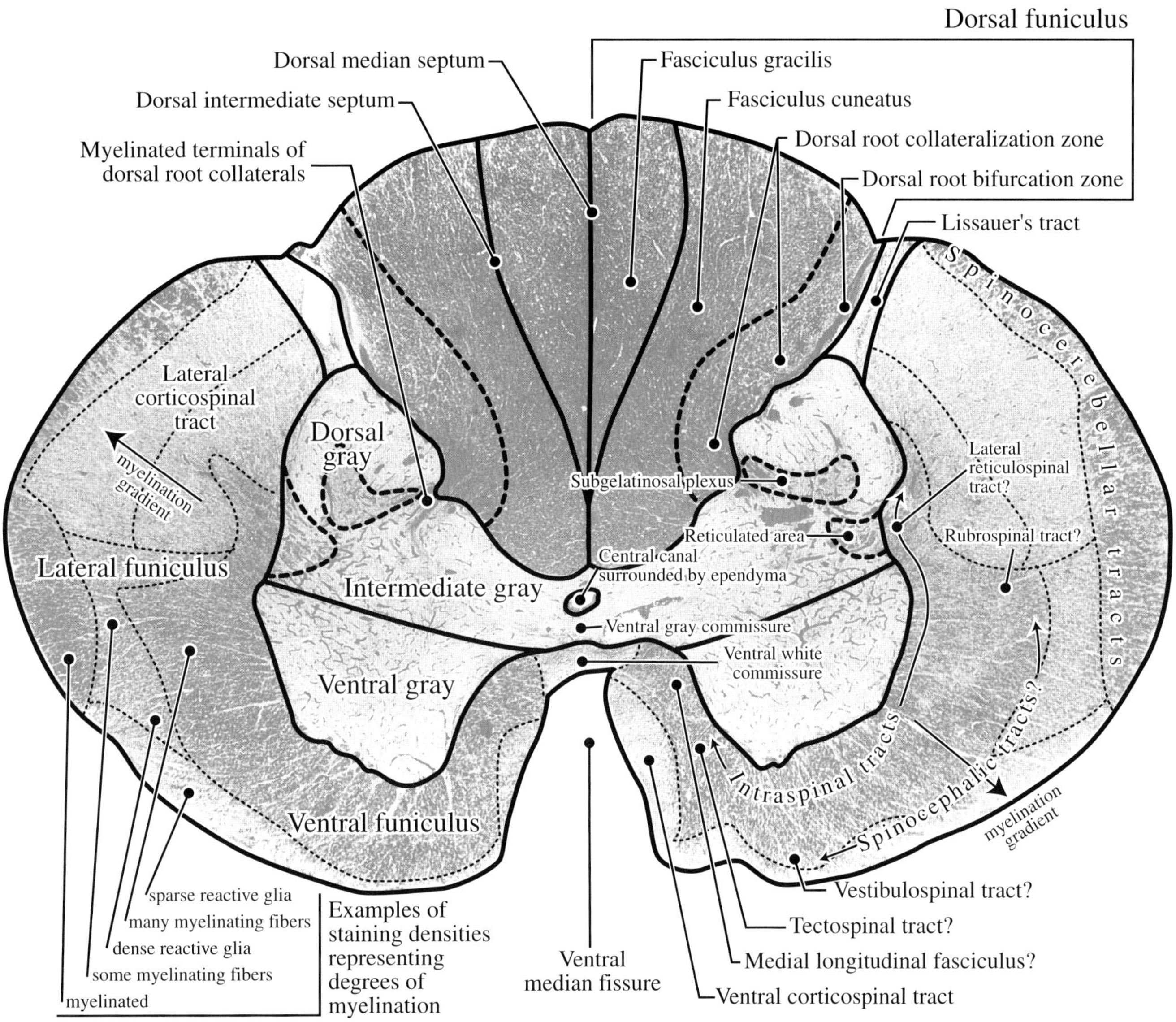

The *fine dashed lines* in the lateral and ventral funiculi segregate regions of differing densities of the myelin stain, not the borders of fiber tracts. These lines may correspond to the borders of some fiber tracts, such as the spinocerebellar tracts. Other lines run across fiber tracts, such as the area surrounding the ventral horn that includes the medial and central parts of the intraspinal tracts and the internal fibers of the spinocephalic tracts. Most tracts in the ventral and lateral funiculi overlap and do not have distinct borders. The lateral corticospinal tract overlaps with the rubrospinal tract. The spinocephalic tracts (*outer curved arrow*) overlap with the vestibulospinal tract and include fibers in the ventrolateral reticulospinal tract (not labeled), the spinotectal tract (not labeled), and the spino-olivary tract (not labeled). The intraspinal tracts (*inner curved arrow*) overlap with the medial longitudinal fasciculus, the tectospinal tract, and the lateral reticulospinal tract. *Straight arrows* in the lateral and ventral funiculi indicate myelination gradients (see comments on these gradients in **Plate 95**).

PLATE 94A

Infant, 4 Weeks
CR 410 mm
Y23-60
Upper Cervical
Cell body stain

Areas (mm^2)

Central canal	.0057
Ependyma	.0151
Gray matter	3.6685
White matter	12.2580

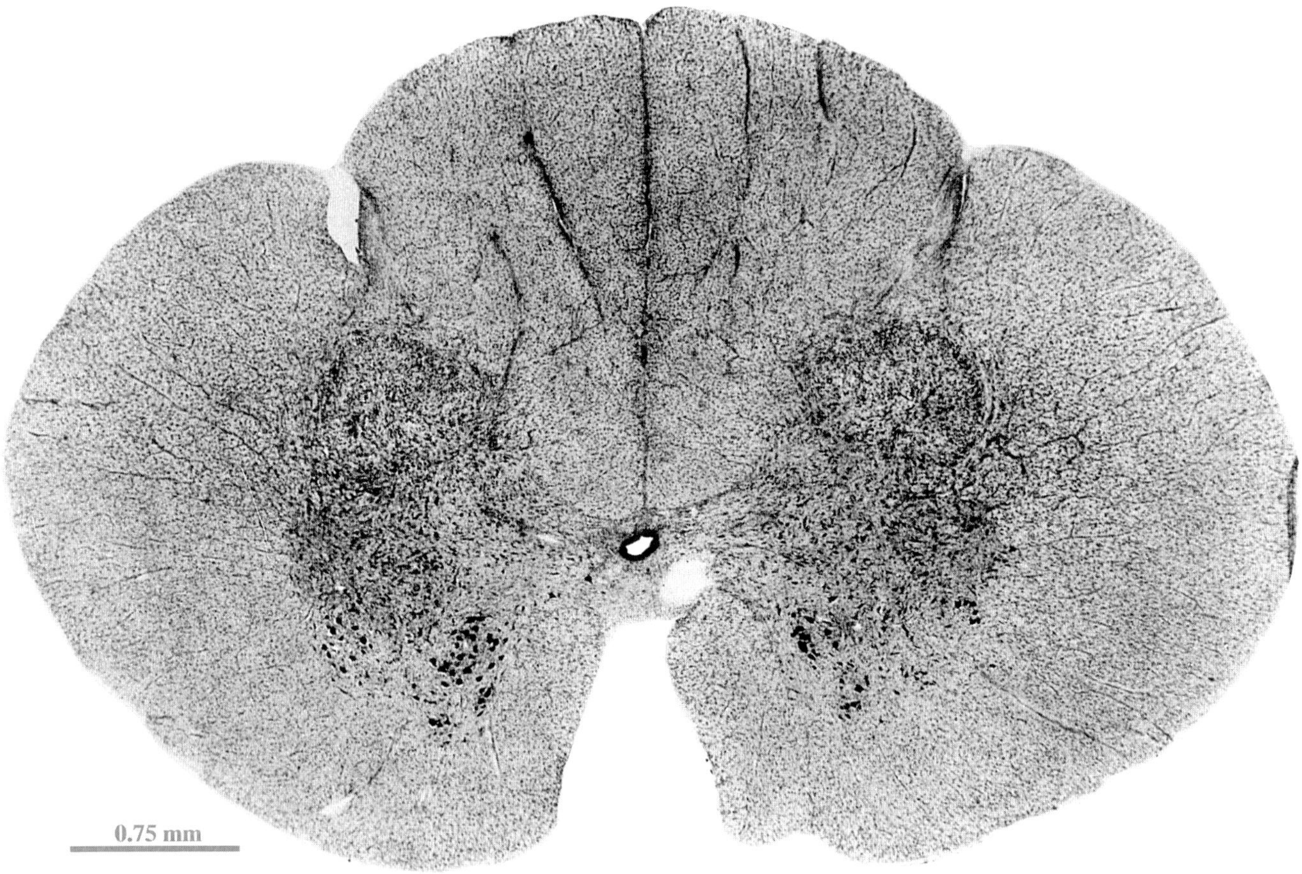

PLATE 94B

See the matched myelin stained section in Plates 93A and B

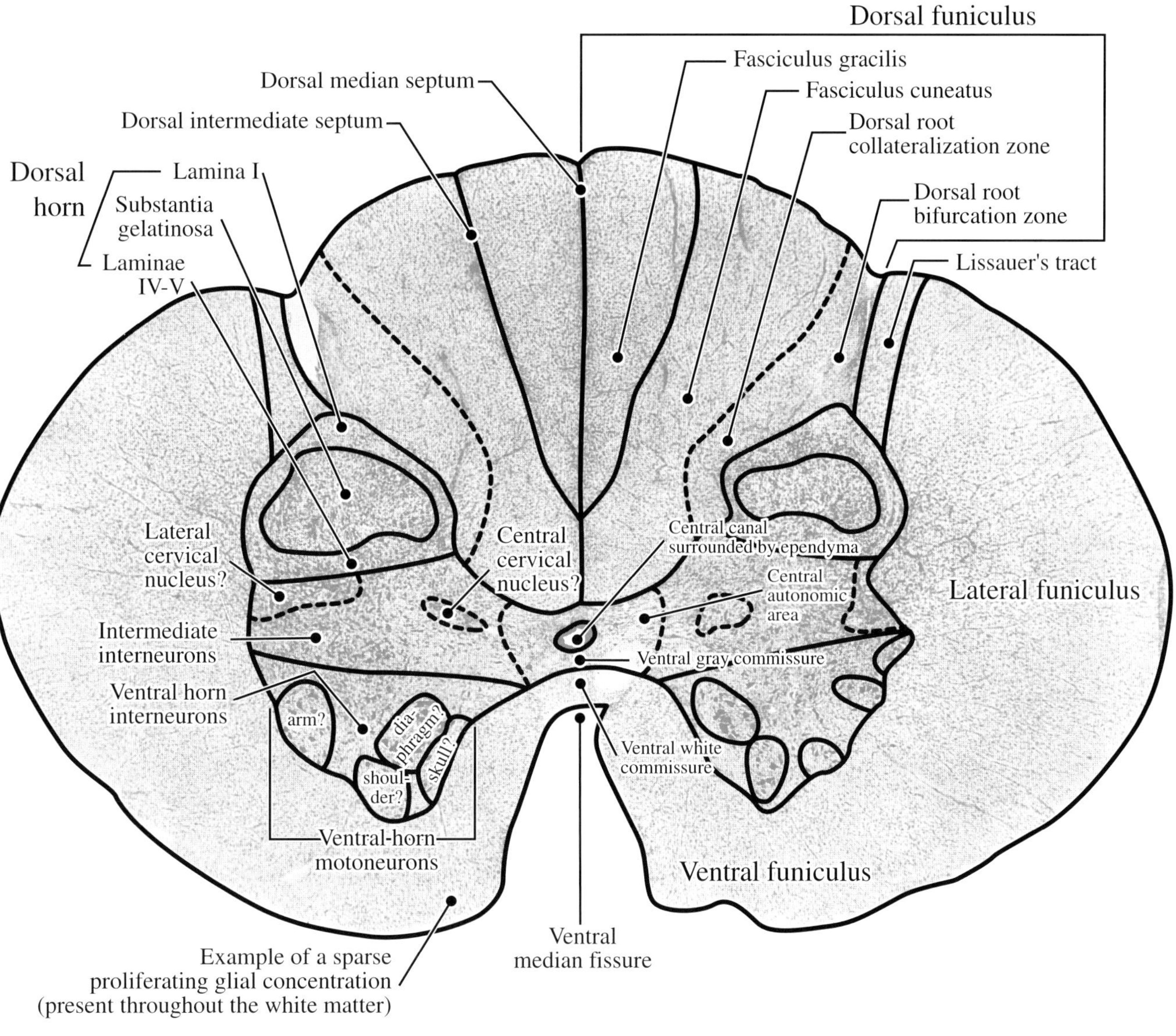

The white matter in this cell body stained section and other sections in this specimen have a uniformly sparse population of proliferating glia. Refer to the matching myelin stained section for the approximate locations of both corticospinal tracts, the medial longitudinal fasciculus, the rubrospinal, lateral reticulospinal, vestibulospinal, tectospinal, intraspinal, and spinocephalic tracts.

PLATE 95A

Infant, 4 Weeks
CR 410 mm
Y23-60
Cervical Enlargement
Myelin stain

Areas (mm^2)

Central canal	.0062
Ependyma	.0143
Gray matter	4.7533
White matter	12.3620

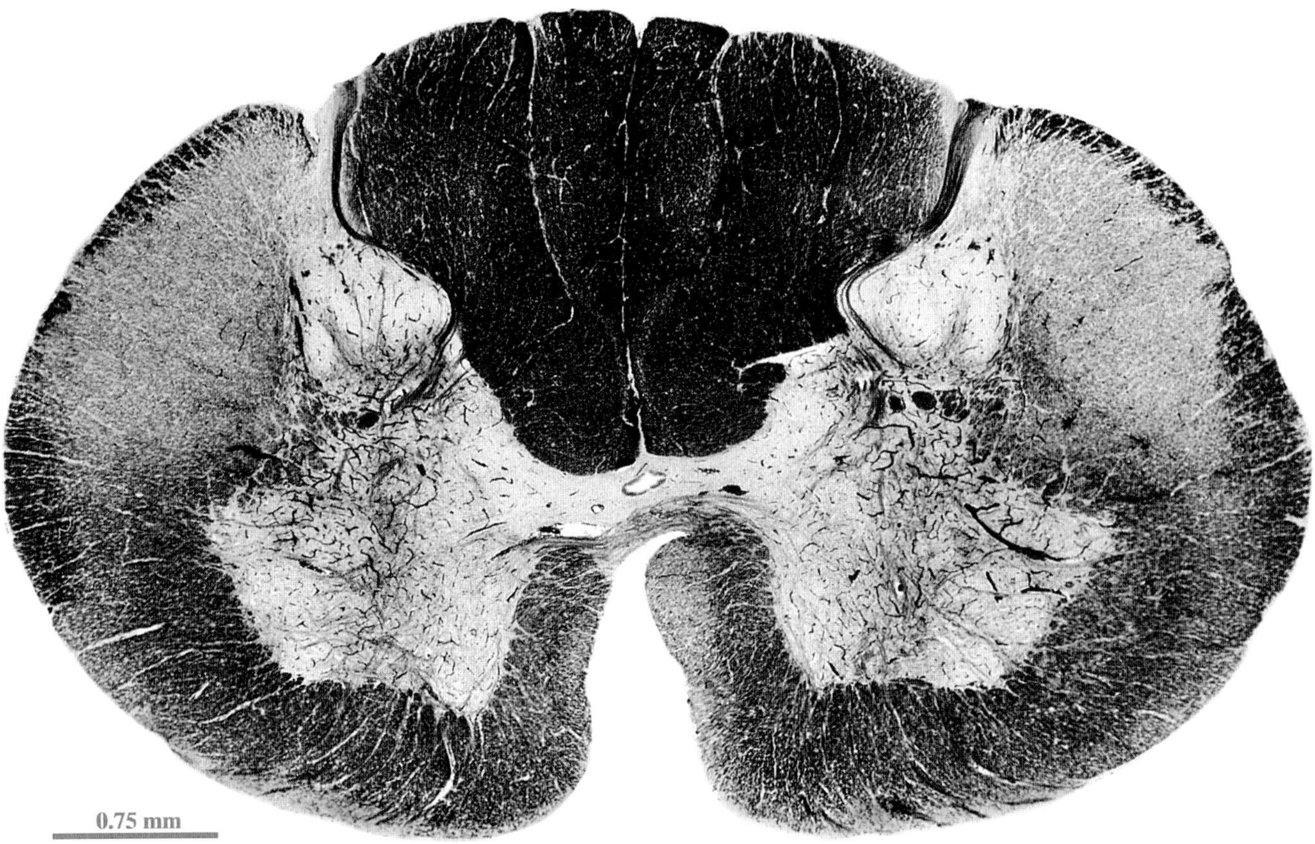

PLATE 95B

This Plate has no matched cell body stained section

As in **Plate 93**, there is lighter staining of the ventral commissure in this section, possibly due to the crossing of unmyelinated axons from the ventral corticospinal tract.

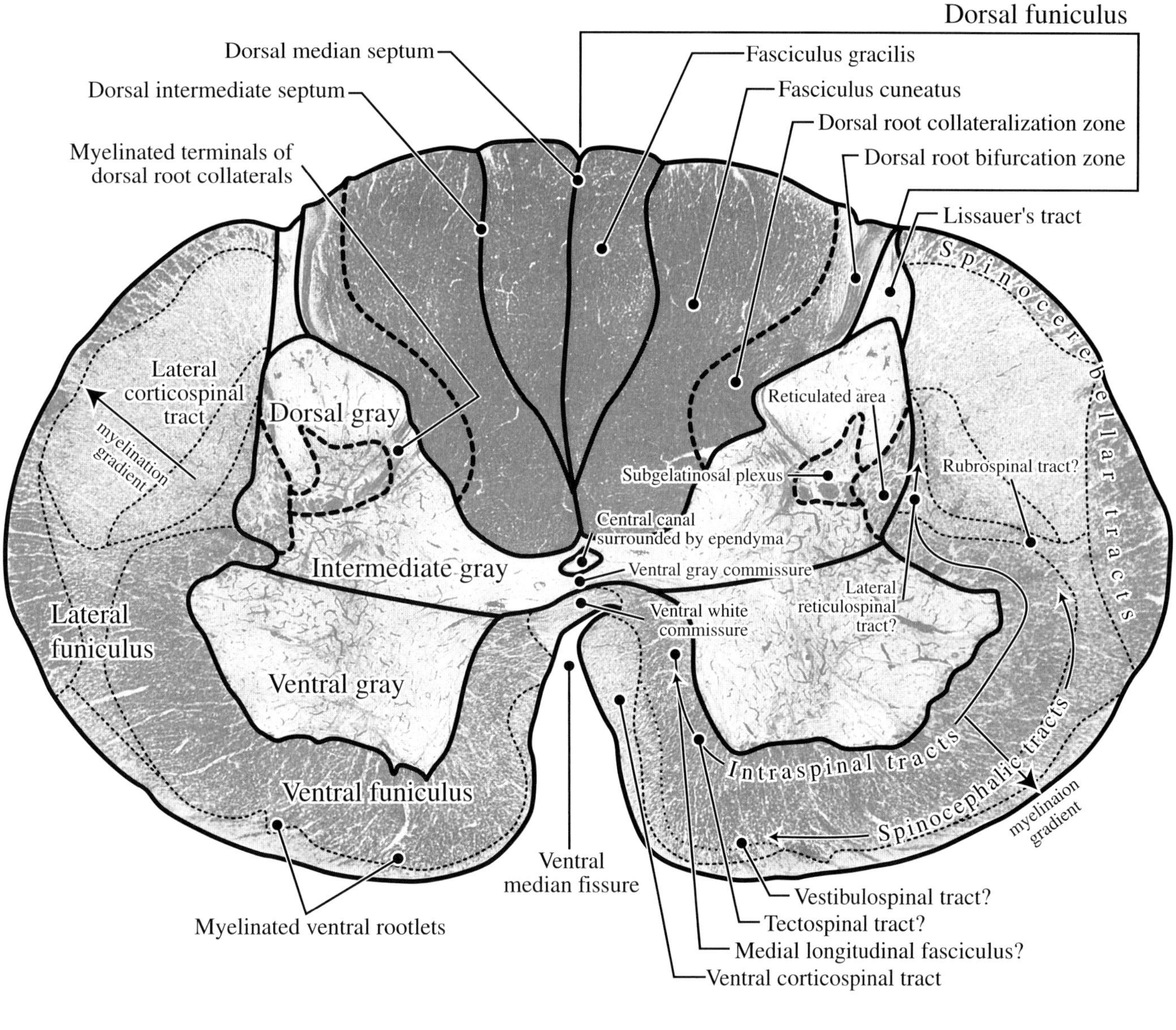

There are two myelination gradients (*straight arrows*) in the ventral and lateral funiculi in this section and also in the section shown in **Plate 93**. The arrow within the lateral corticospinal tract starts ventromedially in the area with a greater density of reactive glia and stops dorsolaterally in the area with sparse reactive glia. Ventromedial fibers enter the gray matter at rostral levels and will be the first to myelinate. Dorsolateral fibers enter the gray matter at caudal levels and will be the last to myelinate. The arrow cutting across the intraspinal and spinocephalic tracts starts close to the ventral gray in a region that contains many myelinating fibers and stops in an area that contains sparse reactive glia just beneath the pial membrane. Fibers in the intraspinal tract myelinate first, then internal fibers in the spinocephalic tracts, then outer fibers in the spinocephalic tracts.

PLATE 96A

Infant, 4 Weeks
CR 410 mm
Y23-60
Upper Thoracic
Myelin stain

Areas (mm^2)

Central canal	.0032
Ependyma	.0094
Gray matter	2.3627
White matter	7.6419

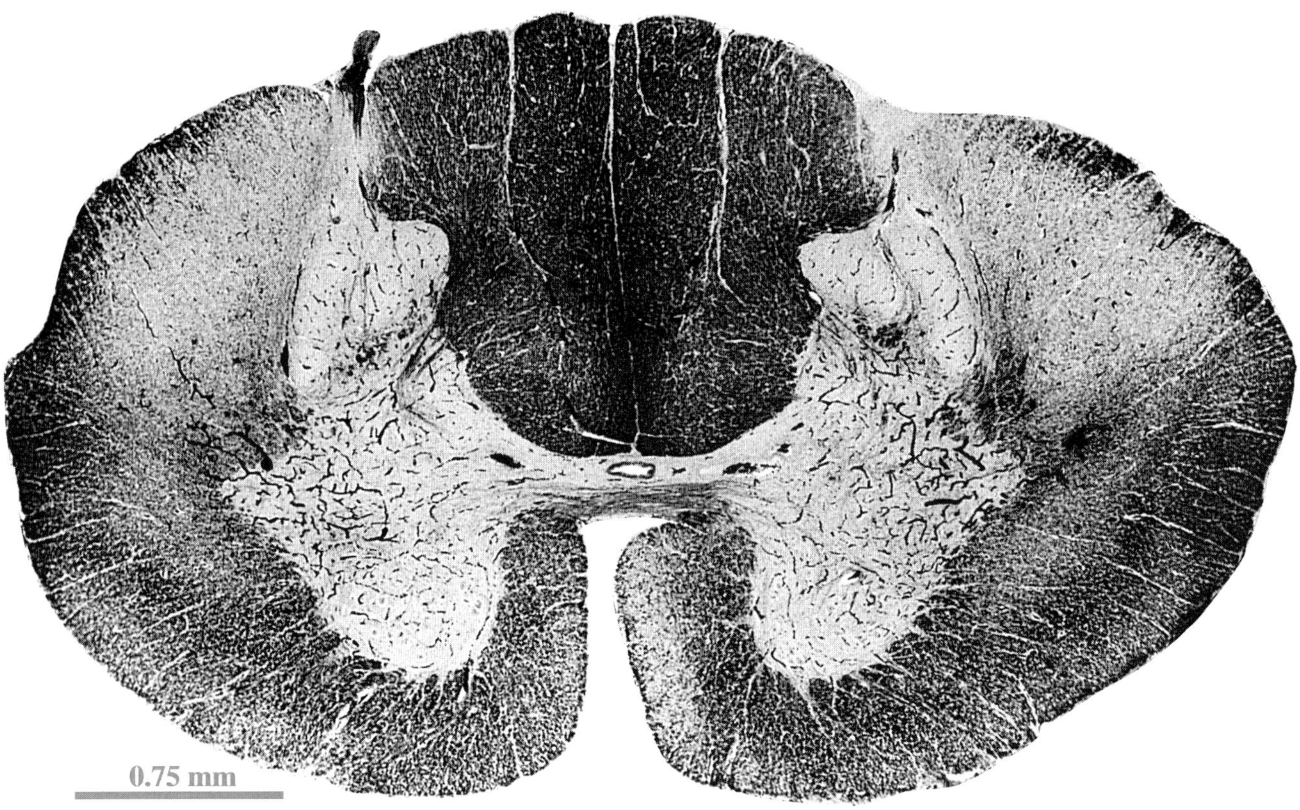

PLATE 96B

See the matched cell body stained section in Plates 97A and B

Note the asymmetry in the size of the ventral corticospinal tract in this section and in **Plates 95 and 97**.

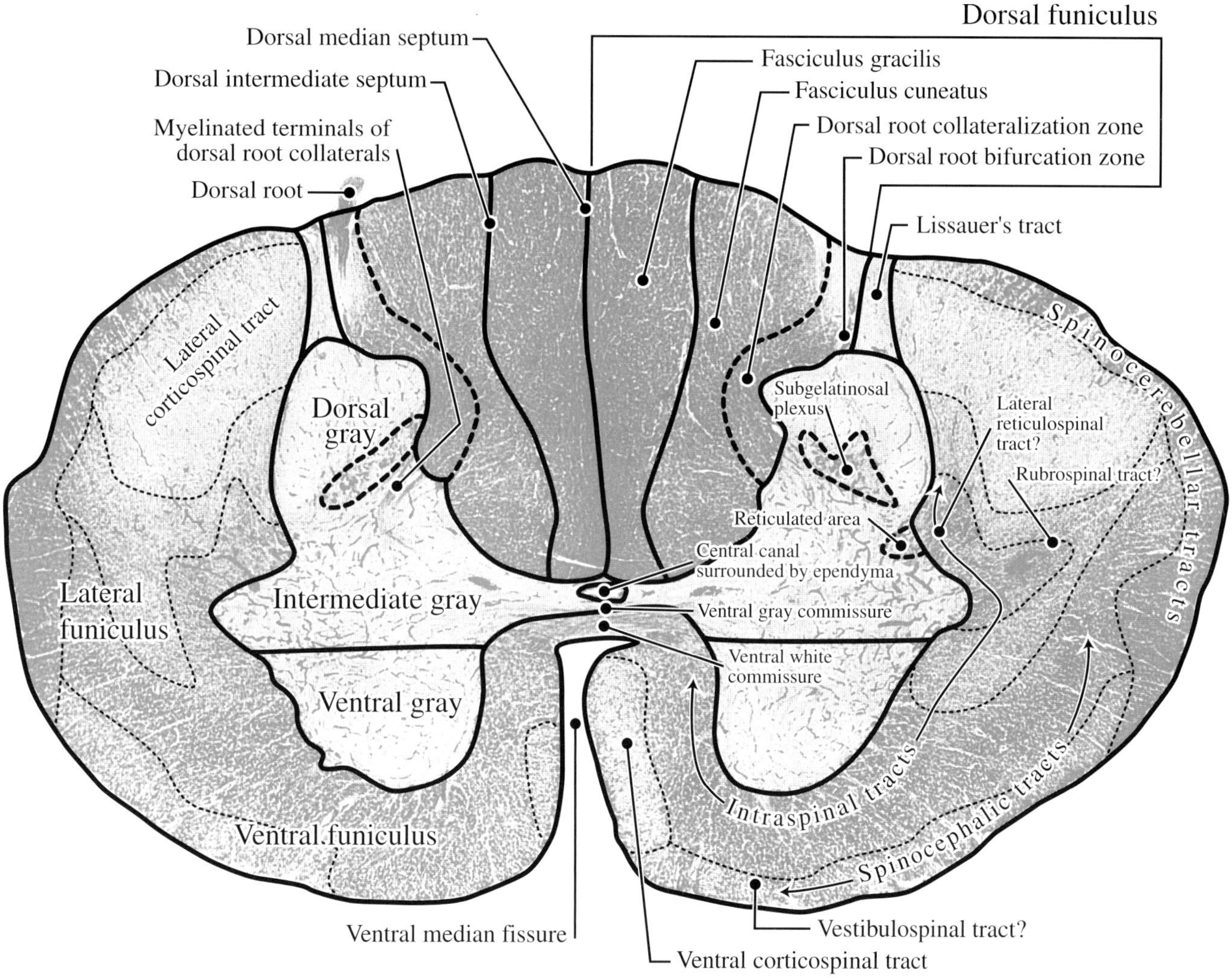

The ventral commissure stains darkly in this section, even though there is a small ventral corticospinal tract. The still unmyelinated axons in the ventral corticospinal tract may not have crossed the midline at this level, indicating a rostral (first) to caudal (last) gradient of maturation.
The myelination gradient in the lateral corticospinal tract is not evident in this section, but note the lighter staining of the superficial fibers in the ventral and lateral funiculi compared to the darker staining of the deep fibers.

PLATE 97A

Infant, 4 Weeks
CR 410 mm
Y23-60
Upper Thoracic
Cell body stain

Areas (mm^2)

Central canal	.0026
Ependyma	.0113
Gray matter	2.2909
White matter	7.4994

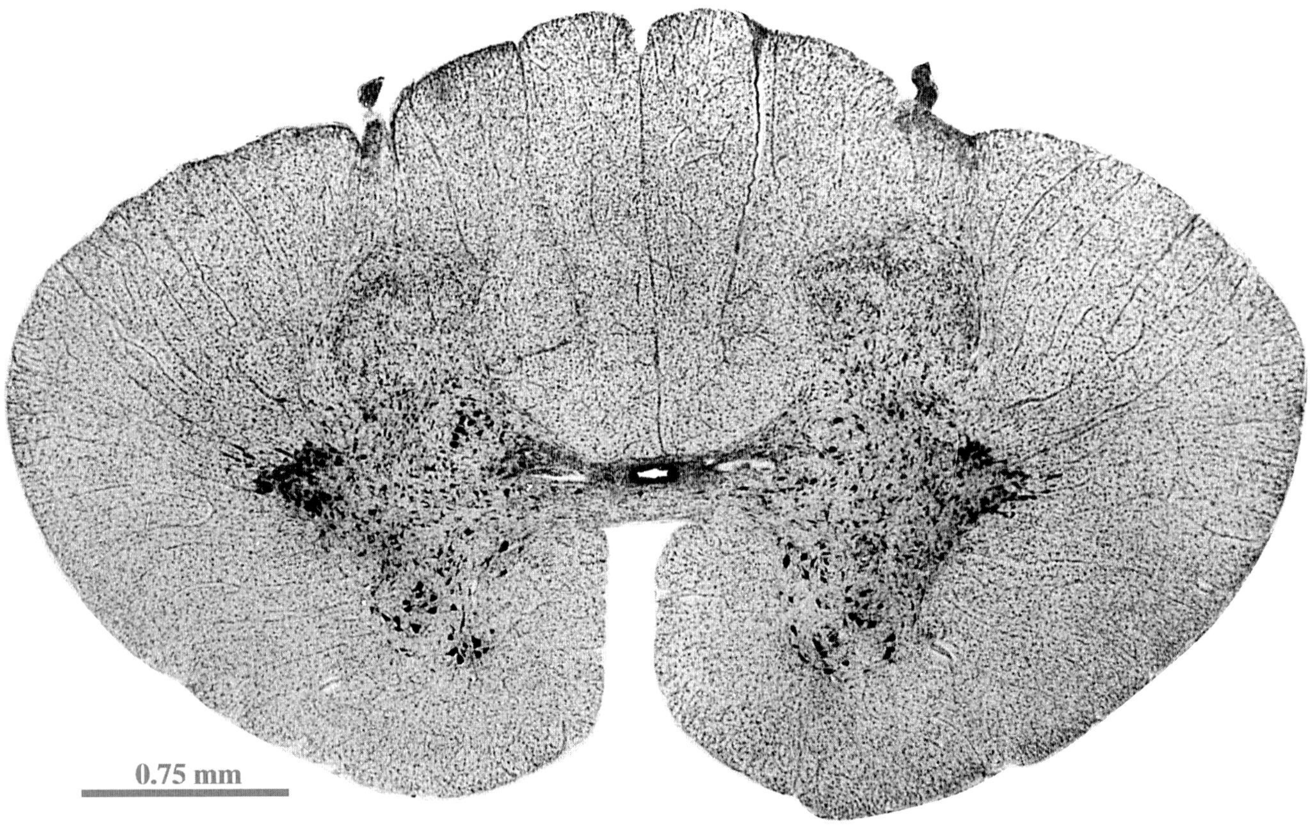

PLATE 97B

See the matched myelin stained section in Plates 96A and B

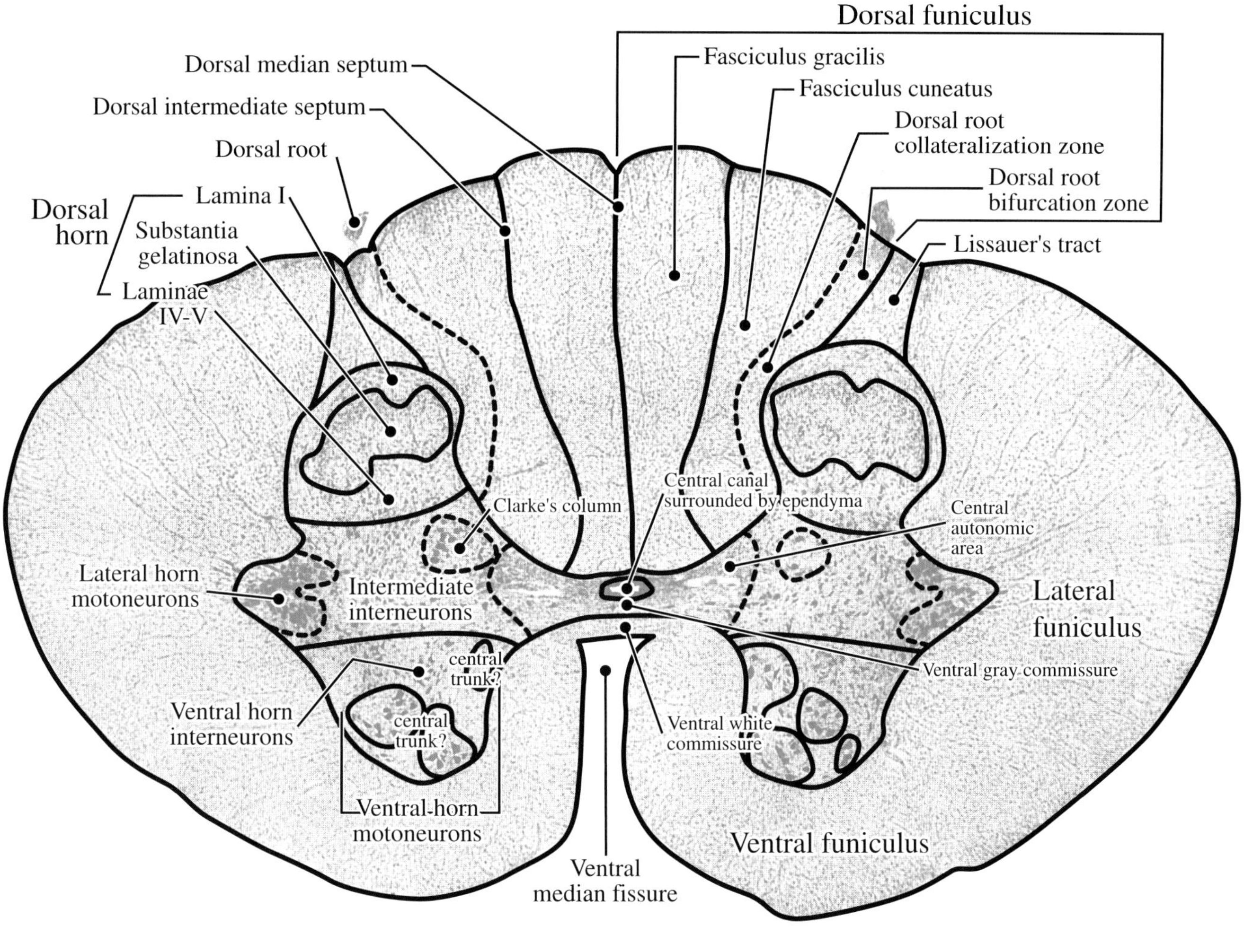

Refer to the matching myelin stained section for the approximate locations of both corticospinal tracts, the rubrospinal, lateral reticulospinal, vestibulospinal, intraspinal, and spinocephalic tracts.

PLATE 98A

Infant, 4 Weeks
CR 410 mm
Y23-60
Lower Thoracic
Myelin stain

Areas (mm^2)

Central canal	.0029
Ependyma	.0083
Gray matter	2.9576
White matter	6.7358

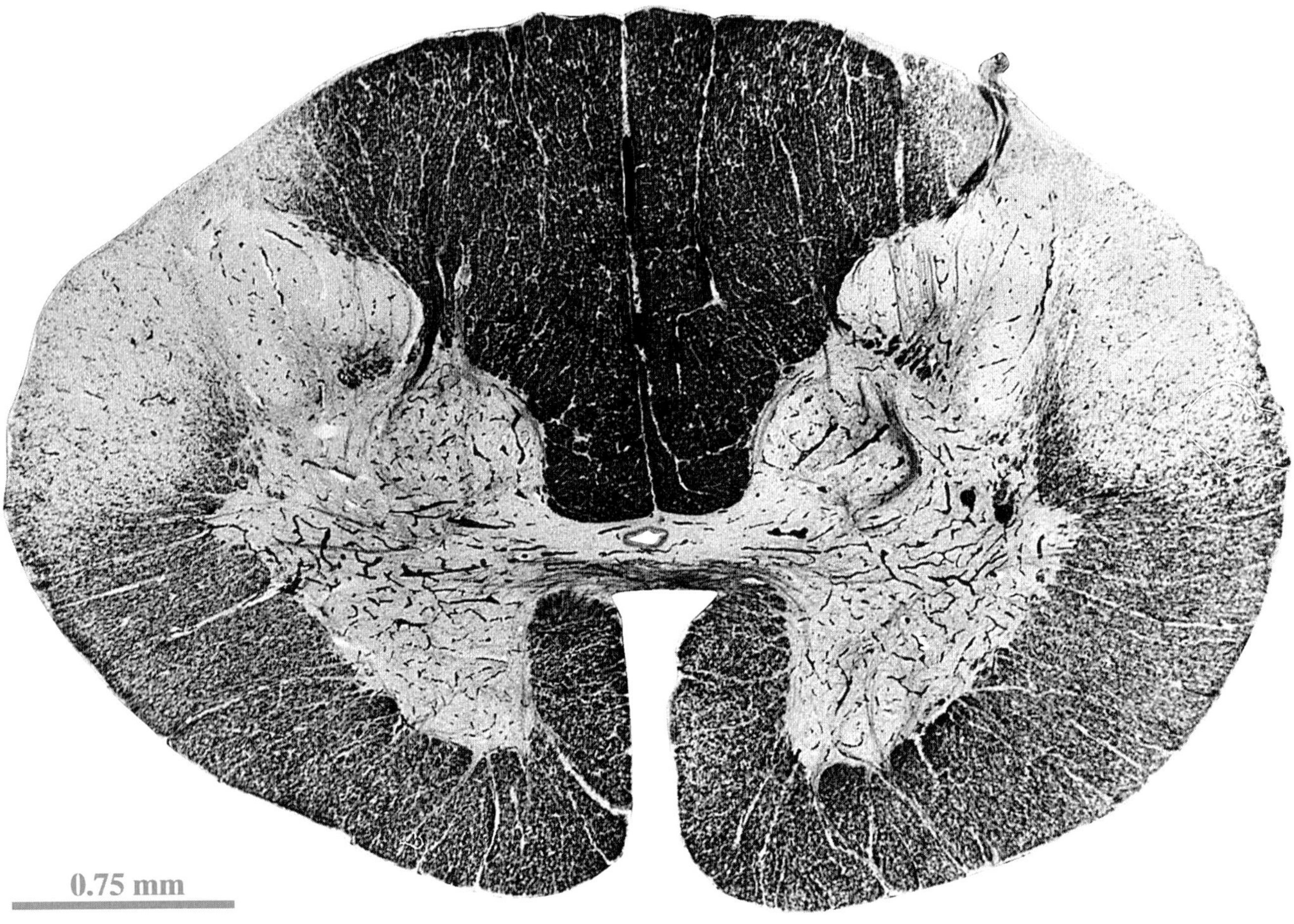

PLATE 98B

See the matched cell body stained section in Plates 99A and B

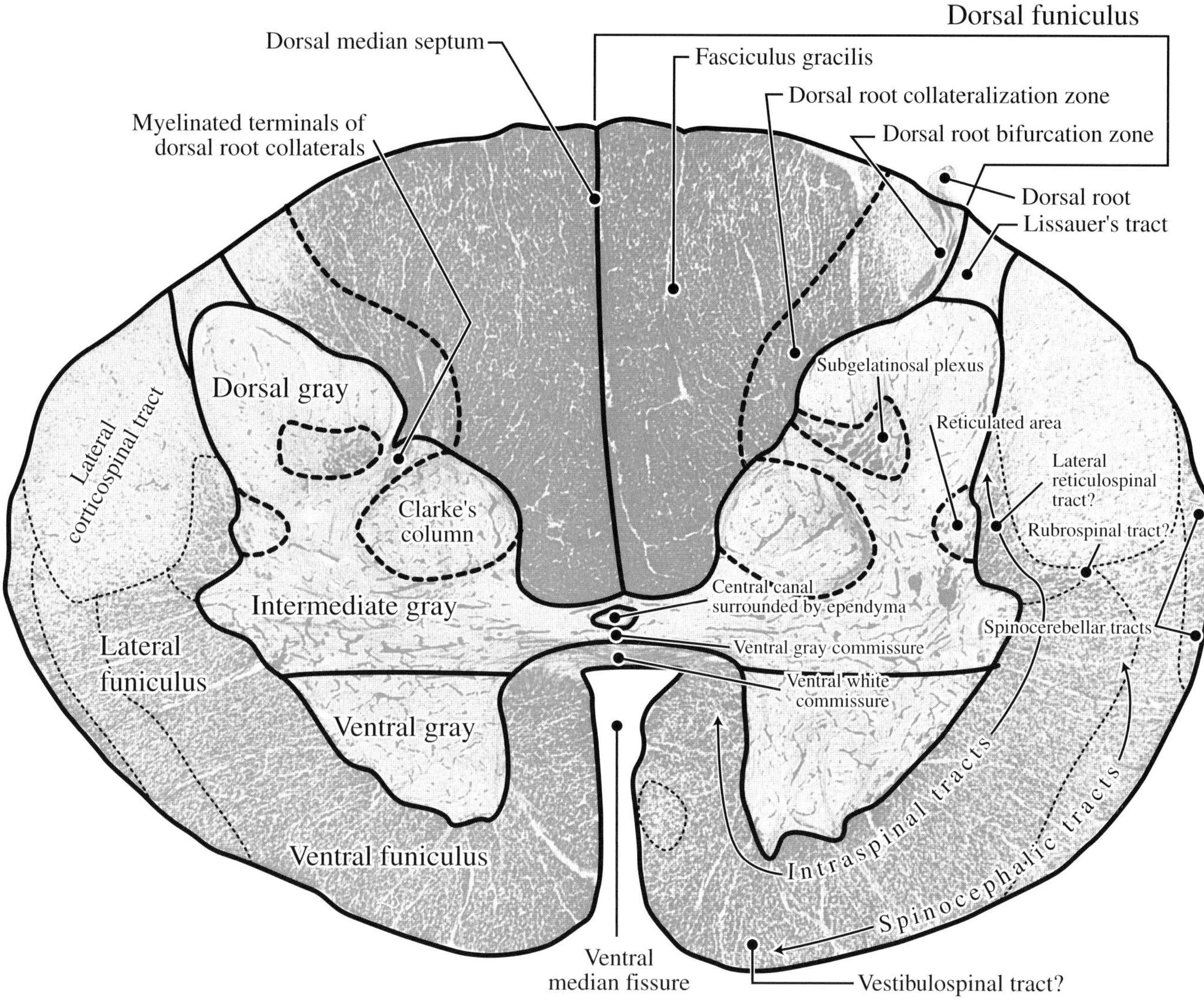

The ventral commissure stains darkly in this section, just as in **Plate 96**. A small ventral corticospinal tract may be present on the right side in the unlabeled circular area bordering the ventral median fissure. All axons in the commissure appear to be myelinated in this section. The ventral part of the lateral funiculus has a lightly stained area (containing late myelinating fibers in the external parts of the spinocephalic tracts) sandwiched by dark staining medially (containing early myelinating fibers in the intraspinal and internal parts of the spinocephalic tracts) and dark staining laterally (containing the early myelinating spinocerebellar fibers).

PLATE 99A

Infant, 4 Weeks
CR 410 mm
Y23-60
Lower Thoracic
Cell body stain

Areas (mm²)

Central canal	.0045
Ependyma	.0132
Gray matter	3.5220
White matter	7.3162

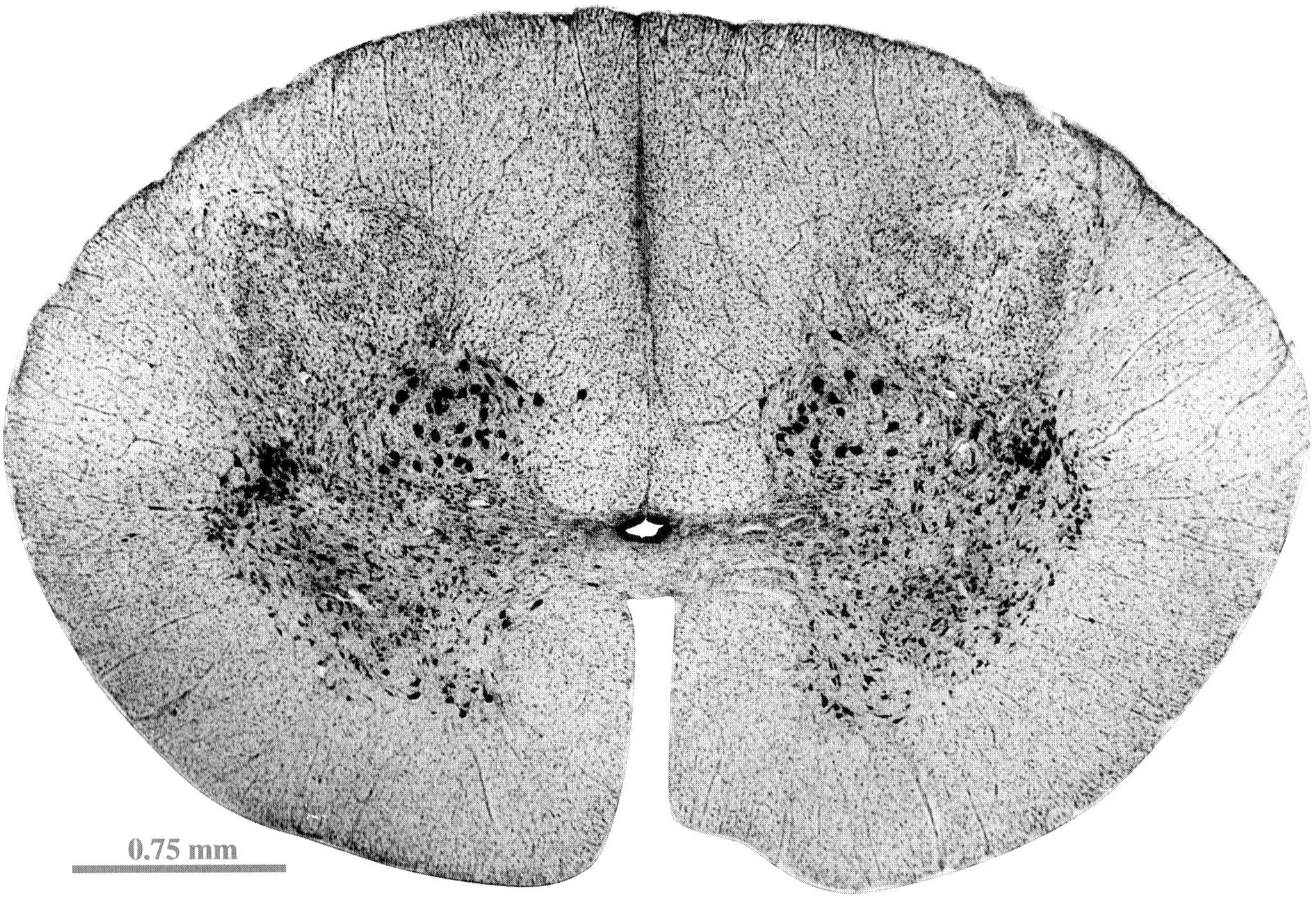

PLATE 99B

See the matched myelin stained section in Plates 98A and B

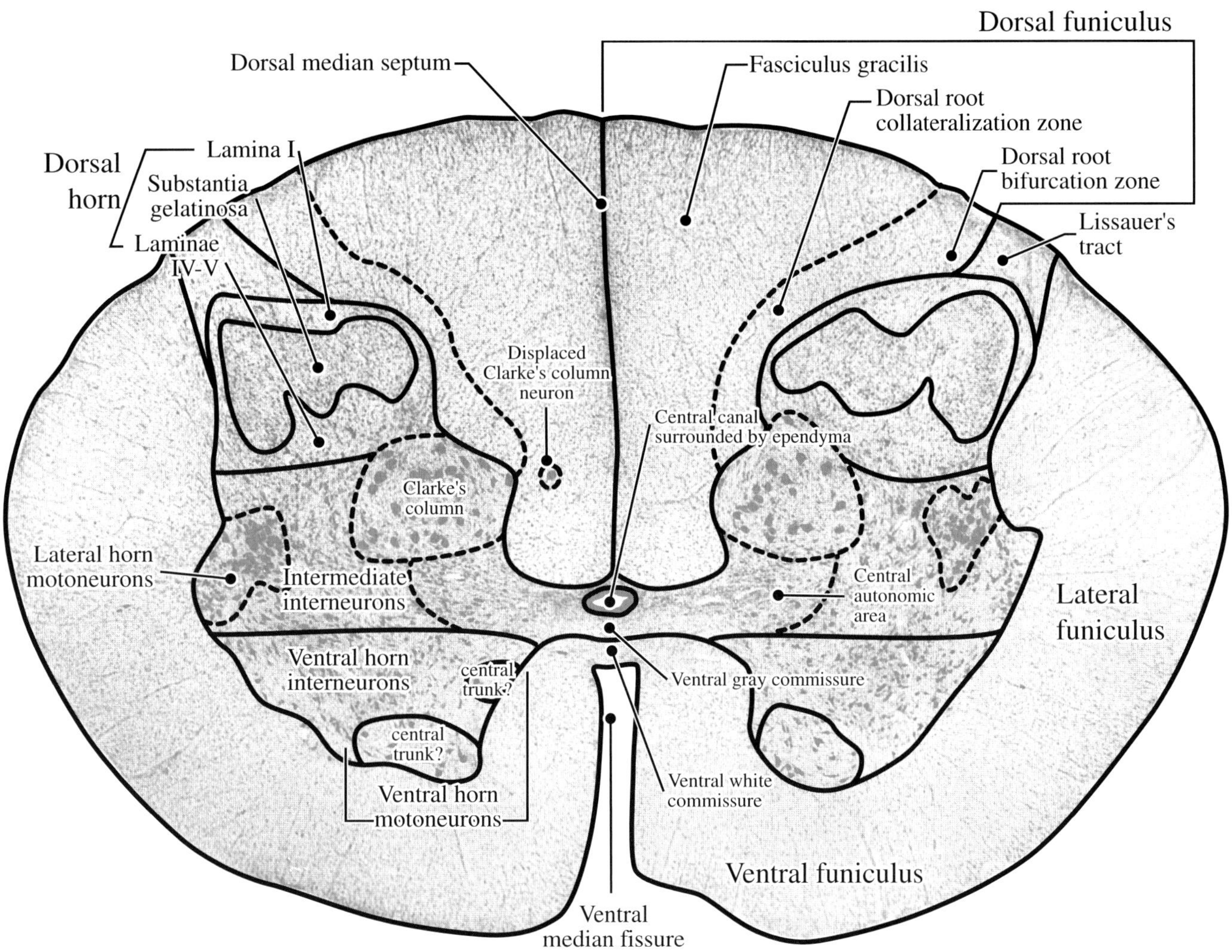

Refer to the matching myelin stained section for the approximate locations of the lateral corticospinal tract, the rubrospinal, lateral reticulospinal, vestibulospinal, intraspinal, and spinocephalic tracts.

PLATE 100A

Infant, 4 Weeks
CR 410 mm
Y23-60
Lumbar
Myelin stain

Areas (mm^2)

Central canal	.0070
Ependyma	.0070
Gray matter	4.1280
White matter	7.0713

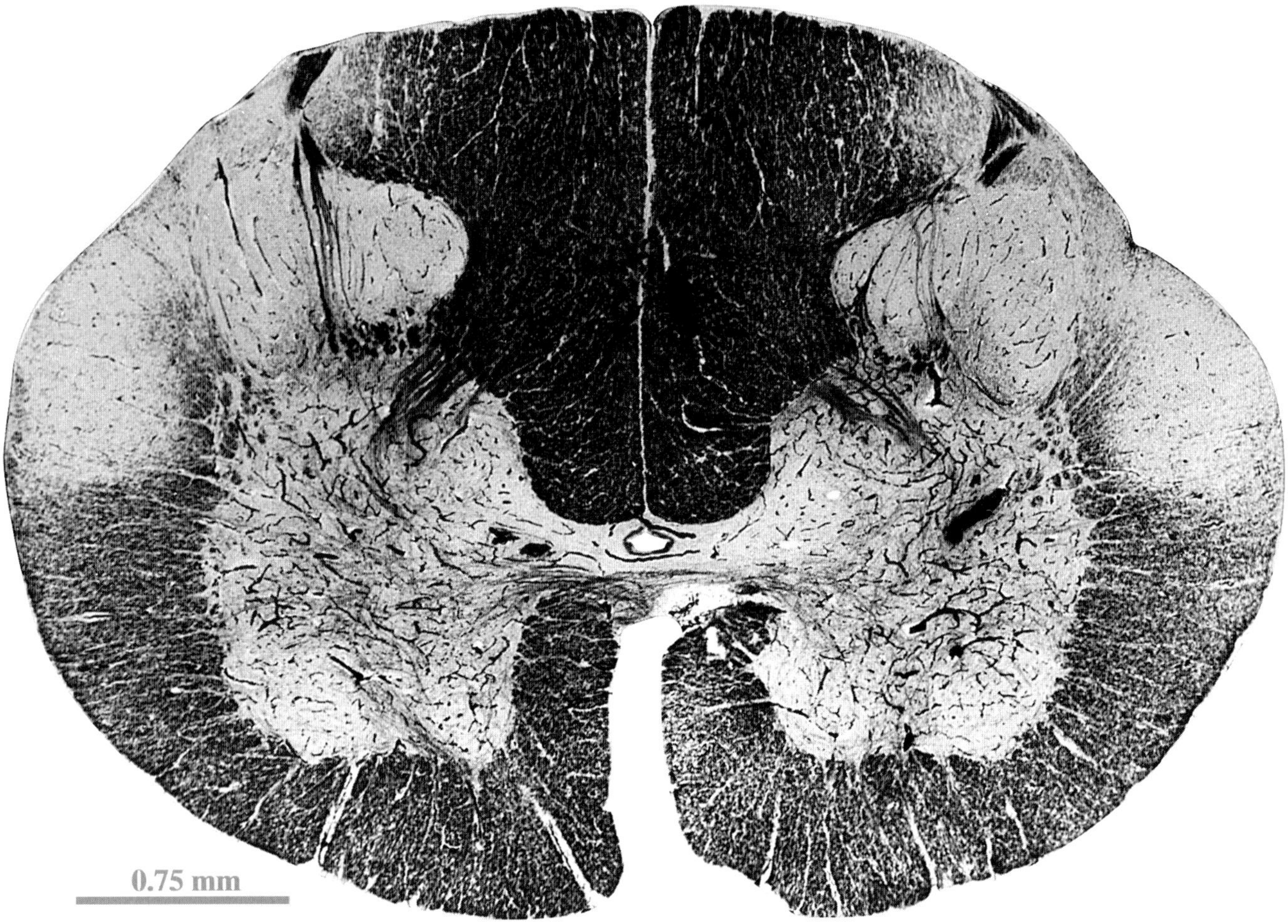

PLATE 100B

This Plate has no matched cell body stained section

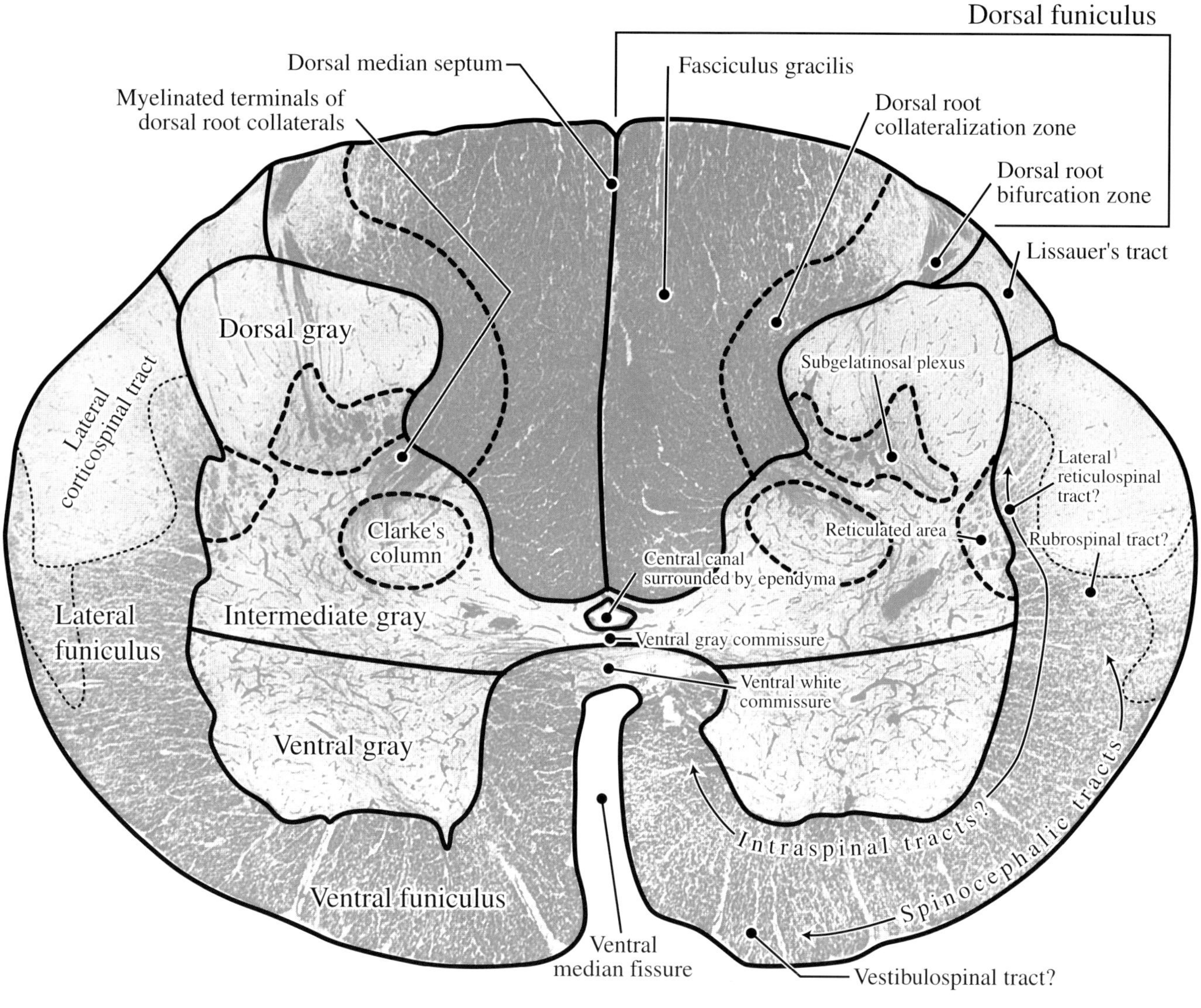

The ventral commissure stains darkly in this section (the light area is a blood vessel). Apparently, all axons in the commissure are myelinated. In contrast to rostral sections, all fibers in the spinocephalic tracts are darkly stained, indicating myelination is well on its way to completion at this level. The external axons in the spinocephalic tracts at this level are more proximal to their cell bodies of origin than at rostral levels, indicating a proximal (first) to distal (second) gradient of myelination.

PLATE 101A

Infant, 4 Weeks
CR 410 mm
Y23-60
Lumbar Enlargement
Myelin stain

Areas (mm²)

Central canal	.0090
Ependyma	.0086
Gray matter	7.8739
White matter	6.4072

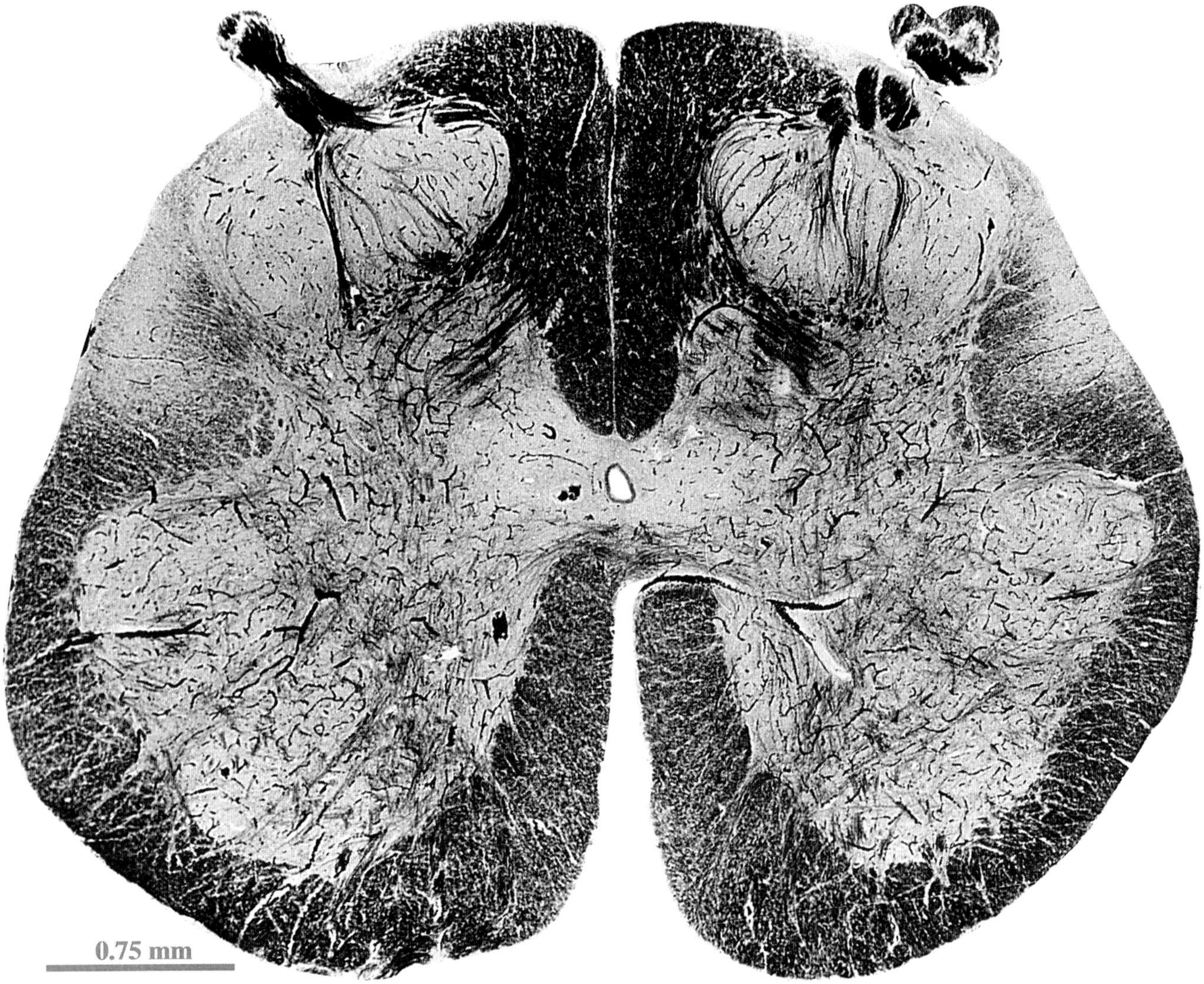

PLATE 101B

See the matched cell body stained section in Plates 102A and B

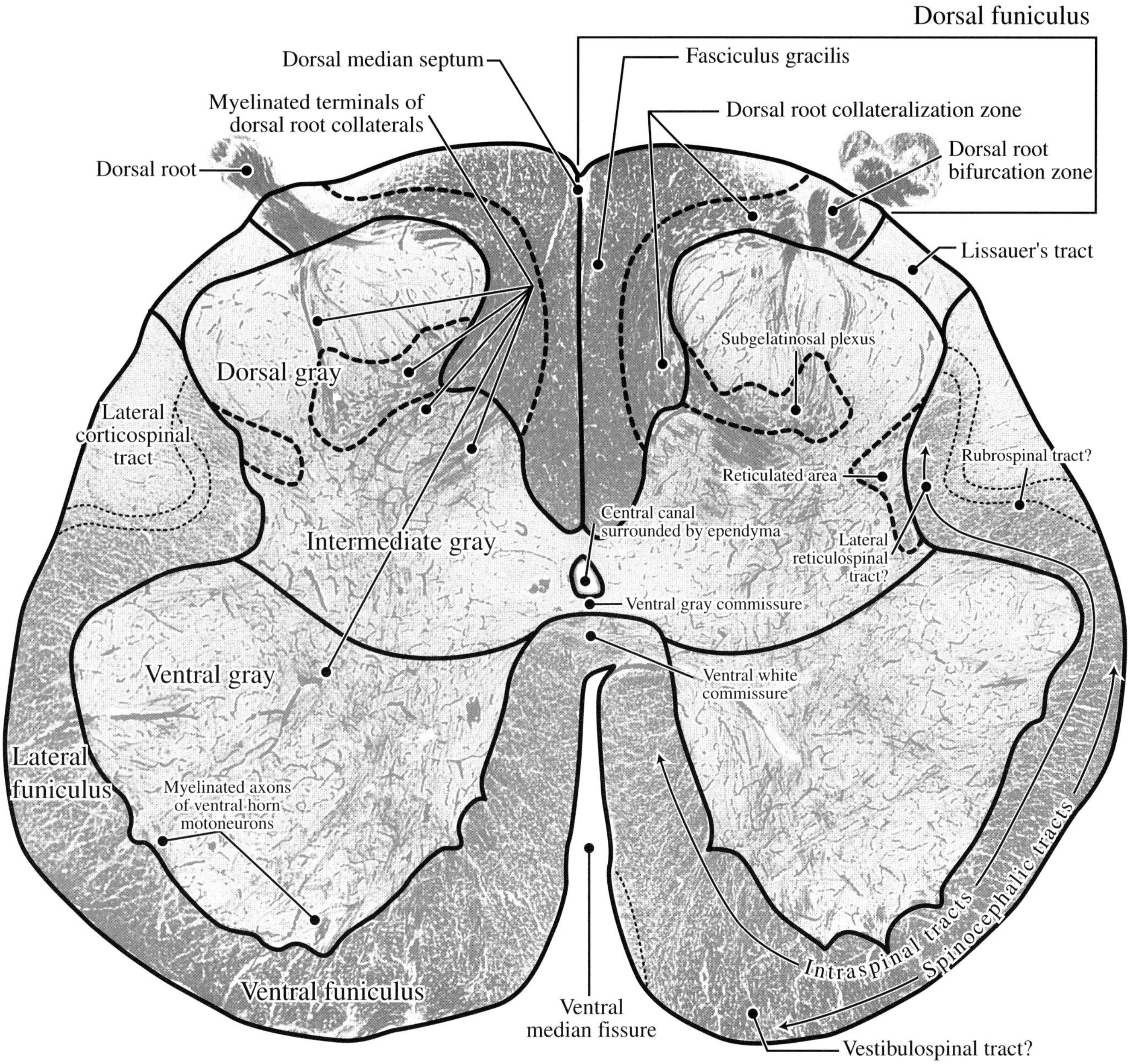

The ventral commissure stains darkly in this section indicating that all axons in the commissure are myelinated. Fibers in the spinocephalic tracts are even more darkly stained than those in **Plate 100**, indicating myelination is more advanced at this level. The axons in the spinocephalic tracts at this level are more proximal to their cell bodies of origin than at rostral levels; hence, myelination follows a proximal (first) to distal (second) gradient.

PLATE 102A

Infant, 4 Weeks
CR 410 mm
Y23-60
Lumbar Enlargement
Cell body stain

Areas (mm^2)

Central canal	.0075
Ependyma	.0148
Gray matter	7.9194
White matter	6.5366

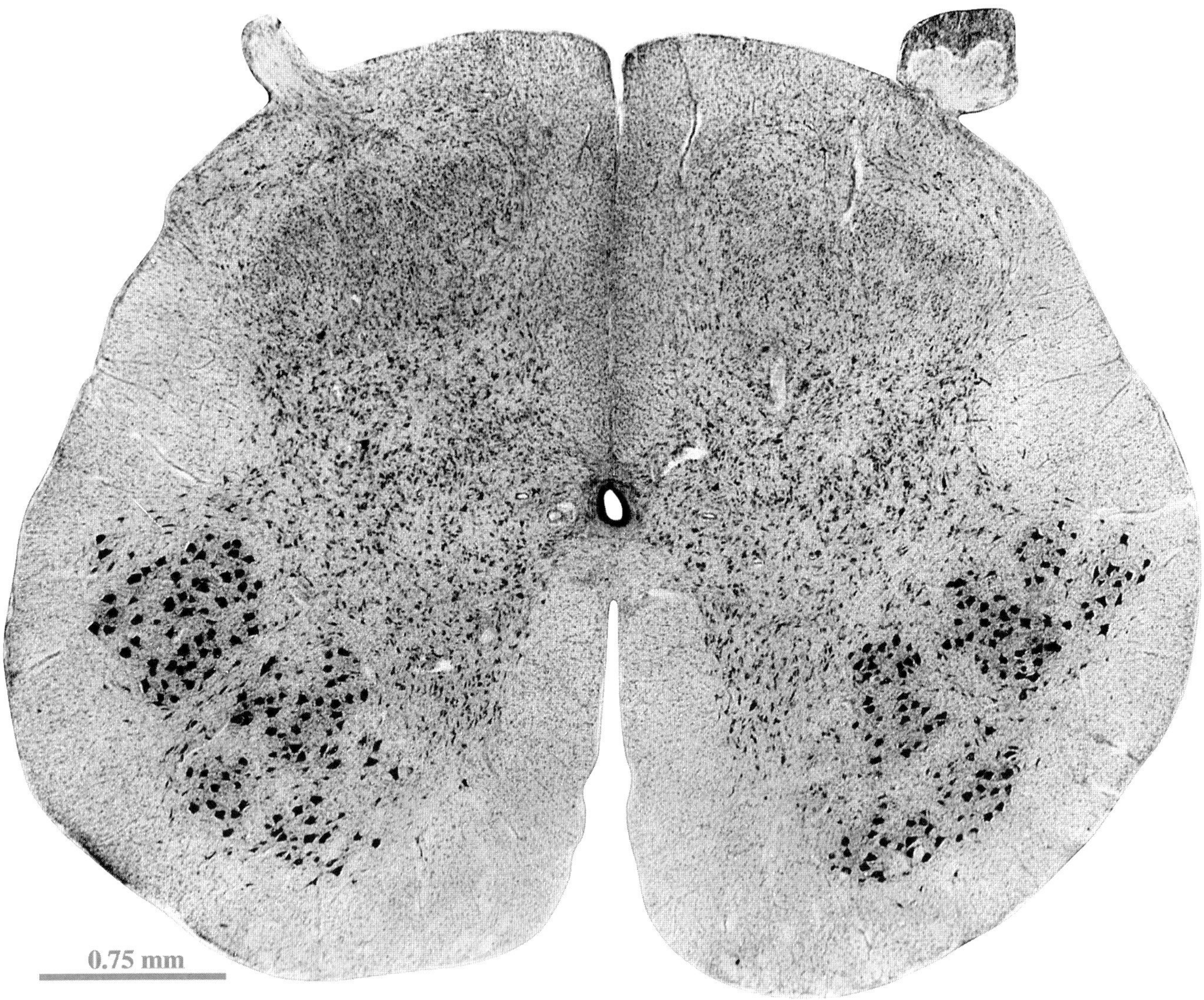

PLATE 102B

See the matched myelin stained section in Plates 101A and B

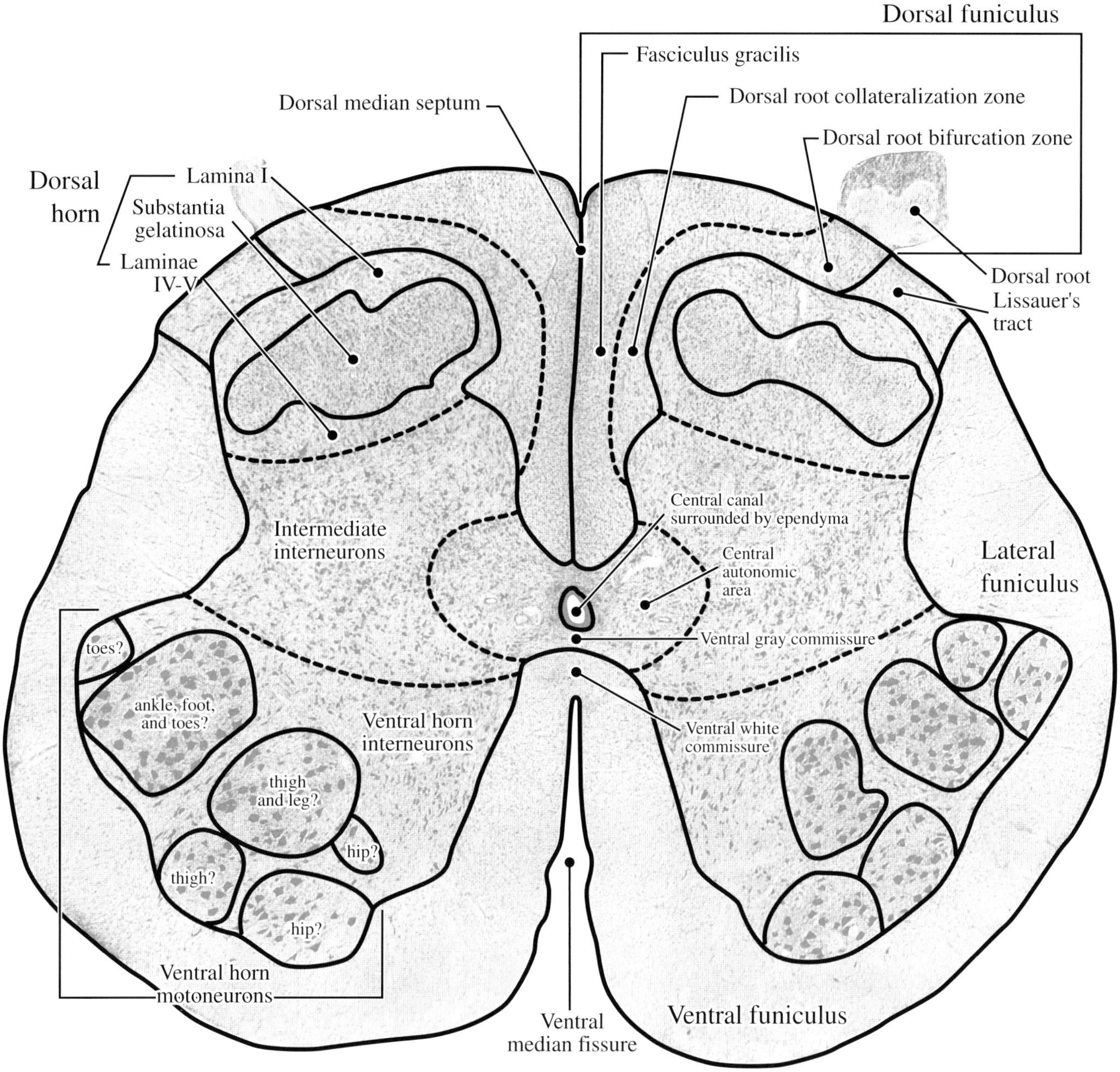

Refer to the matching myelin stained section for the approximate locations of the lateral corticospinal tract, the rubrospinal, lateral reticulospinal, vestibulospinal, intraspinal, and spinocephalic tracts.

Part V: The Early Postnatal Period (concluded)

C. Matched myelin and cell body stained sections in the spinal cord of a 4-month-old infant

Plate 103 is a survey of matched myelin stained and cell body stained sections from Y286-62, a specimen in the Yakovlev Collection with a crown rump length of 440 mm. All sections are shown at the same scale. The boxes enclosing each section list the approximate level and the total area (post-fixation) of the section in square millimeters (mm^2). Full-page normal contrast photographs of each specimen are in **Plates 104A-117A.** Low contrast photographs with superimposed labels and outlines of structural details are in **Plates 104B-117B.** In this specimen, the myelin stained and cell body stained sections were preserved on separate large glass plates without any section numbers. Forty-one myelin stained sections and 38 cell body stained sections were photographed ranging from upper cervical to lumbar enlargement levels. Sections beyond the lumbar enlargement were damaged. The 79 photographic prints were intuitively arranged in order from upper cervical to lumbar enlargement levels, using internal features such as the size of the corticospinal tracts, and the width of the ventral horn. Then, myelin and cell body stained sections were matched and seven different levels were analyzed.

As in the previous specimens, the cross-section area of most myelin stained sections is smaller than the matching cell body stained sections. Using the total areas of the myelin stained sections, the middle thoracic level has the smallest cross-sectional area, 54% smaller than the cervical enlargement and 52% smaller than the lumbar enlargement. The cervical enlargement has the largest cross-sectional area, but it is larger than the lumbar enlargement by only 5%. That difference is much smaller than at earlier ages. The cervical enlargement is 65% larger than the lumbar enlargement in the 4-day-old infant (Page 162) and 20% larger in the 4-week-old infant (Page 188). Growth of the lumbar enlargement lags behind the cervical enlargement indicating a rostral to caudal gradient in maturation.

Myelination in this specimen is nearing completion, and is very similar to that of adults (see **Table V C-1**). All parts of the dorsal funiculus are myelinated except Lissauer's tract, which remains unmyelinated in maturity. Myelination in the lateral corticospinal tract still shows a gradient in the mediolateral and rostrocaudal directions, but it has generally advanced from that seen in the 4-week-old infant. The ventral corticospinal tract may or may not exist in this specimen. If it is present, it is completely myelinated throughout its length (see Chapter 9, Figures 9-25 through 9-32, Altman and Bayer, 2001). The myelination gradient in the spinocephalic tracts is now restricted to the upper cervical level; the outer fibers are surrounded by dense reactive glia, and the inner fibers (those closer to the intraspinal tracts) are myelinating. Caudally, the spinocephalic tract is myelinated. It is important to note that the rostrocaudal myelination gradients in the lateral corticospinal tract and the spinocephalic tracts give clear evidence that axons myelinate first proximal to their site of origin at the cell body and progressively later distally. In addition, the mediolateral myelination gradient in the lateral corticospinal tract is most probably related to the retarded entry of axons from later generated neurons in medial parts of the primary motor cortex (destined to terminate in the ventral horn at lumbar and sacral areas). The gray matter in the myelin stained sections shows heavy fascicles of myelinated fibers penetrating the dorsal horn that were pointed out in the previous specimen. Myelination in the subgelatinosal plexus and the reticulated area continues to progress.

Table V C-1: Glia types and concentration in the white matter in a 4-month-old infant

Name	Myelination	Reactive glia	Proliferating glia
DORSAL ROOT	Myelinated	---	Sparse
VENTRAL ROOT	Myelinated	---	---
DORSAL FUNICULUS:			
dorsal root bif. zone	Many fibers*	---	Sparse
dorsal root col. zone	Myelinated	---	Sparse
fas. gracilis	Myelinated	---	Sparse
fas. cuneatus	Myelinated	---	Sparse
Lissauer's tract	---	none	Sparse
LATERAL and VENTRAL FUNICULI:			
lat. corticospinal tract	Gradient†	Gradient†	Sparse
ven. corticospinal tract	Myelinated**	---	Sparse
spinocerebellar tracts	Myelinated	---	Sparse
ven. commissure	Myelinated	---	Sparse
intraspinal tract	Myelinated	---	Sparse
spinocephalic tract	Gradient††	Gradient††	Sparse
med. long. fasiculus	Myelinated	---	Sparse
vestibulospinal tract	Myelinated	---	Sparse

* intermingled with unmyelinated fibers
† myelinated to myelinating at cervical and thoracic levels, dense reactive glia to sparse reactive glia at upper lumbar level, sparse reactive glia at lumbar enlargement
** the presence of this tract is assumed. If present, it is myelinated.
†† at upper cervical level: myelinating internally, dense reactive glia externally; at other levels: myelinated or myelinating

PLATE 103

INFANT, 4 MONTHS, CR 440 mm, Y286-62

MYELIN STAIN	CELL BODY STAIN
Plates 104A, 104B **Upper Cervical** Total area: 27.489 mm^2 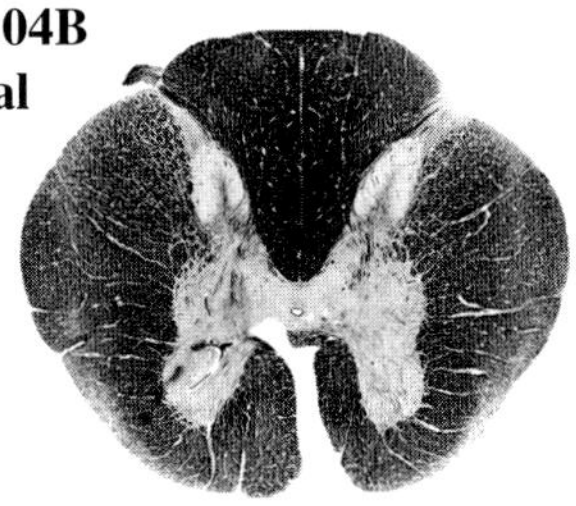	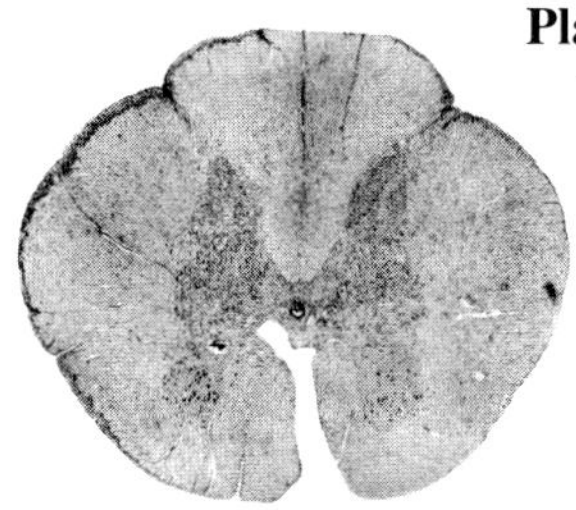 **Plates 105A, 105B** **Upper Cervical** Total area: 28.327 mm^2
Plates 106A, 106B **Cervical Enlargement** Total area: 33.314 mm^2 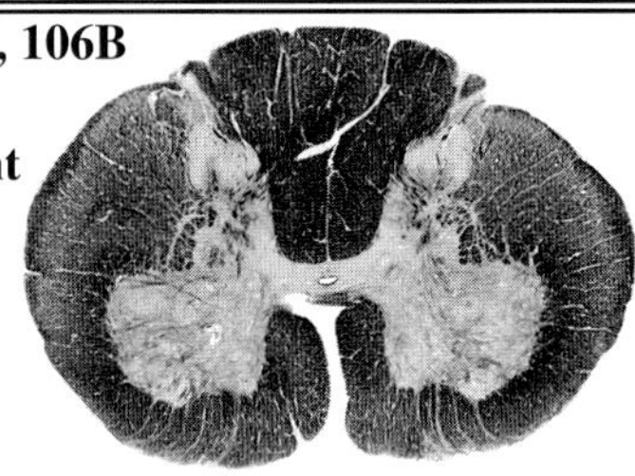	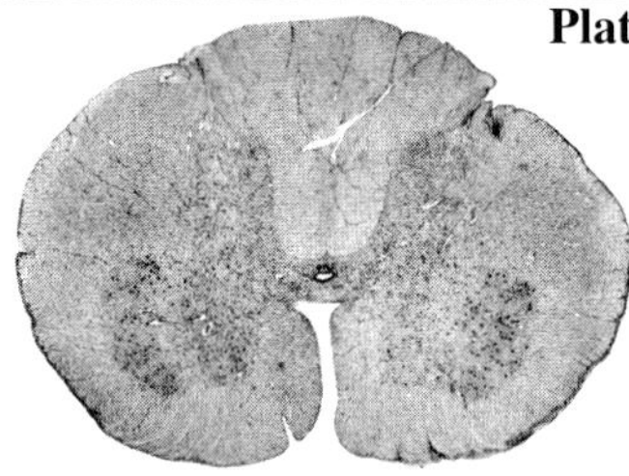 **Plates 107A, 107B** **Cervical Enlargement** Total area: 33.211 mm^2
Plates 108A, 108B **Upper Thoracic** Total area: 22.398 mm^2 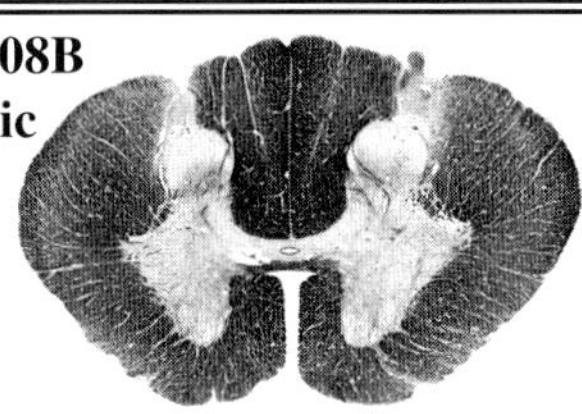	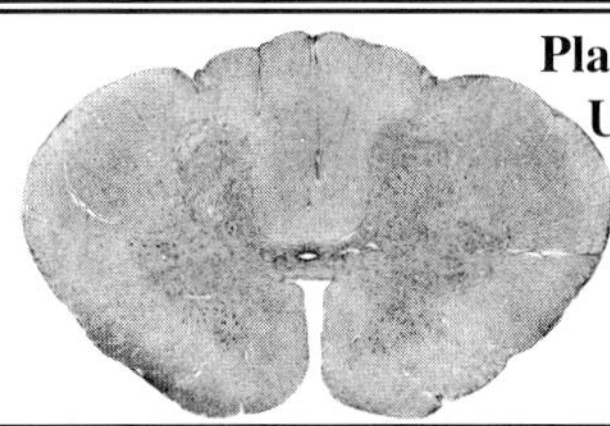 **Plates 109A, 109B** **Upper Thoracic** Total area: 24.760 mm^2
Plates 110A, 110B **Middle Thoracic** Total area: 15.291 mm^2 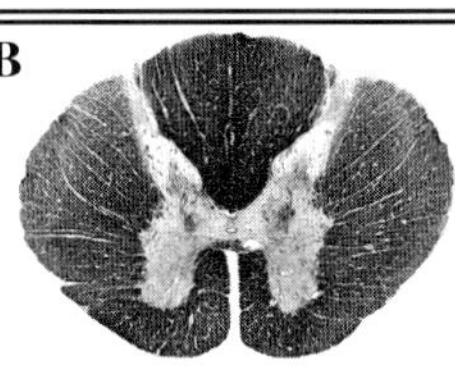	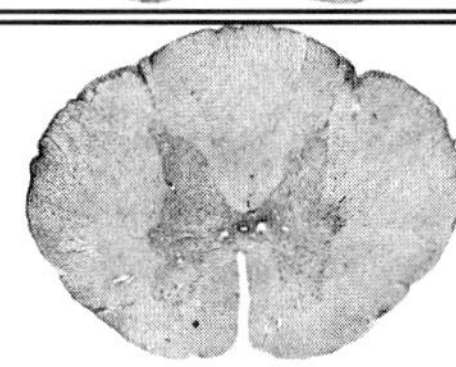 **Plates 111A, 111B** **Middle Thoracic** Total area: 16.112 mm^2
Plates 112A, 112B **Lower Thoracic** 17.098 mm^2 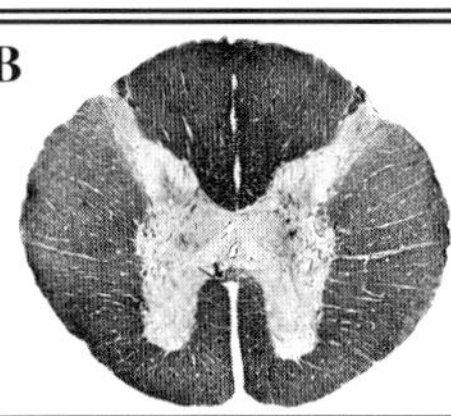	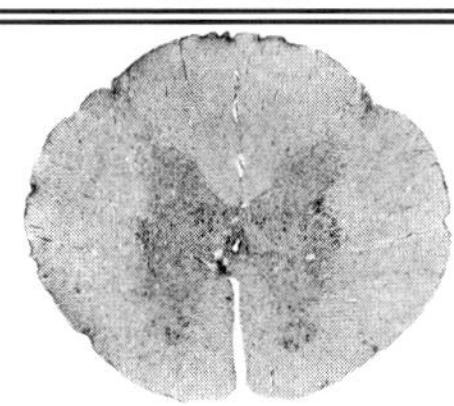 **Plates 113A, 113B** **Lower Thoracic** 17.300 mm^2
Plates 114A, 114B **Upper Lumbar** 24.774 mm^2 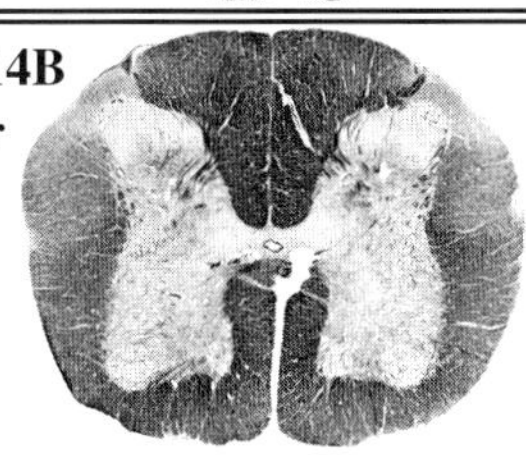	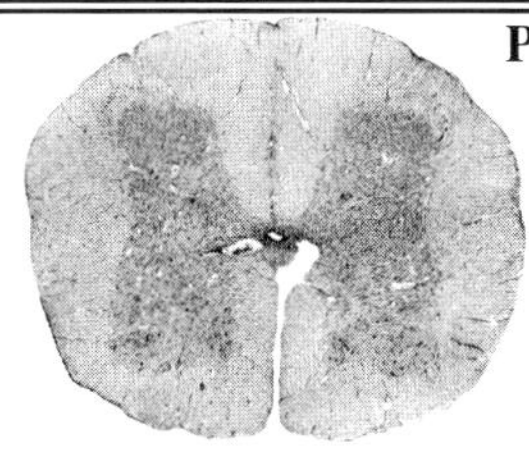 **Plates 115A, 115B** **Upper Lumbar** 24.914 mm^2
Plates 116A, 116B **Lumbar Enlargement** 31.708 mm^2 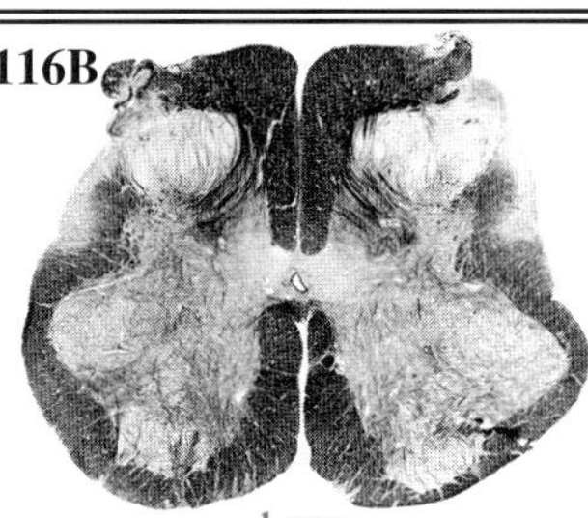 1 mm	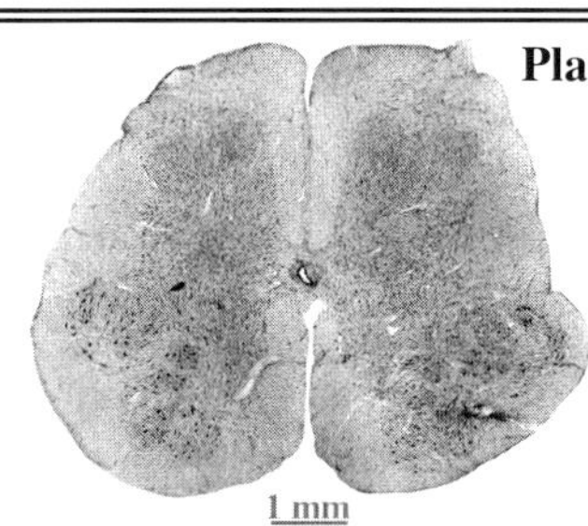 **Plates 117A, 117B** **Lumbar Enlargement** 32.275 mm^2 1 mm

PLATE 104A

Infant, 4 Months
CR 440 mm
Y286-62
Upper Cervical
Myelin stain

Areas (mm^2)

Central canal	.0066
Ependyma	.0082
Gray matter	6.0358
White matter	21.4380

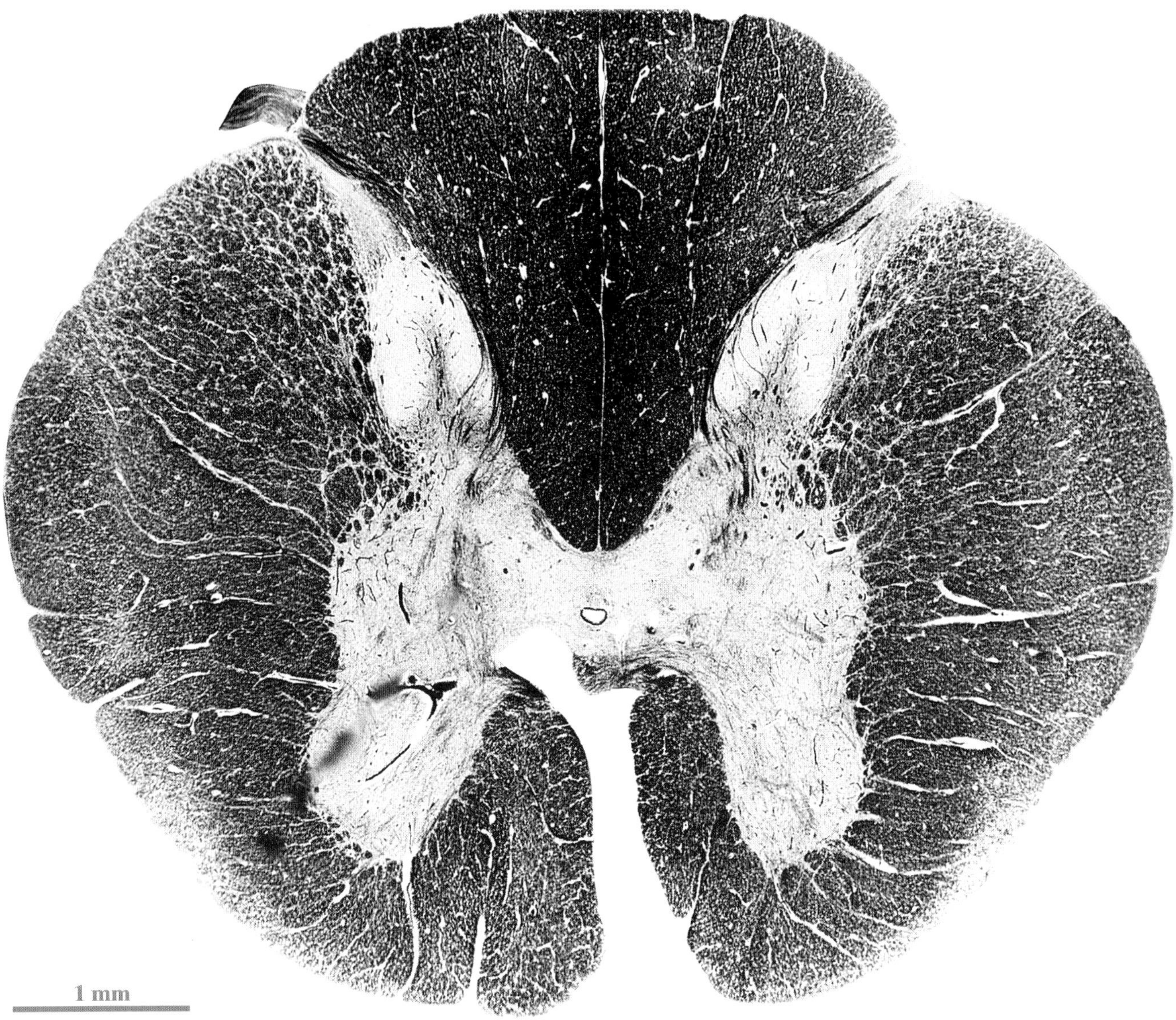

PLATE 104B

See the matched cell body stained section in Plates 105A and B

The myelinating axons in the most superficial region of the lateral corticospinal tract extend to lumbosacral levels where they are *unmyelinated* (see **Plate 116**). The lighter staining in the most superficial region of the spinocephalic tracts is occupied by axons having cells of origin at lumbosacral levels where these axons are *myelinated* (see **Plate 116**). Both gradients indicate that myelination proceeds from proximal to distal in axons.

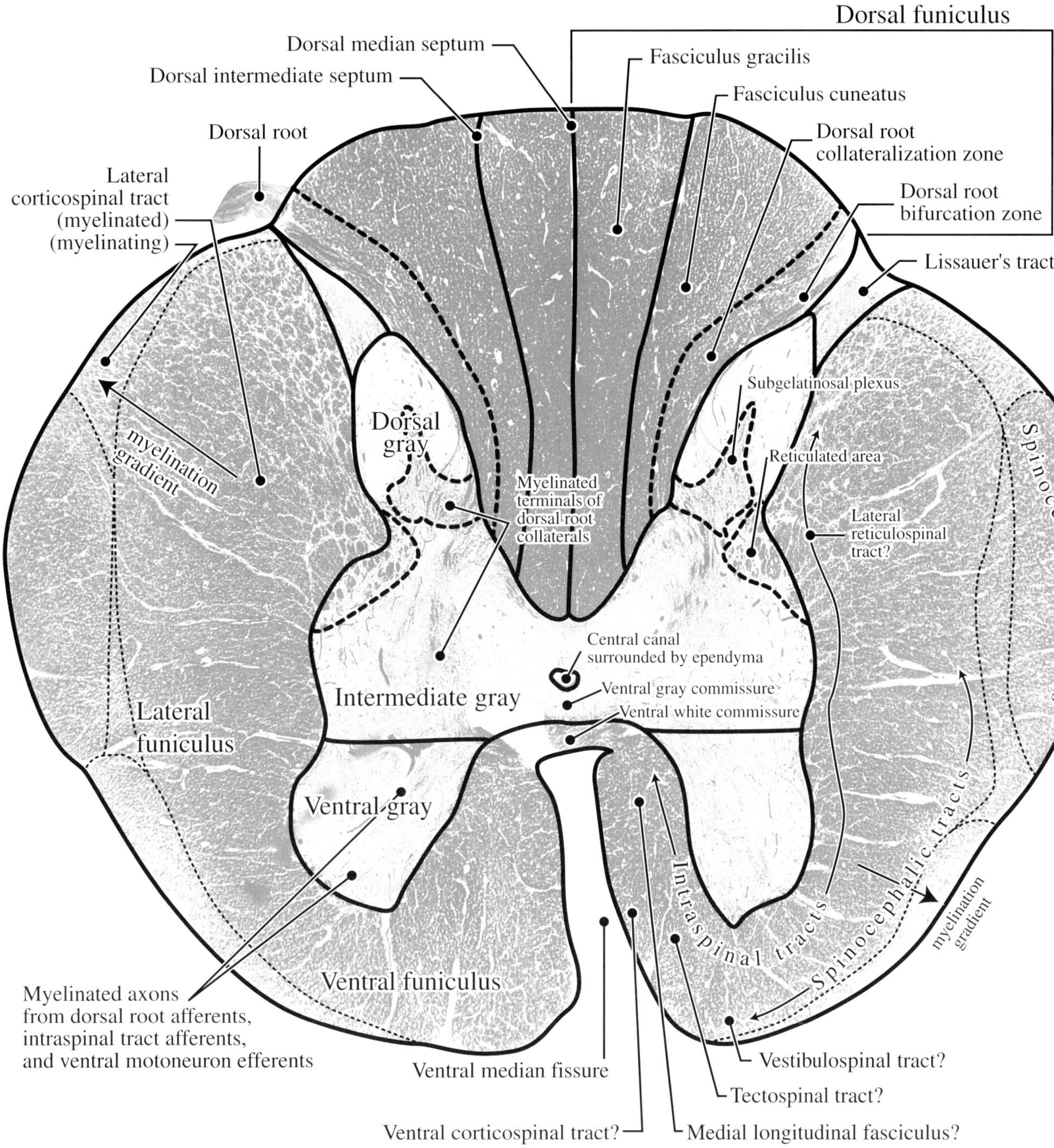

As stated in the previous specimens, the fine dashed lines in the lateral and ventral funiculi segregate regions of differing densities of the myelin stain, not the borders of fiber tracts. In this 4-month-old infant, the ventral and lateral funiculi contain dense staining indicative of advanced myelination. There are three lightly stained areas: sparse reactive glia in the most superficial part of the spinocephalic tracts, a myelinating region in the superficial part of the lateral corticospinal tract, and a thin unidentified myelinating region just medial to the spinocerebellar tracts. Most of the lateral corticospinal tract cannot be distinguished from the rest of the lateral funiculus by its myelination pattern.

PLATE 105A

Infant, 4 Months
CR 440 mm
Y286-62
Upper Cervical
Cell body stain

Areas (mm²)

Central canal	.0045
Ependyma	.0145
Gray matter	6.0685
White matter	22.2390

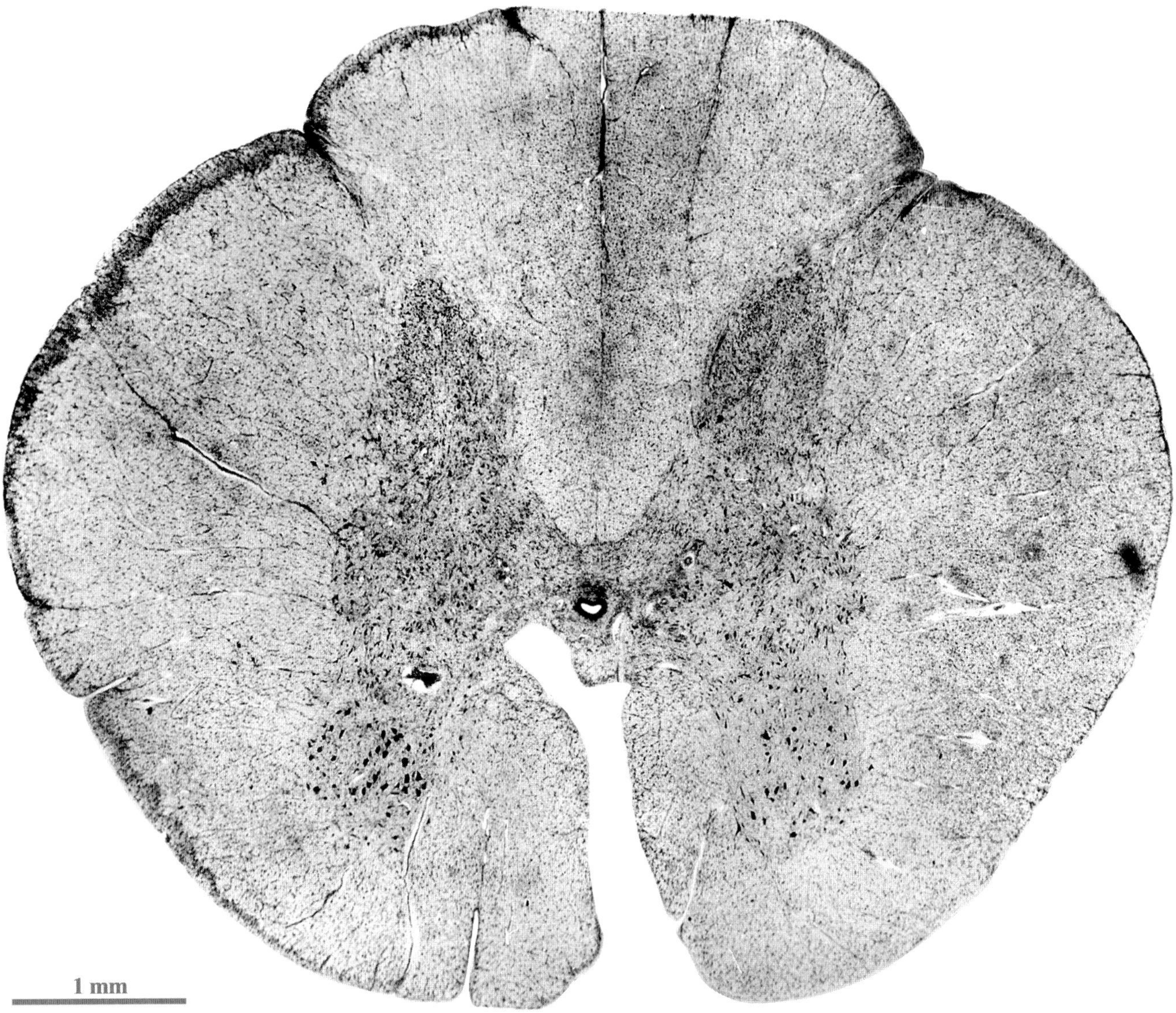

PLATE105B

See the matched myelin stained section in Plates 104A and B

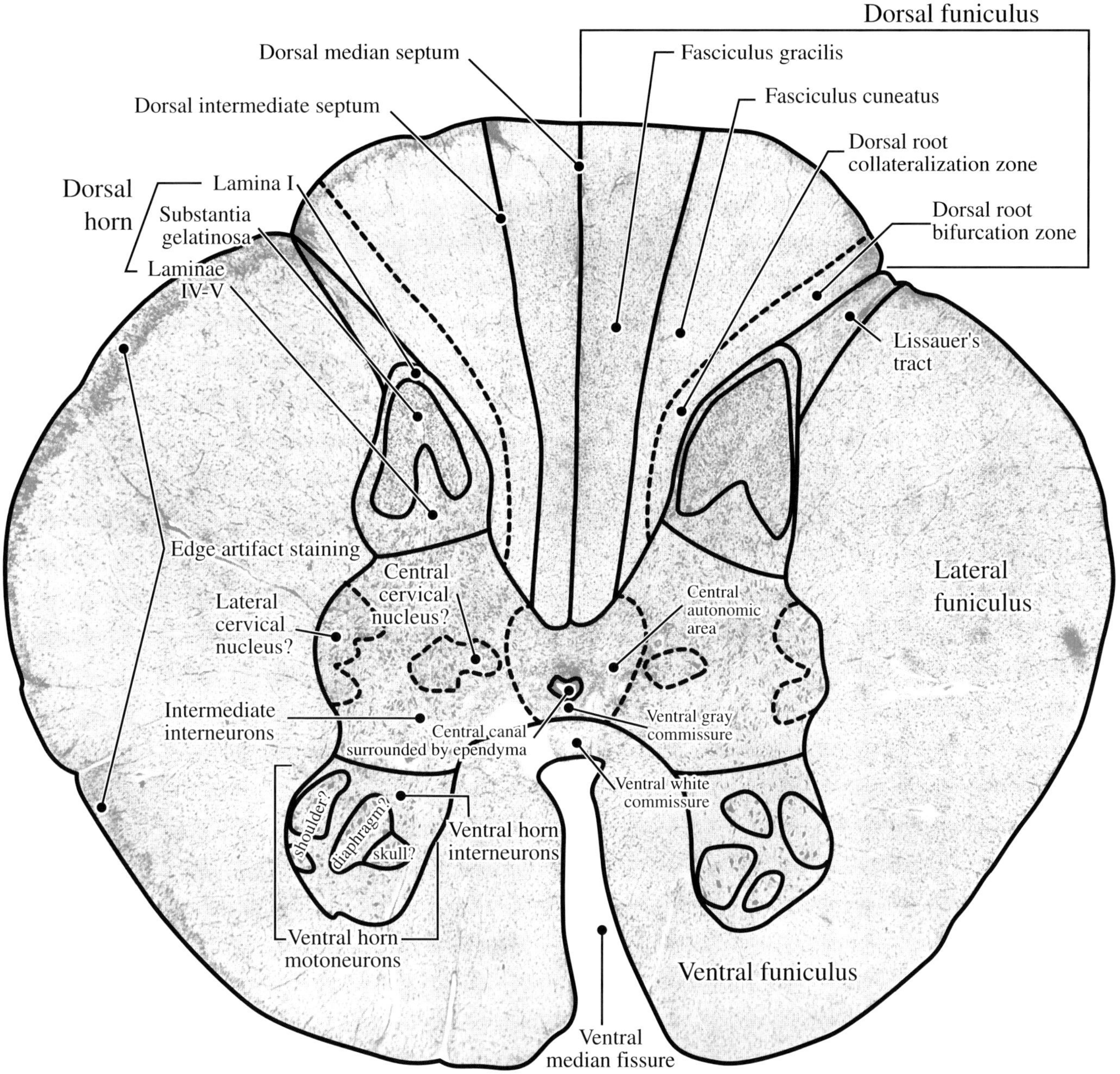

Note the uniform light staining of the white matter in this and all other cell body stained sections in this specimen. That indicates that myelination gliosis is completely finished, even in the lateral corticospinal tract, because nearly all tracts are either myelinated or myelinating. Refer to the matching myelin stained section for the approximate locations of both corticospinal tracts, the medial longitudinal fasciculus, the tectospinal, lateral reticulospinal, vestibulospinal, intraspinal, and spinocephalic tracts.

PLATE 106A

Infant, 4 Months
CR 440 mm
Y286-62
Cervical Enlargement
Myelin stain

Areas (mm^2)

Central canal	.0124
Ependyma	.0160
Gray matter	10.2870
White matter	22.9990

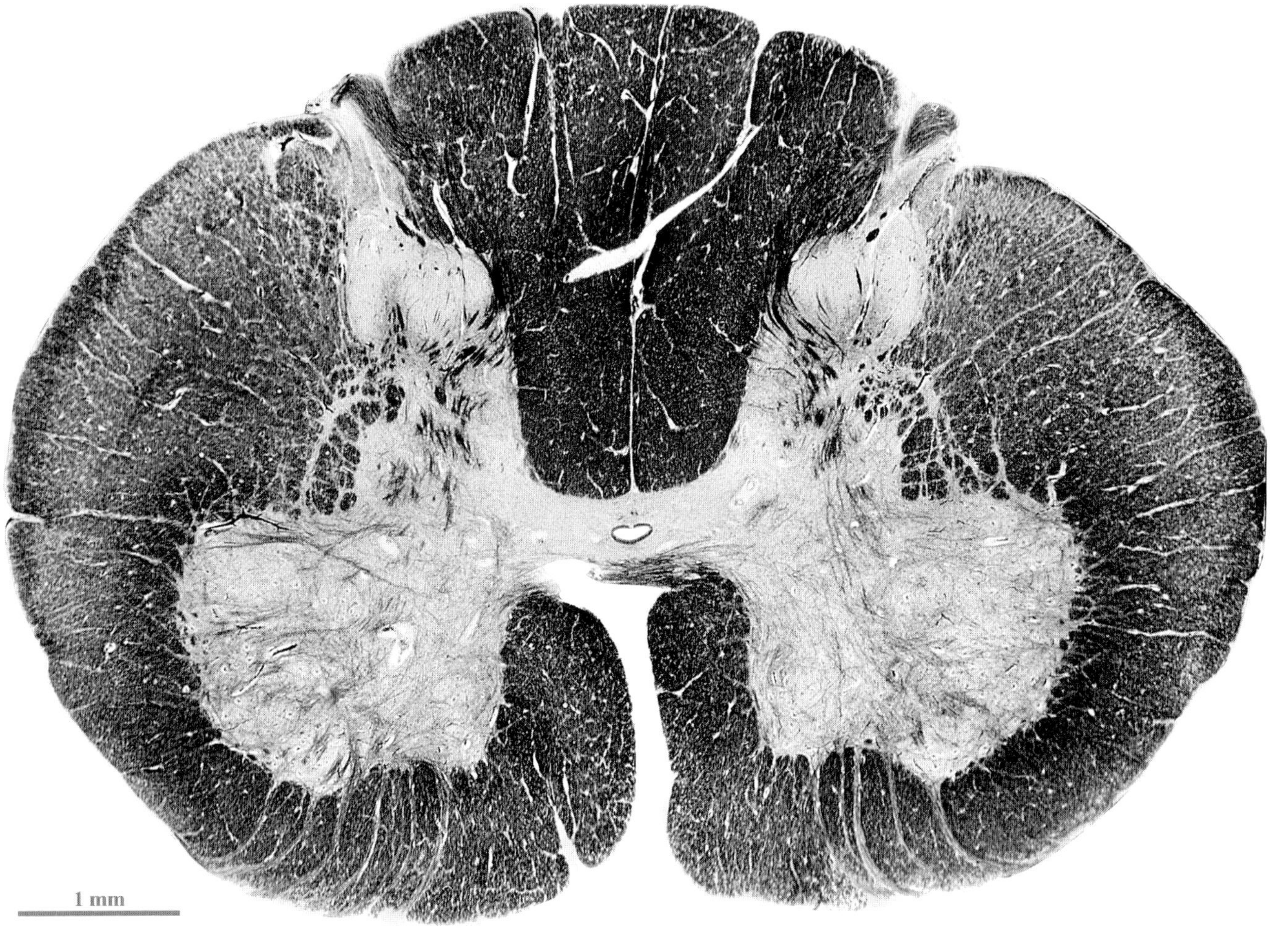

PLATE 106B

See the matched cell body stained section in Plates 107A and B

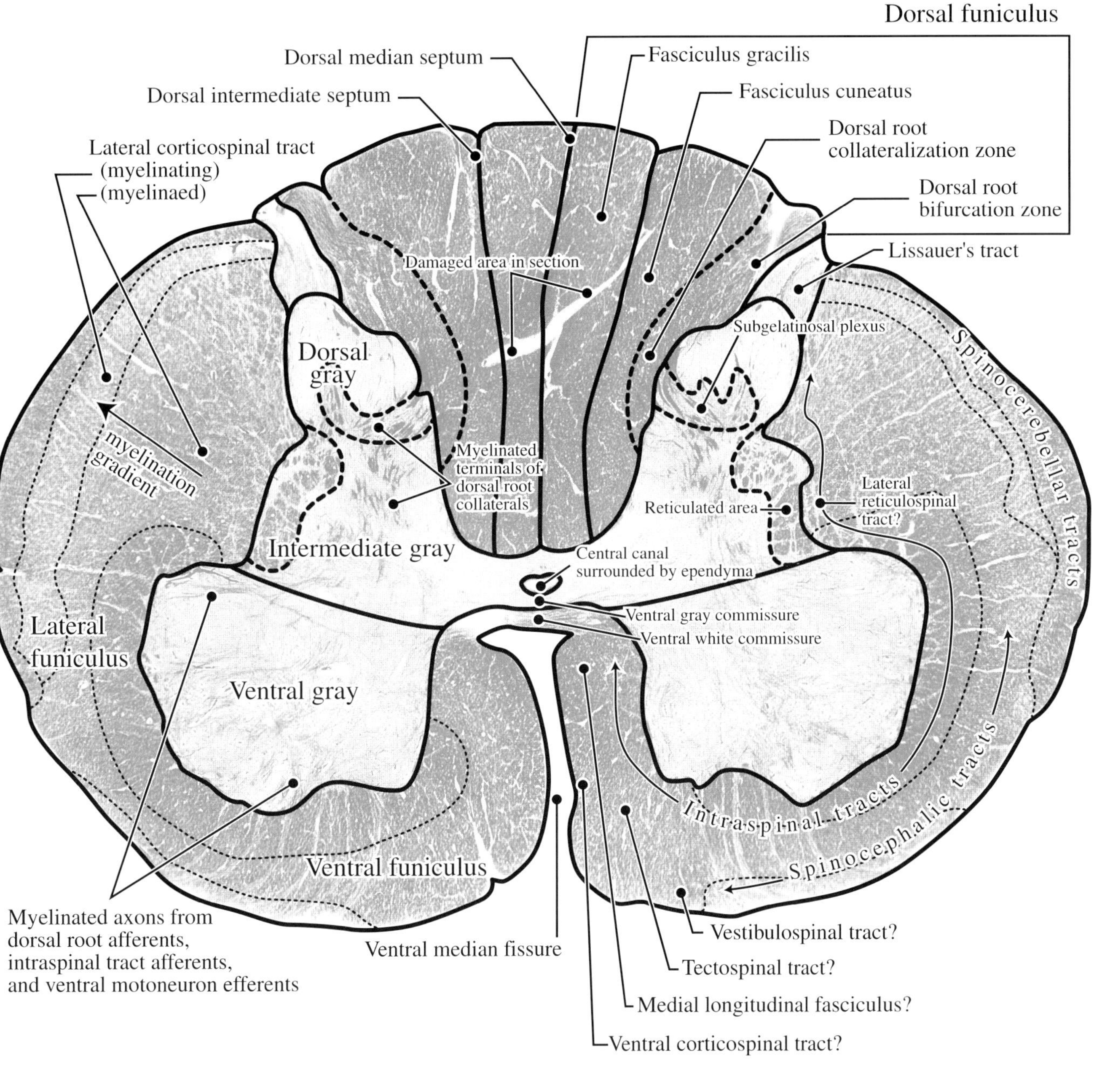

PLATE 107A

Infant, 4 Months
CR 440 mm
Y286-62
Cervical Enlargement
Cell body stain

Areas (mm^2)

Central canal	.0010
Ependyma	.0256
Gray matter	10.1850
White matter	22.9900

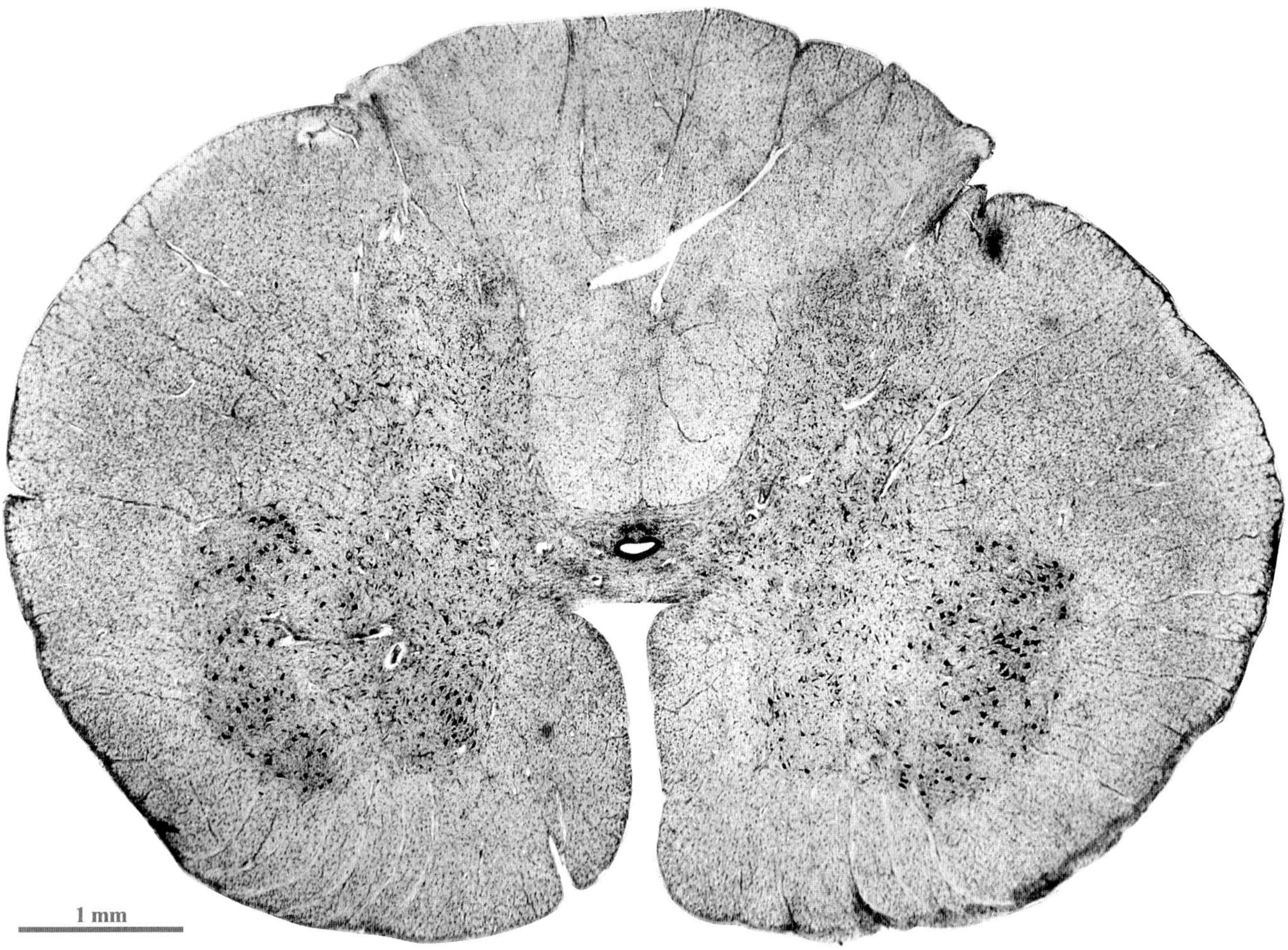

PLATE 107B

See the matched myelin stained section in Plates 106A and B

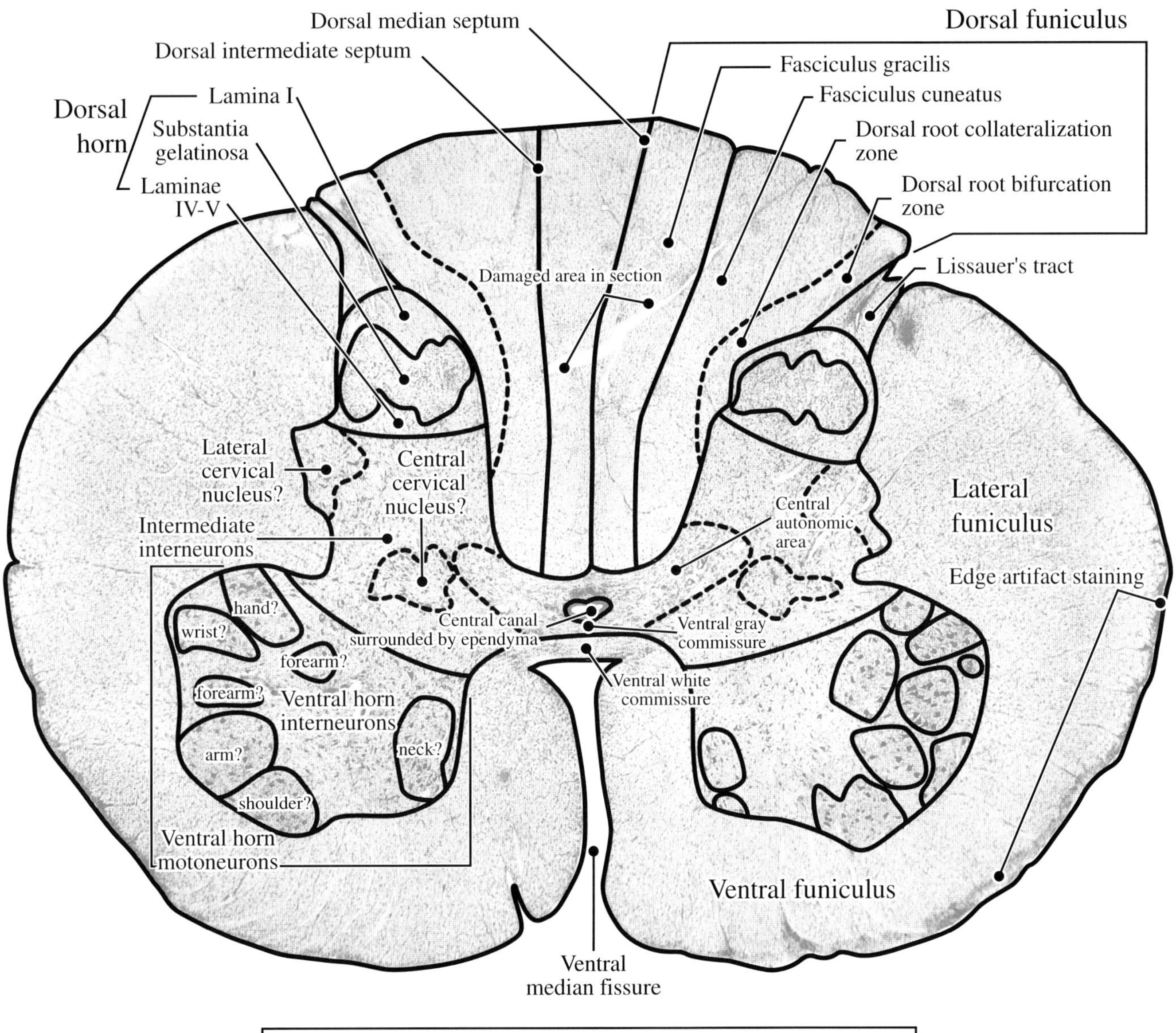

Refer to the matching myelin stained section for the approximate locations of both corticospinal tracts, the medial longitudinal fasciculus, the tectospinal, lateral reticulospinal, vestibulospinal, intraspinal, and spinocephalic tracts.

PLATE 108A

Infant, 4 Months
CR 440 mm
Y286-62
Upper Thoracic
Myelin stain

Areas (mm²)

Central canal	.0088
Ependyma	.0136
Gray matter	5.6424
White matter	16.7340

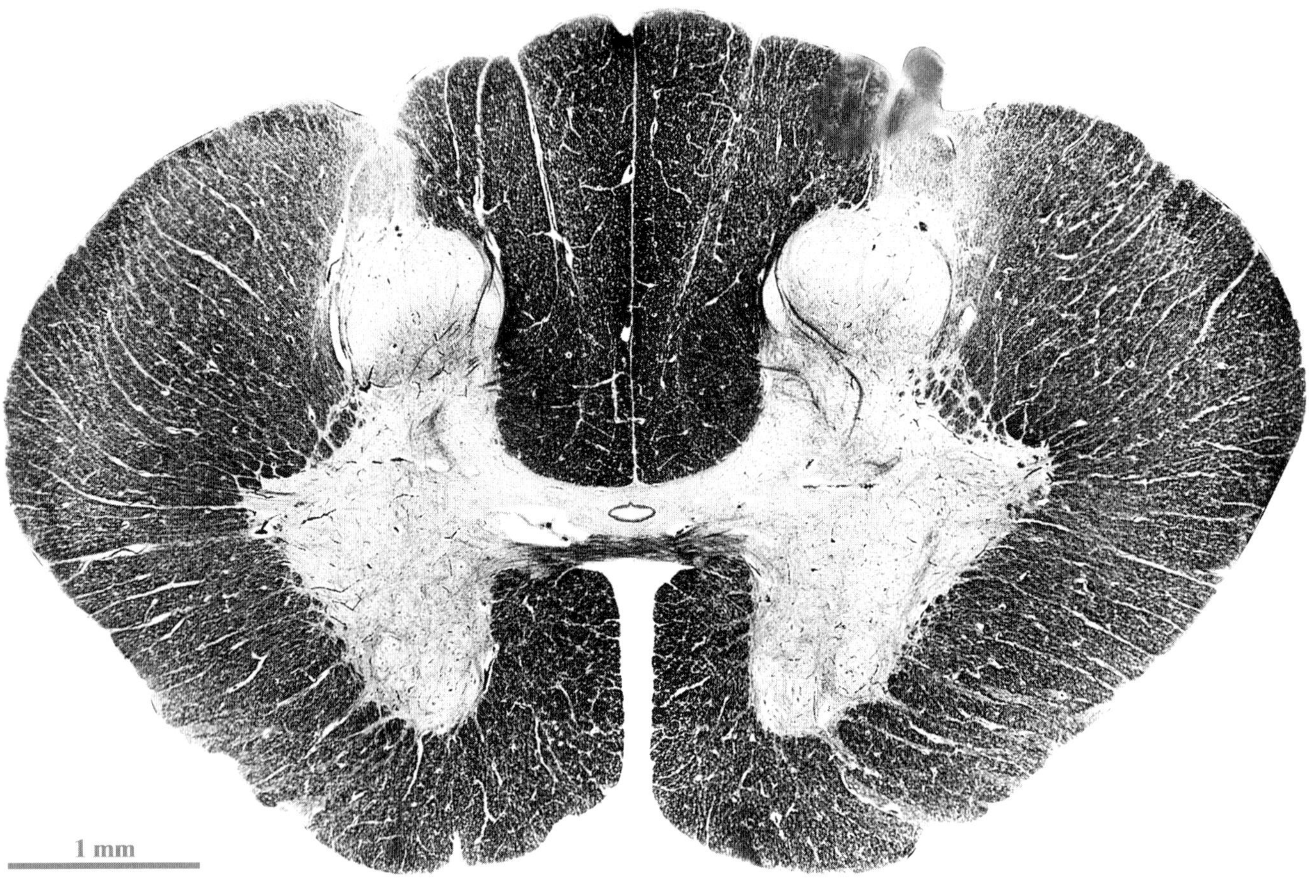

PLATE 108B

See the matched cell body stained section in Plates 109A and B

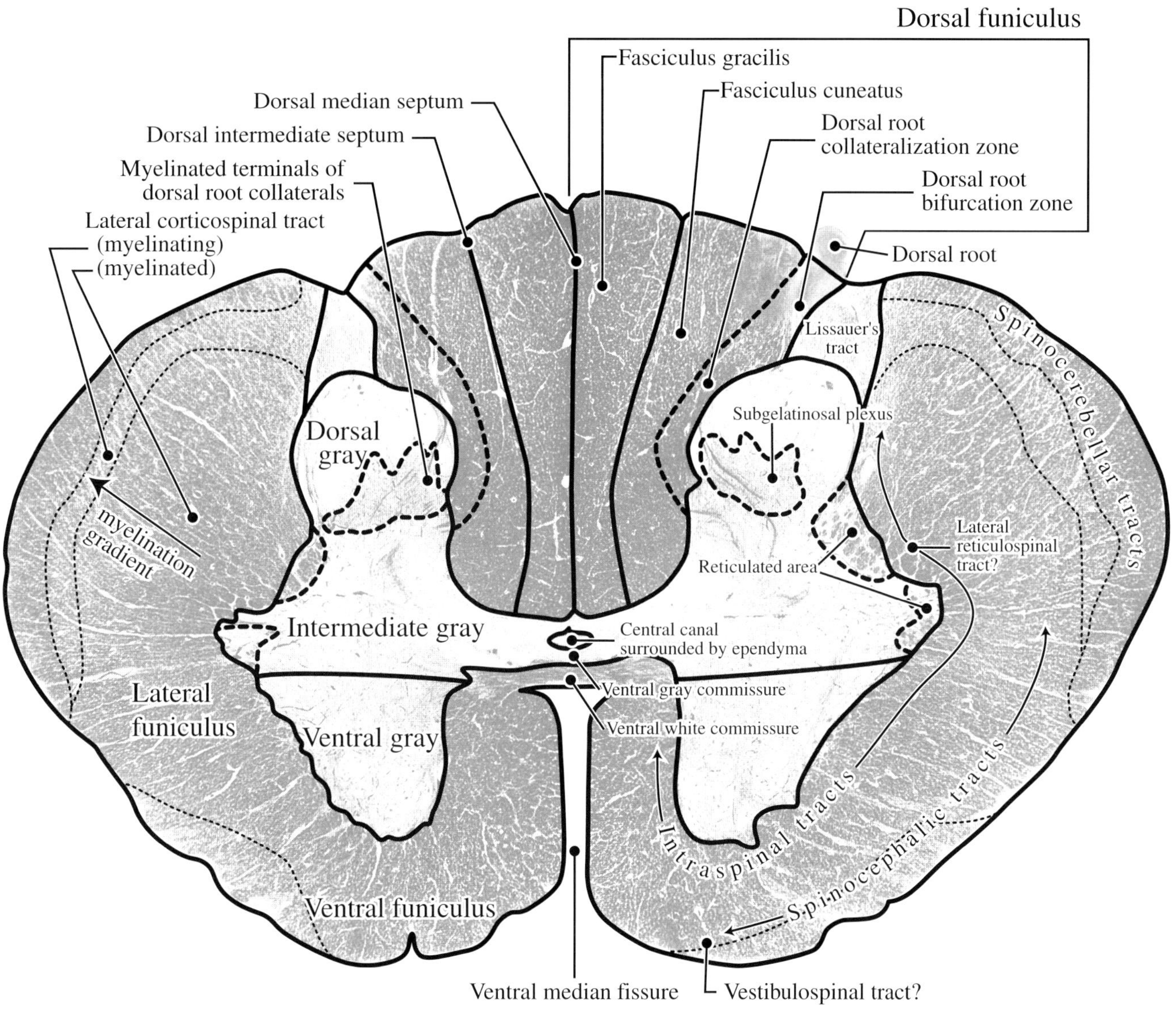

PLATE 109A

Infant, 4 Months
CR 440 mm
Y286-62
Upper Thoracic
Cell body stain

Areas (mm^2)

Central canal	.0099
Ependyma	.0241
Gray matter	6.5545
White matter	18.1720

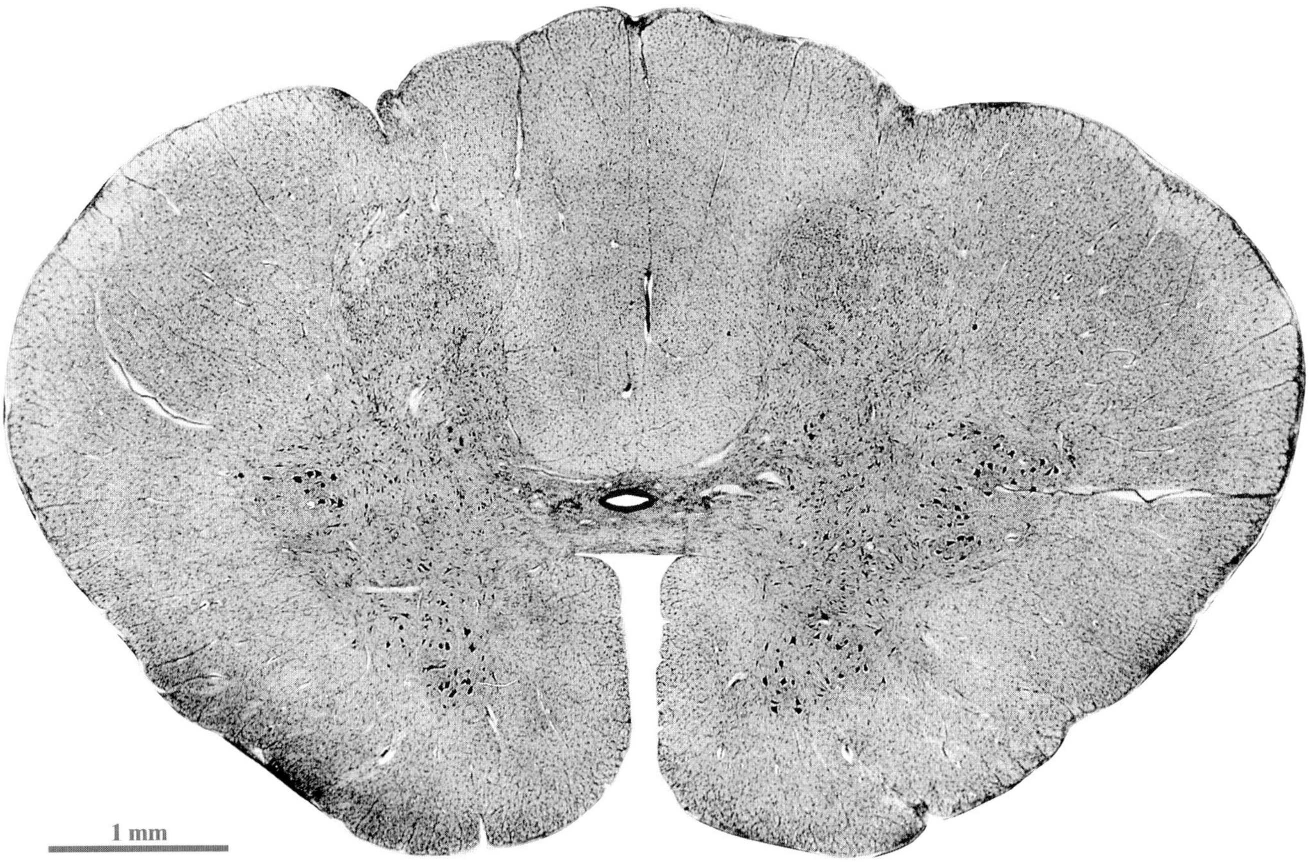

PLATE 109B

See the matched myelin stained section in Plates 108A and B

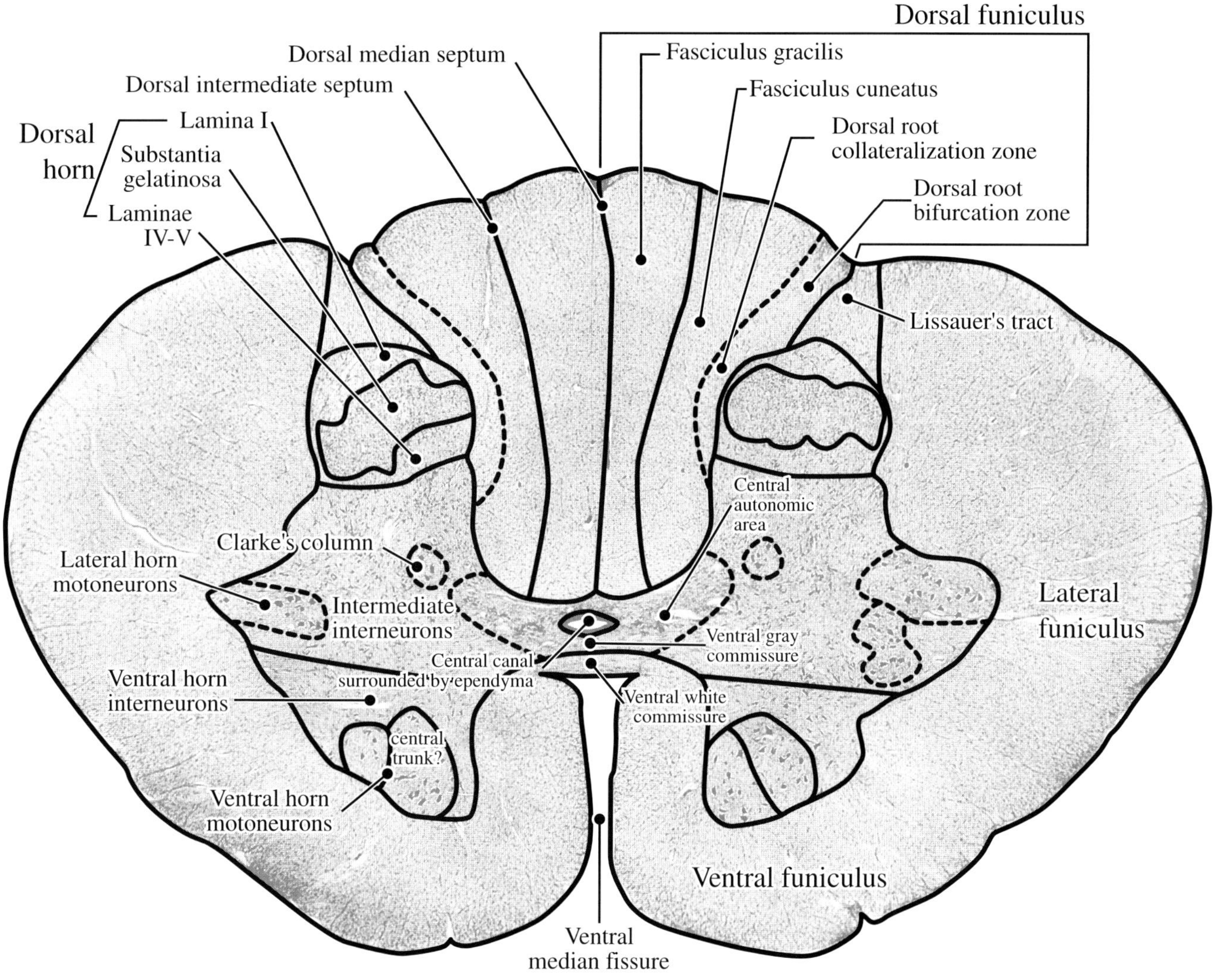

Refer to the matching myelin stained section for the approximate locations of the lateral corticospinal tract, the lateral reticulospinal, vestibulospinal, intraspinal, and spinocephalic tracts.

PLATE 110A

Infant, 4 Months
CR 440 mm
Y286-62
Middle Thoracic
Myelin stain

Areas (mm^2)

Central canal	.0040
Ependyma	.0062
Gray matter	2.8387
White matter	12.4420

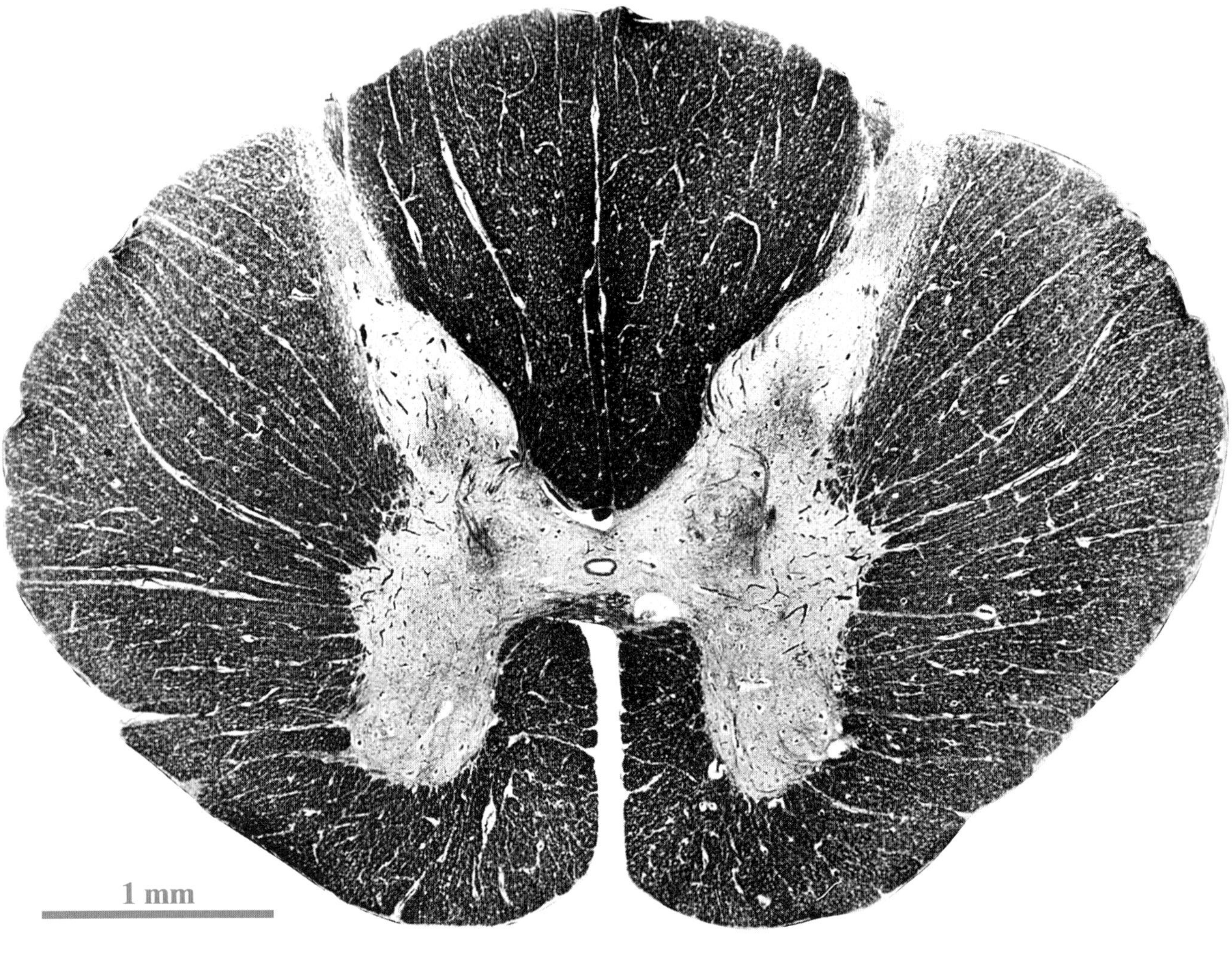

PLATE 110B

See the matched cell body stained section in Plates 111A and B

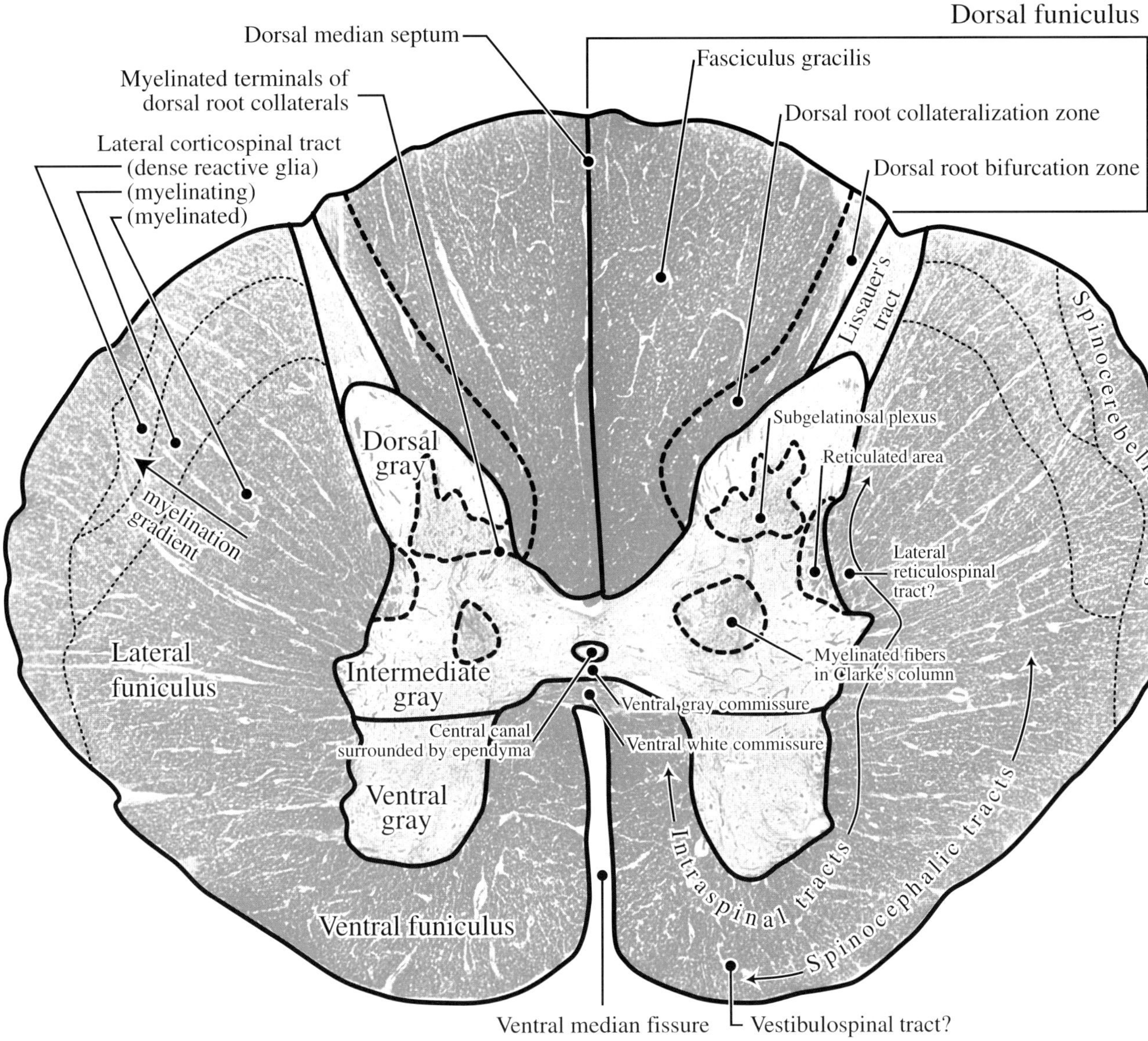

PLATE 111A

Infant, 4 Months
CR 440 mm
Y286-62
Middle Thoracic
Cell body stain

Areas (mm^2)

Central canal	.0031
Ependyma	.0117
Gray matter	2.8799
White matter	13.2170

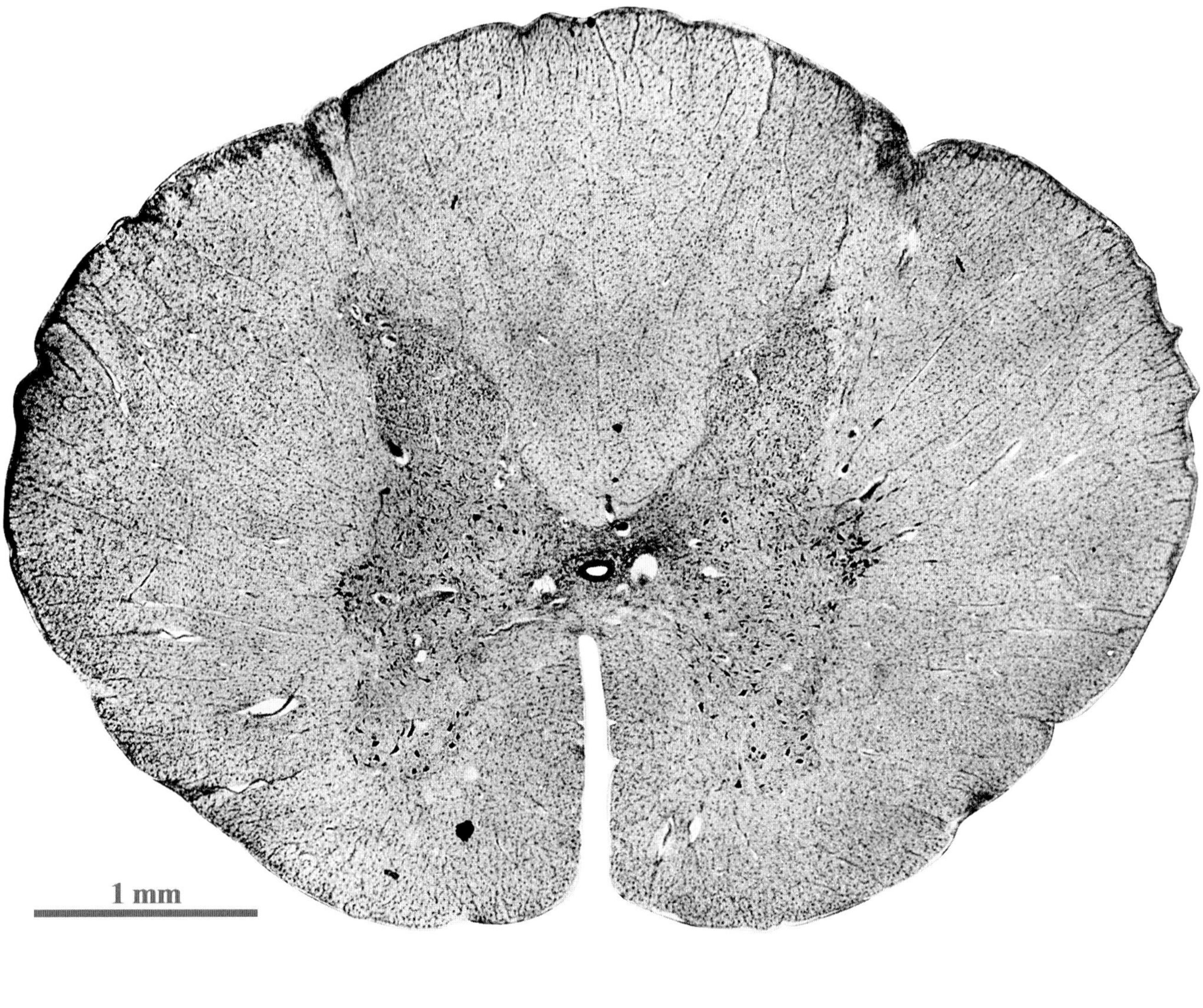

PLATE 111B

See the matched myelin stained section in Plates 110A and B

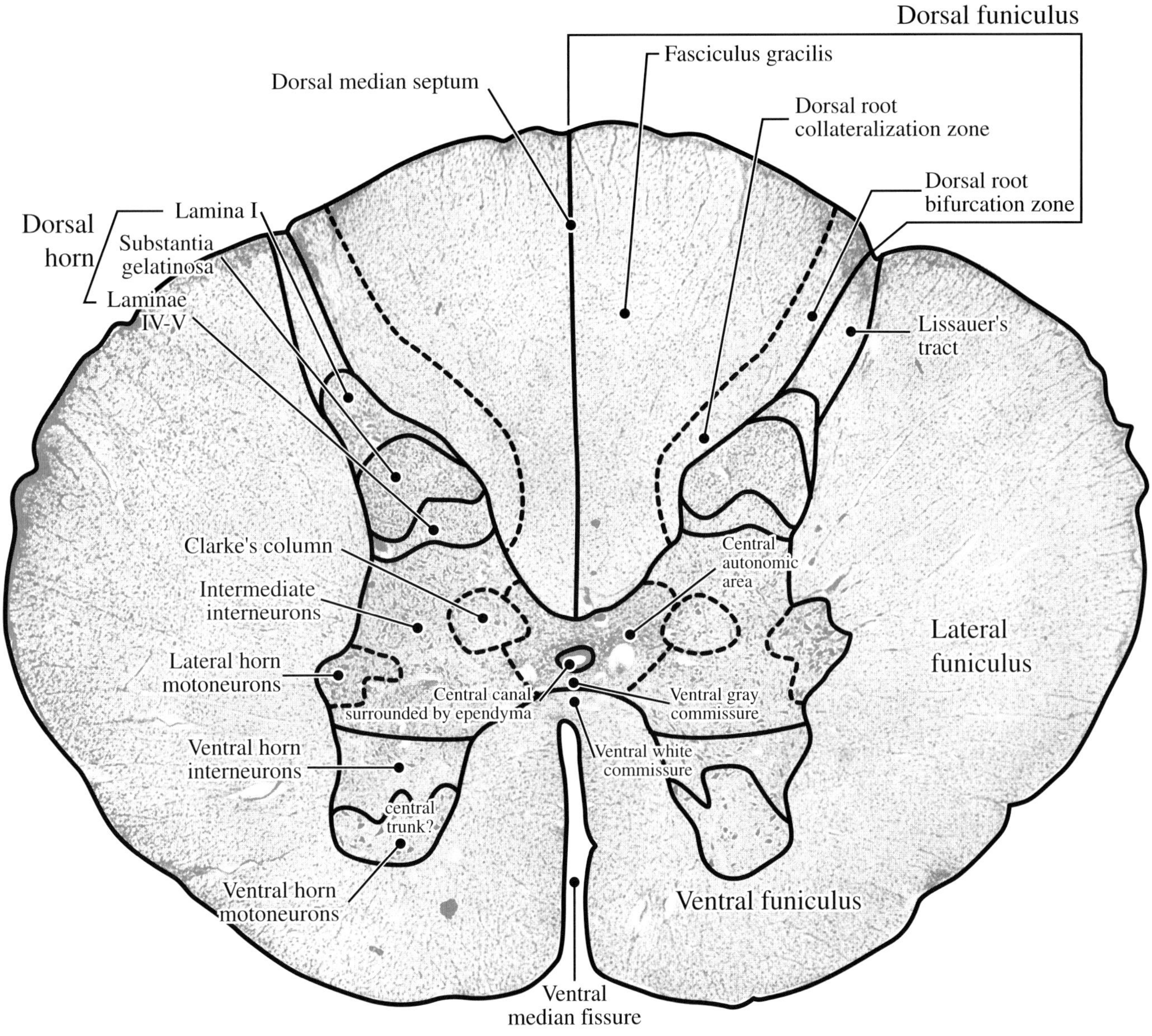

Refer to the matching myelin stained section for the approximate locations of the lateral corticospinal tract, the lateral reticulospinal, vestibulospinal, intraspinal, and spinocephalic tracts.

PLATE 112A

Infant, 4 Months
CR 440 mm
Y286-62
Lower Thoracic
Myelin stain

Areas (mm^2)

Central canal	.0090
Ependyma	.0091
Gray matter	4.0248
White matter	13.0550

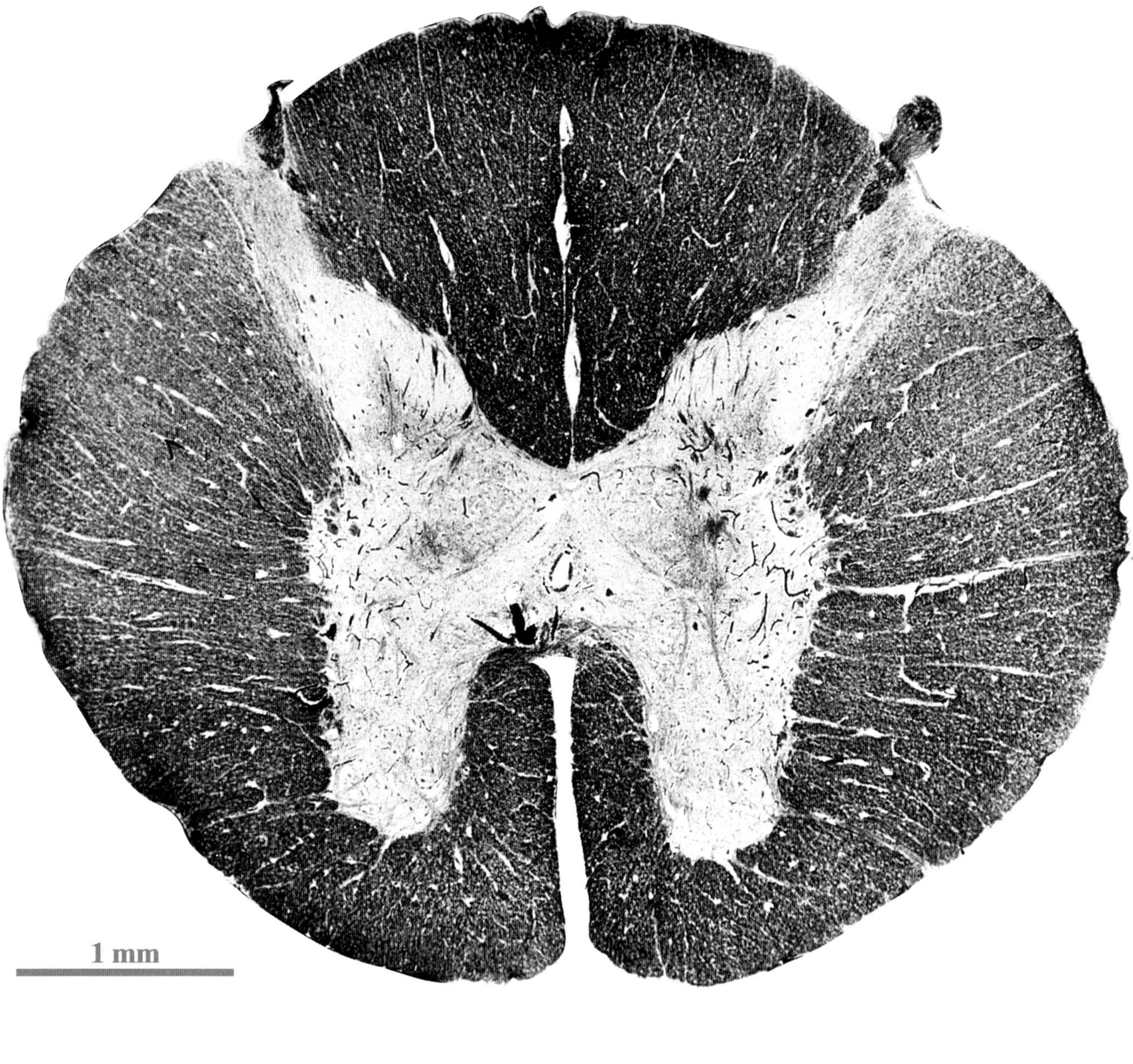

PLATE 112B

See the matched cell body stained section in Plates 113A and B

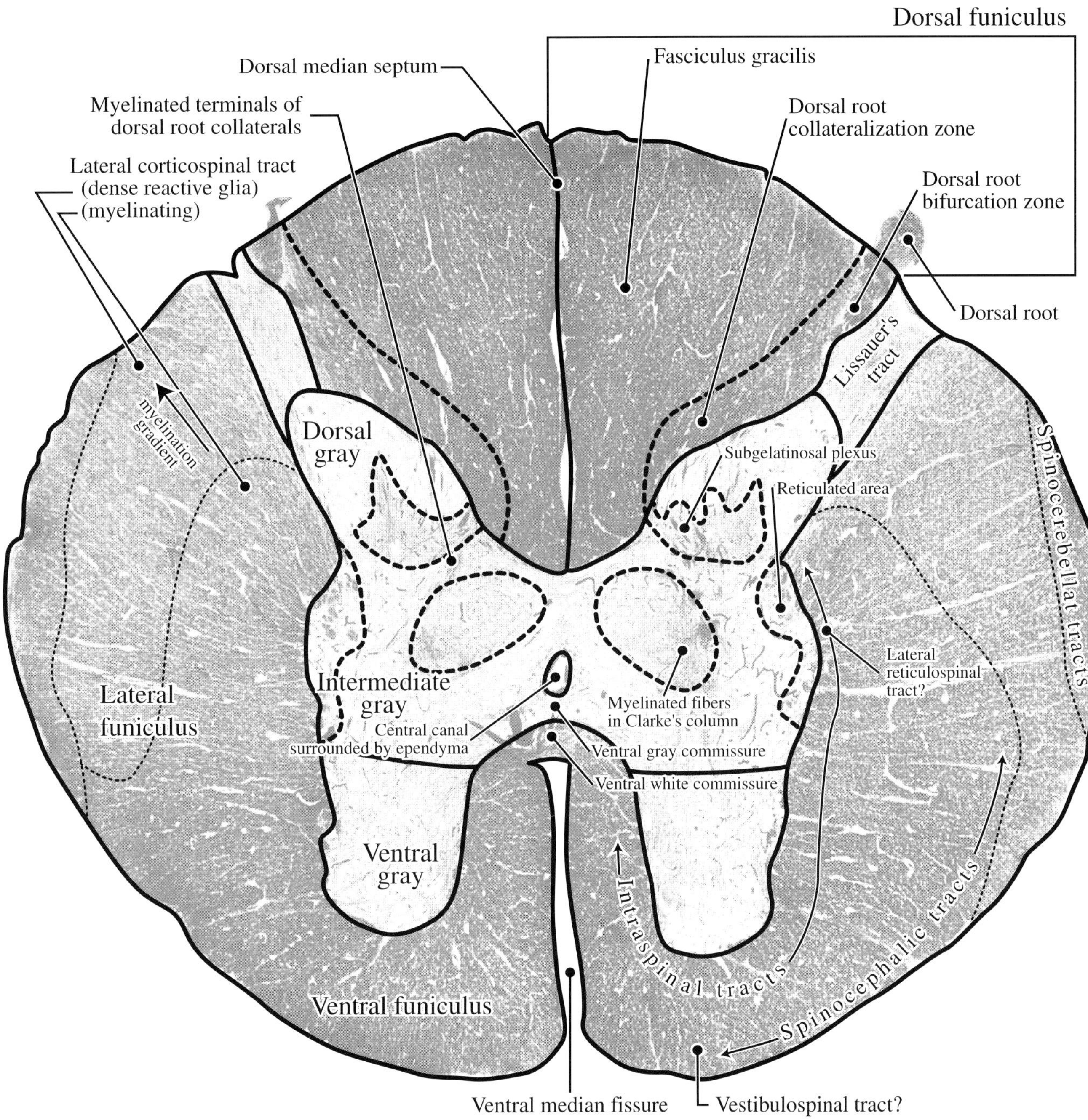

PLATE 113 A

Infant, 4 Months
CR 440 mm
Y286-62
Lower Thoracic
Cell body stain

Areas (mm^2)

Central canal	.0073
Ependyma	.0160
Gray matter	4.2476
White matter	13.0290

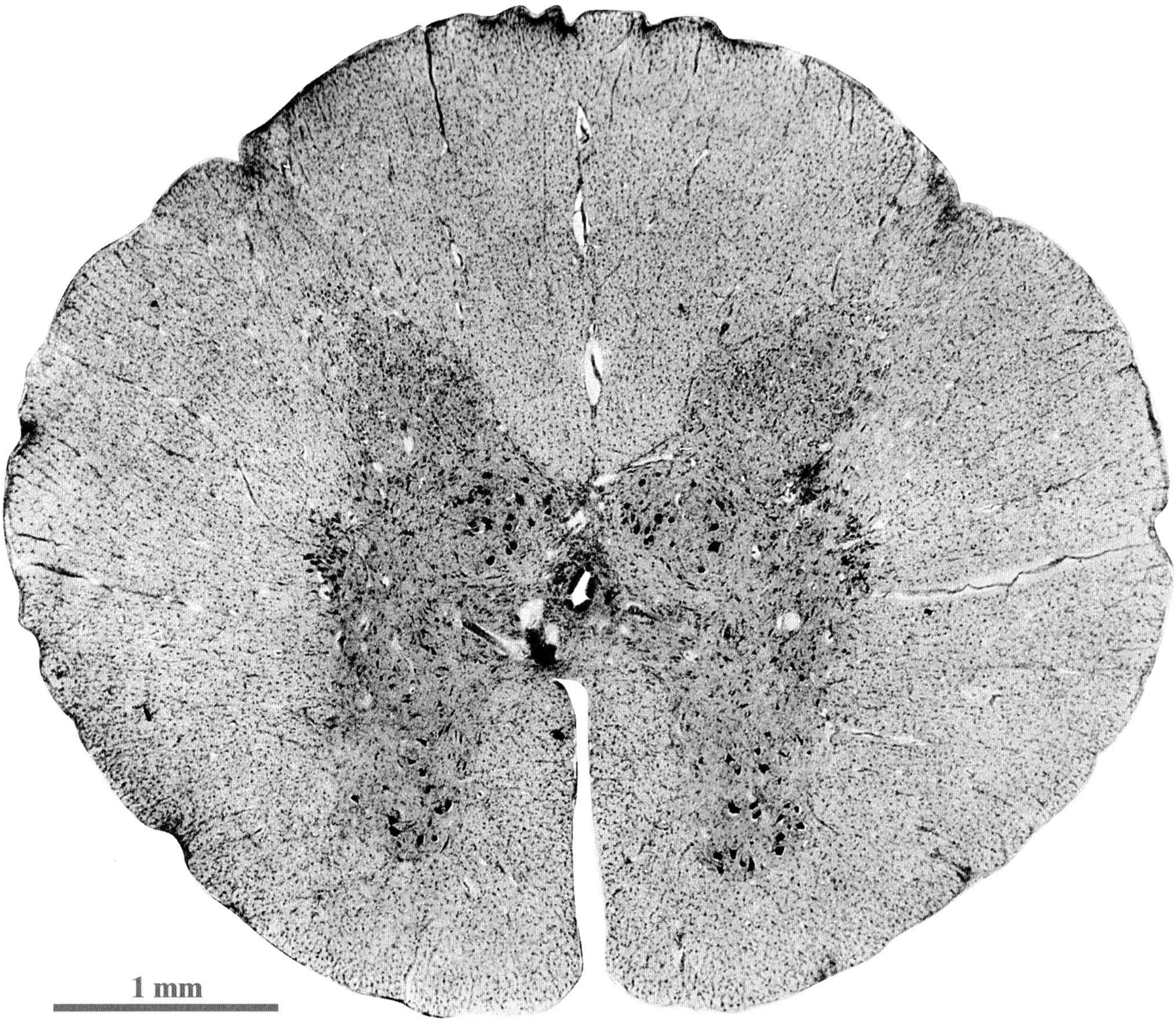

PLATE 113B

See the matched myelin stained section in Plates 112A and B

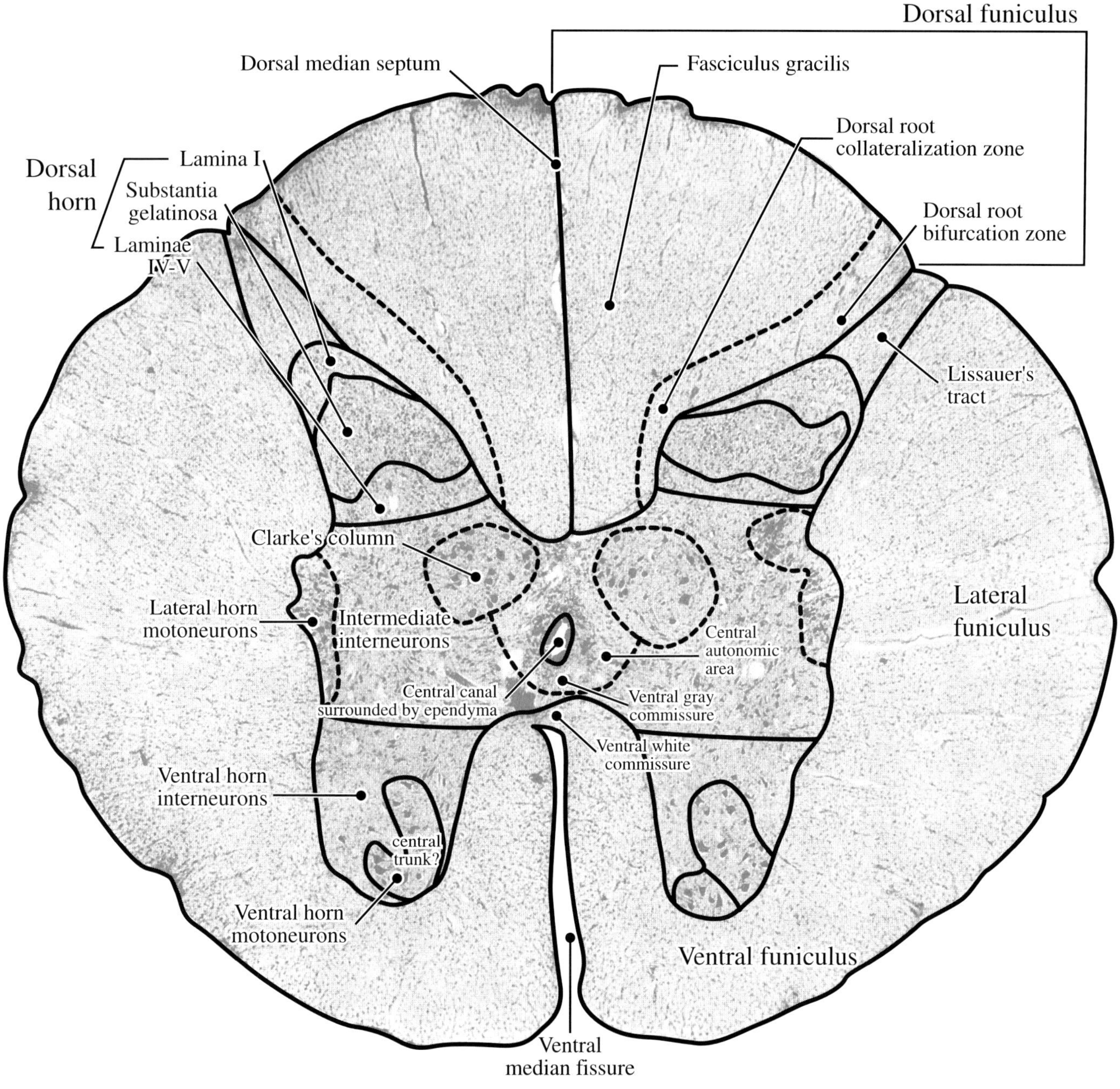

Refer to the matching myelin stained section for the approximate locations of the lateral corticospinal tract, the lateral reticulospinal, vestibulospinal, intraspinal, and spinocephalic tracts.

PLATE 114A

Infant, 4 Months
CR 440 mm
Y286-62
Upper Lumbar
Myelin stain

Areas (mm^2)

Central canal	.0140
Ependyma	.0155
Gray matter	9.5065
White matter	15.2380

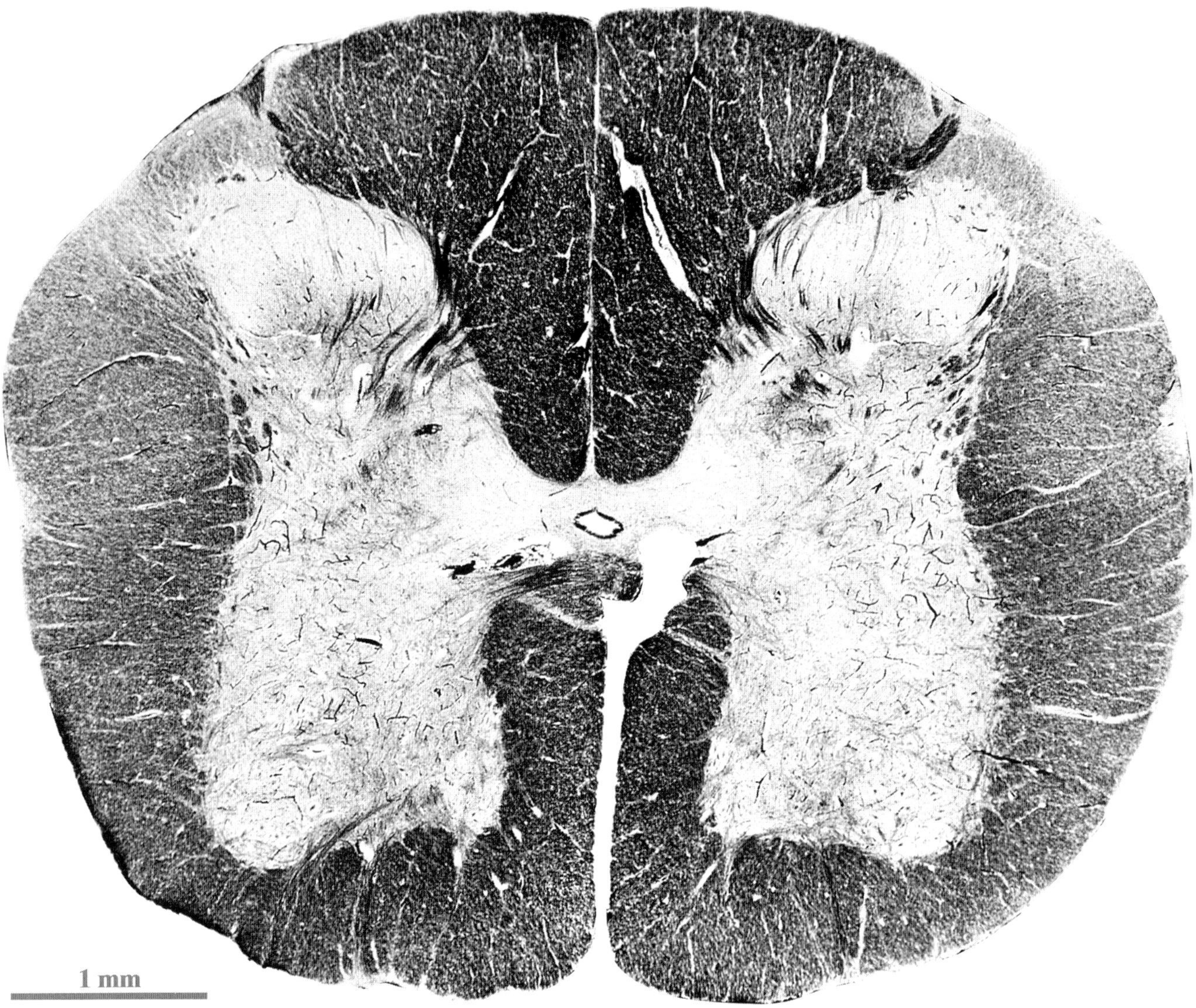

PLATE 114B

See the matched cell body stained section in Plates 115A and B

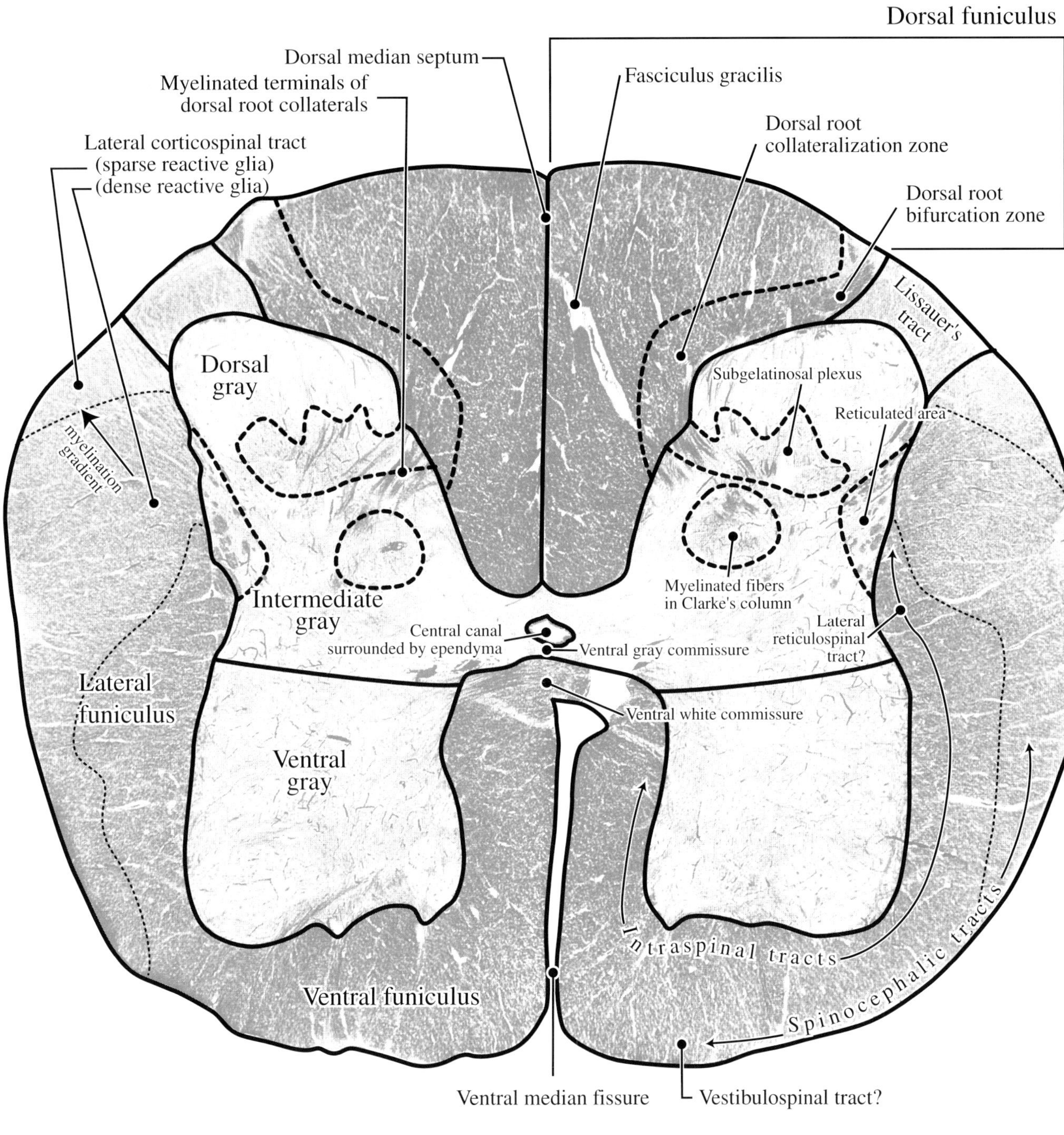

Compare the low staining density of the lateral corticospinal tract in this caudal section with the intense staining in more rostral sections, for instance, **Plate 104**. The progressively decreasing staining intensity indicates a rostral to caudal gradient of myelination in the lateral corticospinal tract.

PLATE 115A

Infant, 4 Months
CR 440 mm
Y286-62
Upper Lumbar
Cell body stain

Areas (mm²)

Central canal	.0086
Ependyma	.0270
Gray matter	9.8606
White matter	15.0180

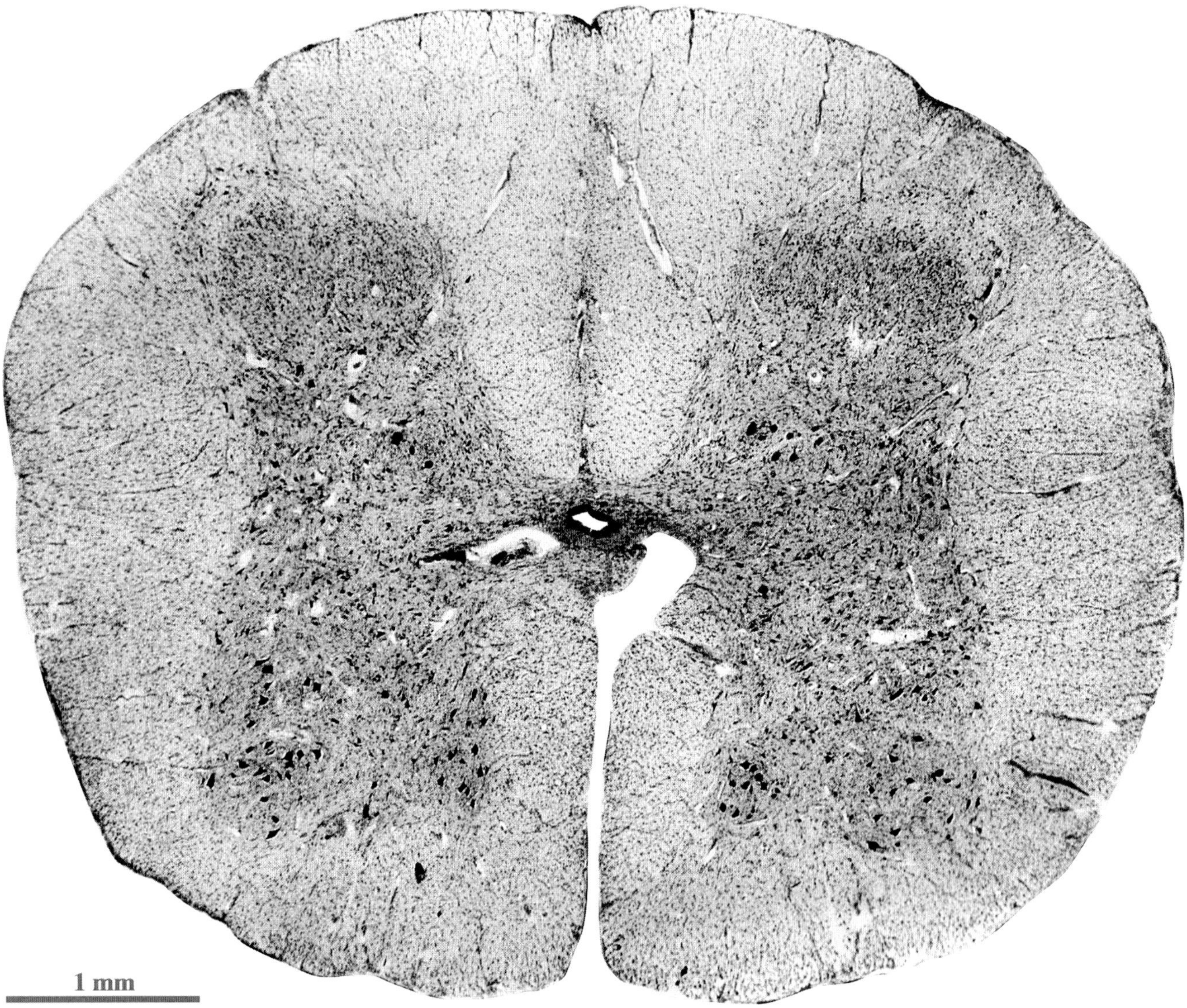

PLATE 115B

See the matched myelin stained section in Plates 114A and B

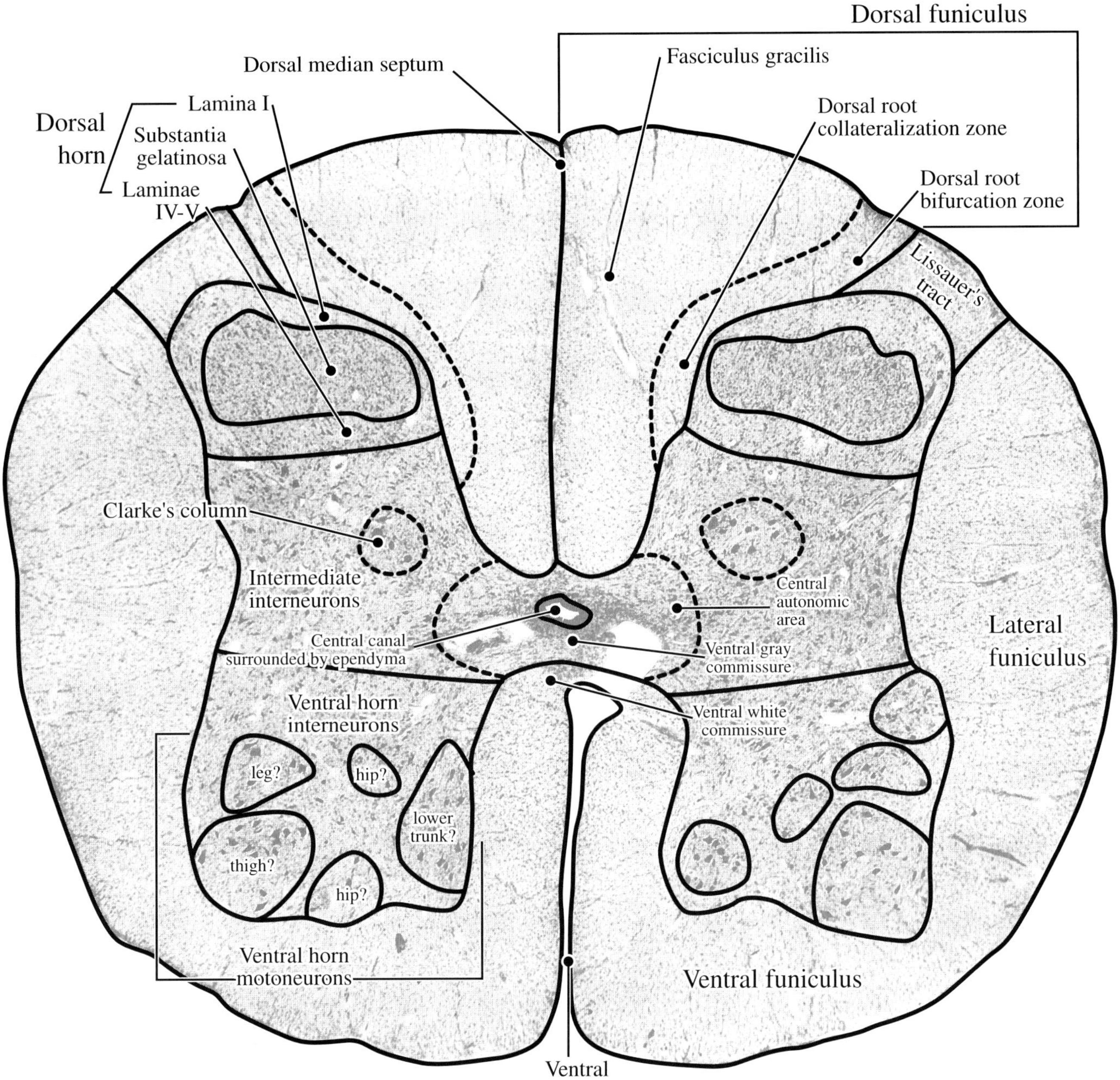

Refer to the matching myelin stained section for the approximate locations of the lateral corticospinal tract, the lateral reticulospinal, vestibulospinal, intraspinal, and spinocephalic tracts.

PLATE 116A

Infant, 4 Months
CR 440 mm
Y286-62
Lumbar Enlargement
Myelin stain

Areas (mm^2)

Central canal	.0180
Ependyma	.0187
Gray matter	17.7370
White matter	13.9340

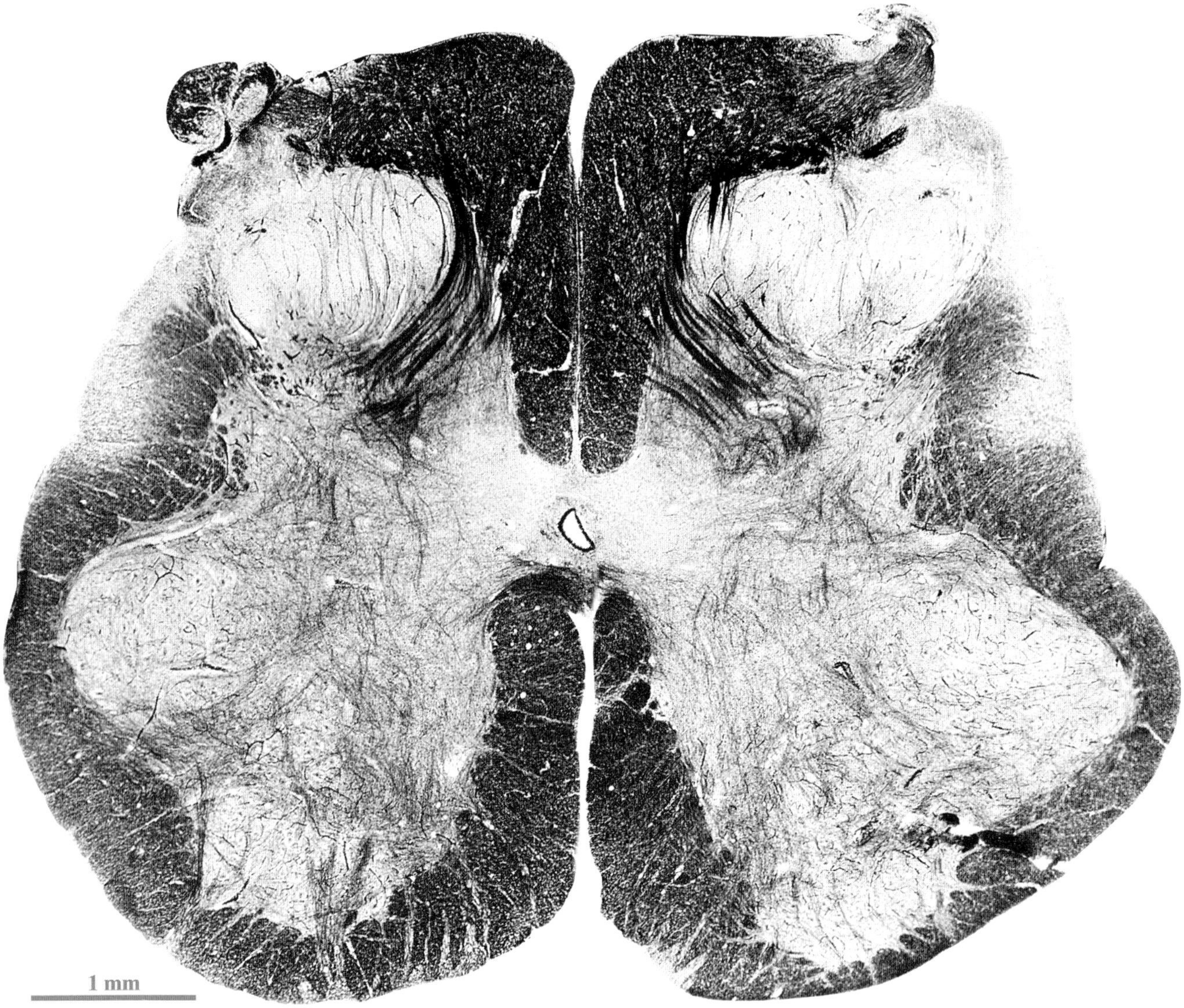

PLATE 116B

See the matched cell body stained section in Plates 117A and B

The axons in the lateral corticospinal tract at this caudal level are *unmyelinated*. These fibers are *more distal* to their cells of origin than they are at cervical and thoracic levels where they are *myelinating* (**Plates 104, 106,** and **108**). The lateral corticospinal tract is small at this level because most of the axons have terminated at more rostral levels.

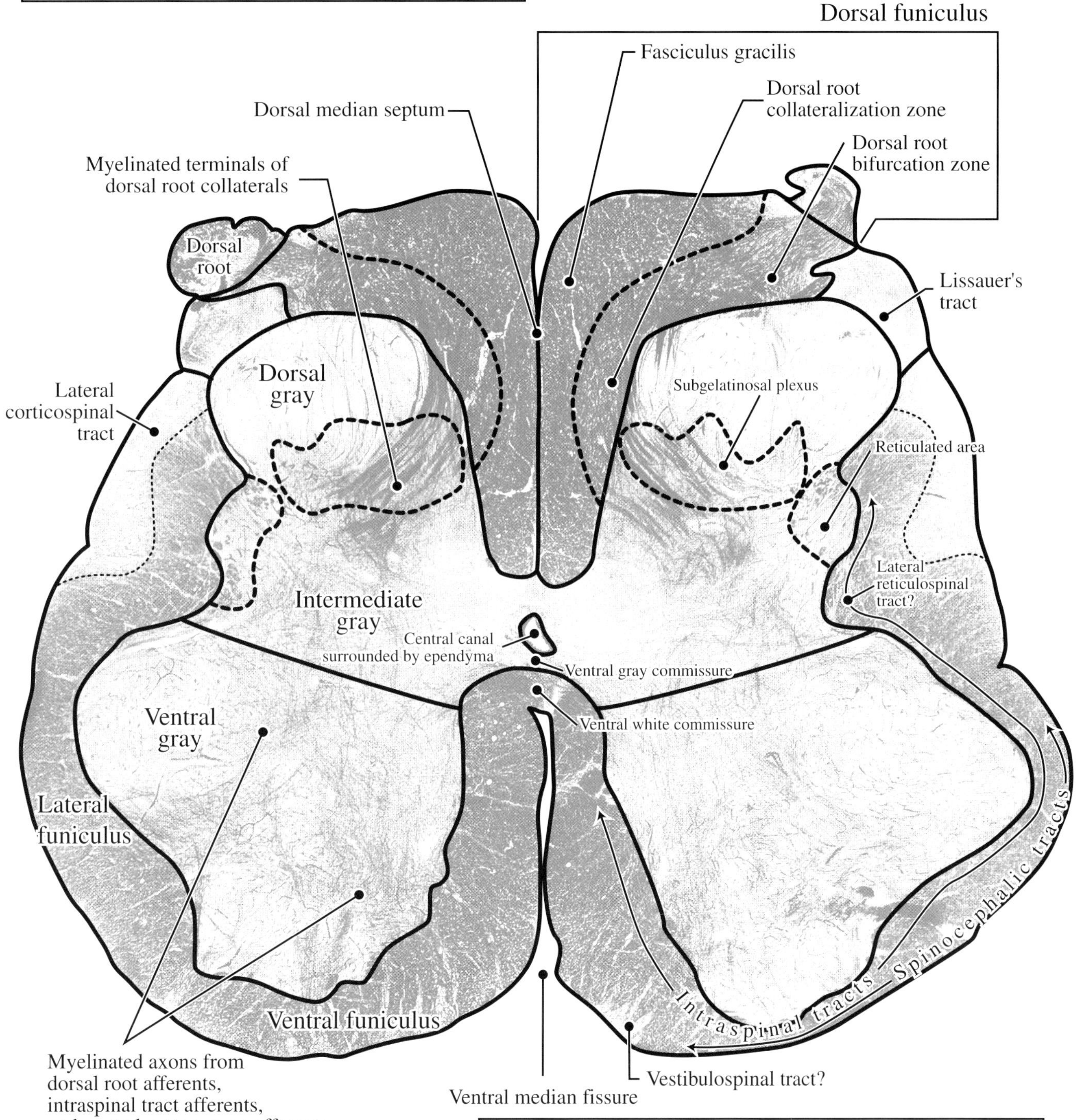

In contrast to the upper cervical level (**Plate 104**), the axons in the spinocephalic tracts at this level are *more proximal* to their cells of origin and are *completely myelinated*. Like the lateral corticospinal tract, the spinocephalic tract is smaller at this level because additional axons join the tract at more rostral levels.

PLATE 117A

Infant, 4 Months
CR 440 mm
Y286-62
Lumbar Enlargement
Cell body stain

Areas (mm^2)

Central canal	.0115
Ependyma	.0231
Gray matter	18.3550
White matter	13.8850

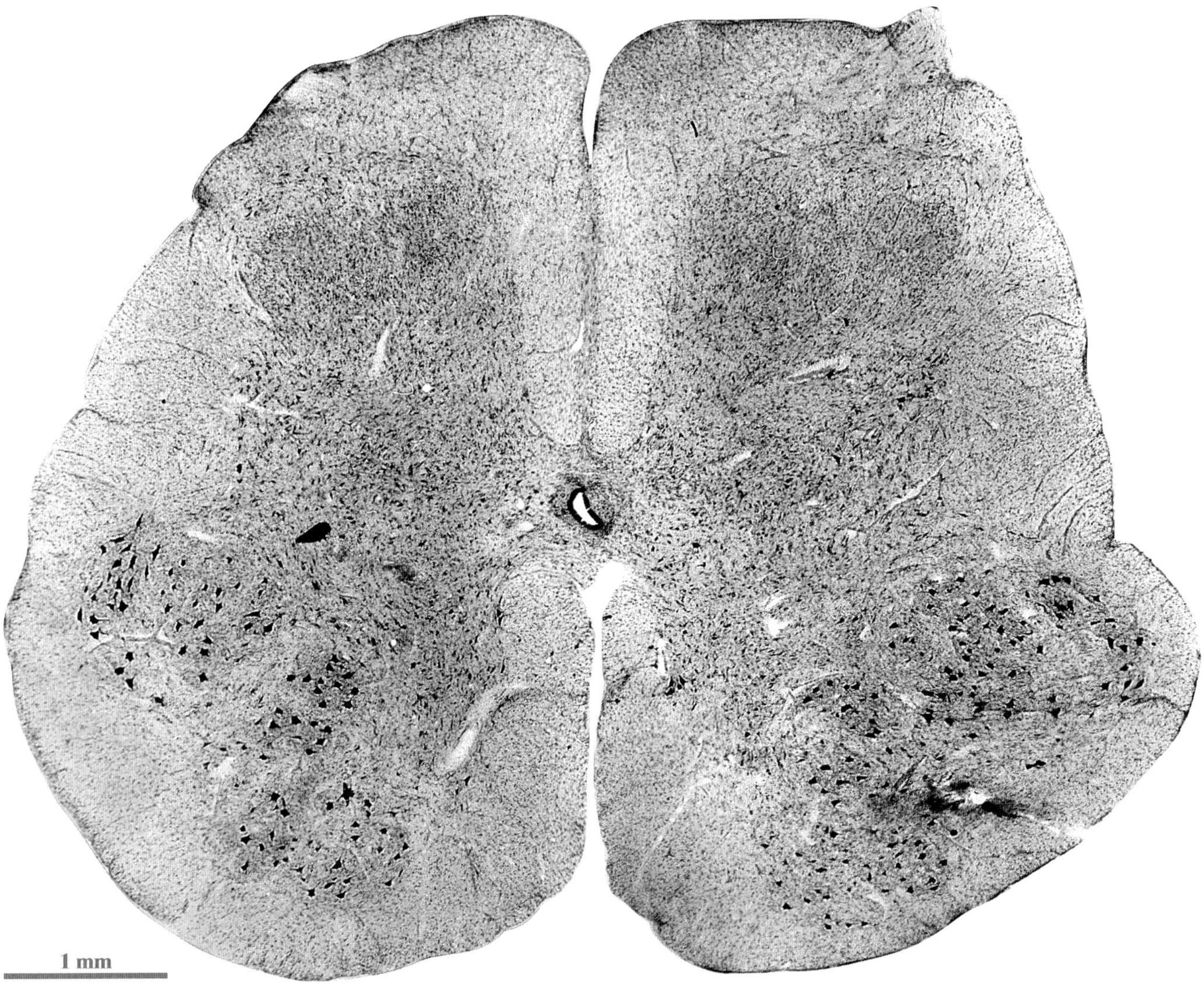

PLATE 117B

See the matched myelin stained section in Plates 116A and B

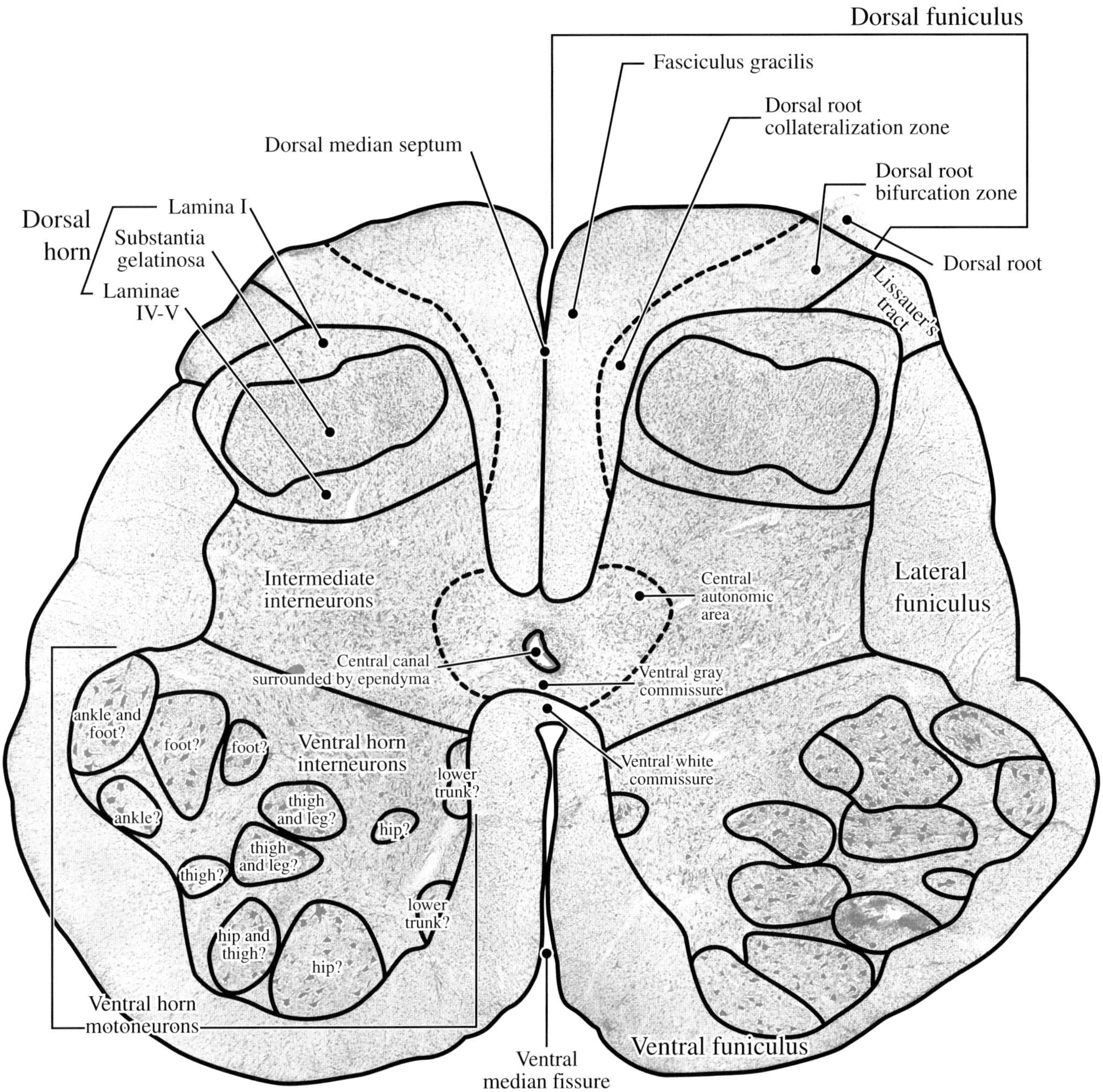

Refer to the matching myelin stained section for the approximate locations of the lateral corticospinal tract, the lateral reticulospinal, vestibulospinal, intraspinal, and spinocephalic tracts.

VI. Three-Dimensional Reconstructions of the Developing Spinal Cord

A. The cervical level of eight first–trimester specimens

Figures 2 through **9** feature computer reconstructed three-dimensional models of the cervical part of the developing spinal cord in eight first-trimester specimens ranging from crown rump (CR) lengths of 3.3 to 56 mm (see **Table VI A–1**). It is during this period that the spinal cord shows the most dramatic morphogenetic changes, starting out as a structure containing mostly neuroepithelium, and ending with a structure where most components of the neuroepithelium have disappeared, neurons have been generated, migration to the gray matter is largely completed, and most major components of the white matter appear and fill with growing axons. **Table VI A–2** lists the colors used to distinguish 13 different structures in the developing spinal cord in all the figures. The reconstructed models in the figures are all shown in the same orientation but differ in the number of structures that are visible to demonstrate various features of spinal cord morphogenetic development.

These figures demonstrate three important relationships during spinal cord development. **First**, before the appearance of a population of neurons in the gray matter, the neuroepithelial stem cells that will produce that population proliferate and expand. Those stem cells recede and ultimately disappear when all the neurons in that population are generated. **Second**, there is a ventral (first) to dorsal (last) gradient of neuroepithelial growth and decline. Ventral horn neurons, especially the large alpha motoneurons, are generated earlier than neurons in the intermediate gray and the latest neurons to be generated are in the dorsal horn. These morphogenetic observations confirm the neurogenetic gradient that has been well documented in the rat spinal cord with ^{3}H–thymidine autoradiography (Altman and Bayer, 1984, 2001). **Third**, the sequential neuroepithelial gradient is followed by a sequential gradient of morphogenesis. The ventral horn appears first, followed by the intermediate gray and finally by the dorsal horn.

Table VI A–1: Specimens Used for 3-D Reconstructions of the Cervical Spinal Cord

NAME	CR (mm)	GW
C6144	3.3	3.5-4.0
C836	4.0	4.0
M2065	8.0	5.25
C6517	10.5	5.5
C8965	19.1	7.0
C8553	22.0	7.2
M2050	36.0	8.5
Y380-62	56.0	14

Table VI A–2: Color Key for Structures in the Spinal Cord

STRUCTURE	COLOR
BOTH SIDES OR IN MIDLINE:	
Entire outside edge	Transparent clear
Roof plate	Brown
Floor plate	Brown
Dorsal funiculus	Pale blue
Ventral funiculus	Pale violet
Lateral funiculus	Pale purple
LEFT SIDE:	
Undivided neuroepithelium	Dark cyan
Gray matter	Bright cyan
RIGHT SIDE:	
Ventral neuroepithelium	Dark red
Ventral horn	Bright red
Intermediate neuroepithelium	Dark yellow-green
Intermediate gray	Bright yellow-green
Dorsal neuroepithelium	Dark yellow
Dorsal horn	Bright yellow

Figure 2. The neuroepithelium, gray matter, and the outer edge of the spinal cord is visible in all eight specimens. The specimens are not shown to scale because structures in the smaller ones (**A-D**) cannot be seen clearly after reduction to the scale of the largest one (**H**). At CR 3.3 mm (**A**), there is no gray matter. The ventral neuroepithelium is larger than the intermediate or dorsal neuroepithelia. At CR 4.0 mm (**B**) a thin sliver of gray matter, the ventral horn, appears adjacent to the ventral neuroepithelium. At CR 8.0 mm (**C**), there is gray matter adjacent to all parts of the neuroepithelium, but the ventral horn is largest. The intermediate and dorsal components of the neuroepithelium are growing larger. At CR 10.5 mm (**D**), the ventral neuroepithelium starts to recede, but the intermediate and dorsal components are still large. The ventral horn is still the largest gray matter component, but neurons are also accumulating in the intermediate gray and dorsal horn. That same process continues at CR 19.1 mm (**E**). The intermediate neuroepithelium is receding at CR 22.0 mm (**F**) but the dorsal neuroepithelium is still prominent. The dorsal horn is now larger than the intermediate gray. By CR 36 mm (**G**), the dorsal neuroepithelium considerably recedes and the dorsal horn is as large as the ventral horn. By CR 56 mm (**H**), the neuroepithelium has been replaced by an ependyma surrounding the shrinking central canal, and the dorsal and ventral horns are prominent components of the gray matter.

Overview of Spinal Cord Development – GW3.5 to GW14

FIGURE 2

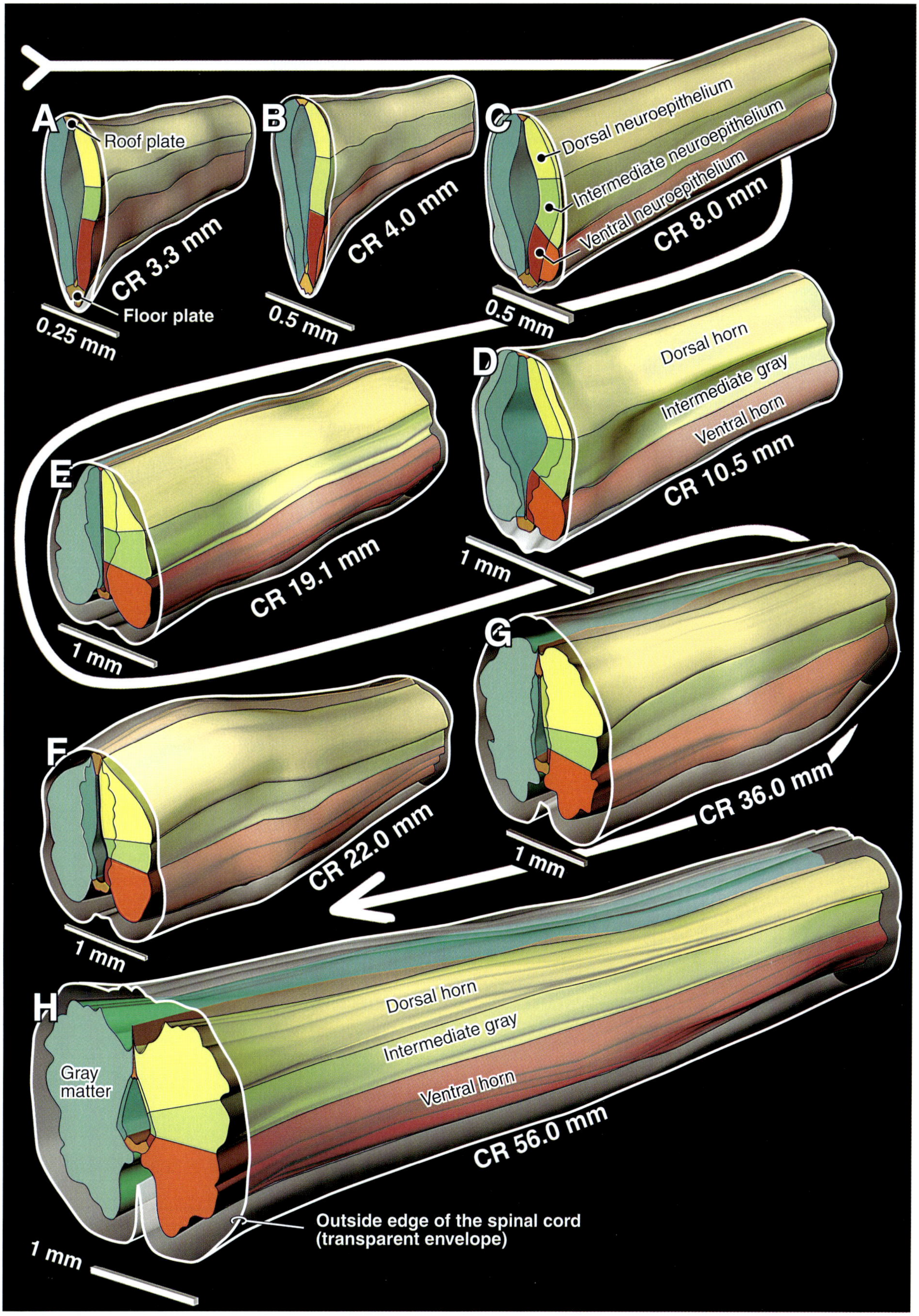

FIGURE 3

The Neuroepithelium – GW3.5 to GW4.0

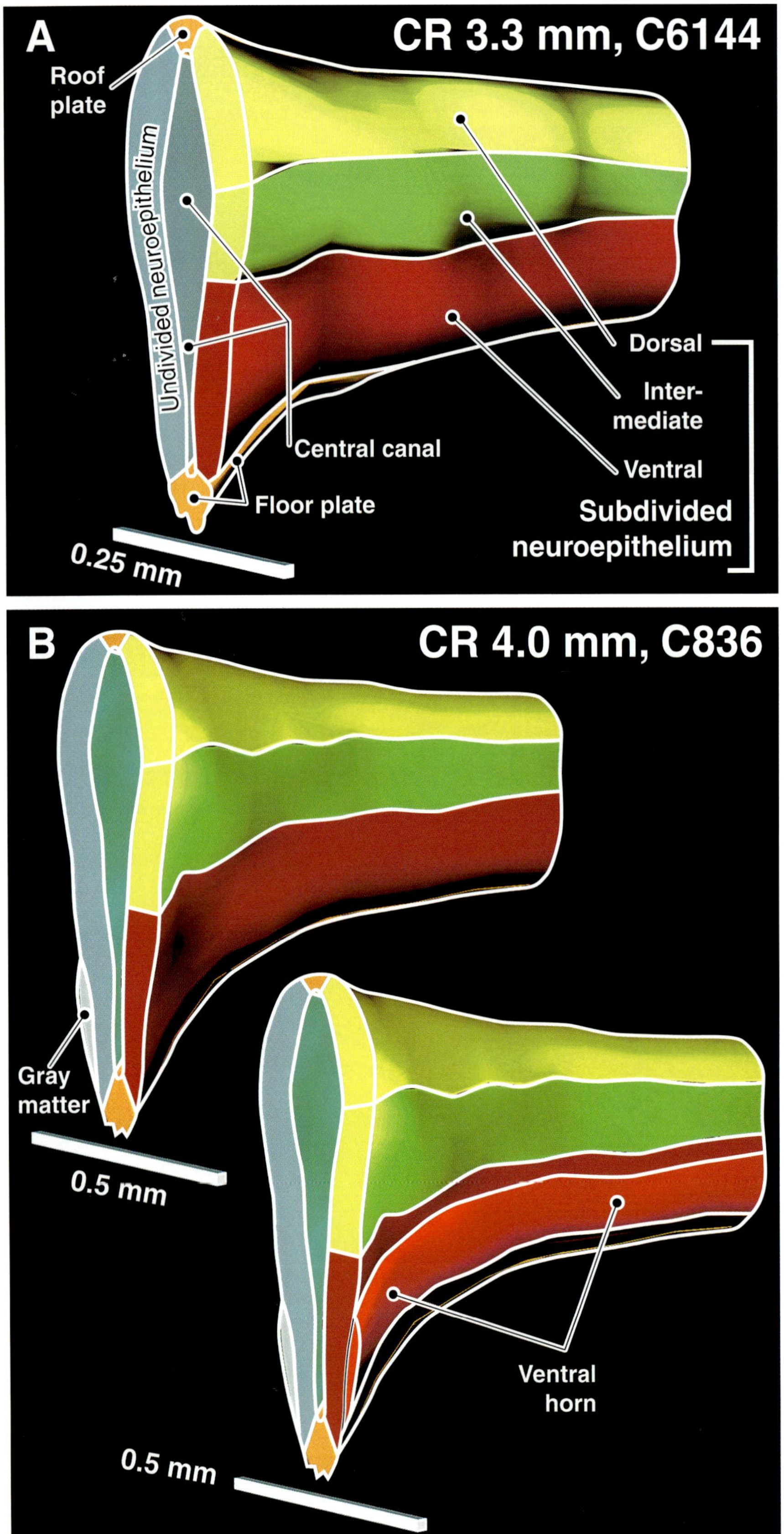

Figure 4. Enlarged views of the CR 8.0 mm (**A**) and CR 10.5 mm (**B**) specimens (note different scales). The **top** views do not show the gray matter on the right side so that surface features of the neuroepithelium can be observed. In **A**, the three components of the neuroepithelium are roughly the same size, and their outer surfaces are smooth. The outer surface represents the *basal* aspect of the neuroepithelium; the *apical* aspect (inner surface) forms the edge of the central canal. The ventral horn bulges outward. There is an intermediate gray that is thicker at its junction with the ventral horn and thinner at its junction with the few postmitotic neurons accumulating in the dorsal horn. The early dorsal horn neurons may be the large Waldeyer cells (Altman and Bayer, 2001). In **B**, the ventral neuroepithelium is beginning to recede in comparison with the still growing intermediate and dorsal neuroepithelia. The **dashed line** indicates a depression in the ventral neuroepithelium that may indicate the region where the greatest concentration of stem cells of large motoneurons are located at earlier stages. Possibly, this depression in the neuroepithelium signals the end of the neurogenetic period for motoneuron generation. The **downward arrowheads** point to undulations in the outer surface of the intermediate and dorsal parts of the neuroepithelium that may be "sojourn zones" of postmitotic neurons. Short survival ^{3}H–thymidine autoradiagraphic studies in the rat spinal cord show no label uptake in these zones, even though cell density indicates that they are still within the neuroepithelium (Altman and Bayer, 2001). It is postulated that the sojourn zones are accumulations of premigratory neurons destined to settle within the intermediate gray, dorsal horn, and possibly the ventral horn. All components of the gray matter are growing larger when **B** is compared to **A**. The ventral horn is still the most prominent gray matter component and features a smooth outer surface. There is a shallow concave area in the ventral part of the outer surface of the dorsal horn. That shallow depression is the region where axons from large cells in the dorsal root ganglion accumulate in the oval bundle of His, or what we call the dorsal root bifurcation zone (see **Figure 9**).

Figure 3. An enlarged view of the specimens at CR 3.3 mm (**A**) and CR 4.0 mm (**B**) with only the neuroepithelium, roof plate, floor plate, and gray matter visible (note scale differences). The specimen in **A** has no gray matter. That indicates that postmitotic neurons have not yet been generated. The neuroepithelium is full of mitotically active stem cells that will eventually produce neurons. The **top** reconstruction in **B** does not show any gray matter on the right side. The **bottom** view in **B** shows a sliver of postmitotic neurons, presumably motoneurons, adjacent to the *central part* of the ventral neuroepithelium. Notice that the surface of the neuroepithelium is smooth in both specimens. In addition, the ventral neuroepithelium is larger than intermediate or dorsal neuroepithelia in both specimens.

Neuroepithelium and Gray Matter – GW5.25 to GW5.5

FIGURE 4

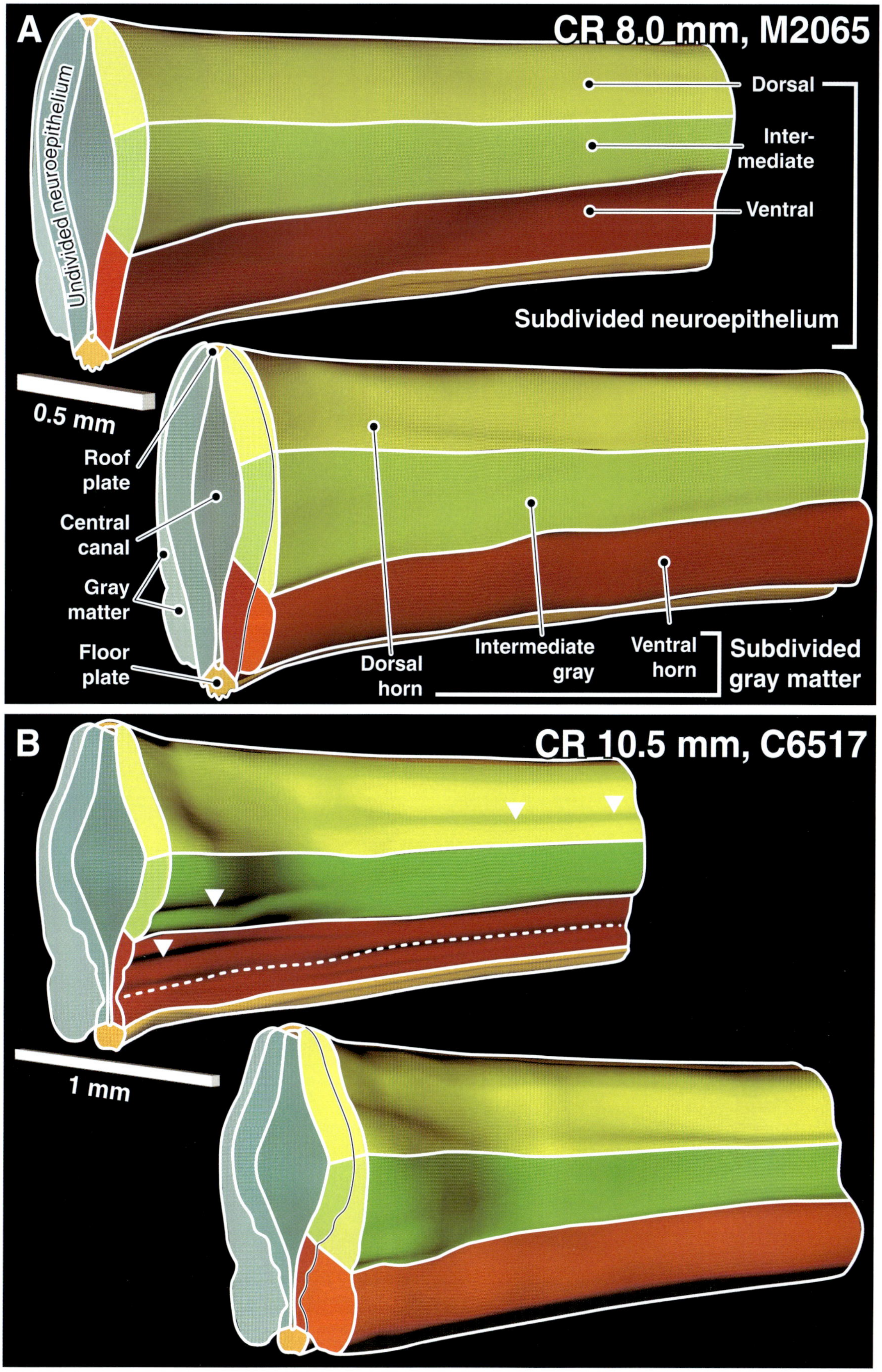

FIGURE 5 Neuroepithelium and Gray Matter – GW7.0 to GW7.2

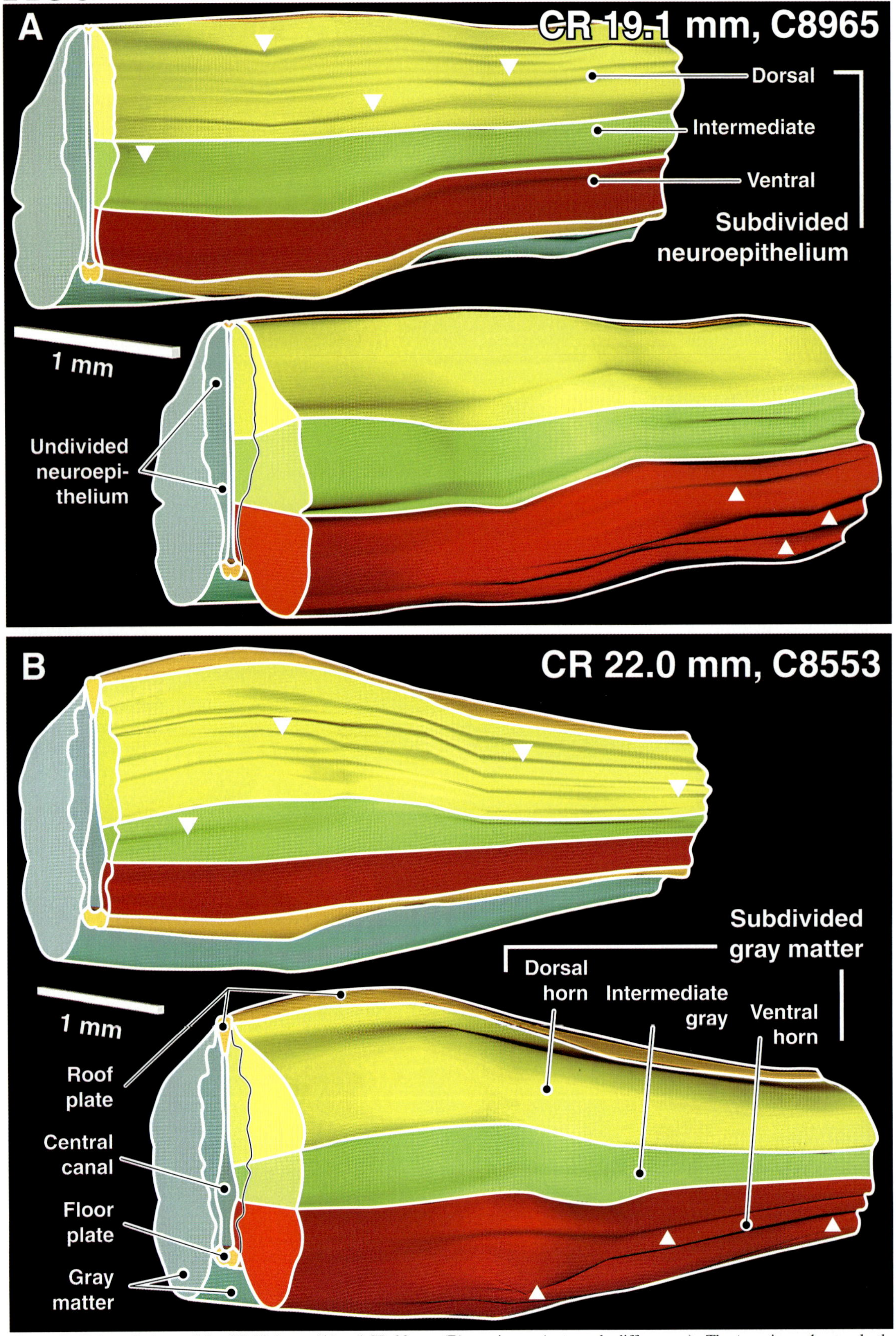

Figure 5. Enlarged views of the CR 19.1 mm (**A**) and CR 22 mm (**B**) specimens (note scale differences). The **top** views show only the surface features of the neuroepithelium on the right side; **bottom** views show the adjacent gray matter on the right side. In **A**, the ventral neuroepithelium has receded but the intermediate and dorsal neuroepithelia are still active (now with many undulations, **downward arrowheads**). The outer surface of the posterior ventral horn has bulges and depressions (**upward arrowheads**) representing the segregation of the ventral motoneurons into columns in the cervical enlargement. In **B**, the same features are present, but now the intermediate neuroepithelium is starting to recede, and the dorsal horn is growing rapidly.

Neuroepithelium and Gray Matter – GW8.5 to GW14

FIGURE 6

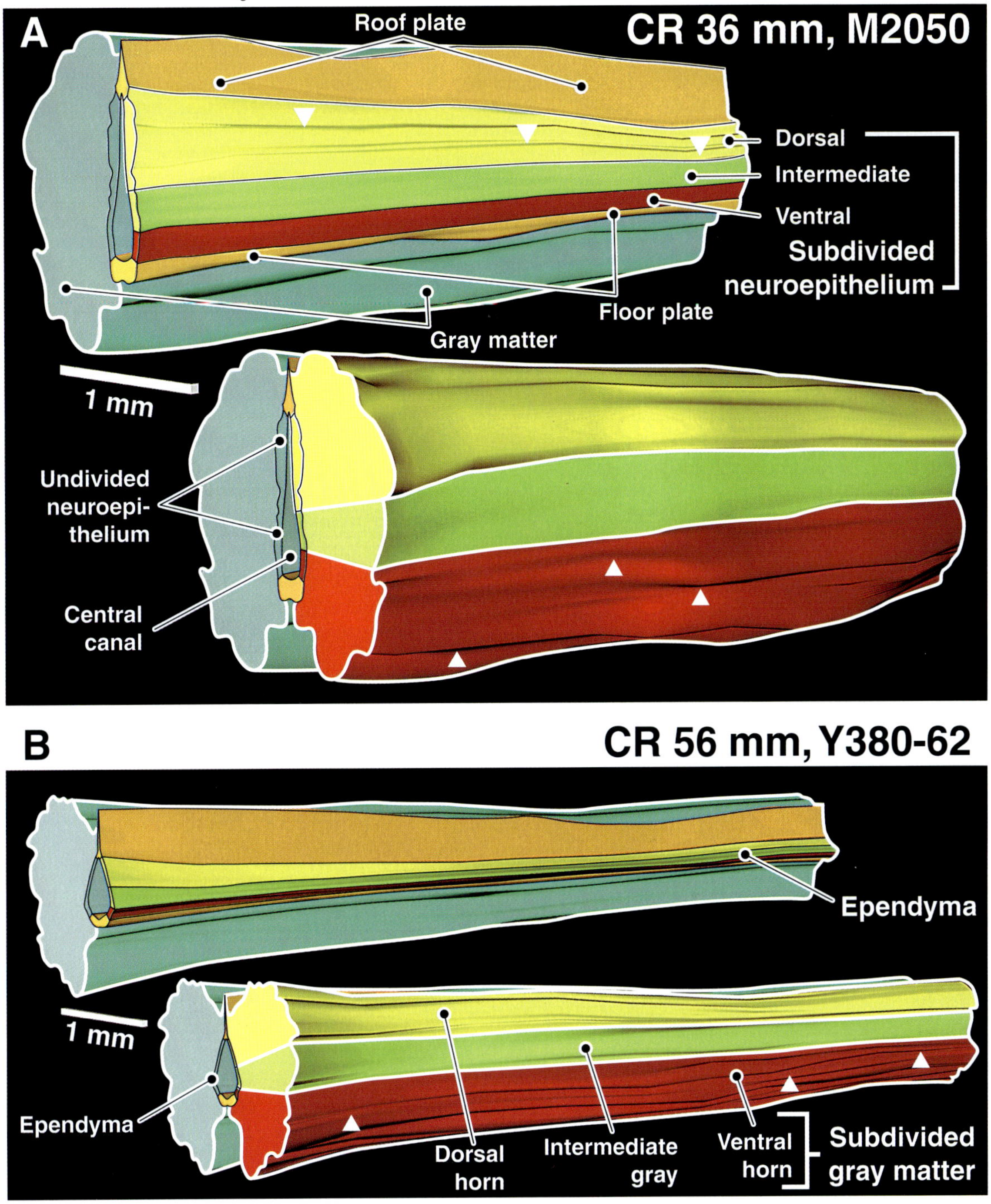

Figure 6. Enlarged views of the CR 36 mm (**A**) and CR 56 mm (**B**) specimens (note scale differences). The **top** views show only the surface features of the neuroepithelium on the right side; **bottom** views show the adjacent gray matter on the right side. In **A**, all parts of the neuroepithelium are relatively smaller than in previous specimens but the dorsal neuroepithelium (still with many undulations, **downward arrowheads**) is active. The bulges in the outer surface of the ventral horn (**upward arrowheads**) extend throughout its entire length representing the advancing segregation of the ventral motoneurons into columns at all cervical levels. In **B**, the neuroepithelium has been transformed into an ependyma that surrounds a considerably smaller central canal. The outer surface of the ependyma is smooth. The dorsal horn has grown relatively larger and reaches the same size as the ventral horn.

FIGURE 7 Neuroepithelium, Roof and Floor Plates – GW3.5 to GW5.5

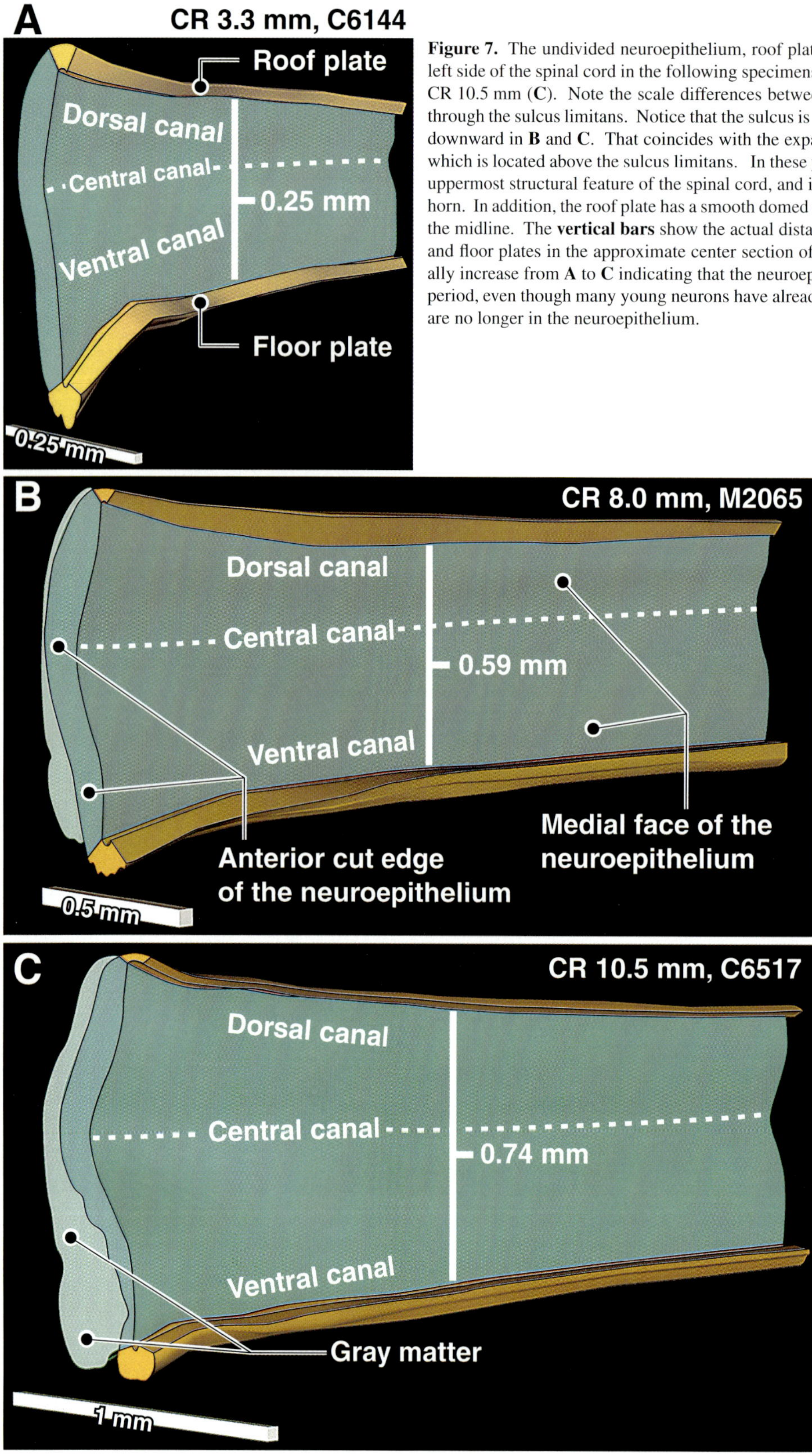

Figure 7. The undivided neuroepithelium, roof plate, floor plate, and gray matter on the left side of the spinal cord in the following specimens: CR 3.3 mm (**A**), CR 8 mm (**B**), and CR 10.5 mm (**C**). Note the scale differences between specimens. The **dashed lines** run through the sulcus limitans. Notice that the sulcus is closer to the roof plate in **A** and shifts downward in **B** and **C**. That coincides with the expansion of the dorsal neuroepithelium, which is located above the sulcus limitans. In these young specimens, the roof plate is the uppermost structural feature of the spinal cord, and is always above the level of the dorsal horn. In addition, the roof plate has a smooth domed surface with no indication of a cleft in the midline. The **vertical bars** show the actual distance (in millimeters) between the roof and floor plates in the approximate center section of each model. The distances continually increase from **A** to **C** indicating that the neuroepithelium has a net growth during this period, even though many young neurons have already been generated and their stem cells are no longer in the neuroepithelium.

Neuroepithelium, Roof and Floor Plates – GW7.0 to GW14

FIGURE 8

Figure 8. A continuation of the series in the following specimens: CR 19.1 mm (**A**), CR 22 mm (**B**), CR 36 mm (**C**), and CR 56 mm (**D**). Note the scale differences between specimens. The sulcus limitans (**dashed line**) gets closer and closer to the floor plate from **A** to **D**, reflecting the regression of the ventral and intermediate neuroepithelia. On the other hand, the dorsal neuroepithelium is expanding between **A** and **B**, indicating that it is still producing neurons. Note that the distance between the roof and floor plates (**vertical bars** with measurements in millimeters) reaches its maximum between **A** and **B** as the dorsal neuroepithelium stretches in the dorsoventral plane. The dorsal neuroepithelium declines between **B** and **C** as neurogenesis wanes, and the distance between the roof and floor plates gets shorter. The dorsal neuroepithelium disappears between **C** and **D**, marking the end of neurogenesis in the cervical spinal cord. By this time, the distance between the roof and floor plates shrinks to a smaller value than the youngest specimen (**A, Figure 7**). In **A**, **B**, and **C**, the roof plate is positioned at the same level as the most dorsal part of the gray matter, then it sinks downward in **D**, and the gray matter begins to arch dorsally, as neurons settle and start to differentiate in the dorsal horn. During that downward movement, the roof plate changes its surface features. In all panels of **Figure 7**, the roof plate surface is a single arch spanning the midline. In **A** of this series, the surface is a shallow double arch with a cleft in the midline. The roof plate still has a double arch in the dorsal midline in **B**, but there is a noticable lengthening in the dorsoventral plane. In **C** and **D**, the double arch is replaced by a sharp vertical spike in the midline that starts out elongated (**C**) and then moves downward (**D**) as the dorsal canal completely disappears to leave only the central canal. Concurrent with these changes in the roof plate, the dorsal funiculus fills with axons in its most medial parts and marks the beginning of the fasciculus gracilis and fasciculus cuneatus in the cervical spinal cord. It is postulated that the roof plate provides some guidance or structural support for axonal recruitment in these fasciculi (Altman and Bayer, 2001).

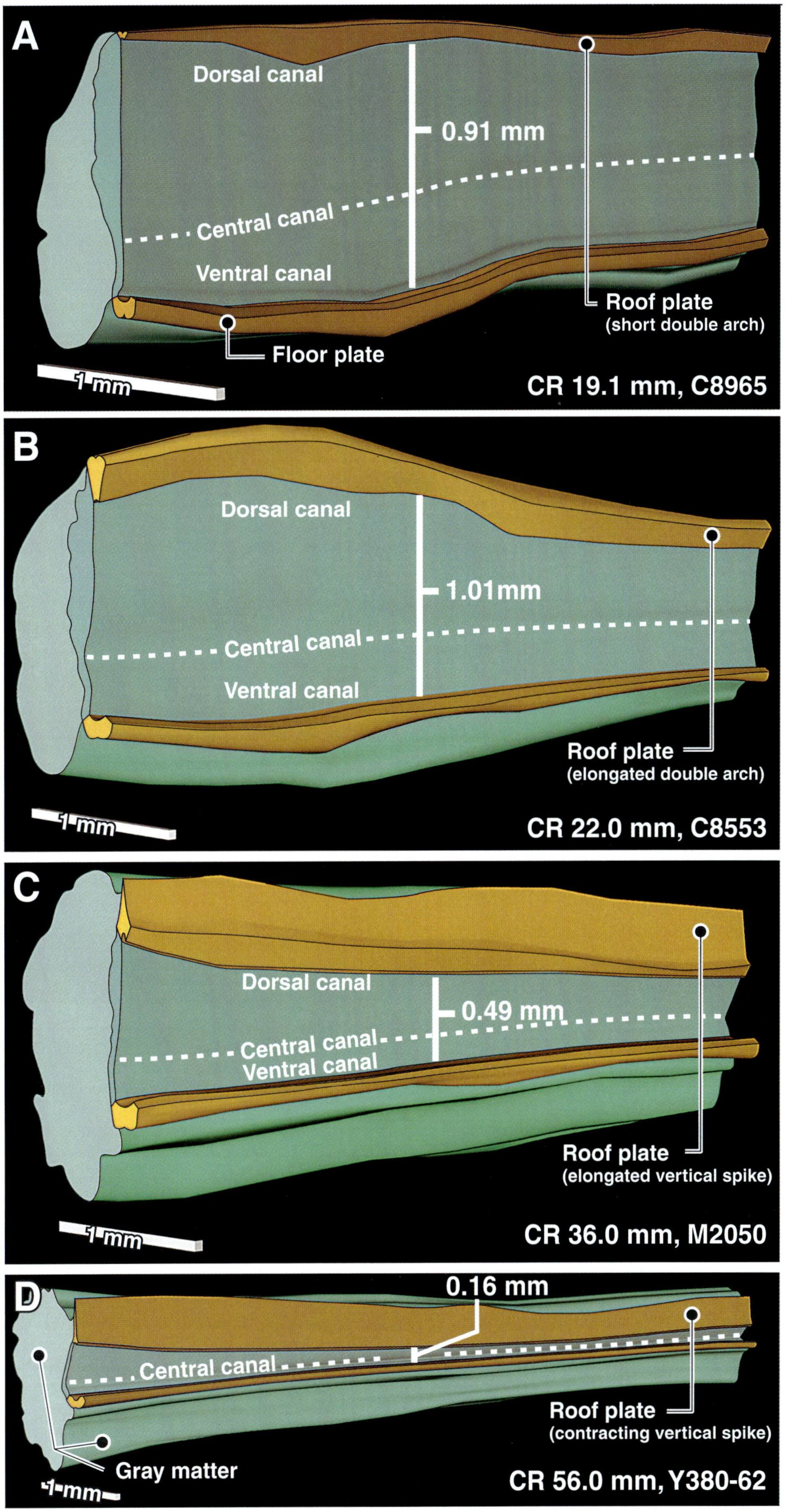

FIGURE 9 Growth of the White Matter – GW5.25 to GW14

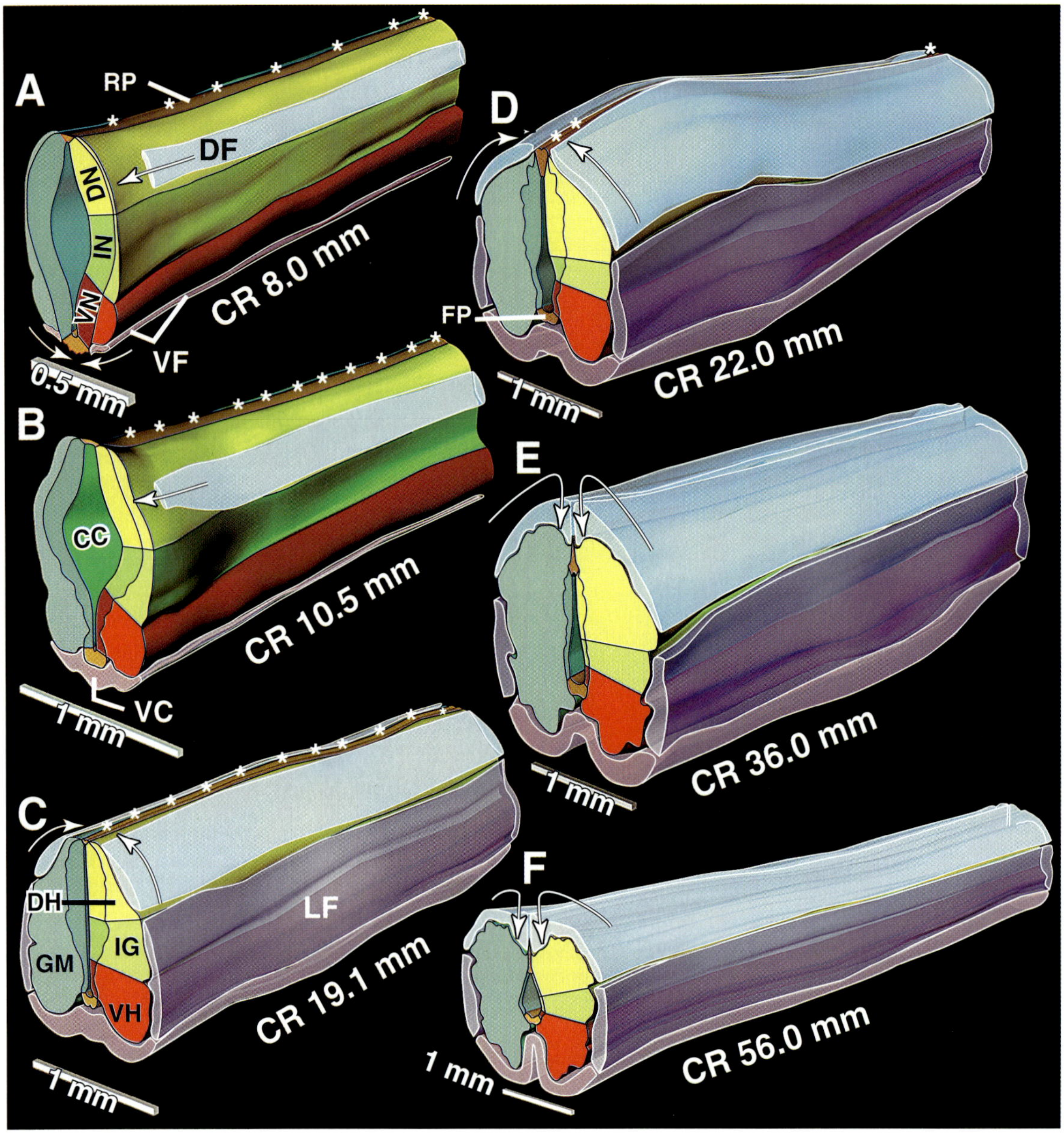

Figure 9. The emergence and growth of the white matter (see **Table VI A–2** for color code) in the following specimens: CR 8 mm (**A**), CR 10.5 mm (**B**), CR 19.1 mm (**C**), CR 22 mm (**D**), CR 36 mm (**E**), and CR 56 mm (**F**). Note scale differences between specimens. All parts of the gray matter and neuroepithelium are visible in the models. **Asterisks (A** to **D)** indicate where the top of the roof plate (**RP**) is visible in the dorsal midline. The dorsal funiculus (**DF**) and ventral funiculus (**VF**) appear simultaneously in **A**; the lateral funiculus (**LF**) is definite in **C**.

The first axons in the ventral funiculus are most likely those of the early generated commissural neurons that accumulate in the space between the lateral edges of the floor plate (**FP**), the most ventral part of the ventral neuroepithelium (**VN**), and the ventromedial edge of the ventral horn (**VH**). In some sections of the model in **A**, the axons are even closer to the midline, so the ventral funiculus first has a medial direction of growth (**bottom arrows in A** pointing toward the midline). In **B**, these fibers cross the midline in the ventral commissure (**VC**), and the ventral funiculus increases in size and fills out laterally as well as medially. The ventral funiculus continues to enlarge throughout the remainder of spinal cord development and contains several descending tracts (medial longitudinal fasciculus, tectospinal tract, and vestibulospinal tract) as well as some of the ascending spinocephalic tracts and the collaterals of axons from neurons throughout the ventral and intermediate gray in the intraspinal tract (also known as the propriospinal tract).

The dorsal funiculus has a complex pattern of growth that features three different gradients. In **A**, the dorsal funiculus contains axons from large neurons in the dorsal root ganglia that enter the spinal cord and bifurcate into ascending and descending branches. This is the dorsal root bifurcation zone (also known as the oval bundle of His). That zone is more obvious at lower than upper cervical levels because the highest dorsal root ganglion is below the highest cervical level of the spinal cord, and axons have not yet reached there. The **forward-facing arrows** in **A** and **B** indicate the advance of the dorsal root bifurcation zone to upper cervical levels, which are filled with axons in **C**. Axons from the dorsal root continue to fill the bifurcation zone, and the next gradient of growth is from ventrolateral to dorsomedial (**arrows arching upward** toward the midline in **C** and **D**). During this time, the bifurcating fibers are growing collaterals (the dorsal root collateralization zone) that will eventually penetrate the gray matter. In **E** and **F**, the dorsal funiculus reaches the midline and expands downward (**downward facing arrows**). These axons are segregated in the midline by a band of glia that extends to the pia from the top of the roof plate. This medial part of the dorsal funiculus contains the long ascending branches of the dorsal root axons in the fasciculus gracilis and fasciculus cuneatus (see Chapter 5, Figures 5-28 and 5-49 in Altman and Bayer, 2001).

Part VI: Three-Dimensional Reconstructions (concluded)

B. The progressive segregation of ventral horn motoneurons into columns

This section features computer reconstructed three-dimensional models of the entire spinal cord in three specimens: CR 36 mm (GW8.5), CR 56 mm (GW10.5), and CR 108 mm (GW14). Throughout the ventral horn, individual sections contain clumps of motoneurons. When the sections are aligned, these clumps form *longitudinal columns*. Each distinguishable motoneuron column has been reconstructed in the three specimens. **Figures 10** and **11** provide an overview of the sections included in the GW8.5 model; the model is shown in **Figures 12** through **15**. **Figures 16** through **18** provide an overview of the sections included in the GW10.5 model; the model is shown in **Figures 19** through **22**. **Figures 23** through **25** provide an overview of the sections included in the GW14 model; the model is shown in **Figures 26** through **31**. The segregation of motoneuron columns is a progressive developmental process. On each side of the ventral horn, there are 7 motor columns in the GW8.5 specimen, 9 in the GW10.5 specimen, and 15 in the GW14 specimen.

Experimental studies in animals show that motoneuron segregation is related to dendritic differentiation (see Chapter 1, Section 1.6 and Chapter 3, Section 3.3 in Altman and Bayer, 2001). When there are fewer columns containing larger groups of neurons, dendrites are growing in many directions. As development progresses, the dendrites are reshaped into well-defined bundles that extend longitudinally in the spinal cord (see Figures 1–68, 1–69, 3–23 and 3–24 in Altman and Bayer, 2001). That process produces progressively better defined motoneuron columns in cell body stained sections. Possibly, motoneuron dendritic bundles develop similarly in humans. Compare, for example, the progressive segregation of motoneurons in the cervical enlargement in **Plates 15, 23,** and **31**.

All parts of the spinal cord contain at least one motoneuron column located in the ventromedial part of the ventral horn. In the thoracic cord, that is the only column present. Progressively more lateral columns are found at the cervical and lumbar levels, especially in their respective enlargements. Traditionally, these columns are named according to their *position*. In our book on development of the human spinal cord (Altman and Bayer, 2001), that locational terminology has been put into a grid system containing up to four panels (medial, central, lateral, and far–lateral) and three tiers (ventral, dorsal, and retrodorsal). The positional labeling system is functionally neutral because little is known about the exact groups of muscles that are innervated by motoneurons in the human spinal cord. The overview figures that introduce each model contain the positional labeling for each reconstructed motoneuron column.

However, there is a wealth of experimental anatomical evidence in animals, including primates, that the locations of motoneuron columns in the ventral horn have functional significance (see Chapter 1, Section 1.6 and Chapter 2, Section 2.3.2 in Altman and Bayer, 2001). The ventromedial columns at all levels (medial panel, ventral tier) contain motoneurons that innervate the axial muscles. More laterally placed columns at upper cervical and upper lumbar levels contain motoneurons that innervate proximal limb muscles (central, lateral, and far-lateral panels, ventral to dorsal tiers). Lateral columns in the cervical and lumbar enlargements contain motoneurons that innervate distal limb muscles (especially those located in the far-lateral panel and retrodorsal tier). The overview figures preceding each model contain a table listing the possible innervations of each three-dimensionally reconstructed motor column.

FIGURE 10

GW8.5, CR 36 mm, M2050

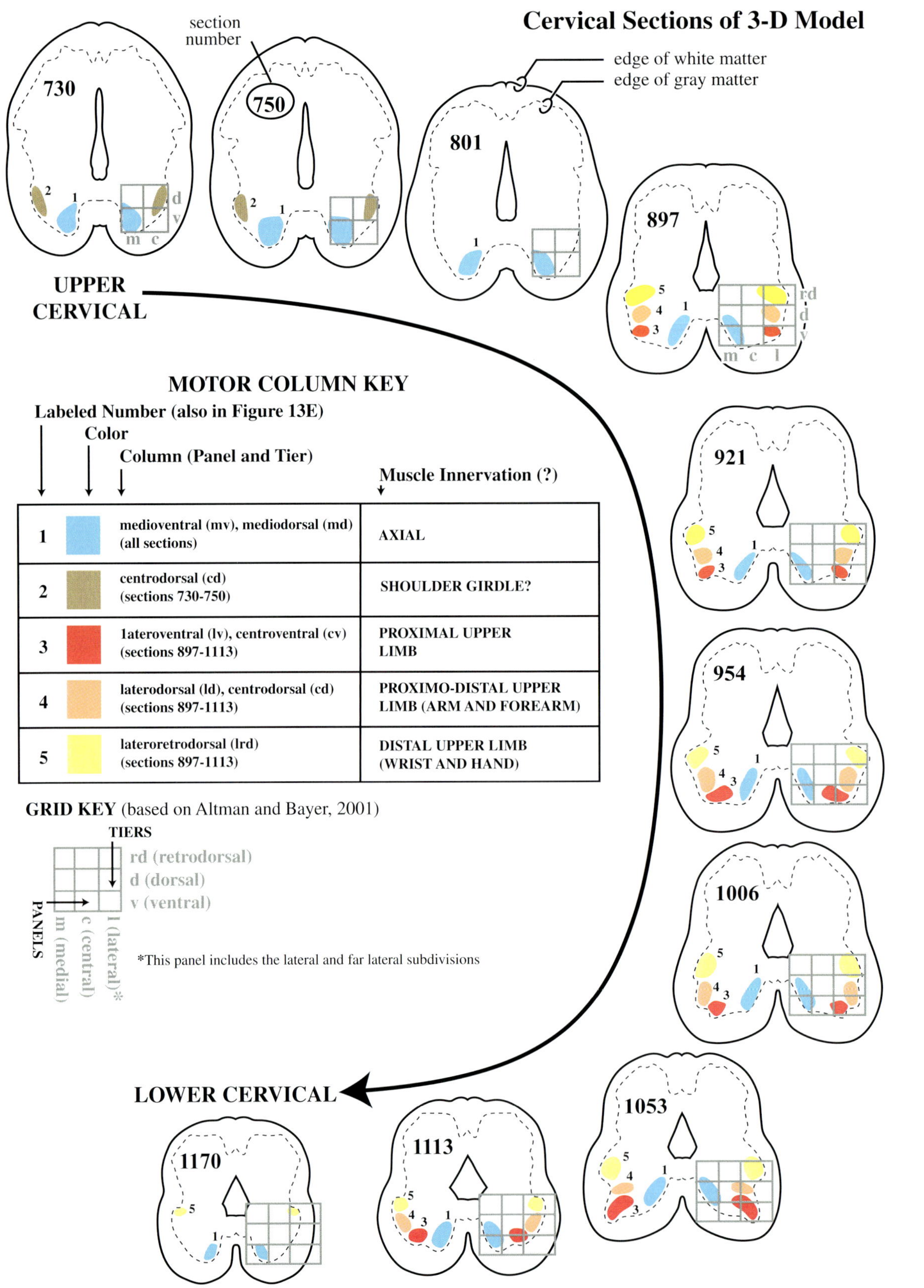

MOTOR COLUMN KEY

Labeled Number (also in Figure 13E) / Color / Column (Panel and Tier) / Muscle Innervation (?)

Labeled Number	Color	Column (Panel and Tier)	Muscle Innervation (?)
1	blue	medioventral (mv), mediodorsal (md) (all sections)	AXIAL
2	brown	centrodorsal (cd) (sections 730-750)	SHOULDER GIRDLE?
3	red	lateroventral (lv), centroventral (cv) (sections 897-1113)	PROXIMAL UPPER LIMB
4	orange	laterodorsal (ld), centrodorsal (cd) (sections 897-1113)	PROXIMO-DISTAL UPPER LIMB (ARM AND FOREARM)
5	yellow	lateroretrodorsal (lrd) (sections 897-1113)	DISTAL UPPER LIMB (WRIST AND HAND)

Figure 10. This specimen from the Minot Collection contains 10-μm thick sections that are consecutively numbered on the slides. All outlines are profiles of the sections used in the cervical spinal cord. Three of these sections are illustrated in the Atlas. **Section 730** is in **Plate 14**; **section 921** is in **Plate 15**, and **section 1113** is in **Plate 16**. After fixation, the total length of the cervical part of the model (shown in **Figure 13**) extends 4.4 mm.

GW8.5, CR 36 mm, M2050

FIGURE 11

Thoracic, Lumbar and Sacral Sections of 3-D Model

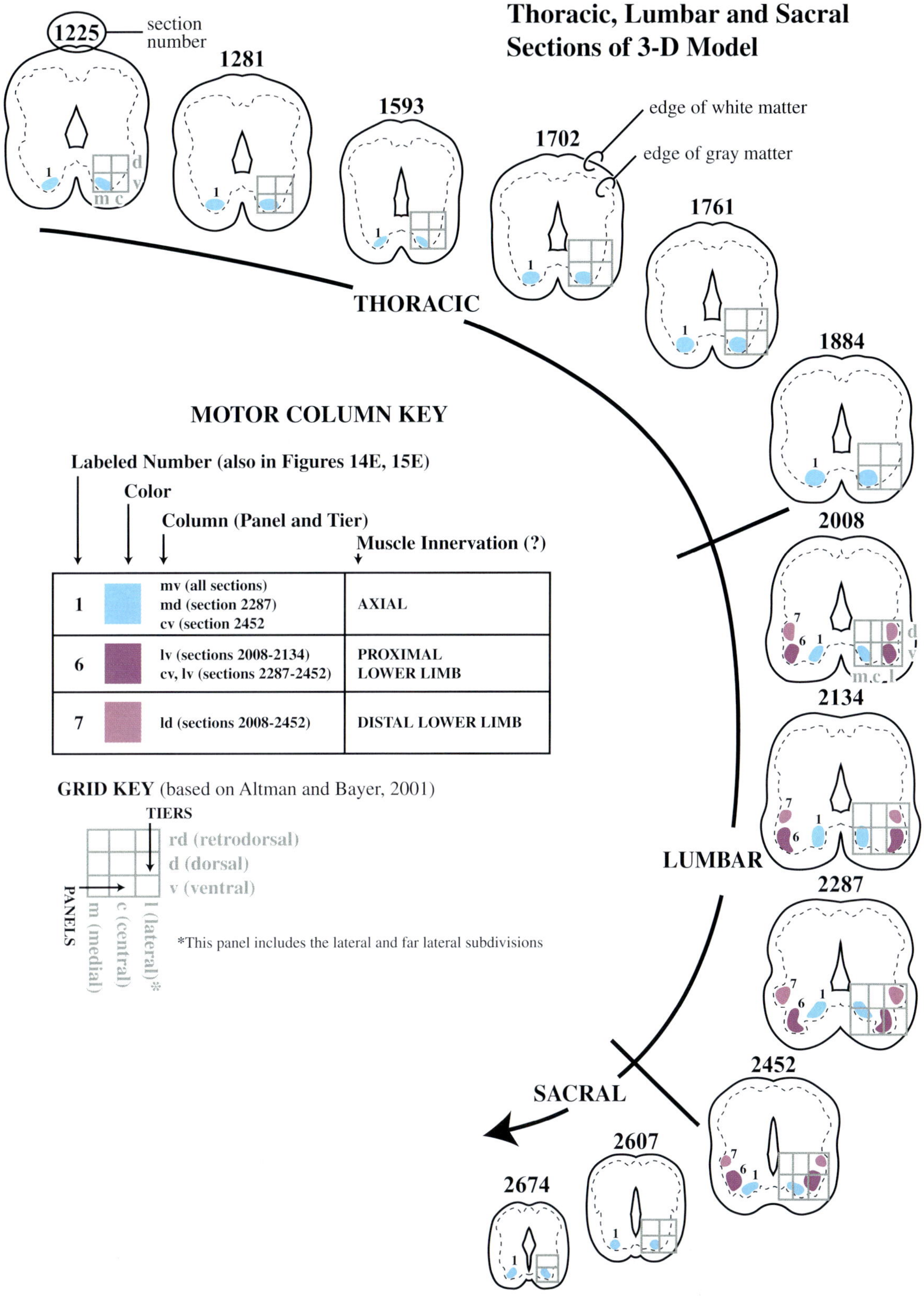

MOTOR COLUMN KEY

Labeled Number (also in Figures 14E, 15E) → Color → Column (Panel and Tier) → Muscle Innervation (?)

Labeled Number	Color	Column (Panel and Tier)	Muscle Innervation (?)
1	(light blue)	mv (all sections) md (section 2287) cv (section 2452	AXIAL
6	(dark purple)	lv (sections 2008-2134) cv, lv (sections 2287-2452)	PROXIMAL LOWER LIMB
7	(pink)	ld (sections 2008-2452)	DISTAL LOWER LIMB

GRID KEY (based on Altman and Bayer, 2001)

TIERS

rd (retrodorsal)
d (dorsal)
v (ventral)

PANELS

m (medial)
c (central)
l (lateral)*

*This panel includes the lateral and far lateral subdivisions

Figure 11. A continuation of the previous figure showing outlines of the sections used in the thoracic, lumbar, and sacral spinal cord. Six of these sections are illustrated in the Atlas. **Section 1593** is in **Plate 17**; **section 1884** is in **Plate 18**, **section 2134** is in **Plate 19,** and **sections 2607** and **2674** are in **Plate 20**. After fixation, the total length of the thoracic part of the model (**Figure 14**) is 6.99 mm and the combined lumbar and sacral regions is 6.66 mm (**Figure 15**).

FIGURE 12

Top Front View GW8.5, CR 36 mm, M2050

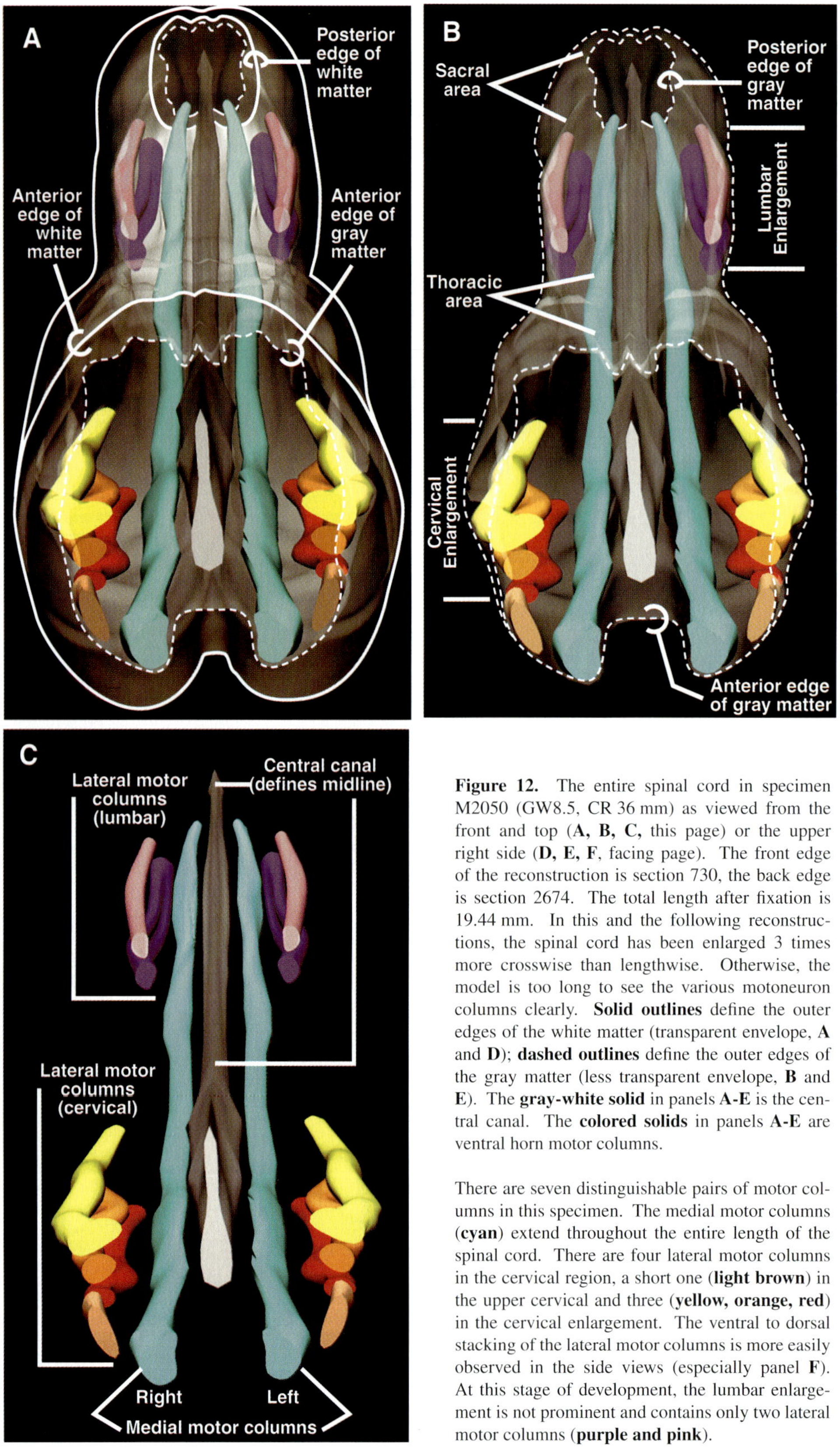

Figure 12. The entire spinal cord in specimen M2050 (GW8.5, CR 36 mm) as viewed from the front and top (**A, B, C,** this page) or the upper right side (**D, E, F**, facing page). The front edge of the reconstruction is section 730, the back edge is section 2674. The total length after fixation is 19.44 mm. In this and the following reconstructions, the spinal cord has been enlarged 3 times more crosswise than lengthwise. Otherwise, the model is too long to see the various motoneuron columns clearly. **Solid outlines** define the outer edges of the white matter (transparent envelope, **A** and **D**); **dashed outlines** define the outer edges of the gray matter (less transparent envelope, **B** and **E**). The **gray-white solid** in panels **A-E** is the central canal. The **colored solids** in panels **A-E** are ventral horn motor columns.

There are seven distinguishable pairs of motor columns in this specimen. The medial motor columns (**cyan**) extend throughout the entire length of the spinal cord. There are four lateral motor columns in the cervical region, a short one (**light brown**) in the upper cervical and three (**yellow, orange, red**) in the cervical enlargement. The ventral to dorsal stacking of the lateral motor columns is more easily observed in the side views (especially panel **F**). At this stage of development, the lumbar enlargement is not prominent and contains only two lateral motor columns (**purple and pink**).

GW8.5, CR 36 mm, M2050 Side View

FIGURE 12

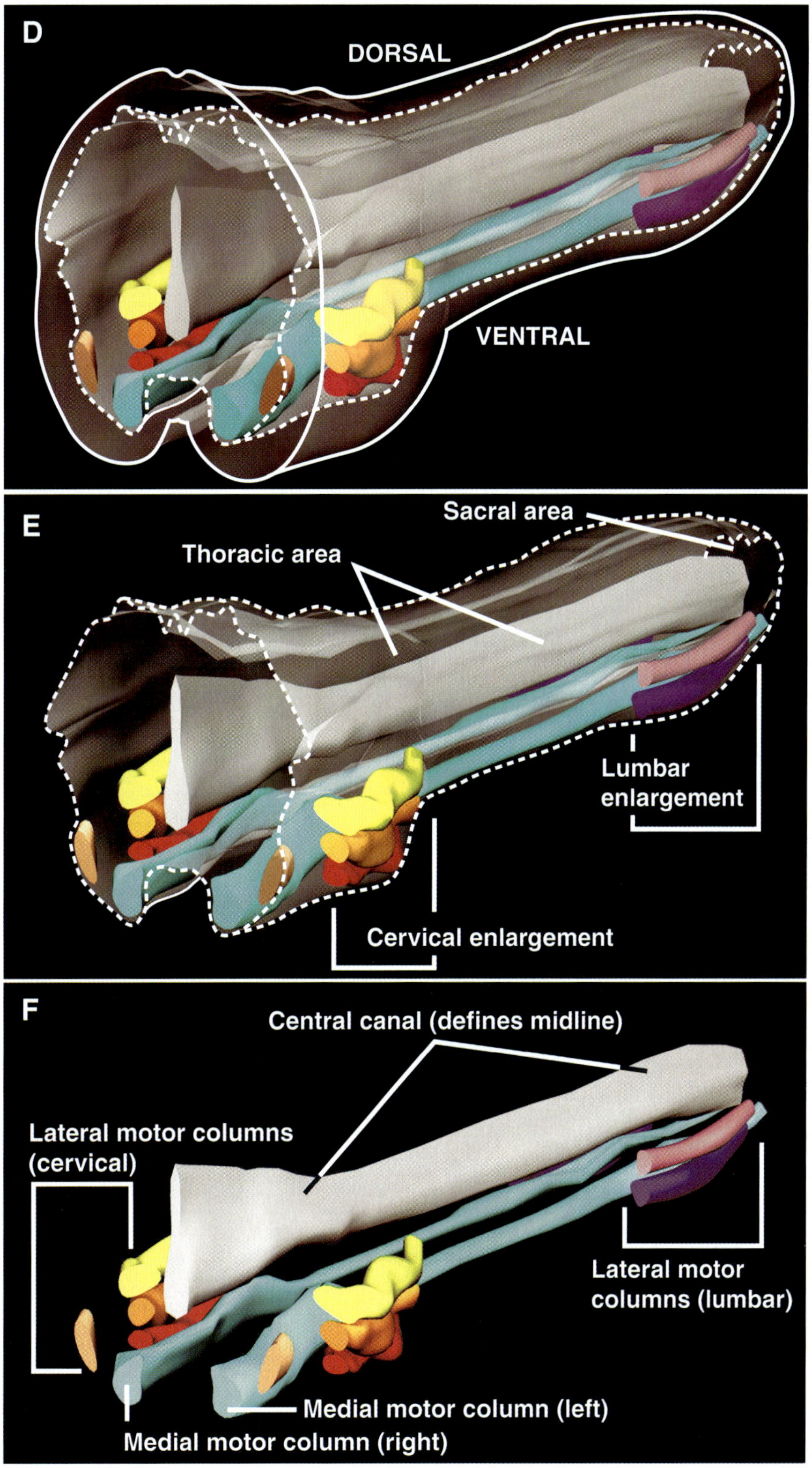

FIGURE 13 — GW8.5, CR 36 mm, M2050

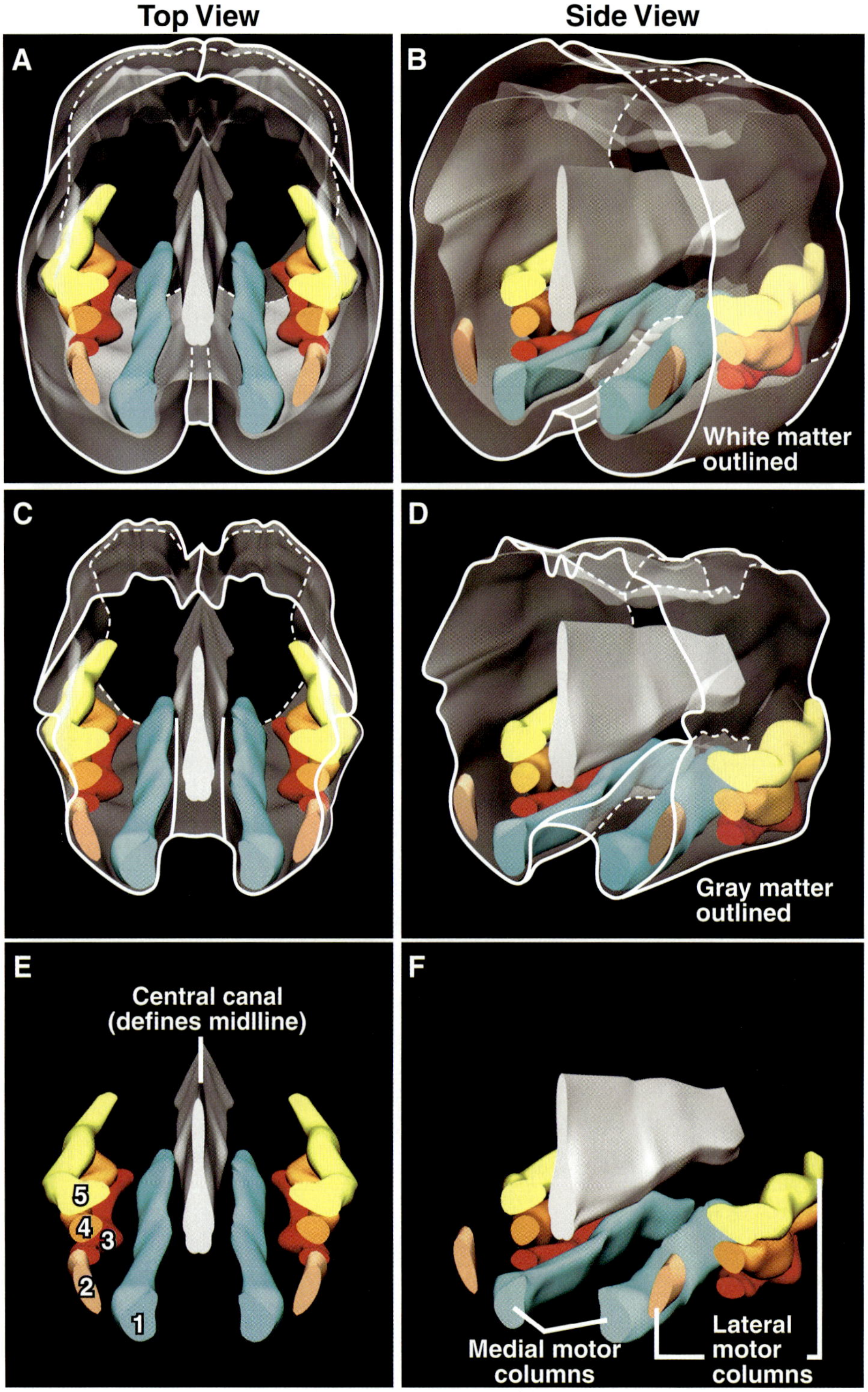

Figure 13. The cervical region in specimen M2050 (GW8.5, CR 36 mm). The individual sections (730–1170) in this model are diagrammed in **Figure 10**. The reconstructed length in the specimen is 4.4 mm after fixation. Panels **A**, **C**, and **E** show the model from the top front, panels **B**, **D**, and **F** from the upper right side. Panels **A** and **B** show both the white matter (**outlined** outer transparent envelope) and gray matter (inner transparent envelope) around the central canal (**gray-white solid**) and motor columns (**colored solids**). Panels **C** and **D** show the only gray matter (**outlined** outer transparent envelope) around the central canal and motor columns. Panels **E** and **F** show the central canal and motor columns alone. Motor column **1** (**cyan**, axial muscle motoneurons?) is located in the medial panel, ventral and dorsal tiers. Motor column **2** (**light brown**, shoulder girdle motoneurons?) is located in the central panel, dorsal tier at high cervical levels. Motor column **3** (**red**, shoulder muscle motoneurons?) in the cervical enlargement is located in the lateral and central panels, ventral tier. Motor column **4** (**orange**, arm and forearm muscle motoneurons?) in the cervical enlargement is located in the lateral and central panels, dorsal tier. Motor column **5** (**yellow**, wrist and hand muscle motoneurons?) in the cervical enlargement is located in the lateral panel, retrodorsal tier.

GW8.5, CR 36 mm, M2050

FIGURE 14

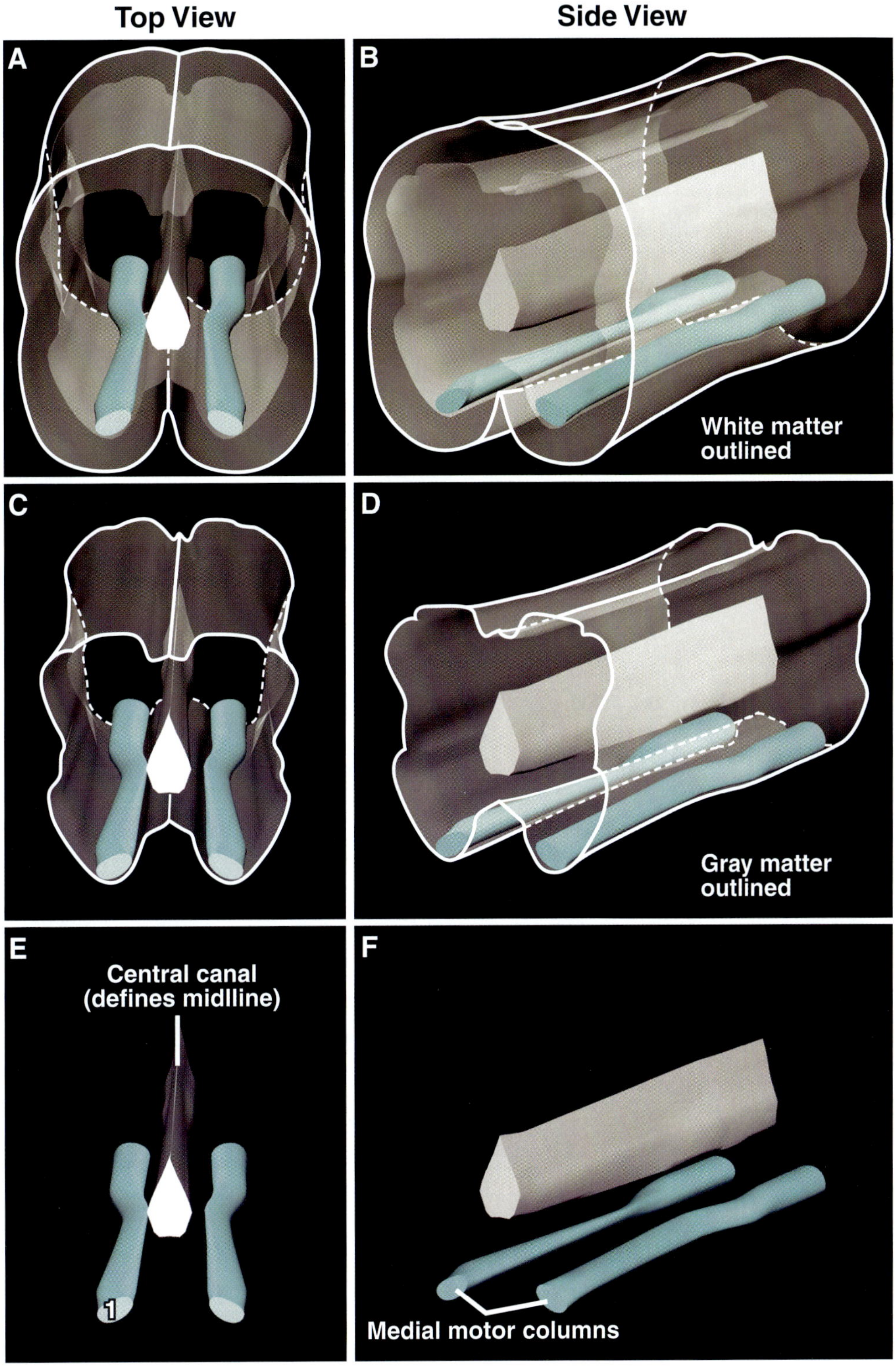

Figure 14. The thoracic region in specimen M2050 (GW8.5, CR 36 mm). The individual sections in this model are diagrammed in **Figure 11** and include sections 1225-1884. The reconstructed length in the specimen is 6.99 mm after fixation. Panels **A**, **C**, and **E** show the model from the top front, panels **B**, **D**, and **F** from the upper right side. Panels **A** and **B** show both the white matter (**outlined** outer transparent envelope) and gray matter (inner transparent envelope) around the central canal (**gray-white solid**) and the medial motor columns (**cyan**). Panels **C** and **D** show the only gray matter (**outlined** outer transparent envelope) around the central canal and the medial motor columns. Panels **E** and **F** show the central canal and the medial motor columns alone. This region of the spinal cord is distinguished from other regions by having only one motor column (**1**) that probably innervates axial muscles associated with the thoracic vertebrae, rib cage and dorsal abdominal wall.

FIGURE 15

GW8.5, CR 36 mm, M2050

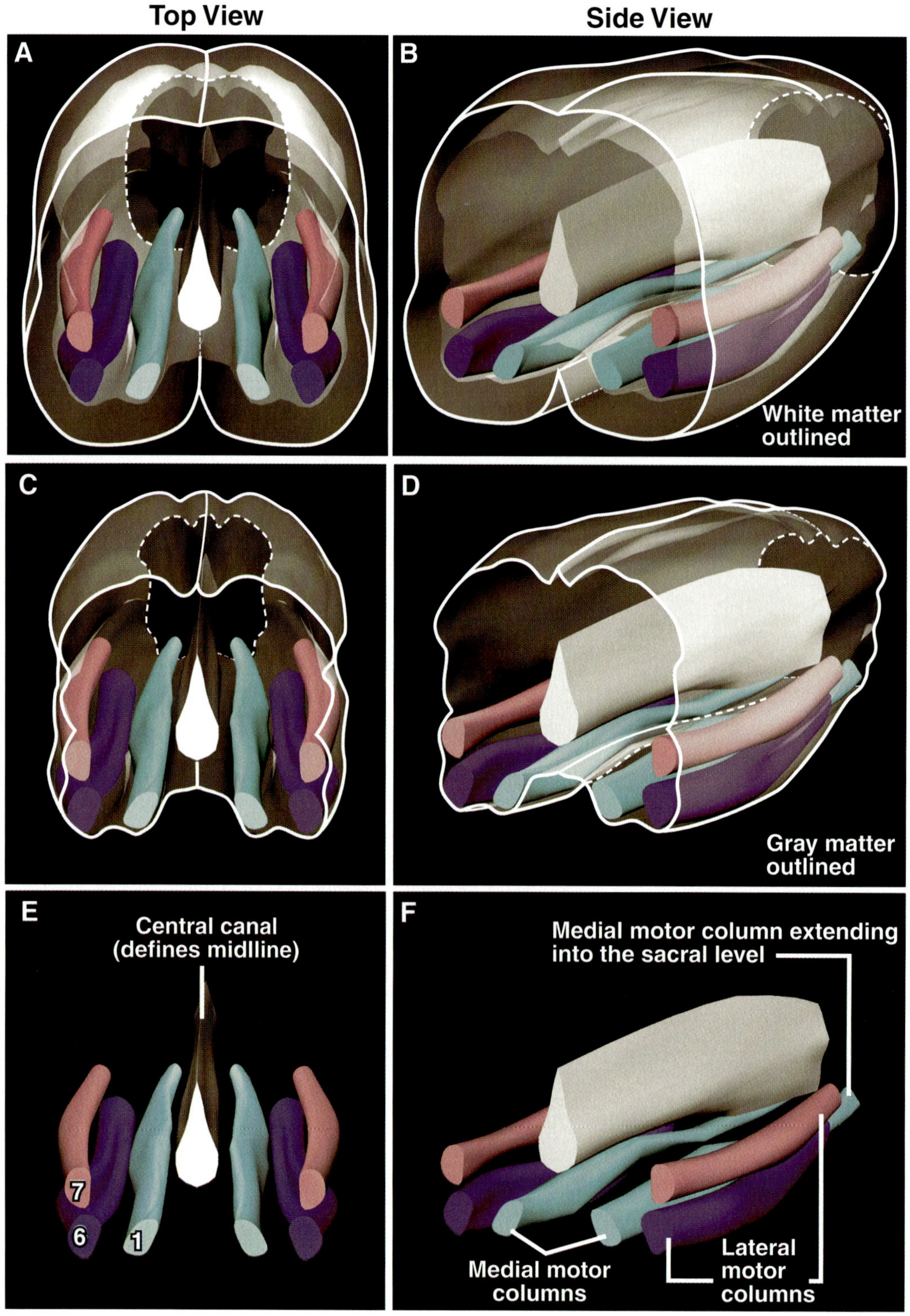

Figure 15. The lumbar and sacral regions in specimen M2050 (GW8.5, CR 36 mm). The individual sections in this model are diagrammed in **Figure 11** and include sections 2008–2674. The reconstructed length in the specimen is 6.66 mm after fixation. Panels **A**, **C**, and **E** show the model from the top front, panels **B**, **D**, and **F** from the upper right side. Panels **A** and **B** show both the white matter (**outlined** outer transparent envelope) and gray matter (inner transparent envelope) around the central canal (**gray-white solid**) and motor columns (**colored solids**). Panels **C** and **D** show the gray matter (**outlined** outer transparent envelope) around the central canal and motor columns. Panels **E** and **F** show the central canal and motor columns alone. Motor column **1** (**cyan**, axial muscle motoneurons?) is located in the medial and central panels, ventral and dorsal tiers. Motor column **6** (**purple**, proximal lower limb muscle motoneurons?) in the lumbar enlargement is located in the central and lateral panels, ventral tier. Motor column **7** (**pink**, distal lower limb muscle motoneurons?) in the lumbar enlargement is located in the lateral panel, dorsal tier.

GW10.5, CR 56 mm, Y380-62

FIGURE 16
Cervical Sections of 3-D Model

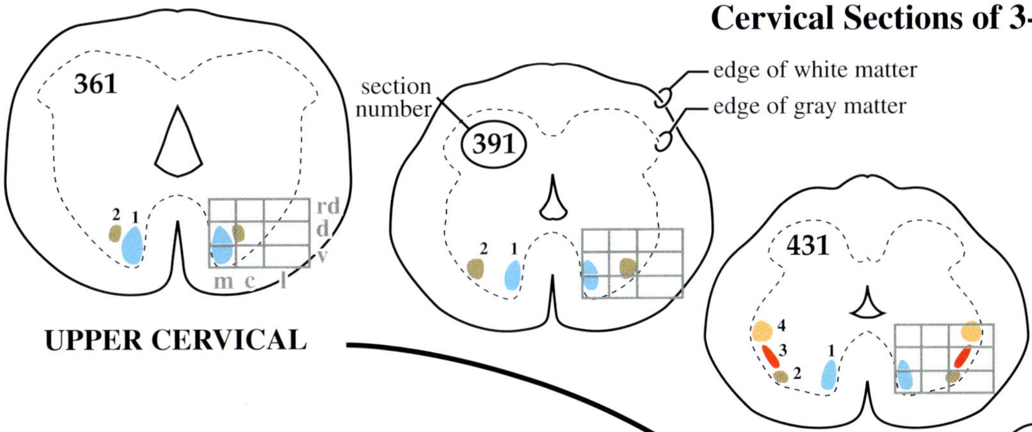

MOTOR COLUMN KEY

Labeled Number (also in Figure 20E) → Color → Column (Panel and Tier) → Muscle Innervation (?)

Labeled Number	Color	Column (Panel and Tier)	Muscle Innervation (?)
1	(blue)	mv, md (all sections)	AXIAL
2	(brown)	cd (sections 361-391) lv, lc (sections 431-551) ld (sections 581-611)	DIAPHRAGM AND PROXIMAL UPPER LIMB (SHOULDER)
3	(red)	1d (sections 431-581)	PROXIMO-DISTAL UPPER LIMB (ARM AND FOREARM)
4	(orange)	lrd (sections 431-551)	DISTAL UPPER LIMB (WRIST AND HAND)
5	(yellow)	cv (sections 551-581) mv, cv (section 611)	A medial subdivision of motor column 2? A lateral subdivision of motor column 1?

GRID KEY (based on Altman and Bayer, 2001)

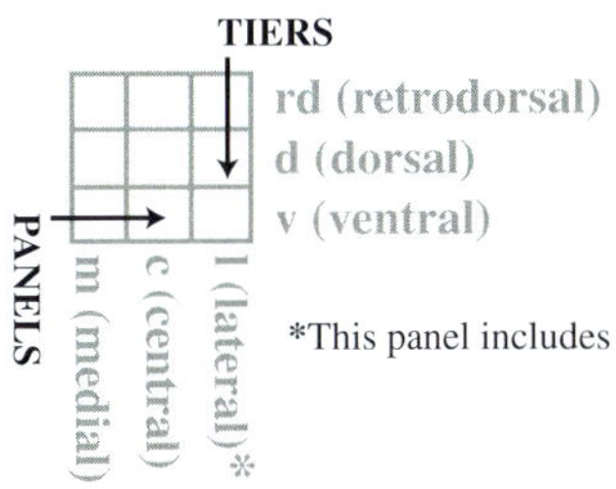

*This panel includes the lateral and far lateral subdivisions

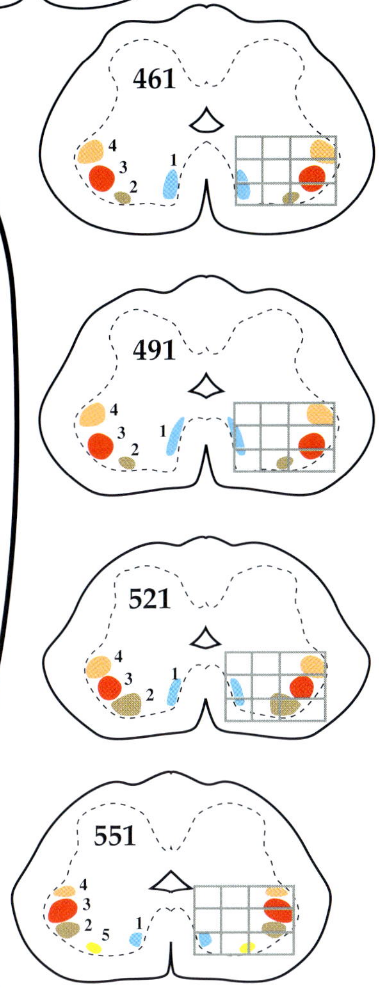

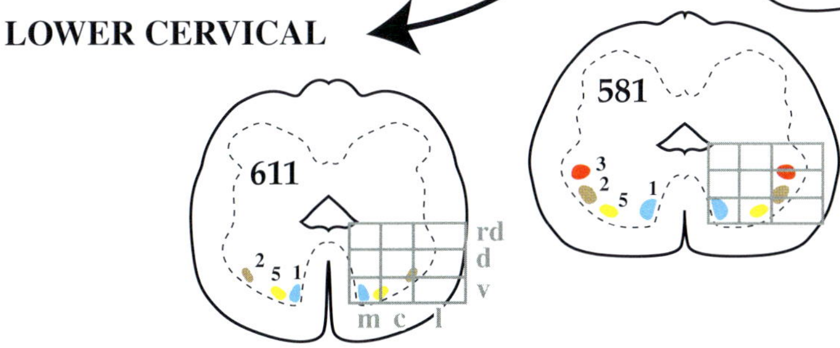

Figure 16. This specimen from the Yakovlev Collection contains serially numbered 35-μm thick sections. All outlines are profiles of the sections used in the cervical region of the spinal cord. These profiles were constructed from the right side of the section, then copied to the left side to make the model symmetrical. After fixation, the cervical region of the model is 8.79 mm long. Two of these sections are illustrated in the Atlas. **Section 361** is in **Plate 22**; **section 521** is in **Plate 23**. The cervical model is shown by itself in **Figure 20**.

FIGURE 17

GW10.5, CR 56 mm, Y380-62

Thoracic Sections of 3-D Model

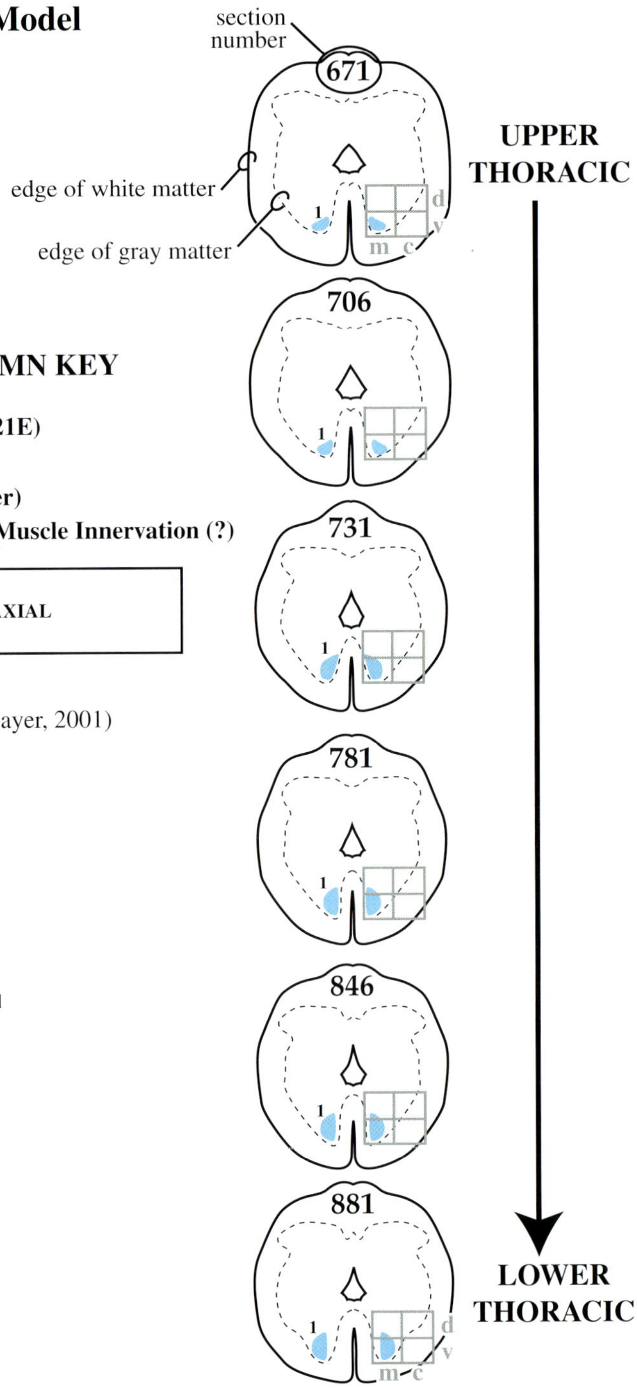

MOTOR COLUMN KEY

Labeled Number (also in Figure 21E) — Color — Column (Panel and Tier) — Muscle Innervation (?)

Labeled Number	Color	Column (Panel and Tier)	Muscle Innervation (?)
1	(blue)	mv (all sections)	AXIAL

GRID KEY (based on Altman and Bayer, 2001)

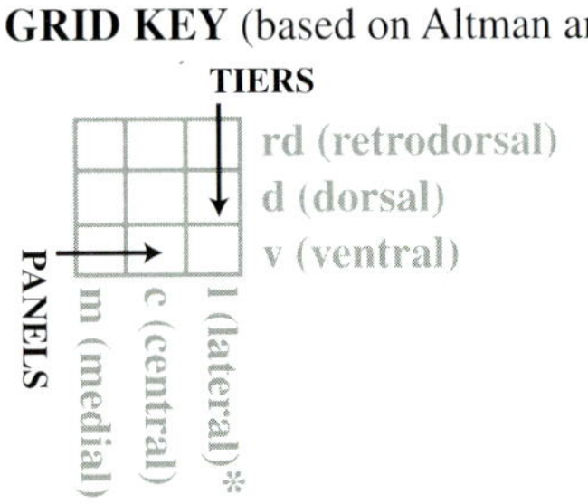

*This panel includes the lateral and far lateral subdivisions

Figure 17. A continuation of the previous figure showing outlines of the sections used in the thoracic region of the spinal cord. Two of these sections are illustrated in the Atlas. **Section 671** is in **Plate 24**; **section 881** is in **Plate 25**. After fixation, the total length of the thoracic part of the model is 7.35 mm. The thoracic model is shown by itself in **Figure 21**.

GW10.5, CR 56 mm, Y380-62

FIGURE 18

Lumbar and Sacral Sections of 3-D Model

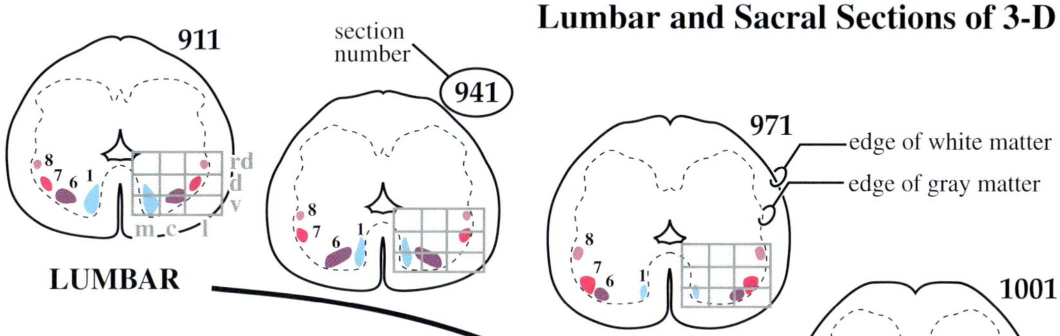

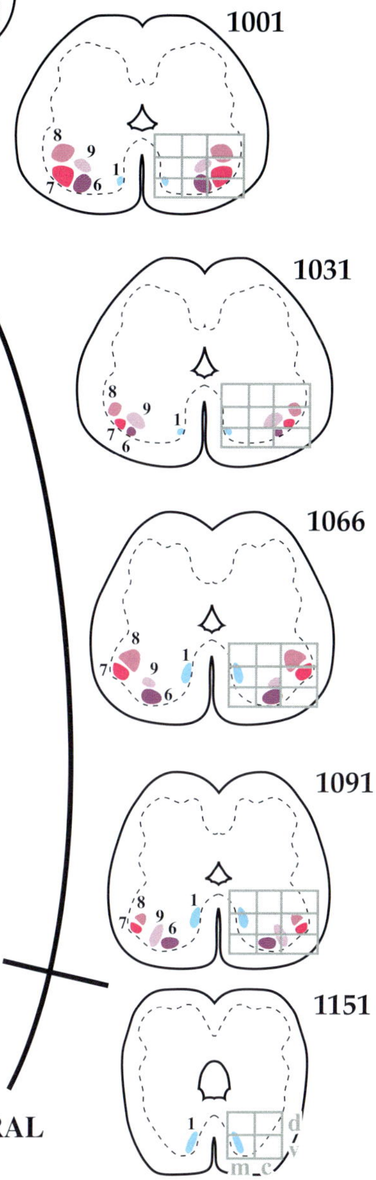

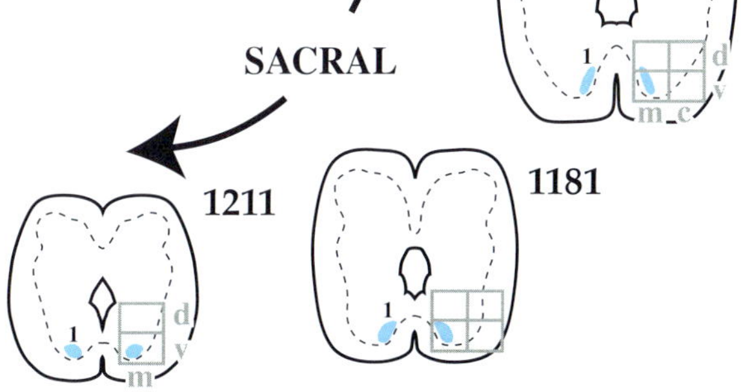

MOTOR COLUMN KEY

Labeled Number (also in Figure 22E) → Color → Column (Panel and Tier) → Muscle Innervation (?)

Labeled Number	Color	Column (Panel and Tier)	Muscle Innervation (?)
1	light blue	mv, md (all sections)	AXIAL
6	purple	cv, lv (sections 911-1091)	PROXIMAL LOWER LIMB (HIP AND THIGH)
7	red	ld (sections 911-1091)	PROXIMO-DISTAL LOWER LIMB (KNEE AND LEG)
8	pink	lrd, ld (sections 911-1091)	DISTAL LOWER LIMB (ANKLE AND FOOT)
9	light pink	cd, cv (sections 1001-1091)	PROXIMAL LOWER LIMB (HIP AND THIGH)

GRID KEY (based on Altman and Bayer, 2001)

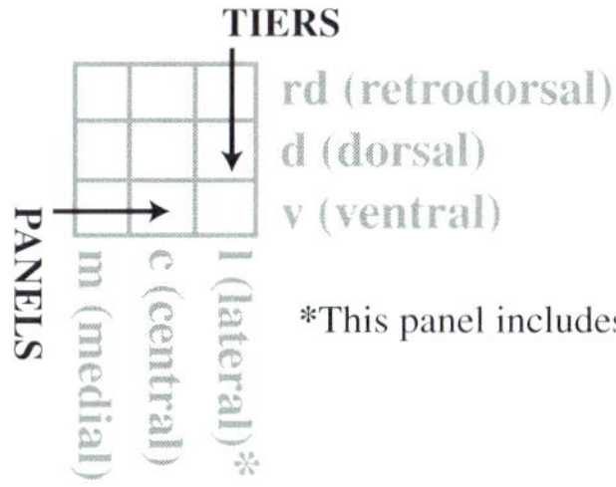

*This panel includes the lateral and far lateral subdivisions

Figure 18. A continuation of the previous figure showing outlines of the sections used in the lumbar (911-1091) and sacral (1151-1211) regions of the spinal cord. Three of these sections are illustrated in the Atlas. **Section 911** is in **Plate 26**; **section 1006** is in **Plate 27**, **section 1211** is in **Plate 28**. After fixation, the total length of the lumbosacral part of the model is 10.5 mm. The lumbosacral part of the model is in **Figure 22**.

FIGURE 19

GW10.5, CR 56 mm, Y380-62

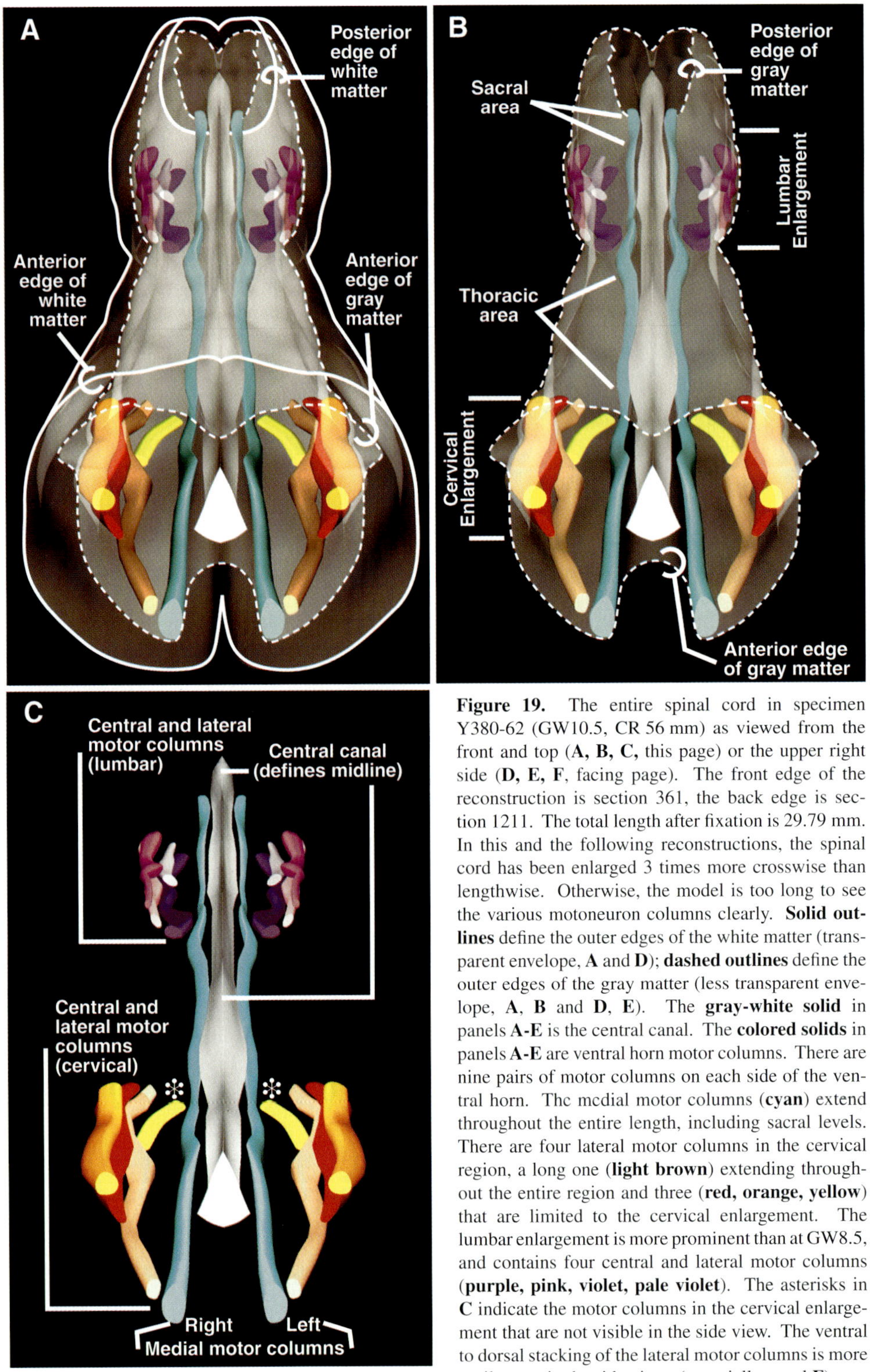

Figure 19. The entire spinal cord in specimen Y380-62 (GW10.5, CR 56 mm) as viewed from the front and top (**A, B, C,** this page) or the upper right side (**D, E, F,** facing page). The front edge of the reconstruction is section 361, the back edge is section 1211. The total length after fixation is 29.79 mm. In this and the following reconstructions, the spinal cord has been enlarged 3 times more crosswise than lengthwise. Otherwise, the model is too long to see the various motoneuron columns clearly. **Solid outlines** define the outer edges of the white matter (transparent envelope, **A** and **D**); **dashed outlines** define the outer edges of the gray matter (less transparent envelope, **A**, **B** and **D**, **E**). The **gray-white solid** in panels **A-E** is the central canal. The **colored solids** in panels **A-E** are ventral horn motor columns. There are nine pairs of motor columns on each side of the ventral horn. The medial motor columns (**cyan**) extend throughout the entire length, including sacral levels. There are four lateral motor columns in the cervical region, a long one (**light brown**) extending throughout the entire region and three (**red, orange, yellow**) that are limited to the cervical enlargement. The lumbar enlargement is more prominent than at GW8.5, and contains four central and lateral motor columns (**purple, pink, violet, pale violet**). The asterisks in **C** indicate the motor columns in the cervical enlargement that are not visible in the side view. The ventral to dorsal stacking of the lateral motor columns is more easily seen in the side views (especially panel **F**).

GW10.5, CR 56 mm, Y380-62

FIGURE 19

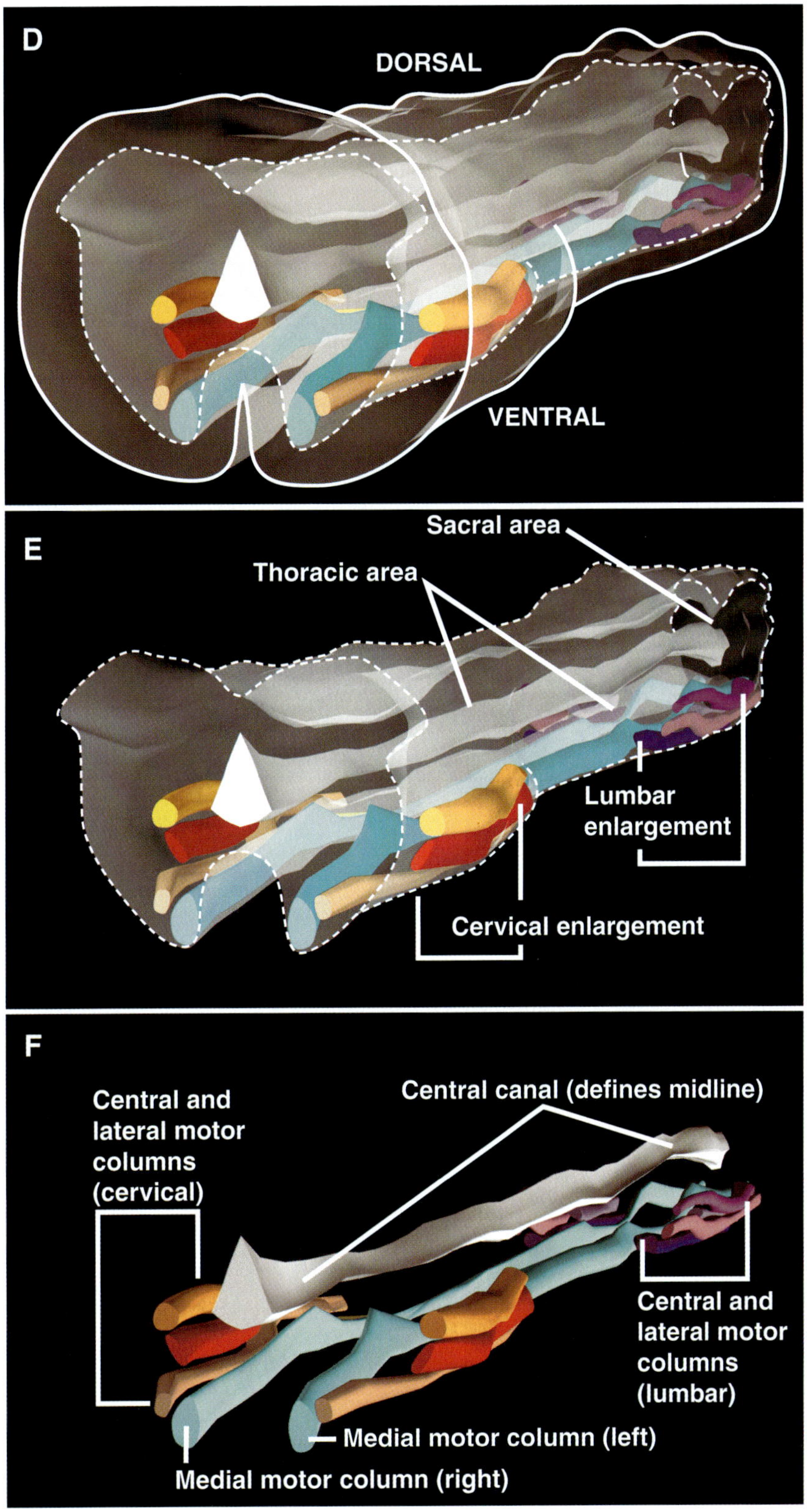

FIGURE 20

GW10.5, CR 56 mm, Y380-62

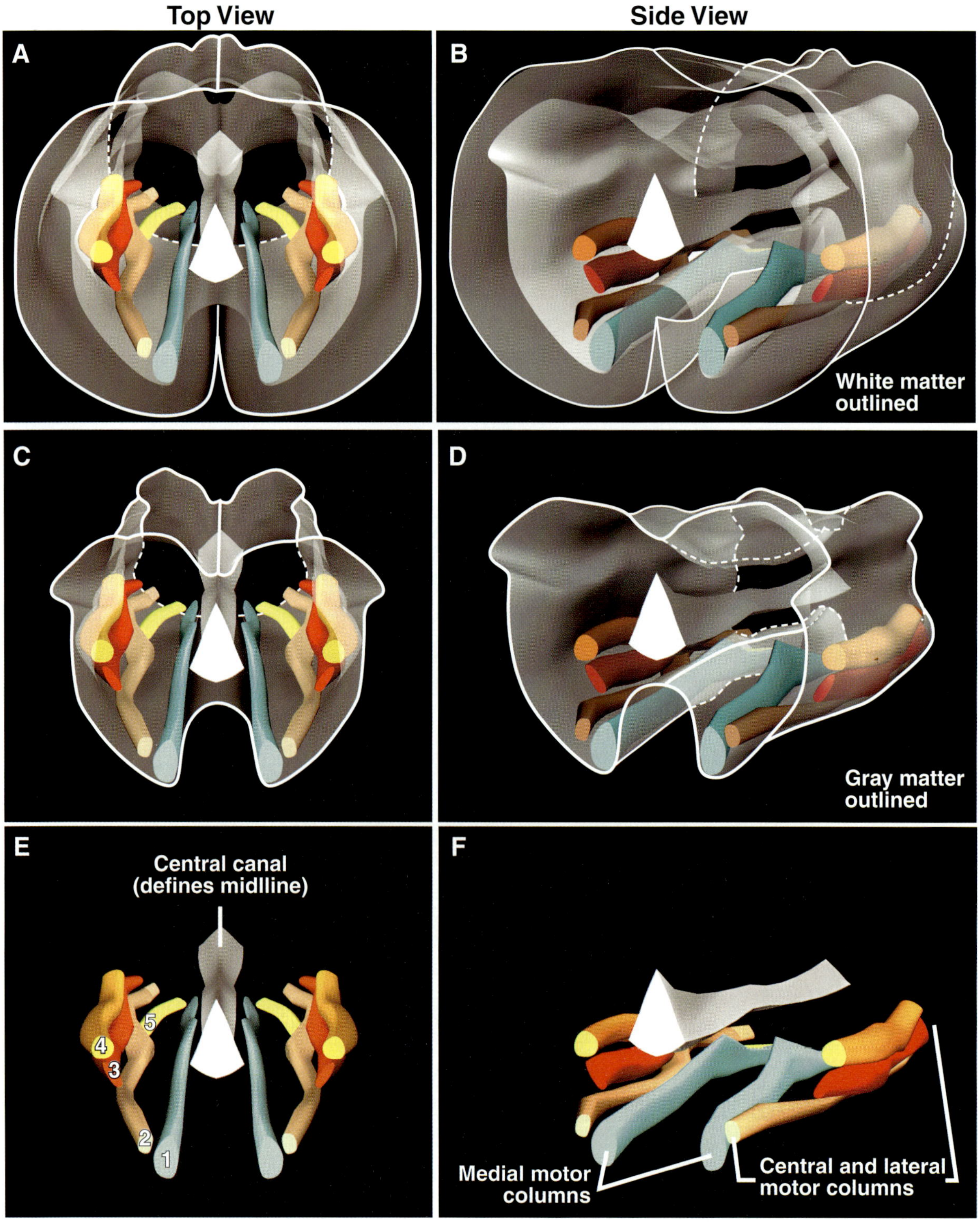

Figure 20. The cervical region in specimen Y380-62 (GW10.5, CR 56 mm). The individual sections in this model are diagrammed in **Figure 16** and include sections 361-611. The reconstructed length is 8.79 mm after fixation. Panels **A**, **C**, and **E** show the model from the top front, panels **B**, **D**, and **F** from the upper right side. Panels **A** and **B** show both the white matter (**outlined** transparent envelope) and gray matter (inner transparent envelope) around the central canal (**gray-white solid**) and motor columns (**colored solids**). Panels **C** and **D** show the gray matter (**outlined** transparent envelope) around the central canal and motor columns. Panels **E** and **F** show the central canal and motor columns alone. Motor column **1** (**cyan**, axial muscle motoneurons?) is located in the medial panel, ventral and dorsal tiers. Motor column **2** (**light brown**, proximal upper limb muscle motoneurons?) is located in the central and lateral panels, ventral and dorsal tiers. Motor column **3** (**red**, arm and forearm muscle motoneurons?) is located in the cervical enlargement in the lateral panel, dorsal tier. Motor column **4** (**orange**, wrist and hand muscle motoneurons?) is located in the cervical enlargement in the lateral panel, retrodorsal tier. Motor column **5** (**yellow**, forearm muscle motoneurons?) in the posterior cervical enlargement (sections 551-611) is located in the medial and central panels, ventral tier.

GW10.5, CR 56 mm, Y380-62

FIGURE 21

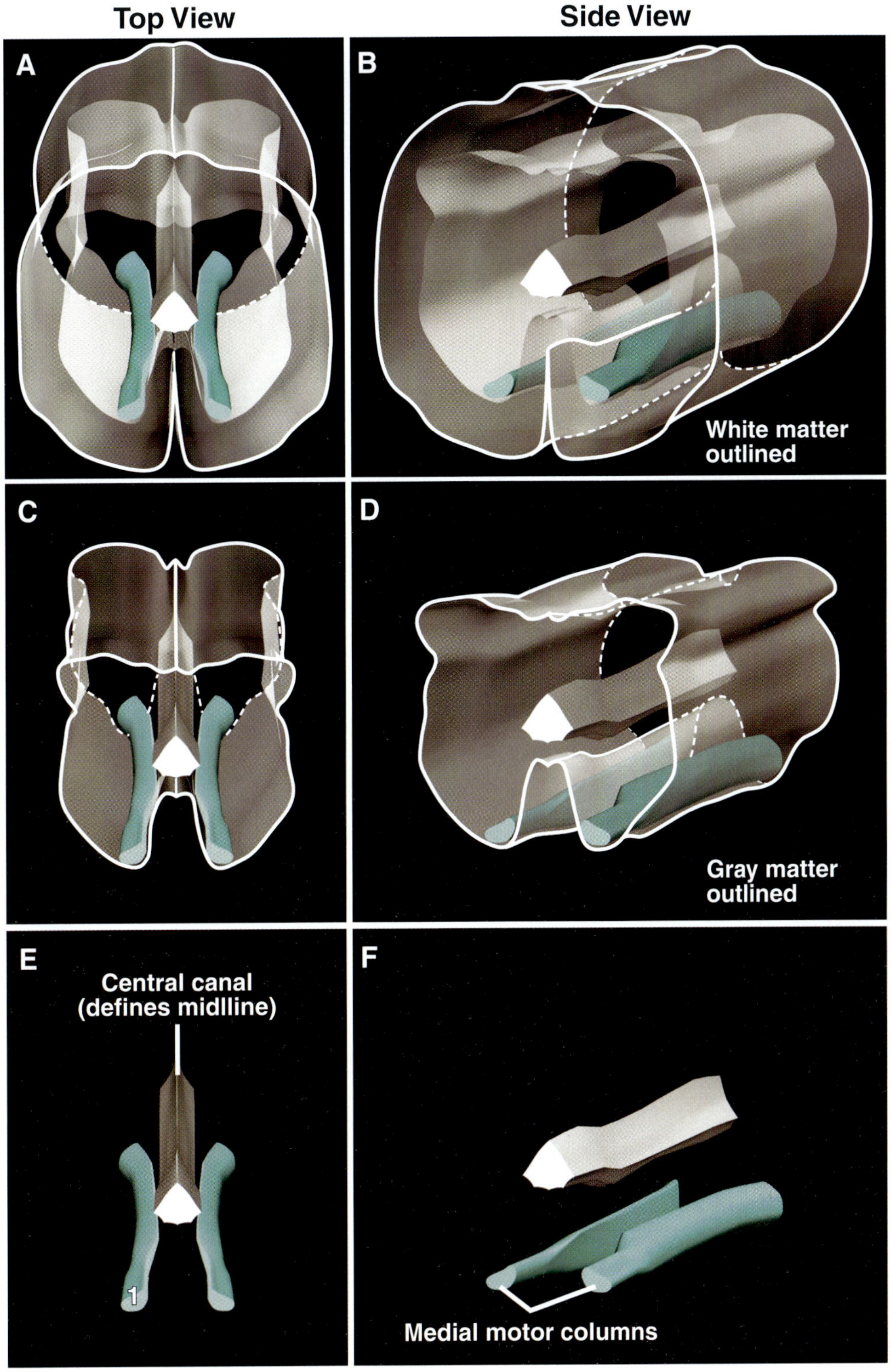

Figure 21. The thoracic region in specimen Y380-62 (GW10.5, CR 56 mm). The individual sections in this model are diagrammed in **Figure 17** and include sections 671-881. The reconstructed length is 7.35 mm after fixation. Panels **A**, **C**, and **E** show the model from the top front, panels **B**, **D**, and **F** from the upper right side. Panels **A** and **B** show both the white matter (**outlined** transparent envelope) and gray matter (inner transparent envelope) around the central canal (**gray-white solid**) and the medial motor columns (**cyan**). Panels **C** and **D** show the gray matter (**outlined** transparent envelope) around the central canal and the medial motor columns. Panels **E** and **F** show the central canal and the medial motor columns alone. This region of the spinal cord is distinguished from other regions by having only one motor column (**1**) that probably innervates axial muscles associated with the thoracic vertebrae, rib cage, and dorsal abdominal wall.

FIGURE 22

GW10.5, CR 56 mm, Y380-62

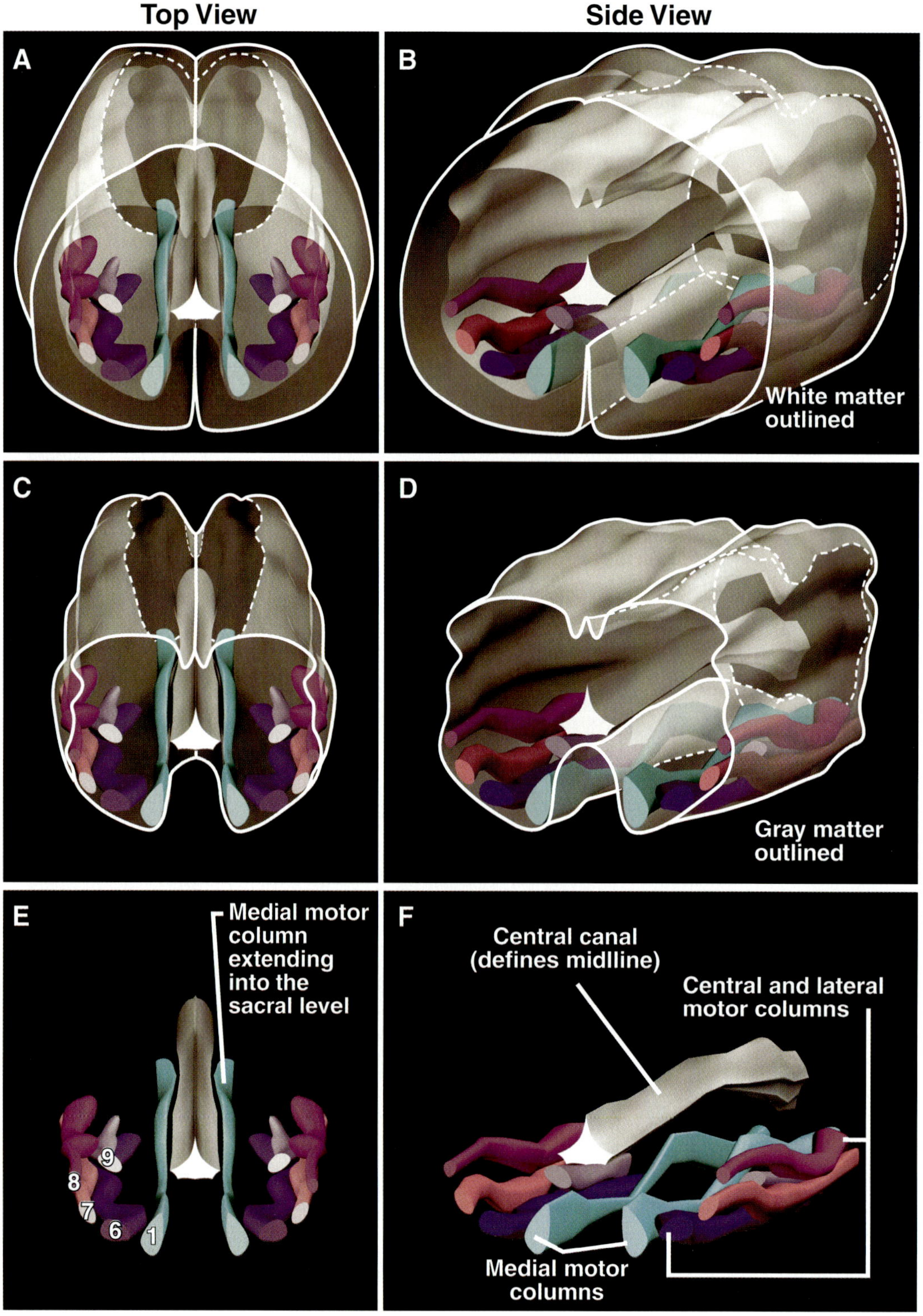

Figure 22. The lumbar and sacral regions in specimen Y380-62 (GW10.5, CR 56 mm). The individual sections (911-1211) in this model are diagrammed in **Figure 18**. The reconstructed length of the specimen is 10.5 mm after fixation. Panels **A**, **C**, and **E** show the model from the top front, panels **B**, **D**, and **F** from the upper right side. Panels **A** and **B** show both the white matter (**outlined** transparent envelope) and gray matter (inner transparent envelope) around the central canal (**gray-white solid**) and motor columns (**colored solids**). Panels **C** and **D** show the gray matter (**outlined** transparent envelope) around the central canal and motor columns. Panels **E** and **F** show the central canal and motor columns alone. Motor column **1** (**cyan**, axial muscle motoneurons?) is located in the medial and central panels, ventral and dorsal tiers. The remaining motor columns (**6-9**) are limited to lumbar levels (sections 911-1091). Column **6** (**purple**, motoneurons controlling hip movements?) is located in the central and lateral panels, ventral tier. Column **7** (**pink**, motoneurons controlling knee movements?) is located in the lateral panel, dorsal tier. Column **8** (**violet**, motoneurons controlling ankle and foot movements?) is located in the lateral panel, dorsal and retrodorsal tiers. Column **9** (**pale violet**, motoneurons controlling hip movements?) is located in the central panel, dorsal and ventral tiers.

GW14, CR 108 mm, Y68-65

FIGURE 23

Cervical Sections of 3-D Model

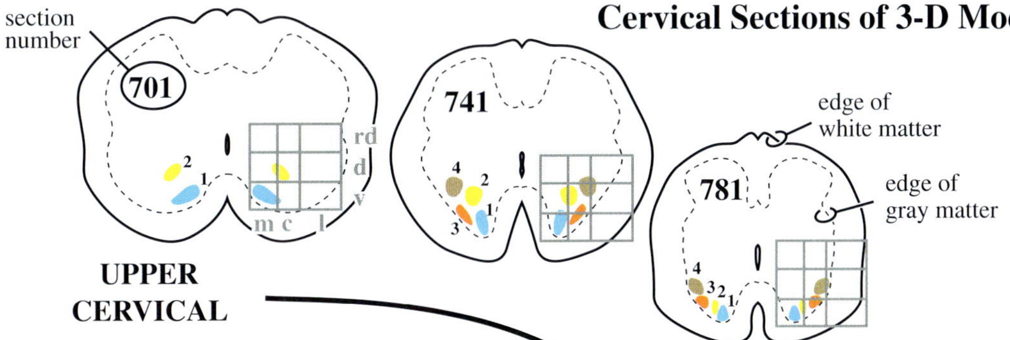

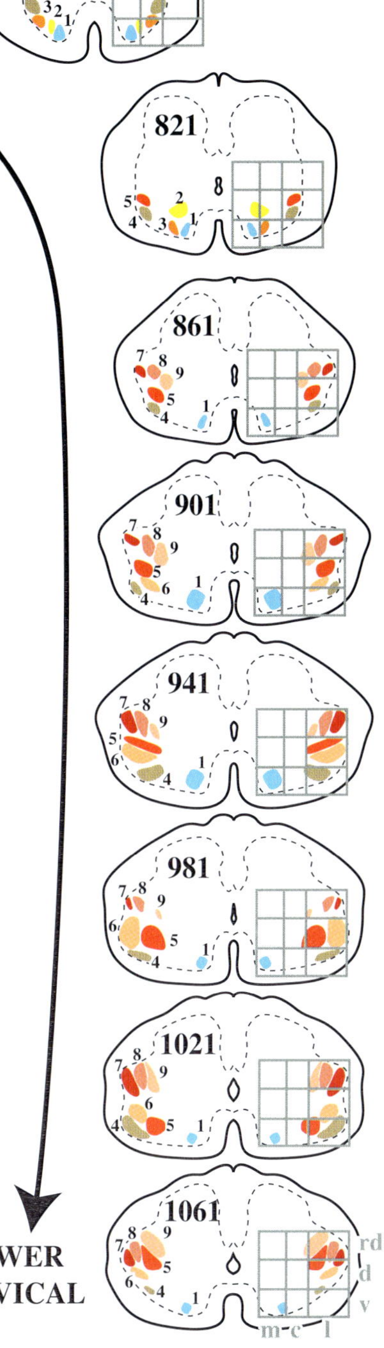

MOTOR COLUMN KEY

Labeled Number (also in Figures 27C-28C) | Color | Column (Panel and Tier) | Muscle Innervation (?)

Labeled Number	Color	Column (Panel and Tier)	Muscle Innervation (?)
1		mv (all sections)	AXIAL
2		cv, cd (sections 701-821)	DIAPHRAGM, PROXIMAL UPPER LIMB (SHOULDER)
3		cv, cd (sections 701-821)	
4		cd (sections 741-781) ld, lv (sections 821-1061)	PROXIMO-DISTAL UPPER LIMB (ARM AND FOREARM)
5		ld (sections 821-981) 1v (section 1021) lrd (section 1061)	
6		ld (sections 901-1061)	
7		lrd (sections 861-1061)	DISTAL UPPER LIMB (WRIST AND HAND)
8		lrd (sections 861-1061)	
9		lrd (sections 861-1061)	

GRID KEY (based on Altman and Bayer, 2001)

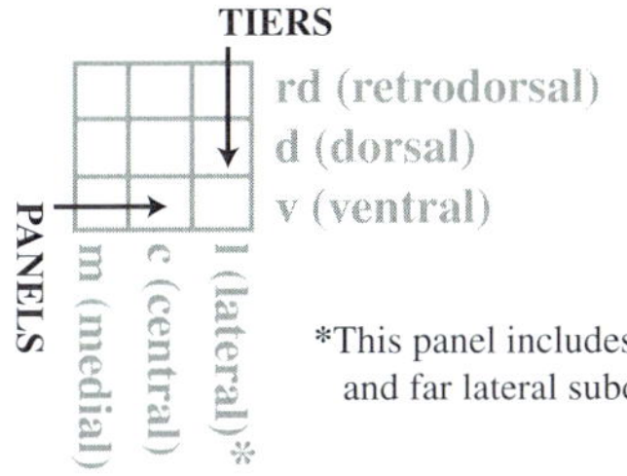

*This panel includes the lateral and far lateral subdivisions

Figure 23. This specimen from the Yakovlev Collection contains serially numbered 35-µm thick sections. All outlines are profiles of the sections used in the cervical region of the spinal cord. These profiles were constructed from the right side of the section, then copied to the left side to make the model symmetrical. Two of these sections are illustrated in the Atlas. **Section 741** is in **Plate 30**; **section 981** is in **Plate 31**. The cervical model is shown by itself in **Figures 27** and **28**. After fixation, the cervical region of the model is 12.6 mm long.

FIGURE 24

GW14, CR 108 mm, Y68-65

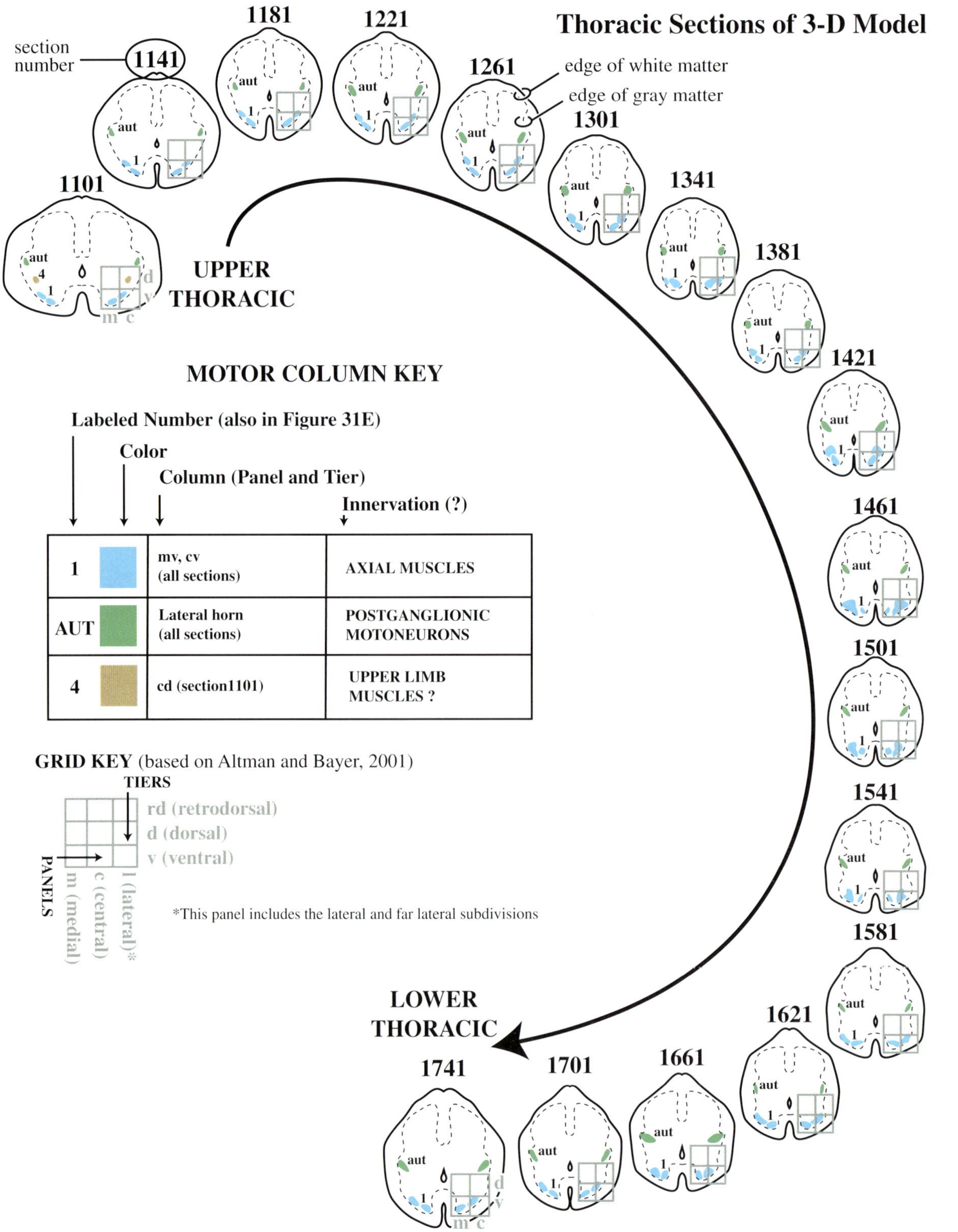

MOTOR COLUMN KEY

Labeled Number (also in Figure 31E) — Color — Column (Panel and Tier) — Innervation (?)

Labeled Number	Color	Column (Panel and Tier)	Innervation (?)
1	(blue)	mv, cv (all sections)	AXIAL MUSCLES
AUT	(green)	Lateral horn (all sections)	POSTGANGLIONIC MOTONEURONS
4	(tan)	cd (section1101)	UPPER LIMB MUSCLES ?

GRID KEY (based on Altman and Bayer, 2001)

*This panel includes the lateral and far lateral subdivisions

Figure 24. A continuation of the previous figure showing outlines of the sections used in the thoracic region of the spinal cord. Three of these sections are illustrated in the Atlas. **Section 1141** is in **Plate 32**; **section 1581** is in **Plate 33**; **section 1741** is in **Plate 34**. After fixation, the total length of the thoracic part of the model is 22.4 mm. The thoracic model is shown by itself in **Figure 31**.

GW14, CR 108 mm, Y68-65

FIGURE 25

Lumbar and Sacral Sections of 3-D Model

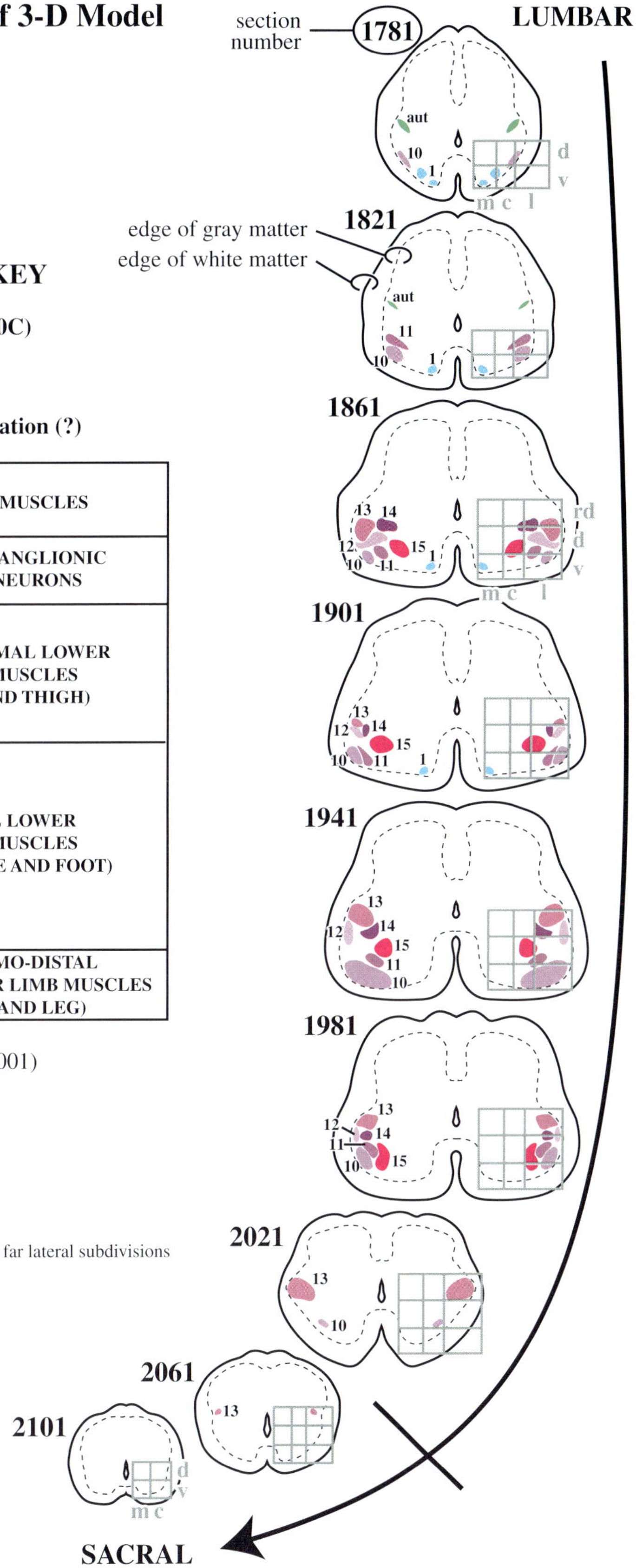

MOTOR COLUMN KEY

Labeled Number (also in Figures 29C-30C)

Color

Column (Panel and Tier)

Innervation (?)

Labeled Number	Color	Column (Panel and Tier)	Innervation (?)
1		mv (sections 1781-1901)	AXIAL MUSCLES
aut		Lateral horn (sections 1781-1821)	POSTGANGLIONIC MOTONEURONS
10		ld, lv (sections 1781-1981) cd (section 2021)	PROXIMAL LOWER LIMB MUSCLES (HIP AND THIGH)
11		1d, lv (sections 1821-1981)	
12		ld (sections 1861-1901) 1rd (sections 1941-1981)	DISTAL LOWER LIMB MUSCLES (ANKLE AND FOOT)
13		ld, lrd (sections 1861-2061)	
14		lrd (sections 1861-1981)	
15		cd, ld (sections 1861-1981)	PROXIMO-DISTAL LOWER LIMB MUSCLES (KNEE AND LEG)

GRID KEY (based on Altman and Bayer, 2001)

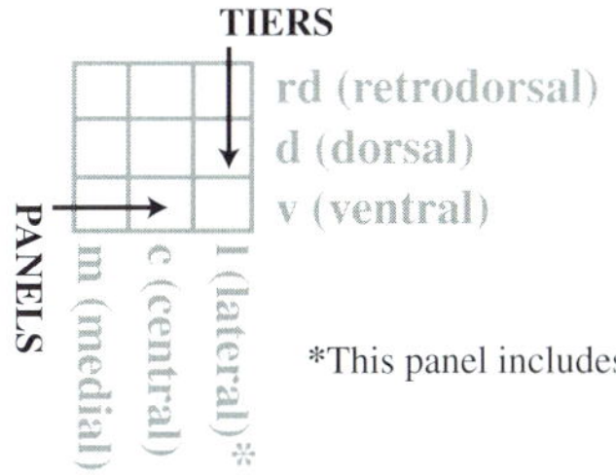

*This panel includes the lateral and far lateral subdivisions

Figure 25. A continuation of the previous figure showing outlines of the sections used in the lumbar (1781-1981) and sacral (2021-2101) regions of the spinal cord. Three of these sections are illustrated in the Atlas. **Section 1821** is in **Plate 35**; **section 1901** is in **Plate 36**, **section 2061** is in **Plate 37**. After fixation, the total length of the lumbosacral part of the model is 11.2 mm. The lumbosacral part of the model is in **Figures 29-30**.

FIGURE 26

GW14, CR 108 mm, Y68-65

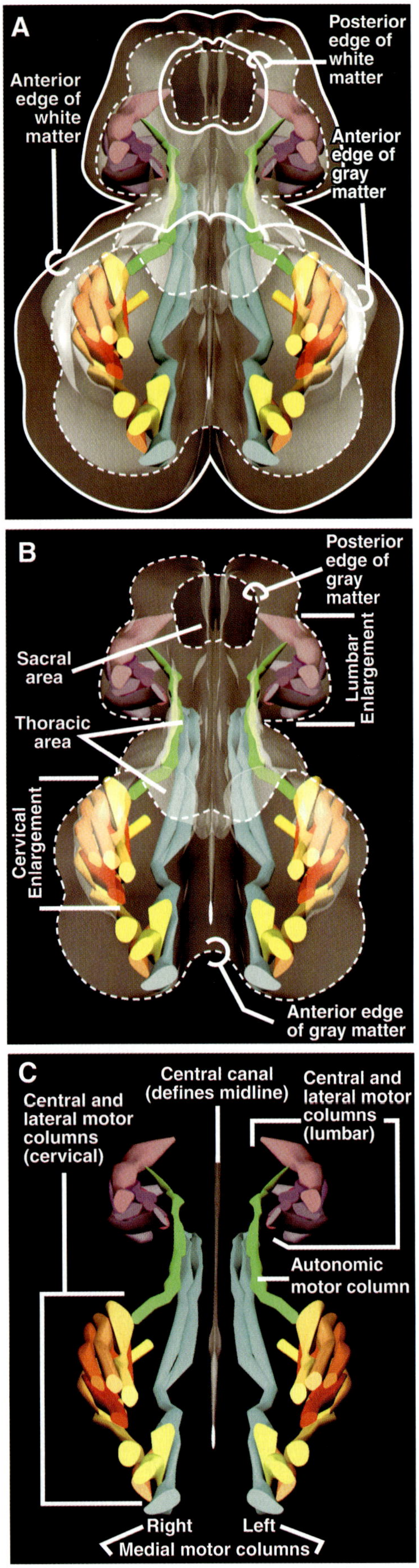

Figure 26. The entire spinal cord in specimen Y68-65 (GW14, CR 108 mm) as viewed from the front and top (**A, B, C,** this page) or the upper right side (**D, E, F**, facing page). The front edge of the reconstruction is section 701, the back edge is section 2101. The total length after fixation is 49 mm. In this and the following reconstructions, the spinal cord has been enlarged 3 times more crosswise than lengthwise. Otherwise, the model is too long to see the various motoneuron columns clearly. **Solid outlines** define the outer edges of the white matter (transparent envelope, **A** and **D**); **dashed outlines** define the outer edges of the gray matter (less transparent envelope, **A**, **B** and **D**, **E**). The **gray-white solid** in panels **A-E** is the central canal. The **colored solids** in panels **A-E** are ventral horn motor columns.

There are 15 pairs of motor columns on each side of the ventral horn in this specimen. The medial motor columns (**cyan**) are split in the thoracic region and extend into the lumbar enlargement up to section 1901 and then disappear. The remaining motor columns are in the ventral, central, and lateral sectors. There are eight central and lateral motor columns in the cervical region (**light brown, orange, yellow shades**). The lumbar enlargement is large in this specimen and contains six central and lateral motor columns (**purple, violet, magenta shades**). The top and side views in panels **C** and **F** show the intertwined arrangement of the motor columns in the cervical and lumbar enlargements. The autonomic motor columns (**green**) in the lateral horn of the thoracic and upper lumbar regions are also reconstructed in this specimen.

GW14, CR 108 mm, Y68-65

FIGURE 26

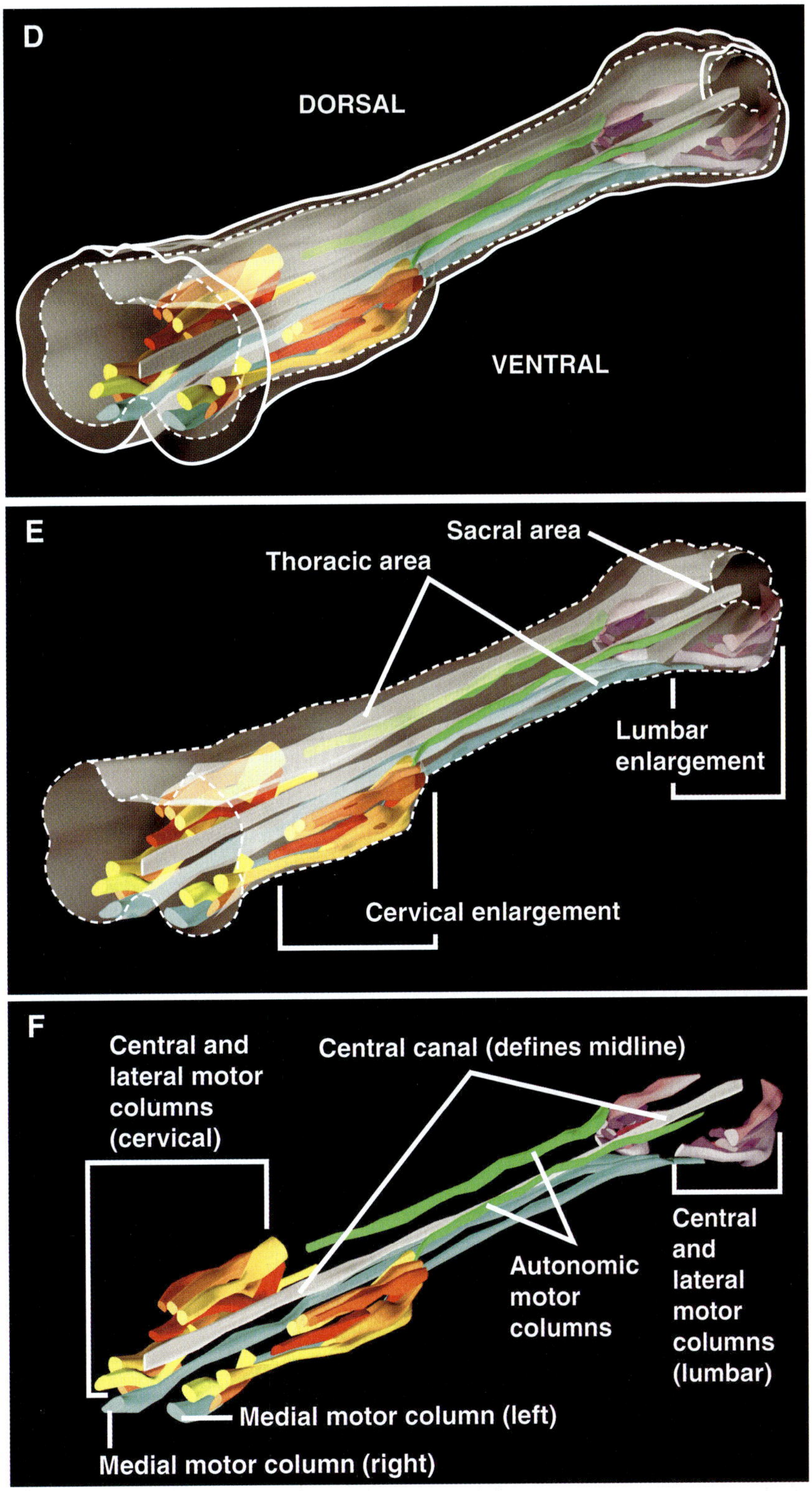

FIGURE 27

GW14, CR 108 mm, Y68-65

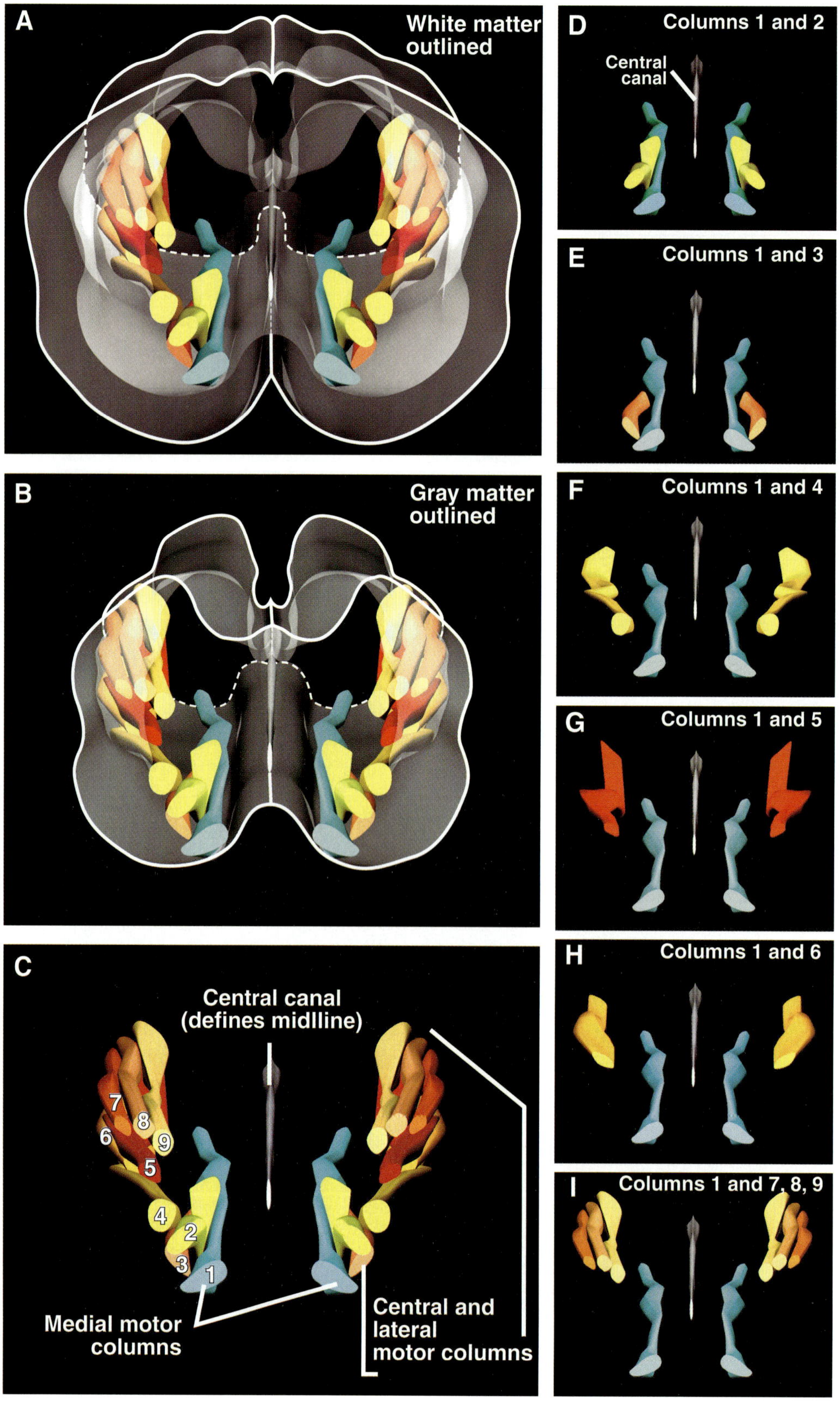

Figure 27. The top front view of the cervical region in specimen Y68-65 (GW14, CR 108 mm). The individual sections (701-1061) in this model are diagrammed in **Figure 23**. The length of the cervical region is 12.6 mm after fixation. All panels show the entire model or parts of it from the top front. Panel **A** shows both the white matter (**outlined** transparent envelope) and gray matter (inner transparent envelope) around the central canal (**gray-white solid**) and motor columns (**colored solids**). Panel **B** shows only the gray matter (**outlined** transparent envelope) around the central canal and motor columns. Panel **C** shows the central canal and all motor columns alone. Subsets of motor columns are shown in the small side panels **D-I**.

GW14, CR 108 mm, Y68-65

FIGURE 28

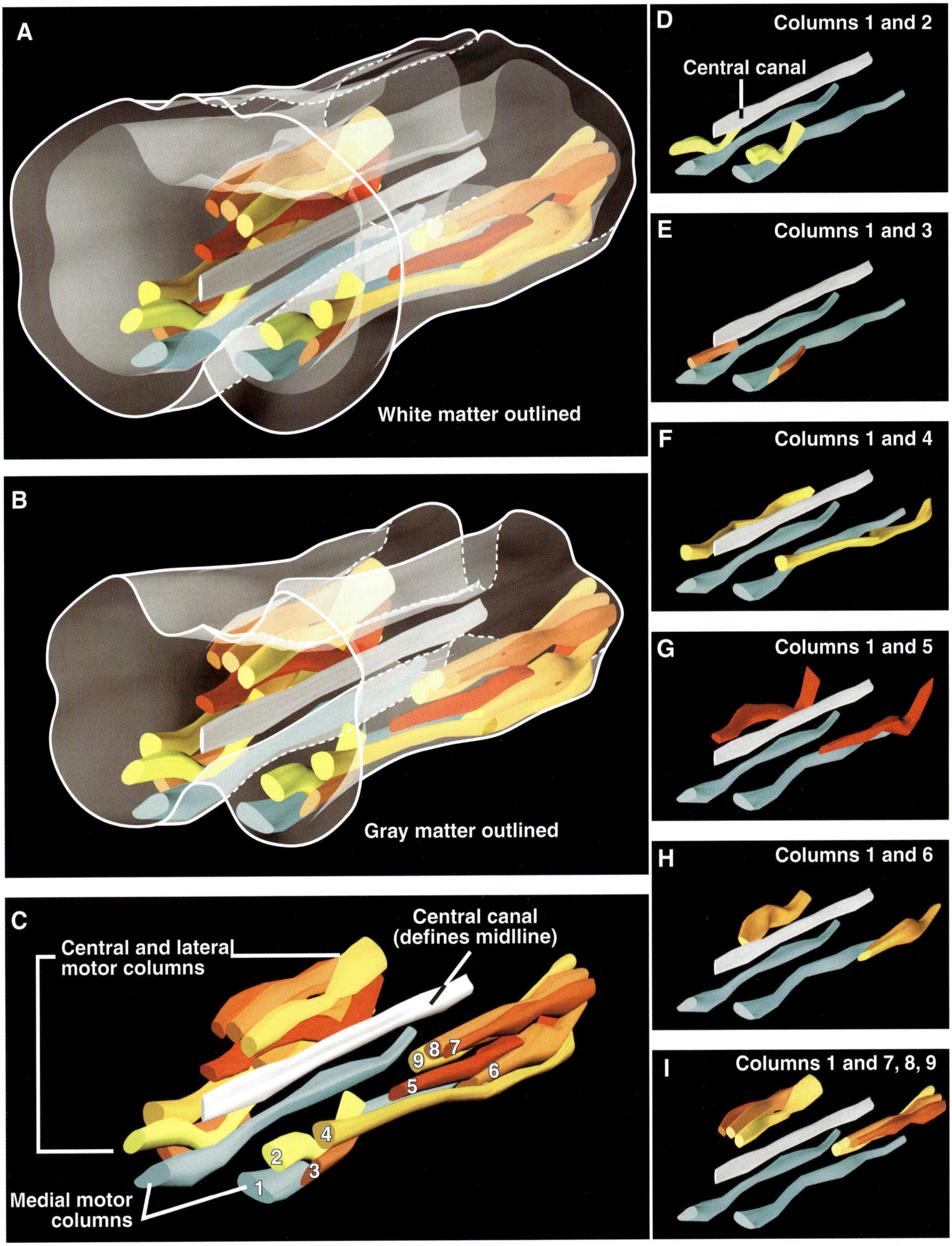

Figure 28. A continuation of the previous figure in the same specimen, except that all panels show either the entire model or parts of it from the side view. Motor column **1** (**cyan**, axial muscle motoneurons?) is located in the medial panel, ventral tier. Motor columns **2** and **3** (**yellow, orange**) at upper cervical levels (sections 701-821) may innervate proximal upper limb muscles, and are located in the central and lateral panels, ventral and dorsal tiers. Motor columns **4** to **6** (**yellow brown**, **red**, **orange**) are located in the cervical enlargement in the central and lateral panels, ventral, dorsal, and retrodorsal tiers and may innervate arm and forearm muscles. Motor columns **7** to **9** (**dark red orange**, **light red orange**, **yellow orange**) are located in the cervical enlargement in the lateral panel, retrodorsal tier and may innervate wrist and hand (digit) muscles.

FIGURE 29

GW14, CR 108 mm, Y68-65

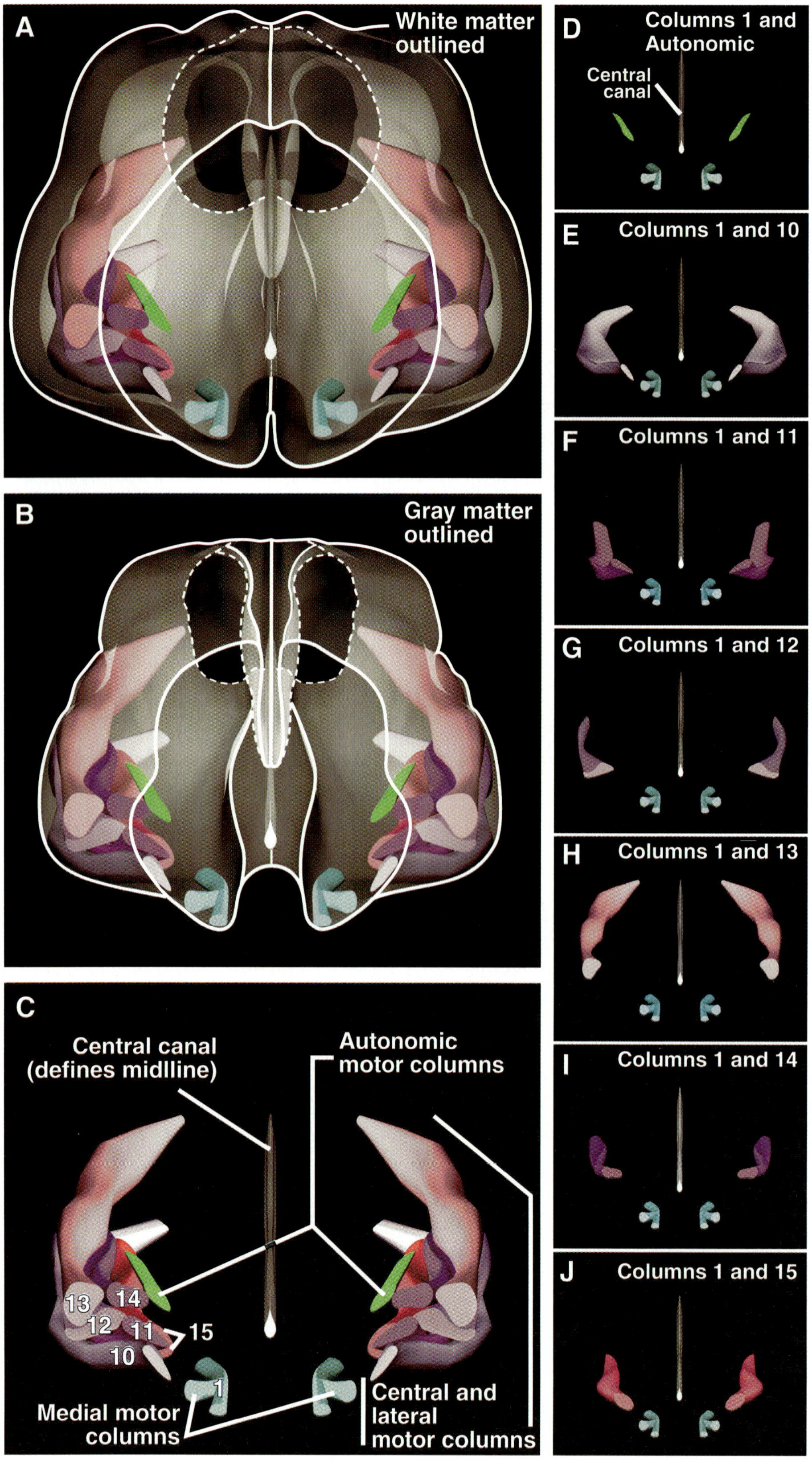

Figure 29. The lumbar and sacral regions in specimen Y68-65 (GW14, CR 108 mm). The individual sections (1781-2101) in this model are diagrammed in **Figure 25**. The reconstructed length is 11.2 mm after fixation. All panels show the model from the top front. Panel **A** shows both the white matter (**outlined** transparent envelope) and gray matter (inner transparent envelope) around the central canal (**gray-white solid**) and motor columns (**colored solids**). Panel **B** shows the gray matter (**outlined** transparent envelope) around the central canal and motor columns. Panel **C** shows the central canal and all the motor columns. The small side panels (**D-J**) show various subsets of motor columns with the central canal.

Figure 30. A continuation of the previous figure in the same specimen except that all panels show either the entire model or parts of it in a side view. Motor column **1** (**cyan**, medial panel, ventral tier) may innervate axial muscles. It is bifurcated in the first section (1781) and single in sections 1821 to 1861. The autonomic motor column (**green**) is in the first two sections of the model (1781-1821) and contains the most posterior neurons that innervate the sympathetic ganglia. The remaining motor columns (**10-15**) are present only at lumbar levels or upper sacral levels (sections 1781-2061). Columns

GW14, CR 108 mm, Y68-65

FIGURE 30

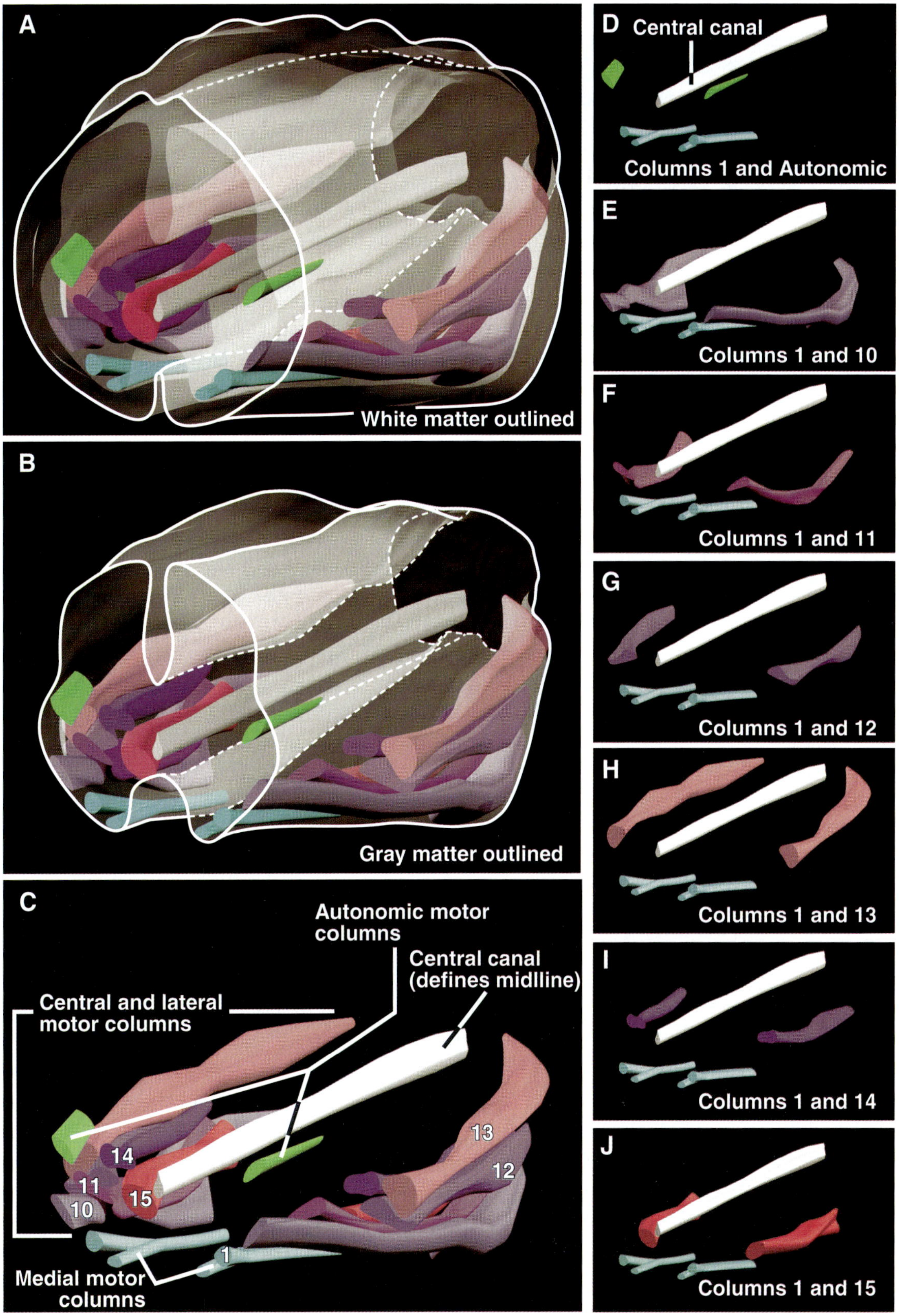

10 (**pale violet**) and **11** (**dark violet**) may innervate proximal lower limb muscles controlling the hip joint and are located in the lateral panel, ventral and dorsal tiers. Column **12**, **13**, and **14** (**pale violet**, **pink**, **purple**) are located in the lateral panel, dorsal and retrodorsal tiers. They may innervate lower limb muscles controlling the ankle and foot. Column **15** (**reddish-magenta**) is located in the central and lateral panels, dorsal tier and may innervate leg muscles.

FIGURE 31

GW14, CR 108 mm, Y68-65

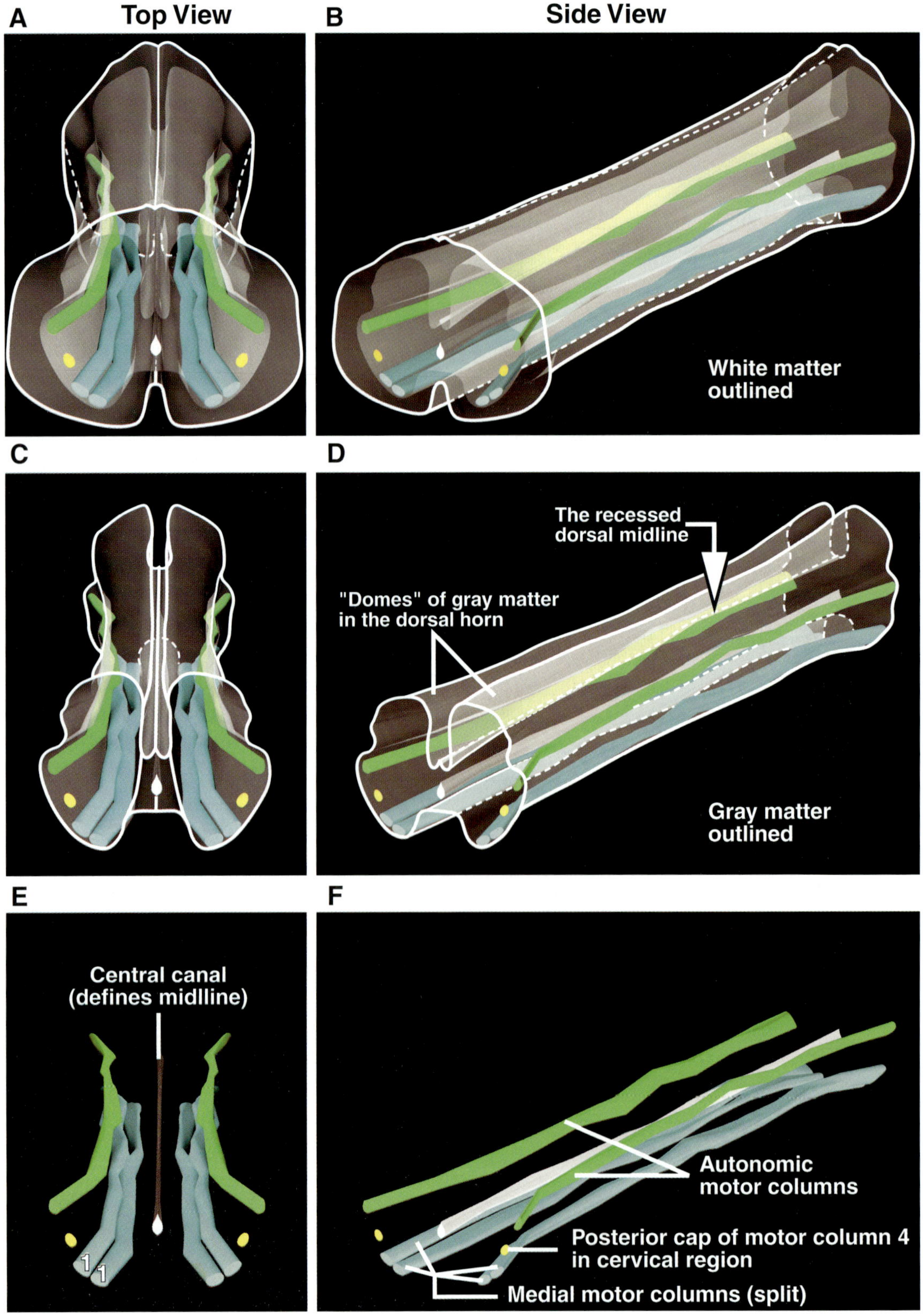

Figure 31. The thoracic region in specimen Y68-65 (GW14, CR 108 mm). The individual sections (1101-1741) in this model are diagrammed in **Figure 24**. The reconstructed length is 22.4 mm after fixation. Panels **A**, **C**, and **E** show the model from the top front, panels **B**, **D**, and **F** from the upper right side. Panels **A** and **B** show both the white matter (**outlined** transparent envelope) and gray matter (inner transparent envelope) around the central canal (**gray-white solid**) and the medial motor columns (**1**, **cyan**), autonomic motor columns (**green**), and the posterior cap of motor column **4** (**yellow-brown**) from the cervical region that is present only in the first section (1101). Panels **C** and **D** show only the gray matter (**outlined** transparent envelope) around the central canal and the motor columns. Panels **E** and **F** show the central canal and the motor columns alone. This region of the spinal cord is distinguished from other regions by having a bifurcated motor column **1** that probably innervates axial muscles associated with the thoracic vertebrae, rib cage, and dorsal abdominal wall. The autonomic motor columns in the lateral horn are prominent and innervate neurons in the sympathetic ganglia.

VII. Summary and Conclusions

This section features a quantitative summary of the major developmental events that have been qualitatively demonstrated in photographs of the spinal cord at immature through mature stages. In order to compare the complete timetable of development, the cervical level is dealt with most completely because that level is illustrated at all ages. First, absolute changes in area are discussed, followed by relative changes in area. Finally, a semi-quantitative approach using myelin staining density is used to determine myelination sequences in various fiber tracts.

Figures 32 and **33** show progressive changes in area of various components of the spinal cord in *absolute values* (mm^2). It is important to note that the absolute values are based on a single specimen at each age; these specimens were collected over a period of several years, were preserved with different fixatives, and were not all cut at the same thickness or in exactly the same plane. All of these factors contribute to "noise" in the data, and the graphs are not smooth progressions from one age to another. The time when most noise in the data exists is between GW4.0 and GW6.8 in the first trimester. Nine specimens from this period are shown in **Plates 2** through **10**. Fortunately, six of these specimens are from the Minot Collection, where the fixative is constant and the cutting plane is more regular between specimens. When all of the specimens in the Carnegie Collection are removed from the data, the changes in absolute areas from one age to the next indicate developmental changes that allow a more accurate interpretation of real events (the first 6 data points in **Figure 32** graphs). These quantitative data, when combined with the qualitative evidence in the photographs, allow the construction of *timetables of developmental events* among neuronal populations in the spinal cord as shown in **Table VII-1**.

Figures 34 through **37** feature pie-charts showing relative changes in the areas of varous components of the spinal cord throughout its development. The problems associated with fixative, cutting planes, section thickness, and staining types are much less relevant here. **Figures 34** and **35** look at the entire range of development from GW4.0 to the 4th postnatal week. **Figures 36** and **37** look at the changes in maturation of cervical through sacral levels in 6 specimens from GW8.5 to the 4th postnatal month.

Finally, **Table VII-2** shows a semi-quantitative analysis of the sequential steps in myelination in several of the major fiber tracts in the spinal cord at the level of the cervical enlargement. These events can be used to make some generalizations about myelination sequences in fiber tracts, not only in the spinal cord, but throughout the central nervous system. These conclusions are explained in the caption of **Table VII-2**.

FIGURE 32

Changes in Area During the First Trimester

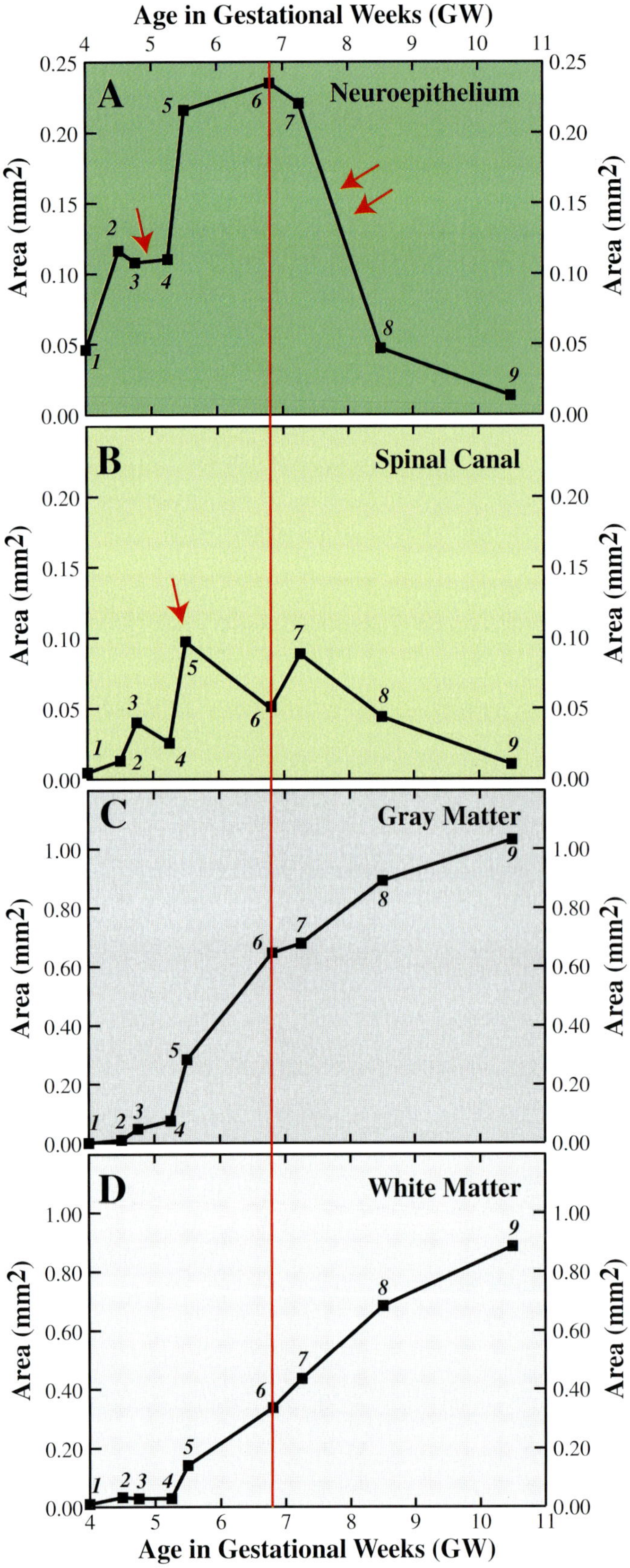

Figure 32. The total area of the neuroepithelium (**A**), spinal canal (**B**), gray matter (**C**), and white matter (**D**) at *cervical levels* during the first trimester. The **italicized numbers** in each graph indicate data points. The first six data points in each graph are from Minot specimens only because the same fixative was used and there is a more constant cutting plane between specimens. Thus, quantitative data more reliably indicate developmental trends.

The **red line** marks the peak area of the neuroepithelium (**A**) and serves as a dividing line. First consider what happens before the peak. Between GW4 and GW4.5 (**data points 1** and **2, A**), the neuroepithelium expands, especially its ventral part (see **Plates 2** and **3**). Neuroepithelial area reaches a plateau (**single red arrow, A**) between GW4.75 and GW5.25 (**data points 3** and **4, A**). That plateau occurs at the time when the area of the ventral neuroepithelium remains stable during the production of the large motoneurons in the ventral horn (see **Plates 4** and **5**). The most rapid growth spurt in the neuroepithelium occurs between GW5.25 and GW5.5 (**data points 4** and **5, A**) when the spinal canal reaches its peak area (**red arrow, B**) as the dorsal neuroepithelium rapidly expands about the dorsal part of the spinal canal. There is a simultaneous growth spurt in the area of the gray matter (**C**) as neurons accumulate in the ventral horn and in the area of the white matter (**D**) as axons accumulate in the ventral funiculus, ventral commissure, and dorsal root bifurcation zone of the dorsal funiculus. These growth spurts are qualitatively visible by looking at the changing appearance of the spinal cord in **Plates 5, 6,** and **7**. After GW5.5, the area of the neuroepithelium grows more slowly to reach a peak at GW6.8 (**data point 6, A**). That slow growth is due to concurrent shrinking of the ventral neuroepithelium and expansion of the dorsal neuroepithelium (see **Plates 8-10**).

Now consider what happens after the peak. The neuroepithelial area gradually declines by GW7.25 (**data point 7, A**), then plummets by GW8.5 (**data point 8, A**). During the rapid decline (**double red arrows, A**; see **Plates 11** and **12**) the dorsal neuroepithelium shrinks dramatically as late-generated neurons migrate into the dorsal horn and the dorsal and ventral parts of the spinal canal disappear to leave only a central canal (**B**). During the decline in the neuroepithelium, the gray matter (**C**) and white matter (**D**) continue to increase. However, throughout the entire first trimester, the area of the gray matter is greater than the area of the white matter.

TABLE VII-1 and FIGURE 33

Table VII–1: Developmental Events in Neuronal Populations during the First Trimester

NAME	GW4.0-4.5	GW4.75-5.25	GW5.5-6.8	GW7.25-8.5	GW10.5 on
Alpha Motoneurons	Stem cells proliferate; First neurons generated	Most neurons generated	Migration and settling	Segregation into columns and maturation	
Ventral Horn Interneurons	Stem cells proliferate		Neurons generated	Migration and settling	Maturation
Intermediate Gray	Stem cells proliferate		Neurons generated	Migration and settling	Maturation
Dorsal Horn Layers I,IV,V	Stem cells proliferate		Neurons generated	Migration and settling	Maturation
Dorsal Horn Layers II,III	Stem cells dormant?		Stem cells proliferate	Neurons generated; Migration and settling	Maturation
Central Auto-nomic Area	Stem cells dormant?		Stem cells proliferate	Neurons generated; Migration and settling	Maturation

Table VII-1. The major events that take place in neuronal populations in the cervical spinal cord during the first trimester. **Dark green** indicates the first step in development when the stem cells rapidly proliferate just prior to neurogenesis. **Light green** indicates the approximate period of neurogenesis. **Pink** indicates the time during which neurons migrate away from the neuroepithelium and settle in the gray matter. **Orange** indicates the period of maturation, which continues throughout the second and third trimesters and well after birth. **Light brown** indicates the postulated time when the stem cells of neuronal populations that are generated late are either dormant or have not yet appeared. The time periods for the various events in each population are based on the data in **Figure 32** and the qualitative changes visible in **Plates 2** through **12**.

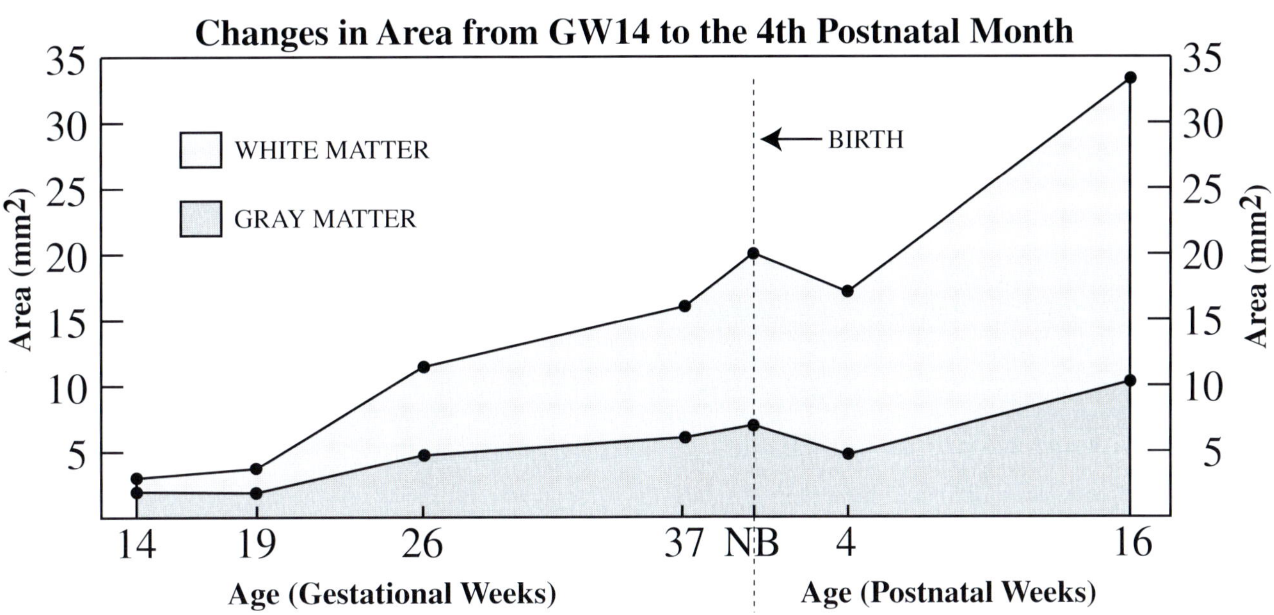

Figure 33. The area of the gray matter (**dark gray**) and white matter (**light gray**) at the cervical enlargement from the beginning of the second trimester (GW14), birth (NB-newborn), to 16 weeks after birth. From GW26 on, data of the myelin stained section area is used. The area graphs are stacked so that the scale on the Y axis indicates the total area of the spinal cord. These areas are measured in fixed tissue, so the actual values in the living spinal cord will be higher by an unknown amount. The areas of the ependyma and spinal canal are so small that they do not register above the baseline. Since neurogenesis finishes in the first trimester (see **Figure 32**), the gradual increase in the area of the gray matter throughout this period is due to the growth and differentiation of the neurons themselves. The white matter increases more rapidly, especially between GW19 and GW26 (compare **Plates 40** and **47**). There is gradual growth from GW 26 to the perinatal period (between GW37 and postnatal week 4), then more rapid growth between postnatal weeks 4 and 16 (compare **Plates 95** and **106**). The first growth spurt is probably due to the accumulation of axons in all of the major fiber tracts, including both parts of the corticospinal tract. The second growth spurt is probably due to the completion of myelination throughout the cervical enlargement at the 4th postnatal month. Only the outermost crescent of the corticospinal tract is still myelinating at this time (see **Plate 106**). From GW26 on, the area of the white matter exceeds the area of the gray matter.

FIGURE 34

Proportional Areas at the Cervical Enlargement

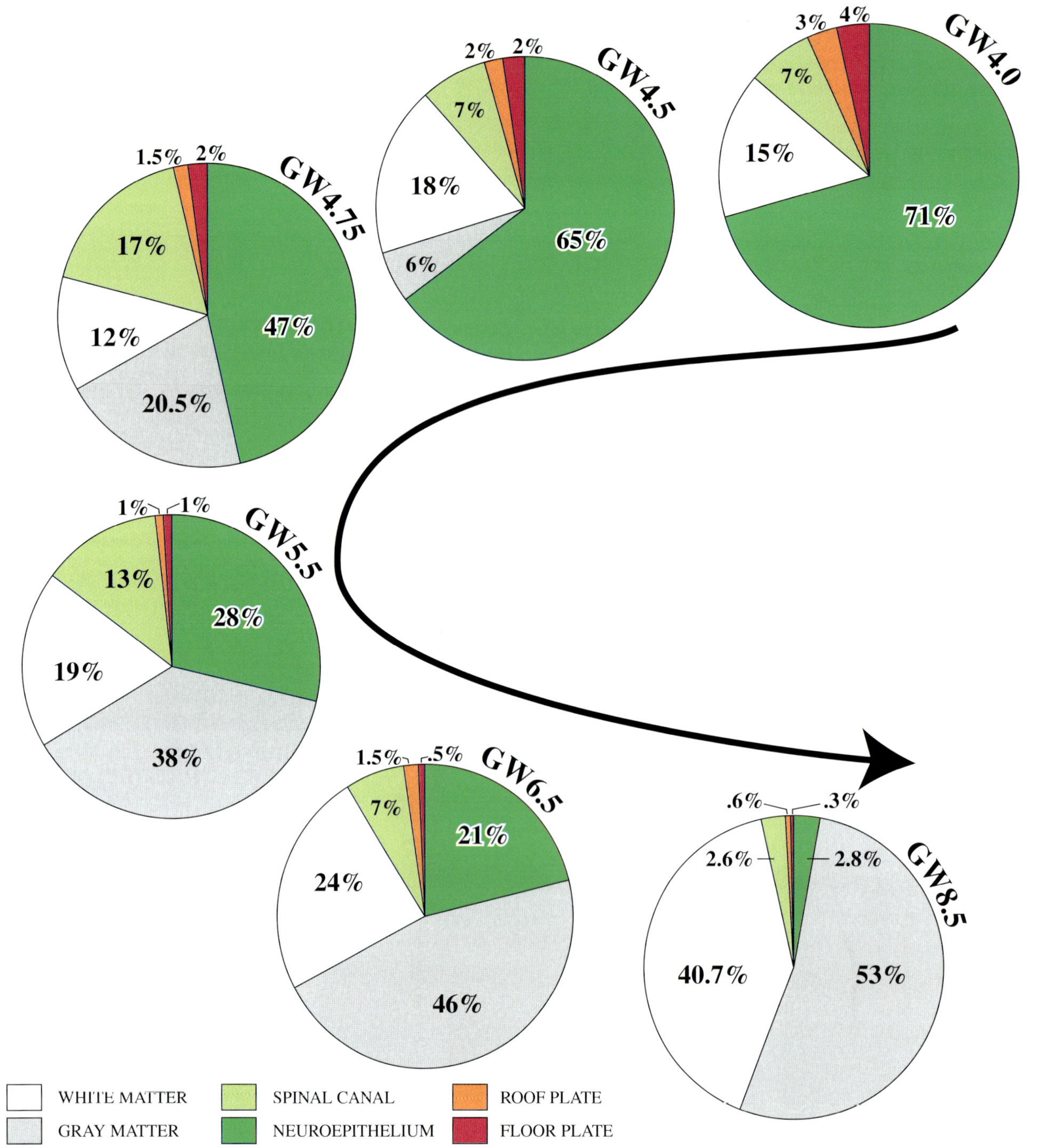

Figure 34. A series of pie-shaped graphs showing the proportional areas of the white matter (**white**), gray matter (**gray**), spinal canal (**light green**), neuroepithelium (**dark green**), roof plate (**orange**), and floor plate (**red**) at the *cervical enlargements* of specimens from gestational week (GW) 4.0 (**upper right graph**) to GW8.5 (**lower right graph**). The section measured from the GW4 specimen is illustrated in **Plate 2**, GW4.5 in **Plate 3**, GW4.75 in **Plate 4**, GW5.5 in **Plate 7**, GW6.5 in **Plate 8**, and GW8.5 in **Plate 12**. The neuroepithelium dominates from GW4-4.75 when stem cells are rapidly increasing in preparation for successive waves of neurogenesis. Even though the neuroepithelium declines in its proportion throughout the entire period, it is actually increasing its absolute area up to GW6.5 (see **Figure 32**). In contrast, the peak proportional areas of the spinal canal (GW4.75-5.5) occur at the same time as the largest absolute area and their decline at GW6.5-8.5 is associated with dramatic shrinkage of the neuroepithelium (**Figure 32**). When neurons are produced and migrate out of the neuroepithelium, the area of the gray matter proportionally increases. The first proportional increase between GW4.75-5.5 is almost exclusively due to growth of the ventral horn. Later proportional increases are due to continual growth of the ventral horn, the emerging intermediate gray, and the rapid expansion of the dorsal horn. The white matter measured at GW4.0-4.75 is primordial and contains only a few axons; that area is the same as the mantle layer described in traditional developmental studies. From GW5.5 onward, the proportional area increases are due to accumulating axons, mainly in the dorsal funiculus, ventral funiculus, and ventral commissure at GW5.5 and GW6.5. By GW8.5 the proportional area of the white matter sharply increases. The dorsal funiculus greatly expands when long ascending collaterals form the gracile and cuneate fasciculi. The roof and floor plates have their largest proportional areas early (GW4.0-4.75) supporting the animal experimental evidence of a promint role in the overall dorsal–to–ventral differentiation of the gray matter and growth of axons parallel to the dorsal midline and across the ventral midline (see Chapter 4 in Altman and Bayer, 2001).

Proportional Areas at the Cervical Enlargement

FIGURE 35

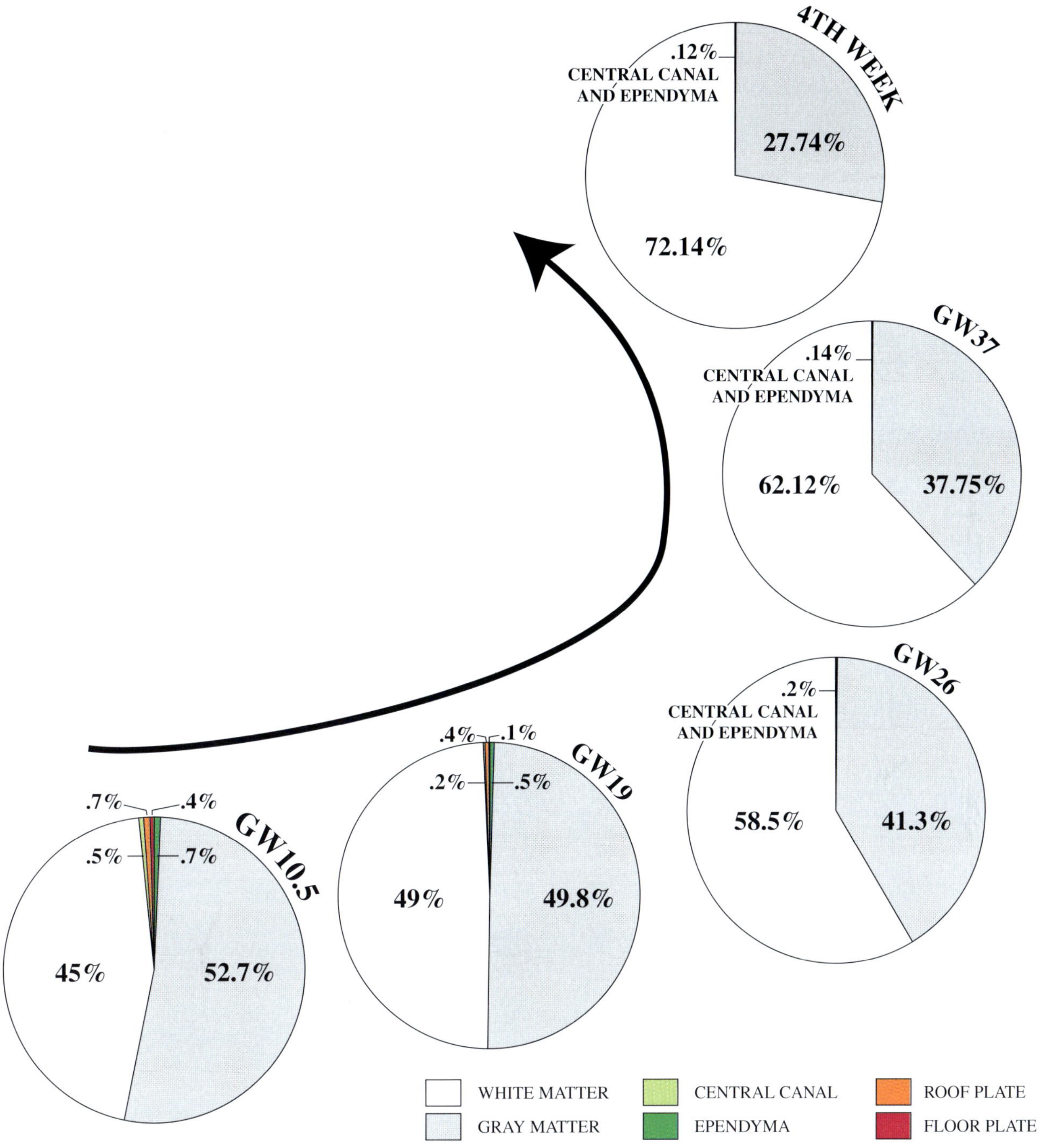

Figure 35. A continuation of the previous figure in specimens at GW10.5 (see section measured in **Plate 23**), GW19 (**Plate 40**), GW26 (**Plate 47**), GW37 (**Plate 65**), and the 4th postnatal week (**Plate 95**). During this period, the proportional areas of the gray and white matter make up most of the spinal cord. The roof and floor plates disappear by GW26, and the proportional areas of the central canal and ependyma are so small that they appear as a slightly thicker black line separating the gray matter and white matter at the 12:00 position of each pie graph. The gray matter continually loses proportional area throughout this period but the absolute area does increase gradually as neurons grow dendrites and some interneurons grow elaborate terminal axon arbors (especially in the dorsal horn). The proportion of the white matter continually increases during this time, and it is the largest proportional area from GW26 onward.

FIGURE 36 Level Differences in Proportional Areas: GW8.5-GW19

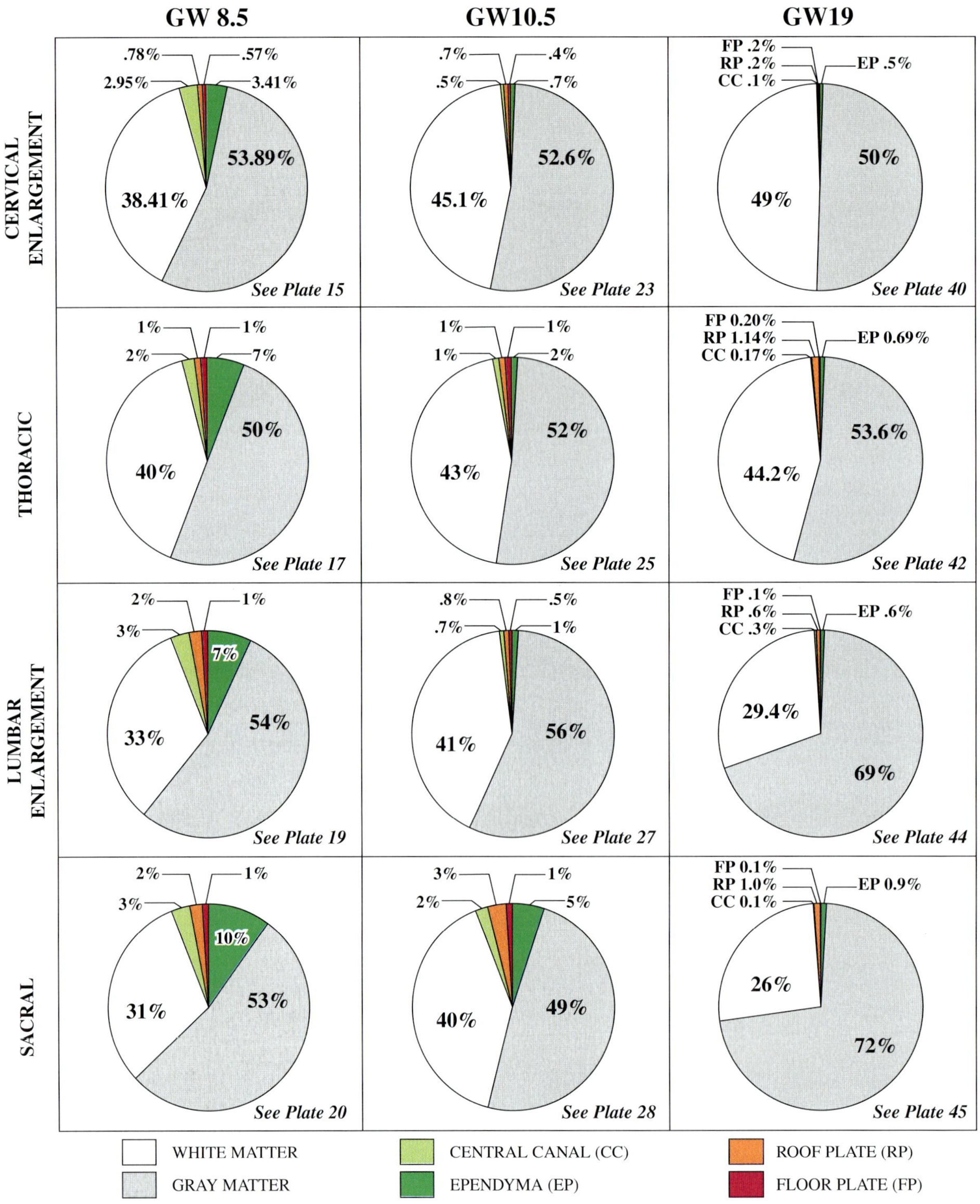

Figure 36. Pie charts showing the proportional areas of the white matter (**white**), gray matter (**gray**), central canal (**light green**), ependyma (**dark green**), roof plate (**orange**), and floor plate (**red**) at the cervical enlargement level (**row 1**), thoracic level (**row 2**), lumbar enlargement level (**row 3**), and the sacral level (**row 4**). Each column contains measured sections from the same specimen: **column 1** is M2050, GW8.5; **column 2** is Y380–62, GW10.5; and **column 3** is Y52–61, GW19. The **plate numbers** in each square refer to the illustration of the measured section in previous parts of this atlas. By following the shifting values of the proportional areas within a column, a *gradient of maturation* is evident between cervical enlargement to sacral levels. The size of the ependyma increases in the first two columns, indicating that the greatest degree of maturation is at the cervical enlargement level followed by declining levels of maturation at thoracic, lumbar enlargement, and sacral levels. By GW19, the central canal, ependyma, roof plate, and floor plate have very small proportional areas indicating the greater maturation of the spinal cord throughout its length at this time. The proportion of gray matter to white matter generally increases from cervical enlargement to sacral levels at all ages. That is because descending fiber tracts from the brain arrive first at cervical levels and progressively later at successively lower levels. Thus from GW8.5 to GW19, the white matter at the lumbar enlargement and sacral levels is less mature than it is at the cervical enlargement level.

Level Differences in Proportional Areas: GW37-4th Month

FIGURE 37

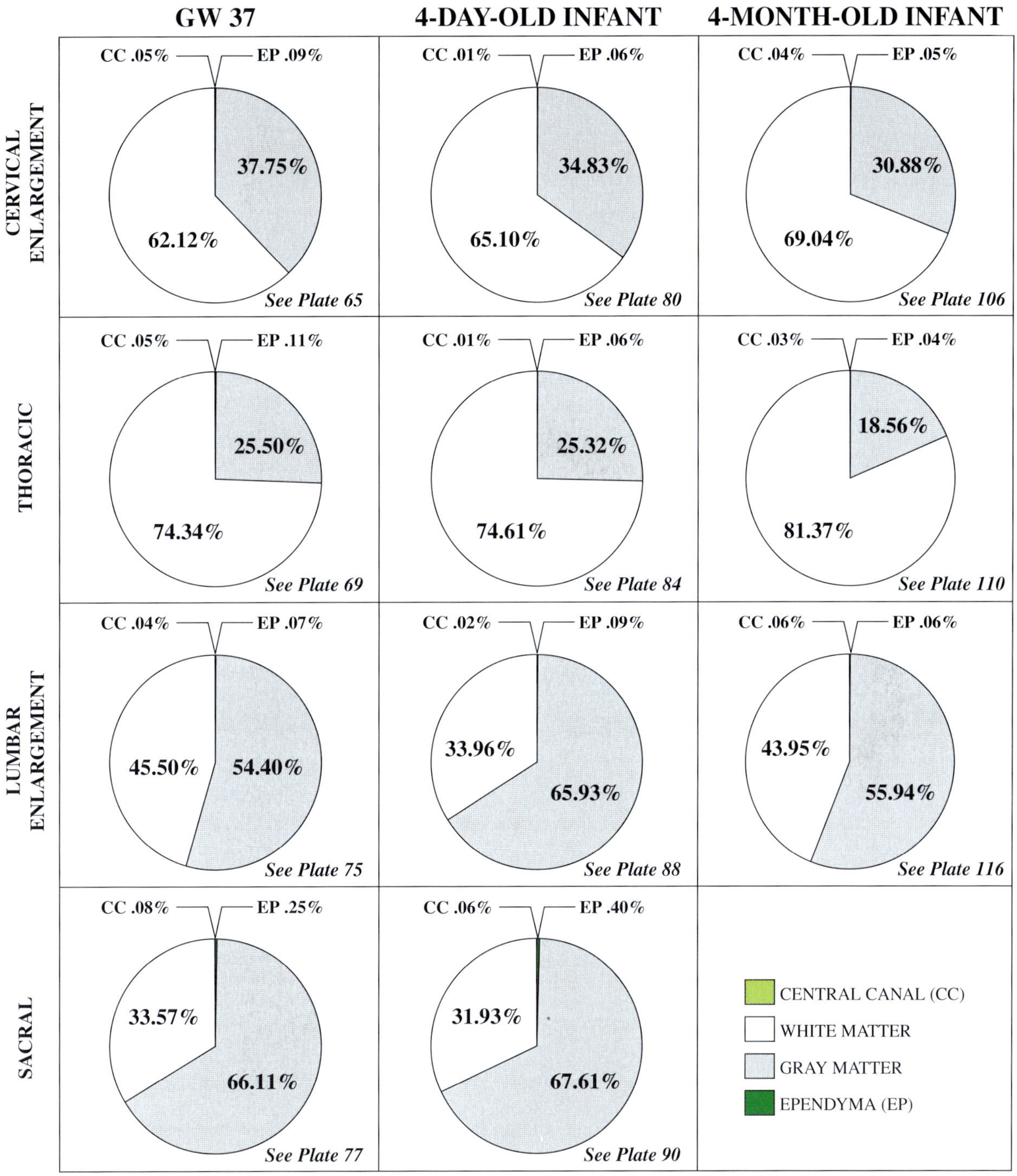

Figure 37. A continuation of the previous figure in specimens at GW37 (**column 1**, Y117–61), a 4-day-old infant (**column 2**, Y299–62), and a 4-month-old infant (**column 3**, Y23–60). A section at the sacral level is not available in the 4-month-old infant. The roof and floor plates are absent at these ages, and the proportional areas of the central canal and ependyma appear on the graphs as slightly thicker black lines separating the gray matter and white matter at the 12:00 position in each chart. The smaller proportional areas of the white matter at the sacral level perinatally (**bottom row, first two columns**) indicate immature white matter. In contrast, the cervical enlargement (**row1**) and thoracic levels (**row 2**) have high to very high proportions of white matter to gray matter indicating that the white matter is quite mature in all these specimens. The very high proportion of white matter at the thoracic level is because the ventral horn is quite small at these levels, just as it is in the adult spinal cord. At the ages shown here, the lower proportional area of white matter at the lumbar enlargement (**row 3**) does not indicate immaturity. That is because all axons from the brain that terminate at lumbar and sacral levels pass through cervical and thoracic parts, increasing the proportional area of the white matter at those levels.

TABLE VII-2

Table VII–2: Myelination Sequences in Major Fiber Tracts (Cervical Enlargement)

TRACT	GW26	GW37	Birth	4th Week	4th Month
Sup. Fas. Gracilis	Sparse reactive glia	Some myelinating fibers	Myelinated	Myelinated	Myelinated
Deep Fas. Gracilis	Dense reactive glia	Myelinated	Myelinated	Myelinated	Myelinated
Sup. Fas. Cuneatus	Sparse reactive glia	Some myelinating fibers	Myelinated	Myelinated	Myelinated
Deep Fas. Cuneatus	Dense reactive glia	Myelinated	Myelinated	Myelinated	Myelinated
D.R. Col. Zone	Dense reactive glia	Myelinated	Myelinated	Myelinated	Myelinated
D.R. Bif. Zone	Some myelinating fibers	Some myelinating fibers	Myelinated	Myelinated	Myelinated
Spino-cerebellar	Dense reactive glia	Some myelinating fibers	Many myelinating fibers	Myelinated	Myelinated
Intraspinal	Dense reactive glia	Some myelinating fibers	Myelinated	Myelinated	Myelinated
Spino-cephalic	Sparse reactive glia	Dense reactive glia	Sparse reactive glia	Some myelinating fibers	Myelinated
Ventral Commissure	Some myelinating fibers	Many myelinating fibers	Many myelinating fibers	Some myelinating fibers	Myelinated
Ventral Corticospinal	No reactive glia	No reactive glia	Sparse reactive glia	Dense reactive glia	Myelinated
Lateral Corticospinal	No reactive glia	No reactive glia	Sparse reactive glia	Dense reactive glia	Many myelinating fibers

Table VII-2. Steps in the myelination of some major fiber tracts in the cervical enlargement of the spinal cord from the second trimester to the fourth postnatal month. The unmyelinated Lissauer's tract and some descending tracts (tectospinal, medial longitudinal fasciculus, and vestibulospinal) are not shown. **Pale violet** indicates the first step when glial cells react with the myelin stain prior to the production of true myelin sheaths; these areas in the myelin stained sections contain a light dusting of punctate stain. **Dark violet** indicates the second step when the punctate staining is more plentiful, presumably due to more glia reacting with the myelin stain; however, the overall density of the stain is still light and few, if any, fibers are myelinating. **Pale purple** indicates step 3 when the density of the stain increases and glia produce myelin sheaths around some fibers. **Medium purple** indicates step four when the density of the stain increases yet again and more glia produce myelin sheaths around many fibers. **Dark purple** indicates the fifth and final step when glia produce myelin sheaths around nearly all the fibers in the tract; the truly myelinated areas appear solid black in the stained sections.

Both the **fasciculus gracilis** and the **fasciculus cuneatus** show gradients in the myelination steps so that deep parts of each fasciculus myelinate before superficial parts. If the sequence of myelination is linked to the time when axons first enter these fasciculi, the conclusion can be drawn that the first axons in the tract occupy the most deep areas, and fibers that enter the tract later occupy progressively more superficial parts.

The **spinocephalic tract** does not have an even sequence of myelination because the dense reactive glial staining at GW37 is bracketed by periods of sparse reactive glial staining. That may be due to the dilution of myelinated axons in the cervical level by the invasion of unmyelinated or less myelinated axons from parts of the spinal cord that are less mature, such as sacral and lumbar levels.

The **ventral commissure** also has an uneven sequence of myelination. From the very first appearance of myelin staining, there are always some fibers that are myelinated. However, more fibers are myelinated at GW37 and postnatal day 4 than at the 4th postnatal week. That apparent regression is possibly due to the growth of unmyelinated axons from the ventral corticospinal tract into the commissure, thus diluting the number of myelinated fibers.

The **lateral corticospinal tract** is the last fiber tract to myelinate. Indeed, some parts of it are still myelinating during the 4th postnatal month, especially the outer crescent closest to the spinocerebellar tracts. That part of the tract contains fibers that will terminate at sacral levels of the cord. The lateral corticospinal tract in thoracic, lumbar, and sacral levels is progressively less mature at the 4th postnatal month. That gradient of maturation suggests that cortical axons myelinate first proximal to the cell body and myelinate last distal to the cell body.

This analysis departs from most previously published accounts of myelination sequences which describe various fiber tracts as either "not myelinated" or "myelinated." It is a novel idea that myelination is a progressive event rather than an "all or none" event within a fiber tract. For a review of the literature on myelination in the spinal cord, see Chapter 8 and Chapter 9, Section 9.4 in Altman and Bayer (2001).

GLOSSARY

Alar plate—According to His' bipartite division of the neural tube, that part of the lateral plate neuroepithelium situated *above* the sulcus limitans. It is hypothesized to be the source of dorsal horn sensory processing neurons and some intermediate gray neurons. In the tripartite division of the spinal cord neuroepithelium in this Atlas, this term is not used. See **Dorsal neuroepithelium** and **Intermediate neuroepithelium**.

Autonomic motoneurons—Preganglionic sympathetic motoneurons in the lateral horn at thoracic and upper lumbar levels; preganglionic parasympathetic motoneurons in the lateral part of the intermediate gray at sacral levels. See **Lateral horn motoneurons**.

Basal plate— According to His' bipartite division of the neural tube, that part of the lateral plate neuroepithelium situated *below* the sulcus limitans. It is hypothesized to be the source of ventral horn motoneurons, ventral horn interneurons, and some intermediate gray neurons. In the tripartite division of the spinal cord neuroepithelium in this Atlas, this term is not used. See **Ventral neuroepithelium** and **Intermediate neuroepithelium**.

Cauda equina—Elongated roots of spinal nerves in the vertebral canal that exit from the spine below the level of their points of attachment to the spinal cord.

Cellular roof plate—Cells in the roof plate that have fibrous-like projections into the roof of the spinal canal. Initially, the cellular roof plate caps the midline of the spinal cord, but it starts to sink downward at the end of the first trimester, apparently leaving behind a string of non-neural cells that form the dorsal median septum. See **Roof plate**.

Central autonomic area—Small- and medium-sized neurons that surround the central canal and provide a bridge between the two wings of the gray matter on either side of the midline. This area is homologous to the central gray surrounding the cerebral aqueduct and fourth ventricle in the brain. It contains the dorsal and ventral gray commissures that bridge the midline. These neurons are activated by nociceptive afferents from small cells in the dorsal root ganglion.

Central cervical nucleus—A cluster of large neurons in the intermediate gray near the central canal at cervical levels that is the continuation of Clarke's column. Axons of these neurons enter the ipsilateral dorsal spinocerebellar tract.

Cervical enlargement—Includes that part of the spinal cord from which the nerves of the brachial plexus arise, from C4 to T1. This part of the spinal cord has a larger diameter because of the greater amount of gray and white matter required to innervate the upper limb.

Cervical region—The part of the spinal cord that is continuous with the medulla of the brain through the foramen magnum of the skull. Eight spinal nerves exit from this region; all nerves except C1 have a dorsal root ganglion. C1 exits above the first vertebra; C8 exits in the transverse foramen of the C7 vertebra.

Clarke's column—A group of large neurons dorsal to the central canal in the intermediate gray at thoracic and upper lumbar levels. Its axons form the ipsilateral dorsal spinocerebellar tract.

Conus medullaris—The cone-shaped terminus of the spinal cord.

Dorsal canal—A transient part of the spinal canal that first expands then recedes in relation to the changing size of the dorsal neuroepithelium during the first trimester.

Dorsal funiculus—Dorsal white matter that caps and fills in the U-shaped space between the paired dorsal horns. Most of the axons in this funiculus are from large dorsal root ganglion cells in four subdivisions: fasciculus gracilis, fasciculus cuneatus, dorsal root bifurcation zone, and dorsal root collateralization zone.

Dorsal horn—Dorsal wings of the butterfly-shaped gray matter that contain sensory processing neurons in laminae I-V. This is the latest differentiating region of the spinal cord gray matter.

Dorsal horn, lamina I—A thin sheet of gray matter that caps the dorsal surface of the dorsal horn. It contains the large Waldeyer cells that process nociceptive sensory information. Axons of the Waldeyer cells cross the midline in the ventral white commissure and enter the ventral and lateral spinothalamic tracts, or the spinocephalic tracts.

Dorsal horn, laminae II-III—See **Substantia gelatinosa**.

Dorsal horn, laminae IV-V—A collection of large and small neurons; also called the nucleus proprius. These neurons process touch, pressure, and nociceptive input. At lumbar and thoracic levels, axons of some of the neurons in lamina IV travel to the medulla in the dorsal funiculus as a postsynaptic–dorsal column system. At all levels, some neurons in lamina IV travel in the dorsolateral region of the lateral funiculus (the spinocervicothalamic pathway) and terminate in a topographic pattern in the lateral cervical nucleus (Haines, 2000).

Dorsal intermediate septum—Aligned non-neural cells that separate the fasciculus gracilis and fasciculus cuneatus. There is a slight depression on the surface of the spinal cord to mark the spot where this septum joins the pia.

Dorsal median septum—Aligned non-neural cells attached to the pia that extend downward in the dorsal midline. This septum separates the two gracile fasciculi on either side of the midline and may be formed by the receding cellular portion of the roof plate during the late first trimester and early second trimester.

Dorsal neuroepithelium—Neuroepithelium surrounding the dorsal canal that produces dorsal horn neurons. This neuroepithelium still produces neurons in the late first trimester.

Dorsal root—Part of a spinal nerve that is formed by axons of dorsal root ganglion cells that bring sensory input to the dorsal funiculus.

Dorsal root bifurcation zone—Region in the dorsal funiculus where incoming axons from the dorsal root divide into ascending and descending branches.

Dorsal root boundary cap—Specialized glial cells that accumulate at the entry points of the dorsal roots of the spinal nerve. These cells may act as targets that direct the growth of dorsal root axons into the spinal cord.

Dorsal root collateralization zone—Region of the dorsal funiculus that contains the local collateral branches of dorsal root ganglion cell axons. These axons curve around the dorsomedial edge of the dorsal horn before penetrating the spinal gray matter. This region also contains axons of the intraspinal tracts.

Dorsal root ganglion—Primary sensory neurons embedded in the dorsal root of a spinal nerve that bring exteroceptive (touch, pressure), proprioceptive (stretch receptors in muscles and tendons), visceroceptive (receptors in the vital internal organs), and nociceptive (pain) input to the spinal cord.

Dorsal spinocerebellar tract—Axons of neurons in Clarke's column and the central cervical nucleus that extend ipsilaterally to the cerebellum in the peripheral dorsal lateral funiculus.

Dorsolateral fasciculus—See **Lissauer's tract**.

Ependyma—Cells lining the central canal in the mature spinal cord. During the late first trimester and early second trimester, these cells gradually appear as the spinal neuroepithelium recedes.

Fasciculus cuneatus—Central fiber tract in the dorsal funiculus above the 6th thoracic level (lateral to the fasciculus gracilis and medial to the dorsal root collateralization zone). Its main component is ipsilateral axons from large neurons in the dorsal root ganglia above the 6th thoracic level that carry exteroceptive information from the upper part of the body to the nucleus cuneatus in the lower medulla. A smaller component is axons of sensory relay neurons in laminae III–V that terminate in the nucleus cuneatus as part of the postsynaptic–dorsal column system.

Fasciculus gracilis—Medial fiber tract in the dorsal funiculus. Its main component is ipsilateral axons from large neurons in the dorsal root ganglia below the 6th thoracic level that carry exteroceptive information from the lower part of the body to the nucleus gracilis in the lower medulla. A smaller component is axons of sensory relay neurons in laminae III–V that terminate in the nucleus gracilis as part of the postsynaptic–dorsal column system.

Fibrous roof plate—The end feet of roof plate cells that extend into the spinal canal in the dorsal midline.

Floor plate—Specialized cells that bridge the ventral midline of the spinal canal during the first trimester. Initially, it is the most ventral structure in the embryonic spinal cord, lying directly above the notochord. The floor plate moves upward at the end of the first trimester, with the receding ventral neuroepithelium. It may be involved in ventral motoneuron induction and may provide directional cues for growing commissural axons.

Intermediate gray—Spinal gray matter that forms a bridge between the dorsal and ventral horns (laminae VI–VII).

Intermediate interneurons—Neurons in the intermediate gray whose axons do not extend to the brain or enter the ventral roots. These neurons may be part of the intrinsic locomotive pattern generator system in the spinal cord. Many of their axons travel up and down the spinal cord in the intraspinal tracts.

Intermediate neuroepithelium—Neuroepithelium surrounding the sulcus limitans that produces intermediate gray neurons.

Intraspinal tracts—Axons of spinal interneurons that travel in the white matter adjacent to the gray matter of the dorsal horn, intermediate gray, and ventral horn; also called the propriospinal tract or the spinospinal tract. These tracts overlap with the medial longitudinal fasciculus, the tectospinal tract, the ventral and lateral reticulospinal tracts, and the dorsal root collateralization zone.

Laminae VI–VII—Gray matter of the intermediate gray.

Lamina VIII—Gray matter of the ventral horn interneurons.

Lamina IX—Gray matter of the ventral horn motoneurons.

Lateral cervical nucleus—Large neurons scattered adjacent to and within the reticulated area at cervical levels. These neurons get input from lamina IV cells at all levels of the spinal cord. Their axons cross the midline in the ventral white commissure and travel to the thalamus in the medial lemniscus.

Lateral corticospinal tract—Contralateral axons in the dorsal part of the lateral funiculus from upper motoneurons in layer V of the primary motor cortex (pyramidal cells of Betz in area 4) and other cortical areas. These axons are somatotopically arranged (cervical terminating axons medial, lumbosacral terminating axons lateral). Axons from area 4 terminate primarily in laminae VI-IX in the ventral horn.

Lateral funiculus—White matter between Lissauer's tract and the ventral funiculus. The border between the lateral and ventral funiculi is indistinct. The lateral funiculus contains the following tracts: dorsal and ventral spinocerebellar, lateral corticospinal, rubrospinal, lateral reticulospinal, dorsolateral spinocephalic, and dorsolateral intraspinal.

Lateral horn motoneurons—Preganglionic (visceral) motoneurons of the autonomic nervous system. These neurons are generated in the ventral neuroepithelium during the early first trimester and migrate dorsolaterally during the late first trimester to settle in the lateral horn. There are no lateral horn motoneurons at cervical levels. At thoracic and upper lumbar levels, the axons of these motoneurons leave the spinal cord in the ventral roots and terminate in the chain of sympathetic ganglia located adjacent to the anterolateral vertebral column. At sacral levels, the axons of these motoneurons leave the spinal cord in the ventral roots and terminate in parasympathetic ganglia located in the abdominopelvic cavity.

Lateral plate—Dorsal, intermediate, and ventral parts of the spinal cord neuroepithelium.

Lateral reticulospinal tract—Axons from large neurons in the medullary reticular formation that terminate primarily in laminae VII at all levels of the spinal cord. Axons in this tract travel close to the gray matter in the lateral funiculus and overlap with axons in the intraspinal tracts.

Lateral spinothalamic tract—See **Spinocephalic tracts**.

Lissauer's tract—Unmyelinated axons of small neurons in dorsal root ganglia that carry pain input to the dorsal horn of the spinal cord; also called the dorsolateral fasciculus. This tract forms the boundary between the dorsal funiculus and the lateral funiculus.

Lumbar enlargement—The part of the spinal cord that has a larger diameter and contains the gray matter and white matter associated with the lumbosacral plexus, the nerves that innervate the lower limb.

Lumbosacral region—The most caudal part of the spinal cord that contains the lumbar enlargement. There are five lumbar spinal nerves, five sacral spinal nerves, and one coccygeal spinal nerve.

Mantle zone or layer—See **Primordial gray matter**.

Marginal zone or layer—See **Primordial white matter**.

Medial longitudinal fasciculus—Axons from neurons in the medial vestibular nucleus; also called the medial vestibulospinal tract. This fasciculus is most prominent at cervical levels adjacent to the medial ventral horn below the ventral white commissure in the ventral funiculus. Axons in this fasciculus overlap with those in the intraspinal tracts.

Medulla—also called medulla oblongata. The part of the brain that is continuous with the spinal cord and contains the posterior part of the fourth ventricle.

Motoneurons—A general term for neurons whose axons leave the spinal cord in the ventral roots of a spinal nerve. Axons of these neurons terminate either on skeletal muscle (somatic motoneurons) or on postganglionic neurons in autonomic ganglia (preganglionic visceral motoneurons).

Neuroepithelium—The primordial nervous system composed of neural stem cells (NEP cells). The ultimate source of neurons, neuroglia, ependymal cells, tanycytes, and other specialized cells in the central nervous system.

Notochord—A solid core of cells that lie in the midline beneath the neural tube floor plate that will form the spinal cord. These cells provide "induction" signals that establish "ventralization" of the neural tube so that motoneurons are only generated by stem cells in the ventral neuroepithelium.

Oval bundle of His—See **Dorsal root bifurcation zone**.

Primordial gray matter—A thin layer of young neurons between the spinal cord neuroepithelium and the primordial white matter present only during the early first trimester. The primordial gray matter differentiates into specific parts of the spinal gray. This area is traditionally called the mantle zone or layer.

Primordial white matter—A thin cell-sparse layer beneath the pial membrane in the embryonic spinal cord. The main components of this layer are the end feet of specialized glial cells and spaces where axons of spinal cord fiber tracts will form the dorsal, lateral, and ventral funiculi. This area is traditionally called the marginal zone or marginal layer.

Propriospinal tract—See **Intraspinal tracts**.

Reticulated area—A region in the lateral funiculus adjacent to the intermediate gray where white matter and gray matter intermingle.

Roof plate—Non-neural cells that cap the dorsal midline of the spinal canal during the first trimester. Initially, it is the most dorsal structure in the embryonic spinal cord, but it sinks downward at the end of the first trimester as the dorsal neuroepithelium recedes. As that happens, the dorsal funiculus fills with axons separated in the midline by aligned non-neural cells that form the dorsal median septum.

Rubrospinal tract—Somatotopically arranged axons of the contralateral red nucleus in the lateral funiculus that intermingle with ventral fibers of the lateral corticospinal tract and terminate in laminae V-VIII of the spinal gray. Medial axons terminate at cervical levels, lateral axons terminate at lumbosacral levels.

Sojourn zone—The peripheral parts of the dorsal and intermediate neuroepithelia that contain dense accumulations of premigratory young neurons that stay ("sojourn") in the neuroepithelium prior to entering the gray matter. These zones have ruffled edges that are difficult to distinguish from the immediately adjacent gray matter. The ventral neuroepithelium may also contain a sojourn zone, but its ruffled outer edges are less obvious.

Somite—Paired oval clusters of cells outside the embryonic spinal cord that contain a dermatome (precursor of the dermis), myotome (precursor of axial skeletal musculature), and sclerotome (precursor of the axial skeleton).

Spinal canal—The enlarged fluid-filled core of the spinal cord present during the embryonic and early fetal period. It is surrounded laterally by the dorsal, intermediate, and ventral parts of the neuroepithelium. It is sealed dorsally by the roof plate and ventrally by the floor plate.

Spinal cord—That part of the central nervous system that is continuous with the medulla of the brain and occupies the vertebral canal in the spine.

Spinal nerve—A mixed nerve that separates into dorsal and ventral roots at its points of entry (dorsal root) or exit (ventral root) from the spinal cord. The incoming axons are those of the dorsal root ganglion. The outgoing axons are those of the ventral horn somatic motoneurons and lateral horn visceral motoneurons. A total of 31 spinal nerves are connected to the spinal cord.

Spinocephalic tracts—A complex of several tracts occupying a crescent-shaped area in the peripheral ventral and lateral funiculi. The traditional anatomists named this region the ventral and lateral spinothalamic tracts. Modern research indicates that the majority of these axons terminate in the brainstem reticular formation (spinoreticular tract), the inferior olive nucleus (spino–olivary tract), the deep layers of the superior colliculus (spinotectal tract), the central gray in the midbrain, and the hypothalamus. Less than 50% of these axons terminate in the thalamus in the ventroposterolateral nuclear complex and in the intralaminar nuclei.

Spinocerebellar tracts—Combined dorsal and ventral spinocerebellar tracts in the peripheral lateral funiculus.

Spino-olivary tract—Axons in the spinocephalic tracts that terminate in the accessory nuclei associated with the inferior olive complex in the lower medulla.

Spinoreticular tract—Axons in the spinocephalic tracts that terminate in the reticular formation of the medulla and pons.

Spinotectal tract—Axons in the spinocephalic tracts that terminate in the deep layers of the superior colliculus.

Subgelatinosal plexus—Intermingled axon terminal branches (lightly myelinated) and dendrites at the base of the substantia gelatinosa.

Substantia gelatinosa—Component of the dorsal horn that contains interneurons in laminae II and III. These neurons include the central cells, tufted cells, stalked cells, and islet cells that receive sensory input from dorsal root ganglion cells.

Sulcus limitans—A shallow depression in the spinal neuroepithelium traditionally a landmark that separates the alar and basal plates. This sulcus marks the area of the spinal canal that will be retained as the central canal in the maturing and adult spinal cord.

Tectospinal tract—Contralateral axons from neurons in the deep layers of the superior colliculus that travel below and intermingle with the medial longitudinal fasciculus in the ventral funiculus. All axons in this tract terminate at cervical levels in laminae VI–VIII in the intermediate gray. This tract overlaps with the intraspinal tracts.

Thoracic region—Part of the spinal cord that is the target and source of 12 pairs of spinal nerves associated with the 12 thoracic vertebrae. The ventral horn gray matter is narrow here because there are only axial muscles to innervate.

Ventral canal—A transient part of the spinal canal that first expands then shrinks in relation to the changing size of the ventral neuroepithelium during the first trimester.

Ventral corticospinal tract—Ipsilateral axons in the medial part of the ventral funiculus from upper motoneurons in layer V of the primary motor cortex (pyramidal cells of Betz in area 4). These axons are somatotopically arranged (cervical terminating axons dorsal, lumbosacral terminating axons ventral). Axons cross the midline in the ventral white commissure before terminating in laminae VI-IX in the ventral horn.

Ventral funiculus—White matter between the lateral funiculus and the ventral white commissure. The border between the lateral and ventral funiculi is indistinct. The ventral funiculus contains the following tracts: medial longitudinal fasciculus, tectospinal, ventral corticospinal, ventral reticulospinal, vestibulospinal, ventromedial spinocephalic, and ventromedial intraspinal.

Ventral gray commissure—A bridge of small neurons in the midline beneath the central canal.

Ventral horn—Ventral wings of the butterfly-shaped gray matter that contain motoneurons (lamina IX) and ventral horn interneurons (lamina VIII). This is the earliest differentiating region of the spinal cord gray matter.

Ventral horn interneurons—Neurons in the ventral gray (lamina VIII) whose axons do not extend to the brain or enter the ventral roots. These neurons are part of the intrinsic locomotive pattern generator system in the spinal cord. Many of their axons travel up and down the spinal cord in the intraspinal tracts.

Ventral horn motoneurons—Neurons in lamina IX whose axons leave the spinal cord in the ventral roots of the spinal nerves and innervate skeletal muscle. These neurons are generated in the ventral neuroepithelium during the early first trimester. During the late first trimester and second trimester, the motoneurons segregate into smaller and smaller clusters (in cross-sections) that are aligned longitudinally to form columns. Each column supplies specific groups of muscles. The most medial columns supply axial muscles. Ventrolateral columns supply proximal limb muscles in the cervical and lumbar enlargements. Dorsolateral columns supply distal limb muscles in the cervical and lumbar enlargements.

Ventral median fissure—A deep cleft in the ventral midline that separates the medial walls of the ventral funiculus on the right and left sides of the spinal cord. The dorsal border of this fissure is formed by the ventral white commissure.

Ventral neuroepithelium—Neuroepithelium surrounding the ventral canal that produces ventral horn motoneurons, autonomic motoneurons, and ventral horn interneurons. This is the first neuroepithelium to produce neurons, and the first to recede in the middle first trimester. A portion of this neuroepithelium may be transformed to produce glia.

Ventral root—Part of the spinal nerve that exits the central nervous system near the junction of the ventral and lateral funiculi. It is composed of the axons of ventral horn motoneurons that project to skeletal muscle and lateral horn motoneurons that project to autonomic ganglia.

Ventral rootlets—Fine bundles of axons visible in the white matter of the ventral funiculus prior to coalescing and exiting the spinal cord in the ventral root.

Ventral spinocerebellar tract—Axons of proprioceptive sensory processing neurons in the intermediate zone and spinal border cells in the intermediate gray that project to the cerebellum.

Ventral spinothalamic tract—See **Spinocephalic tract**.

Ventral reticulospinal tract—Axons from large neurons in the pontine reticular nucleus that terminate primarily in laminae VIII at all levels of the spinal cord. Axons in this tract travel close to the gray matter in the ventral funiculus and overlap with axons in the intraspinal tracts.

Ventral white commissure—Axons that cross the midline beneath the ventral gray commissure. This is one of the earliest identifiable components of the white matter that contains early and late myelinating axons.

Vestibulospinal tract—Axons of the lateral vestibular nucleus (Deiter's) in the medial part of the peripheral ventral funiculus. These axons terminate in laminae VII and VIII throughout the entire length of the spinal cord. These axons overlap with medial axons in the spinocephalic tracts.